T0202445

Green Chemistry

V. K. Ahluwalia

Green Chemistry

Environmentally Benign Reactions

Third Edition

Ane Books
Pvt. Ltd.

 Springer

V. K. Ahluwalia
Department of Chemistry
University of Delhi
New Delhi, Delhi, India

ISBN 978-3-030-58515-0   ISBN 978-3-030-58513-6   (eBook)
https://doi.org/10.1007/978-3-030-58513-6

Jointly published with ANE Books India
The print edition is not for sale in South Asia (India, Pakistan, Sri Lanka, Bangladesh, Nepal and Bhutan) and Africa. Customers from South Asia and Africa can please order the print book from: ANE Books Pvt. Ltd.

This Springer imprint is published by the registered company Springer Nature Switzerland AG
The registered company address is: Gewerbestrasse 11, 6330 Cham, Switzerland

Foreword

I feel happy to congratulate Prof. V. K. Ahluwalia on writing *Green Chemistry—Environmentally Benign Reactions*. The book is replete with basic principles of Green Chemistry and requisite details that are necessary to obtain a desirable organic reactions (which earlier needed anhydrous conditions and used volatile organic solvents) under green conditions. It is hoped that this development will go a long way in not only reducing environmental pollution but also affecting atom economy.

The book has been very well written and presented in a lucid manner. The book is so comprehensive that it can serve as a practical guide to the researchers (including M.Sc., M.Phil. and Ph.D) in various industries, universities and college laboratories.

Dr. R. K. Suri
Additional Director
Ministry of Environment and Forests
Government of India
New Delhi, India

Preface to the Third Edition

On the suggestion of the teachers, researchers and students, the book has been completely revised and enlarged, particularly Chaps. 1 and 3. In fact, the book has also been updated as per the requirements of Choice Based Credit System (CBCS). Besides, multiple choice questions and short answer questions have also been included.

It is hoped that this addition will be extremely helpful to all concerned.

The author takes this opportunity of thanking Mr. Sunil Saxena of Ane books Pvt. Ltd. for his help in the publication of this book.

New Delhi, India   V. K. Ahluwalia

Preface to the Second Edition

The enthusiastic response of the teachers, students and researchers for the first edition of the book encoursed for the second edition. The book has been completely revised and enlarged. A large number of reactions have been included under benign conditions like using aqueous phase (including super critical water), super critical carbon dioxide, ionic liquids, polymer supported reagents, polyethylene glycol and its solutions and perfluorous liquids as solvents. Most of the reactions have been carried out under microwave irradiations and sonication. The use of catalysts like phase transfer catalysts, crown ethers, biocatalysts have also been included. It is hoped that the second edition will be extremely helpful to all concerned. The author takes the opportunity of expressing his thanks to Mr. Sunil Saxena of Ane Books Pvt. Ltd., for help in the publication of this book.

New Delhi, India   V. K. Ahluwalia

Preface to the First Edition

Green chemistry is basically environmentally benign chemical synthesis and is helpful to reduce environmental pollution. A large number of organic reactions were earlier carried out under anhydrous conditions and using volatile organic solvents like benzene, which cause environmental problems and are also potentially carcinogenic. Also, the by-products are difficult to dispose off.

With the advancements of knowledge and new developments, it is now possible to carry out large number of reactions in aqueous phase, using phase transfer catalysts, using sonication and microwave technologies. Some reactions have also be performed enzymatically and photochemically. It is now possible to carry out a number of reactions using the versatile liquids and also in solid state.

The book is divided into three chapters. Introduction to Green Chemistry is described in Chap. 1. The second chapter deals with those reactions which are now performed under the so-called green conditions. Such reactions are now referred to as Green Reactions. Finally in Chap. 3 are described a number of preparations in aqueous phase, using phase transfer catalysis using sonication and microwave technologies. Also, some preparation are carried out enzymatically and photochemically. It is now possible to perform by using ionic liquids as solvents are also described.

The author expresses his sincere thanks to Dr. Pooja Bhagat and Dr. Madhu Chopra for all the help they have rendered.

Grateful thanks are due to Prof. Sukh Dev FNA, INSA Professor, New Delhi; Prof. J. M. Khurana, Department of Chemistry; and Dr. R. K. Suri, Additional Director, Ministry of Forests, Government of India.

Finally, I take the opportunity to thank Prof. Ramesh Chandra, Director, Dr. B. R. Ambedkar Centre for Biomedical Research University of Delhi, Delhi, for all the help rendered.

New Delhi, India   V. K. Ahluwalia

Contents

About the Author

V. K. Ahluwalia was a Professor of Chemistry at Delhi University for more than three decades teaching Graduate, Postgraduate and M.Phil. Students. He was also a Postdoctoral Fellow between 1960 and 1962 and worked with renowned global names from prestigious international universities. He was a Visiting Professor of Biomedical Research at the University of Delhi. V. K. Ahluwalia is widely regarded as a leading subject expert in chemistry and allied subjects along with being a "Choice Award for an Outstanding Academic Title" winner. He has published more than 100 titles. Apart from books, he has published more than 250 research papers in national and international journals.

Chapter 1
Green Chemistry

1.1 Introduction

There is absolutely no doubt that green chemistry has brought about medical revolution (e.g., synthesis of drugs etc.). The world's food supply has increased many folds due to the discovery of hybrid varieties, improved methods of farming, better seeds and use of agro chemicals, like fertilizers, insecticides, herbicides and so on. Also, the quality of life has improved due to the discovery of dyes, plastics, cosmetics and other materials. All these developments increased the average life expectancy from 47 years in 1900 to about 80 years in 2010. However, the ill-effects of all the development became pronounced. The most important effect is the release of hazardous by-products of chemical industries and the release of agro chemicals in the atmosphere, land and water bodies; all these are responsible for polluting the environment, including atmosphere, land and water bodies. Owing to all these, green chemistry assumed special importance.

1.2 What Is Green Chemistry?

Green chemistry is defined[1] as environmentally benign chemical synthesis with a view to minimize the environmental pollution. This is possible by using non-toxic starting materials from renewable sources and reduction in pollution by the prevention of hazardous by-products or substances. The chemists and research scientists all over the globe have been trying for the development of benign green synthesis of not only new products but also development of green synthesis for its existing chemicals.

© The Author(s) 2021
V. K. Ahluwalia, *Green Chemistry*,
https://doi.org/10.1007/978-3-030-58513-6_1

1.3   Need for Green Chemistry

A number of chemical accidents were reported from different parts of the globe. Some of these are:

- **Minamata disease**. It resulted due to the effects of mercury poisoning. It was reported that in 1950, more than 50 people died and a number of people were affected in a sea coast village in Japan. The reason was consuming of fish contaminated with mercury. On investigation, it was found that the water of the Minamata Bay was polluted for more than 30 years (1932–1968) by approximately 27 tonnes of mercury compounds which were dumped by the Chisso Chemical Company, a company located in Kumamoto, Japan. On investigation, it was found that the Chisso Chemical Company used mercuric chloride as a catalyst for the manufacture of acetaldehyde, and so only non-toxic mercury was released in the effluents. However, the sediments from the Minamata Bay were found to be rich in methyl mercury chloride. In the sediments of the lake, microorganisms help the biomethylation of mercury to form methyl mercury chloride. This is lipid soluble and found its way in the tissues of fish. Consumption of these fish caused birth defects and affected neural tissues, mainly the brain. This disease caused by mercury poisoning is called Minamata disease as it was found to occur in Minamata Bay in Japan.
- **Itai-Itai Disease**. It resulted due to the effects of cadmium poisoning. This disease occurred around 1912 in Japan due to consumption of rice affected by cadmium metal. The rice fields were found to be irrigated with effluents released by zinc smelters. On consumption of such rice, particularly the women suffered from acute pain in the entire body. In some cases, the women suffered from broken bones on trying to move. The clinical features were osteomalacia accompanied with osteoporosis and multiple renal tubular dysfunctions. Owing to acute pain, the victims cried "Itai-Itai" and so the disease was called "Itai-Itai". On investigation it was found that the cause of the disease was cadmium poisoning. About 200 people were found to be having this disease. The disease cause cancer of liver and lungs.
- **Bhopal Gas Tragedy**. On 3 December 1984, world's worst air-pollution episode occurred at Union Carbide in Bhopal. Approximately 33 tonnes of methyl isocyanate (MIC), a deadly poisonous gas used in the synthesis of a pesticide "seven", leaked after midnight from a storage tank and spread mist and cloud over the city of Bhopal. A large number of people were exposed to MIC while in sleep. On an estimation about 22,000 people died and more than 120,000 people suffered from diseases. The survivors of Bhopal Gas Tragedy suffered from a number of problems, such as permanent respiratory illness, impairment of vision, damage to lungs, kidneys and muscles, gastrointestinal and reproductive problem combined with low response to the immune system. In a number of women, menstrual abnormalities and abortion were reported.

In Bhopal, MIC was made to react with I-naphthol to make a carbonate insecticide "seven" (Scheme 1.1).

The required MIC was prepared by the reaction of phosgene (another deadly poisonous gas) with methylamine, a primary amine (Scheme 1.2).

The MIC thus prepared is invariably associated with about 2% of $COCl_2$. The threshold limiting value (TLV) for MIC and $COCl_2$ is 50.02 and 0.1 ppm, respectively. The toxic effects of MIC are considerably enhanced by $COCl_2$. Exposure to MIC causes tightness in the chest, breathing problems and eye ache. It also generates cyanide in the body, which is responsible for instantaneous death.

The du Pont Co. developed a method to produce MIC whenever needed so that it need not be stored, as was done in Bhopal [J. R. Thomen, Chem. Eng. News, Feb 6, 1955, 2; V. N. P. Rao and G. E. Heinsohn, U.S. Patent 4,537,726 (1985); L. E. Manzer, Catal. Today, 1993 18(2) 199] (Scheme 1.3).

It has been found that the insecticide "seven" could be replaced by the well-known integrated pest management involving *Bacillus thuringiensis* (cited in Introduction to Green Chemistry, Albert S. Matlack, CRC Press, P-29).

Flixborough Disaster
An explosion occurred in a chemical plant near the village at Flixborough, North Lincolnshire, England, on Saturday, the 1 June 1974. This plant was used to oxidize cyclohexane into cyclohexanone by air at about 155 °C.

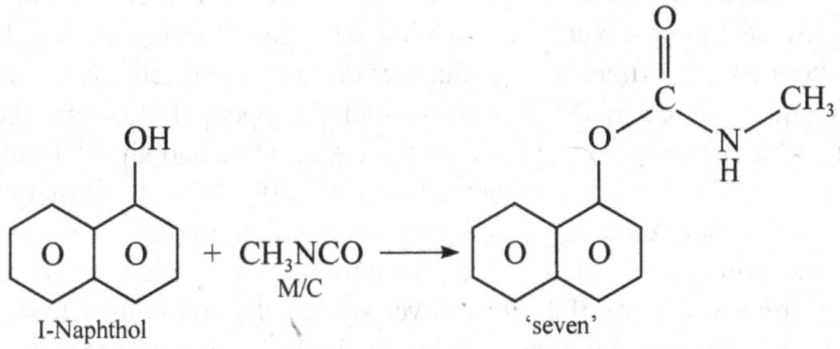

Scheme 1.1 (Synthesis of a carbonate insecticide)

$$CH_3NH_2 + COCl_2 \longrightarrow CH_3N-C=0 + 2HCl$$
Methylamine   Phosgene   MIC

Scheme 1.2 Preparation of MIC

$$CH_3NH_2 + CO \longrightarrow CH_3NHCHO \xrightarrow[>240°C]{O_2} CH_3NCO$$
M/C
84-89 of catal

Scheme 1.3 A convenient method for the preparation of MIC

About 28 people were killed due to the explosion and 36 were seriously injured out of a total of 72 people on the site at that time.

It is believed that the explosion may be due to a hasty modification in the plant. Cyclohexanone (needed to produce caprolactam, which was used to produce nylon 6) was originally produced by the hydrogenation of phenol. Subsequently, additional capacity was added by using a DSM design in which hot liquid cyclohexane was oxidized partially by compressed air. In this process as many as six steel reactors were used. About two months before the explosion, reactor number 5 leaked due to developmental of a crack extending about six feet. It was decided to install a temporary pipe in order to bypass the leaking reactor so that there will be continued operation of the plant while the repairs were carried out. However, on 1 June 1974, a massive release of hot cyclohexane occurred. This was followed by a release of hot cyclohexanone, which got ignited and a cloud of flammable vapor and a massive explosion occurred.

The reason for the explosion was attributed to the change in the design of the original plant.

Cyclohexanone is now made by the hydrogenation of phenol.

Seveso Disaster

In 1976, an explosion occurred at Seveso, Italy in a plant manufacturing herbicide. A dense white cloud of a poisonous gas consisting of 2, 3, 7, 8-tetra-chlorodibenzo-*p*-dioxan (TCDD) (Dioxan) was discharged in the atmosphere. Approximately an area of 150 km^2 with a population of about 40,000 was affected. This caused skin injuries to a large number of people who were exposed to the gas. The skin injuries, however, healed in about a month time. A large number of children suffered from chloracne, a condition characterized by skin blotches which disappeared in several months. A number of children born after the accident were premature and also deformed.

As per the records, during the Vietnam war (1970s) the U.S. military sprayed more than 40 l of herbicide agent orange (which was a mixture of 2, 4-dichloro-phenoxyacetic acid (2, 4–D) and 2, 4, 5 trichloro phenoxyacetic acid (2, 4, 5–T), a defoliant in order to destroy the forest cover so that the communist forces do not find place to hide themselves. Since the herbicide was contaminated with dioxans, it caused havoc in the life of millions of victims of the war and other residents of the area. It is known that dioxans are most important carcinogenic tested so far. The dioxans were absorbed in the skin of the victims. The children born to such women were deformed with low IQ and were mentally retarded. Skin tumours were reported in number of cases.

Dioxans are more dangerous than synthetic carcinogens and are most deadly chemicals known. As these are fat soluble, they bioaccumulate in the food chain. As per EPA (1984), dioxan exposure may cause a number of health problems, including diabetes, lowering immune system and cancer. According to WHO, the tolerable daily intake of dioxan is 1–4 picograms per kg body weight (one picogram is one-trillionth of a gram). The dioxans are known to be produced by the burning of chlorinated compounds, for example, garbage, medical waste and toxic chemicals. These are also produced during bleaching of paper with chlorinated compounds, manufacture of PVC and chlorinated pesticides. The discharged dioxans contaminate air, water and food.

Manufacture of DDT
Dichlorodiphenyltrichloroethane (DDT) is a common insecticide insoluble in water, but easily soluble in ethanol and acetone. This insecticide is useful against agricultural pests, flies, lice and mosquitoes. Widespread use of DDT has resulted in pollution of crop lands, and a large number of pests have become resistant to it. When it enters the food chain, DDT accumulates in the fatty tissues of animals. The long-term effects of DDT stored in body fat made the US Environmental Protection Agency to ban DDT. However, in developing countries it is still in use, particularly in those regions where malaria is still endemic.

DDT was introduced during World War II. It saved millions of lives through malaria control programs. It was discovered by a Swiss chemist Paul Muller, who won the Noble Prize in 1948. In spite of its tremendous service to humanity, DDT was banned in USA. It is known that many species of hunting birds, particularly those having high level of DDT, are threatened with extinction. It is found that the eggs of such species became too thin and fragile possibly due to interference with the hormones which control calcium deposition.

Love Canal Incident
In Niagara Falls, the Love Canal neighbourhood was built on an estimated 22,000 sq. ft. Chemical waste was buried in the abandoned canal. Subsequently, the residents of the love canal found strange fluids seeping into their basements which were responsible for health problems. The entire neighbourhood was finally abandoned and cordoned off.

Manufacture of Adipic Acid
It is well known that adipic acid is used for manufacture of nylon, polyurethane, lubricants and plasticizers. Approximately 2 billion kg of adipic acid are needed each year. The normal standard way of making adipic acid involves the use of benzene, a carcinogen. The procedure has been changed by the development of a process aided by biocatalysts and replacing benzene by simple sugar glucose.

As seen, all the episodes mentioned above were responsible for environmental problems that are mostly caused by the discharge of harmful substances into the environment. All such episodes could be controlled by the use of basic principles of green chemistry. A discussion on the basic principles of green chemistry forms the subject matter of a subsequent section.

1.4 Obstacles in the Pursuit of the Goals of Green Chemistry

As already stated, the environmental pollution can be eliminated or considerably reduced by following the principles of green chemistry. The most important principles (as we will see subsequently) include using renewable resources as starting materials in a chemical synthesis, using safer chemicals, economizing on atoms, using minimum energy for a process and discharging only the safe substances (or by-products) into the environment. However, a number of obstacles are there that hamper the goals of green chemistry. Some of these include:

- It is not always possible to procure starting materials for a reaction from renewable resources.
- Use of benign or safer solvents is not always possible. If feasible, a particular reaction could be conducted without using any solvent, in solid state.
- It is not always possible to economise on atoms. This means that all the atoms of the starting materials cannot be incorporated into the final products. It is well known that only rearrangement reactions and addition reactions are 100% atom-economical. All other reactions are not atom-economical.
- It is not always possible to discharge only the safer by-products into the environment.

It is very important to formulate guidelines and pass strict rules for the practising chemists. But the most important is to bring about changes at the grassroot level, which can be achieved by bringing about necessary changes in the chemistry curriculum in the colleges and the universities, as well as also in the secondary schools. A concerted and pervasive effort is needed to reach the widest audience. Bringing green chemistry to the classroom and the laboratory will have the desired effect in educating the students at various levels about green chemistry.

1.5 Principles of Green Chemistry

As already stated, green chemistry basically involves benign chemical synthesis. This objective can be achieved by following the 12 principles of green chemistry as suggested by Anastas and Warner [1]. The 12 principles of green chemistry are:

1. It is better to prevent waste than to treat or clean up waste after it is formed.
2. Synthetic materials should be designed to maximize the incorporation of all materials used in the process of the final product.
3. Wherever practicable, synthetic methodologies should be designed to use and generate substances that possess little or no toxicity to human health and the environment.
4. Chemical products should be designed to preserve efficacy of function while reducing toxicity.

5. The use of auxiliary substances (solvents, separation agents, etc.) should be made unnecessary whenever possible and, when used, innocuous.
6. Energy requirement should be recognized for their environmental and economic impacts, and should be minimized.
7. A raw material or feedstock should be renewable rather than depleting, whenever technically and economically practicable.
8. Unnecessary derivatization (blocking group, protection/deprotection, temporary modification of physical/chemical processes) should be avoided whenever possible.
9. Catalytic reagents (as selective as possible) are superior to stoichiometric reagents.
10. Chemical products should be so designed that at the end of their function they do not persist in the environment and break down into innocuous degradation products.
11. Analytical methodologies need to be further developed to allow for real-time, in-process monitoring, and control prior to the formation of hazardous substances.
12. Substances and the forms of a substance used in a chemical process should be chosen so as to minimize the potential for chemical accidents, including releases, explosions and fires.

1.6 Explanation of the 12 Principles of Green Chemistry

1. **It is better to prevent waste than to treat or clean up waste after it is formed**. It is best to carry out a synthesis by following a pathway so that formation of waste (by-products) is minimum or absent. It must be kept in mind that in most of the cases, the cost involved in the treatment and disposal of waste adds to the overall cost of production. The unreacted starting materials (which may or may not be hazardous) form part of the waste. The basic principle "prevention is better than cure" should be followed. If the waste is discharged into the atmosphere, sea or land, it not only causes pollution but also requires expenditure for cleaning up.
2. **Synthetic materials should be designed to maximize the incorporation of all materials used in the process into the final product**. It has so far been believed that if the yield in a particular reaction is about 90%, it is considered to be good. The percentage yield is calculated by

$$\% \text{ yield} = \frac{\text{Actual yield of the product}}{\text{Theoretical yield of the product}} \times 100$$

The above calculation implies that if one mole of a starting material produces one mole of the product, the yield is 100%. However, such a synthesis may generate significant amount of waste or by-products which is not visible in the above calculation. Such a synthesis, even though 100% (by above calculation) is not considered to be a green synthesis. For example, reactions like Grignard reactions and Wittig reaction may proceed with 100% yield, but they do not take into account the large amount of by-products obtained (Schemes 1.4 and 1.5).

A reaction or a synthesis is considered to be green if there is maximum incorporation of the starting materials or reagents in the final product. One should take into account the percentage atom utilization, which is determined by the following equation:

$$\% \text{ atom utilization} = \frac{\text{MW of desired product}}{\text{MW of desired product} + \text{MW of waste products}} \times 100$$

This concept of atom economy was developed by Trost [2] in a consideration of total amount of the reactants end up in the final product. The same concept was also determined by Sheldon [3] as given below.

$$\% \text{ atom economy} = \frac{\text{FW of atoms utilized}}{\text{FW of the reactants used in the reaction}} \times 100$$

The most common reactions we generally come across in organic synthesis are rearrangement, addition, substitution and elimination reactions. Let us find out which of the above reactions are more atom-economical.

(a) **Rearrangement Reactions**

These reactions involve rearrangement of atoms that make up a molecule. For example, allyl phenyl ether on heating at 200 °C gives o-allyl phenol (Scheme 1.6).

The rearrangement reaction (in fact all rearrangement reactions) is 100% atom-economical reaction, since all the reactants are incorporated into the product.

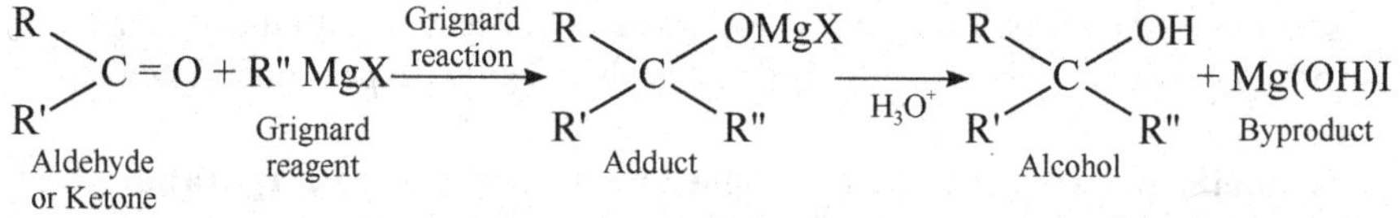

Scheme 1.4 Grignard reaction

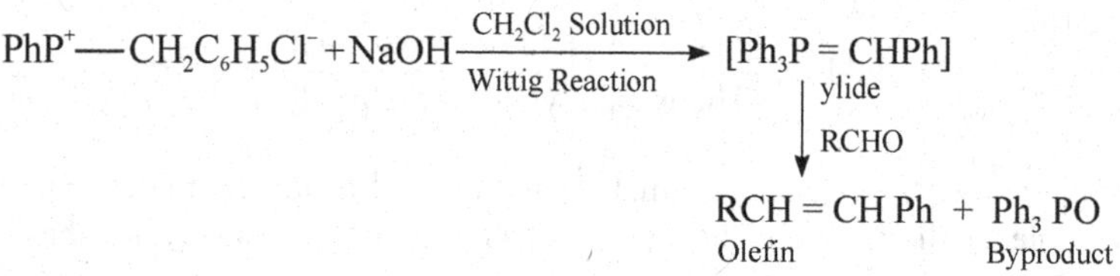

Scheme 1.5 Wittig reaction

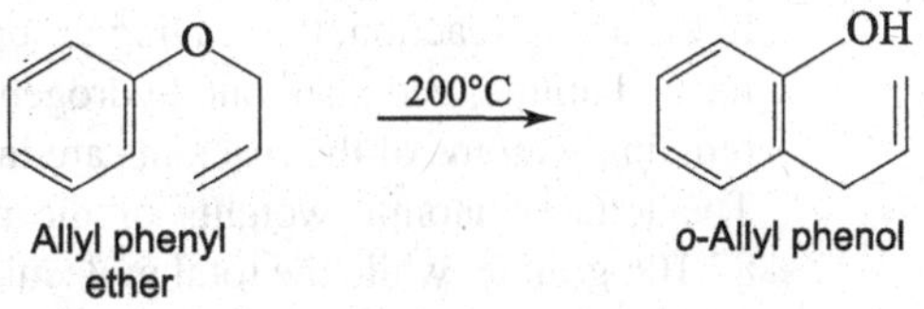

Scheme 1.6 A
rearrangement reaction

Allyl phenyl
ether

$\xrightarrow{200°C}$

o-Allyl phenol

Scheme 1.7 An addition
reaction

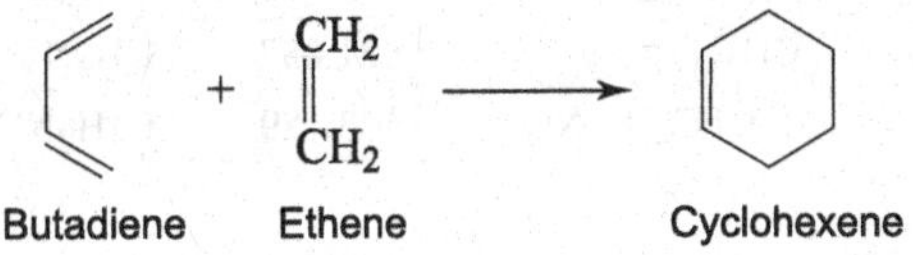

$$H_3C\,CH{=\!=}CH_2 + Br_2 \xrightarrow{CCl_4} H_3C\,CH\,Br\,CH_2Br$$

Propene   1,2-Dibromopropane

Scheme 1.8 Cycloaddition
reaction

Butadiene   Ethene   Cyclohexene

$$+ \begin{array}{c} CH_2 \\ \| \\ CH_2 \end{array} \longrightarrow$$

(b) **Addition Reactions**

Consider the bromination of propene (Scheme 1.7).

Here also, all the elements of the reactants (propene and bromine) are incorporated into the final product (1,2-dibromopropane). So, this reaction is also 100% atom-economical reaction.

In a similar way, cycloaddition reaction of butadiene and ethene (Scheme 1.8) and addition of hydrogen to an olefin (Scheme 1.9) is 100% atom-economical reaction.

(c) **Substitution Reactions**

In substitution reactions, one atom (or group of atoms) is replaced by another atom (or group of atoms). The atom or group that is replaced is not used in the final product. So, the substitution reactions are less atom-economical than rearrangement or addition reactions.

Let us consider the reaction of ethyl propionate with methyl amine (Scheme 1.10).

$$H_3C{-}CH{=\!=}CH_2 + H_2 \xrightarrow{Ni} H_3C{-}CH_2{-}CH_3$$

Propene   Propane

Scheme 1.9 Hydrogenation of ethene

$$\underset{\text{Ethyl propionate}}{CH_3CH_2\overset{O}{\overset{\|}{C}}C_2H_5} + \underset{\text{Methyl amine}}{H_3CNH_2} \longrightarrow \underset{\text{N-Methyl propamide}}{CH_3CH_2\overset{O}{\overset{\|}{C}}NHCH_3} + \underset{\text{Ethyl alcohol}}{CH_3CH_2OH}$$

Scheme 1.10 Substitution reaction

In the above reaction, the leaving group (OC_2H_5) is not incorporated in the formed amide, and also, one hydrogen atom of the amine is not used. The remaining atoms of the reactants are incorporated into the final product.

The total of atomic weights of the atoms in reactants that are used is 87.106 g/mole, while the total molecular weight including the reagent used is 133.189 g/mole. Thus, a molecular weight of 46.069 g/mole remains unused in the reaction.

Reactants		Utilized		Unutilized	
Formula	FW	Formula	FW	Formula	FW
$C_5H_{10}O_2$	102.132	C_3H_5O	57.057	C_2H_5O	45.061
CH_5N	31.057	CH_4N	30.049	H	1.008
Total $C_6H_{15}NO_2$	133.189	C_4H_9NO	87.106	C_2H_5OH	46.069

$$\text{Therefore, the atom economy } (\%) = \frac{87.106}{133.189} \times 100 = 65.40\%$$

(d) **Elimination Reactions**

In an elimination reaction, two atoms or groups of atoms are lost from the reactant to form a π bond. Consider the following Hofmann elimination reaction (Scheme 1.11).

The above elimination reaction is not very atom-economical. The percentage atom economy is 35.30% and is the least atom-economical of all the above reactions.

In the above reaction, only three carbon atoms and six hydrogen atoms are used in the formation of propene. The rest of the atoms are not used. The total atomic weight of all the atoms in the reagent that are used in the final product is 42.080 g/mole (see table below) and the total formula weight of all the reagents used in the reaction is 119.205 g/mole. This implies that 77.125 g/mole of the reactants are not used in

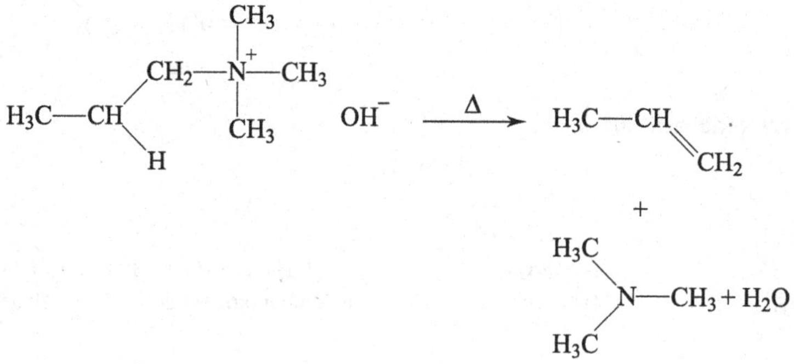

Scheme 1.11 Elimination Reaction

this reaction. By dividing the formula weight of the atoms used by the total formula weight of all the reactants used and multiplied by 100, the percentage atom economy is low at 35.30

$$\frac{42.080}{119.205} \times 100 = 35.30$$

Reagent formula	FW	Utilized formula	FW	Unutilized	
				Formula	FW
$C_6H_{17}NO$	119.205	C_3H_6	42.080	$C_3H_{11}NO$	77.125
Total $C_6H_{17}No$	119.205	C_3H_6	42.080	$C_3H_{11}NO$	77.125

Consider another elimination reaction involving dehydrohalogenation of 2-bromo-2-methylpropane with base to give 2-methylpropene (Scheme 1.12).

The above dehydrohalogenation reaction (an elimination reaction) is also not very atom-economical. The percentage atom economy is 27%, which is even less than the Hofmann elimination reaction.

3. **Wherever practicable, synthetic methodologies should be designed to use and generate substances that possess little or no toxicity to human health and the environment.**
    One of the most important principle of green chemistry is to prevent or at least minimize the formation of hazardous products which may be toxic and or environmentally harmful. In case hazardous products are formed, their effects on the workers must be minimized by the use of protective clothing, respirator and so on. This, of course, will add to the cost of production. At times, it is found that the controls may fail and there may be more risk involved. Green chemistry, in fact, offers a scientific option to deal with such situations.

4. **Chemical products should be designed to preserve efficacy of function while reducing toxicity.**
    It is extremely important that the chemicals synthesized or developed (e.g., dyes, paints, cosmetics, pharmaceuticals etc.) should be safe to use. A typical example of an unsafe drug is thalidomide (introduced in 1961) for reducing the effects of nausea and vomiting during pregnancy (morning sickness). The children born to women taking thalidomide suffered birth defects. Subsequently, the use of

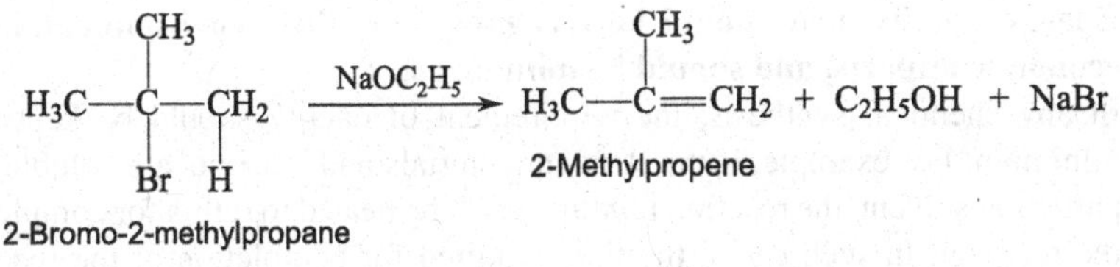

Scheme 1.12 Another elimination reaction

thalidomide was banned, the drug withdrawn and strict regulations passed for testing all new drugs.

It was later found that thalidomide is a chemical drug and exists in the enantiomeric forms. One of the enantiomer had the desired effect of curing morning sickness and the other enantiomer caused birth defects.

With the advancement of technology, the designing and production of safer chemicals has become possible. In fact, it is possible to manipulate the molecular structure to achieve this goal.

5. **The use of auxiliary substances (solvents, separation agents, etc.) should be made unnecessary whenever possible and, when used, innocuous**.

A number of solvents like methylene chloride, chloroform, perchloroethylene, carbon tetrachloride, benzene and other aromatic hydrocarbons have been used (in a large number of reactions) due to their excellent solvent properties. However, the halogenated solvents (mentioned above) have been identified as suspected human carcinogens. Also, benzene and other aromatic hydrocarbons are believed to promote cancer in humans and other animals.

The solvent selected for a particular reaction should not cause any environmental pollution and health hazard. The use of liquid carbon dioxide should be explored. If possible, the reaction should be carried out in aqueous phase or without the use of a solvent in solid phase.

A lot of concerns have been expressed about the use of solvents that have direct hazardous effect on the environment. One such example is chlorofluorocarbons (CFCs) which have been widely used as cleaning agents, blowing agents and as refrigerants. These CFCs are responsible for depleting the ozone layer, which in turn has disastrous effect on human survival. Even the volatile organic compounds (VOCs) like carbon tetrachloride, methylene chloride, chloroform and so on, which have been used as solvents in a number of applications, cause disastrous effects in the atmosphere. In view of all these effects, regulations have been made under the Clear Air Act in some of the advanced countries like USA to control many classes of chemical used as solvents.

It has already been stated that a major problem with many solvents is their volatility that may damage human health and the environment. To avoid this, a lot of work have been carried out on the use of immobilized solvents. These solvents maintain the solvency of the material, but are non-volatile and do not expose humans or the environment to the hazards of that substance.

As far as possible, the pathway for a reaction should be such that there is no need for separation or purification. By this procedure, the energy requirement is kept to a minimum.

6. **Energy requirement should be recognized for their environmental and economic impacts, and should be minimized**.

In any chemical synthesis, the requirement of energy should be kept to a minimum. For example, if the starting materials and reagents are soluble in a particular solvent, the reaction mixture has to be heated to reflux for completing the reaction. In such cases, the time required for completion of the reaction should be minimum so that least amount of energy is required. The use of a catalyst has the great advantage of lowering the requirement of energy of a reaction.

In case the final product is not pure, it has to be purified by distillation, recrystallization or ultrafiltration. All these steps require energy. The process should be designed in such a way that there is no need for separation or purification.

It is possible that the energy to a reaction can be supplied photochemically, by using microwave or sonication.

7. **A raw material or feedstock should be renewable rather than depleting, whenever technically and economically practicable**.

The starting materials can be obtained from renewable or non-renewable material. For example, petrochemicals are mostly obtained from petroleum oil, which is a non-renewable source since its formation take millions of years from animal and vegetable remains. The starting materials which can be obtained from agricultural or biological processes are referred to as renewable starting materials; however, these cannot be obtained in continuous supply due to factors like failure of crops and so on.

Substances like carbon dioxide (generated from natural sources or synthetic routes) and methane gas (obtained from natural sources such as marsh gas, natural gas etc.) are available in abundance; these are considered as renewable starting materials.

8. **Unnecessary derivatization (blocking group, protection/deprotection, temporary modification of physical/chemical processes) should be avoided whenever possible**.

A commonly used technique in organic synthesis is the use of protecting or blocking group. These groups are used to protect a sensitive moiety from the conditions of the reaction, which may make the reaction to go in an unwanted way if it is left unprotected. A typical example of this type of transformation would be protection of amine by making benzyl ether in order to carry out a transformation of another group present in the molecule. After the reaction is complete, the NH_2 group can be regenerated through cleavage of the benzyl ether (Scheme 1.13).

Derivatization of this type is quite common in the synthesis of fine chemicals, pharmaceuticals, pesticides and certain dyes. In the above example, benzyl chloride (a known hazard) needs to be handled with care and used in the preparation of the desired material and then regenerated as waste upon deprotection.

In the above procedure, the protecting group is not incorporated into the final product, and their use makes a reaction less atom-economical. Thus, as far as possible, the use of protecting groups be avoided. Though atom economy is a valuable criterion in evaluating a particular synthesis as "green", other aspects of efficiency must also be considered.

9. **Catalytic reagents (as selective as possible) are superior to stoichiometric reagents**.

In some reactions, the reactants (A and B) react to form a product (C), in which all the atoms contained within A and B are incorporated in the product (C). In

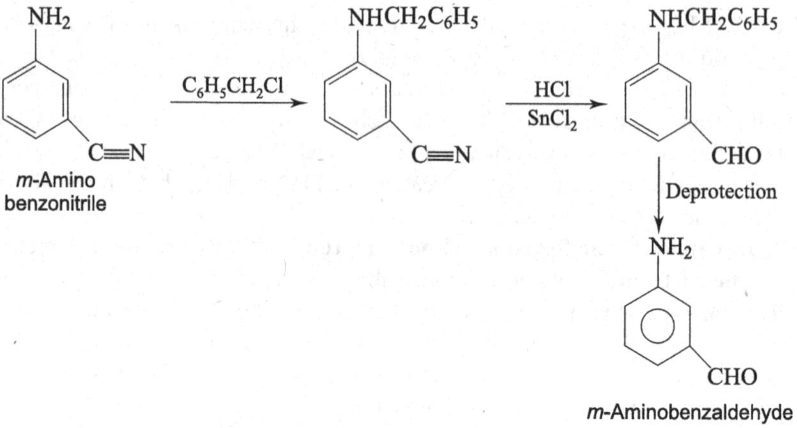

Scheme 1.13 Use of a protection group

such cases, stoichiometric reactions are equally environmentally benign from the point of material usage as any other type of reactions. However, if one of the starting materials (A or B) is a limiting reagent, in such cases, even if the yield is 100%, some unreacted starting material will be left over as waste. In other cases, if the reagents A and B do not give 100% yield of the product (C), both the excess of unreacted reagents will form part of waste. It is found that because of the reason mentioned above, catalysts wherever available offer distinct advantages over typical stoichiometric reagents. The catalyst facilitates the transformations without being consumed or without being incorporated into the final product.

Catalysts are selective in their action, in that the degree of reaction that takes place is controlled, for example, mono addition vs multiple addition; also, the stereochemistry is controlled (e.g., R vs. S enantiomer). By using catalysts, both starting material utilization is enhanced and formation of waste is reduced. An additional advantage of the use of catalyst is that the activation energy of a reaction is reduced, so the temperature necessary for the reaction is also lowered. This results in saving the energy.

It should be understood that in stoichiometric processes, the product obtained is one mole for every mole of the reagent used. However, a catalyst will carry out thousands of transformations before being exhausted.

Following are some of the applications of the use of catalysts:

(i)   Hydrogenation of olefins in the presence of nickel catalyst gives much better yields (Scheme 1.14).

(ii) Conversion of benzyl chloride into benzyl cyanide results in much better yields on using phase transfer catalysts (Scheme 1.15).

(iii) Oxidation of toluene with KMnO$_4$ in the presence of crown ether gives much better yield (Scheme 1.16).

$$H_3C—CH=CH_2 + H_2 \xrightarrow{\text{Ni}} H_3C—CH_2—CH_3$$

Propene   Propane

Scheme 1.14 Hydrogenation of olefin in the presence of Ni

Scheme 1.15 Use of a PTC

$$C_6H_5CH_2Cl + \text{aq. KCN} \xrightarrow{\text{PTC}} C_6H_5CH_2CN$$

Benzyl chloride   > 90% yield
Benzyl cyanide

Scheme 1.16 Oxidation of toluene by $KMnO_4$

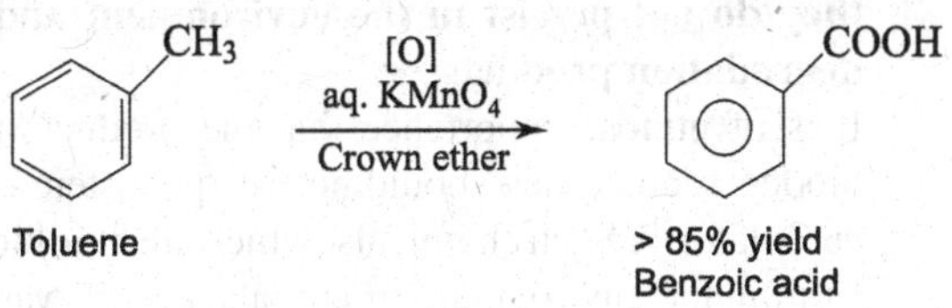

$$\xrightarrow[\text{Crown ether}]{\substack{[O] \\ \text{aq. } KMnO_4}}$$

Toluene   > 85% yield
Benzoic acid

(iv) Even in those cases where no reaction occurs usually, the reaction becomes feasible. An example is the hydration of alkynes to give aldehydes or ketones (Scheme 1.17).

(v) The selectivity enhancement takes place as shown by reduction of a triple bond to double bond (Scheme 1.18).

(vi) Selectivity in C-methylation versus O-methylation (Scheme 1.19).

$$HC\equiv CH + H_2O \xrightarrow[\text{H}_2\text{SO}_4]{\text{HgSO}_4} CH_3CHO$$

Acetylene   Acetaldehyde

$$CH_3C\equiv CH + CO + CH_3OH \xrightarrow{\text{Pd}} CH_3—\overset{\displaystyle O}{\underset{\displaystyle CH_2}{\overset{\| }{\underset{\|}{C}}}}—C—OCH_3$$

Propyne   Methyl methacrylate
(shell corporation)

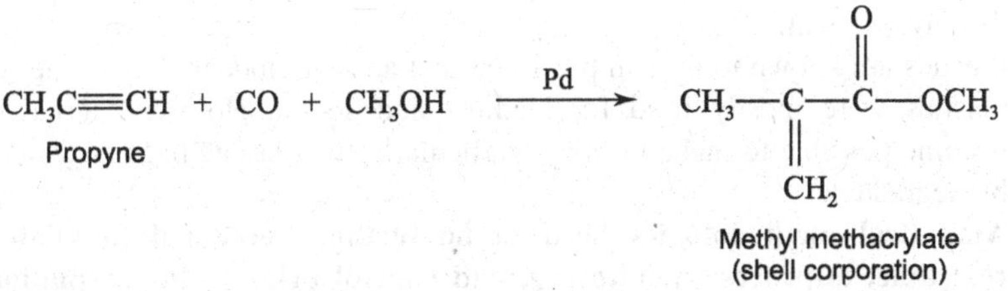

Scheme 1.17 Hydration of alkenes

$$H_3C—C\equiv CH + H_2 \xrightarrow[\text{Mono addition}]{\text{Pd-BaSO}_4} H_3C—CH=CH_2$$

Propyne   Propene

Scheme 1.18 Reduction of triple bond to double bond

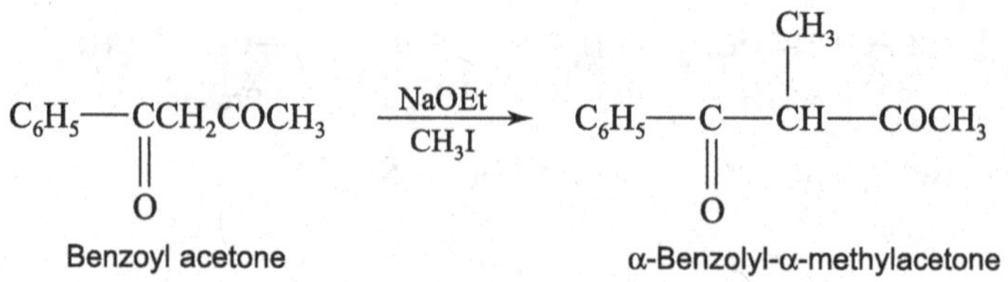

$$C_6H_5-\underset{\underset{O}{\|}}{C}CH_2COCH_3 \xrightarrow[CH_3I]{NaOEt} C_6H_5-\underset{\underset{O}{\|}}{C}-\overset{\overset{CH_3}{|}}{CH}-COCH_3$$

Benzoyl acetone   α-Benzolyl-α-methylacetone

Scheme 1.19 Selectivity in methylate vs O-alkylation

10. **Chemical products should be so designed that at the end of their function they do not persist in the environment and break down into innocuous degradation products.**

It is of utmost importance that the products that are synthesized should be biodegradable; they should not be "persistent chemicals" or "persistent bioaccumulators". Such chemicals (which are non-biodegradable) remain in the same form in the environment or are taken up by various plants and animal species, and accumulate in their systems; this is detrimental to the concerned species. The problem of non-biodegradability is generally associated with pesticides, plastics and a host of other organic molecules.

Most of the pesticides in use are organohalogen-based compounds. These pesticides though effective tend to bioaccumulate in plants and animals. The pesticide DDT was one of the first pesticide which bioaccumulated in plants and animals. Whenever a chemical is being designed, it should be made sure that it will be biodegradable. It is now possible to place functional groups and other features in the molecule which will facilitate its degradation. Functional groups which are susceptible to hydrolysis, photolysis or other cleavage have been used to ensure that products will biodegrade. It is equally important to make sure that the degradation products should not possess any toxicity and be detrimental to the environment.

Plastics are known to remain persistent and are not biodegradable. The waste plastics were mostly used for landfills and so on. However, it has now become possible to make plastics (particularly for garbage bags etc.) that are biodegradable.

11. **Analytical methodologies need to be further developed to allow for real-time, in-process monitoring, and control prior to the formation of hazardous substances.**

Analytical methodologies and technology have been developed which allow the prevention and minimization of the generation of hazardous substances in chemical processes. One need to have accurate and reliable sensors, monitors and analytical techniques to assess the hazards that are present in the process stream. Using various techniques, a chemical process can be monitored for generation of hazardous by-products and side reactions. These procedures can prevent any accident which may occur in chemical plants.

12. **Substances and the forms of a substance used in a chemical process should be chosen so as to minimize the potential for chemical accidents, including releases, explosions and fires.**

The occurrence of accidents in chemical industry must be avoided. The accidents in Bhopal (India) and Seveso (Italy) and many others have resulted in the loss of thousands of lives.

At times, it is possible to increase accident potential inadvertently in an attempt to minimize the generation of waste in order to prevent pollution. It has been noticed that in an attempt to recycle solvents from a process (in order to be economical and also prevent escape of solvent to the atmosphere) increases the potential for a chemical accident or fire. In fact, a process must balance the accident prevention with a desire for preventing pollution. A possible course is not to use volatile solvents, instead solids or low vapour pressure substance can be used.

1.7   Planning a Green Synthesis in a Chemical Laboratory

Following are some of the points that should be kept in mind for carrying out a synthesis in a chemical laboratory.

1.7.1   Percentage Atom Utilization

There should be maximum incorporation of the starting materials and reagents into the final product. This implies that a reaction should be 100% atom-economical. All reactions are not 100% atom-economical. As already stated, only the rearrangement reactions and addition reactions are 100% atom-economical. The substitution reactions and the elimination reactions are about 60–65% and 30–35% atom-economical. The latter two types of reaction generate wastes and by-products, which are not desirable.

1.7.2   Evaluating the Type of the Reaction Involved

The reaction involved must be evaluated with regard to its environmental impact or consequences. For this purpose, the nature of the starting material and the by-products (if formed) must be examined. Following are the different types of reactions which may be involved in a particular synthesis:

(a) **Rearrangement Reactions**

These are the reactions, as the name indicates, in which the atoms that comprise a starting molecule change its orientation relative to one another, including their connectivity and bonding pattern. Such reactions can be performed using a variety of procedures including thermal, photo and chemical means. From the point of view of green chemistry, in such reactions, both the starting materials and the end products contain the same atoms, so there is no waste generated. In fact, a rearrangement reaction is 100% atom-economical and fully efficient.

(b) **Addition Reactions**

In these reactions, a reagent adds to a substrate. All reagents and the substrates are consumed during the reaction. No additional by-products are generated and such reactions are very efficient, and like rearrangement reaction they are also 100% atom-economical. Some typical addition reactions include the addition of bromine to an olefin, Grignard reagent to a carbonyl compound and hydrogen cyanide to an α, β-unsaturated carbonyl compounds (Scheme 1.20).

(c) **Substitution Reactions**

In these reactions, the functional group of a substrate is replaced with another functional group. Typical examples include the well-known S_N1 and S_N2 reactions. In these reactions, nucleophilic reagents displace a leaving group in an aliphatic carbon atom; the product formed incorporates the nucleophile with the removal of the leaving group. Typical examples are given in Scheme 1.21.

In some cases, the leaving group is the desired product. For example, potassium iodide demethylation of a carboxylic acid methyl ester gives free carboxylate salt and methyl iodide (Scheme 1.22).

The usefulness of the methods depends on the nature of the leaving group generated. This pathway can be convenient and efficient if a substitution reaction sequence can be designed where the leaving group has been carefully selected.

(d) **Elimination Reactions**

These are reverse of addition reactions and there are procedures to generate unsaturation in the molecule. Examples of this type include dehydration of an

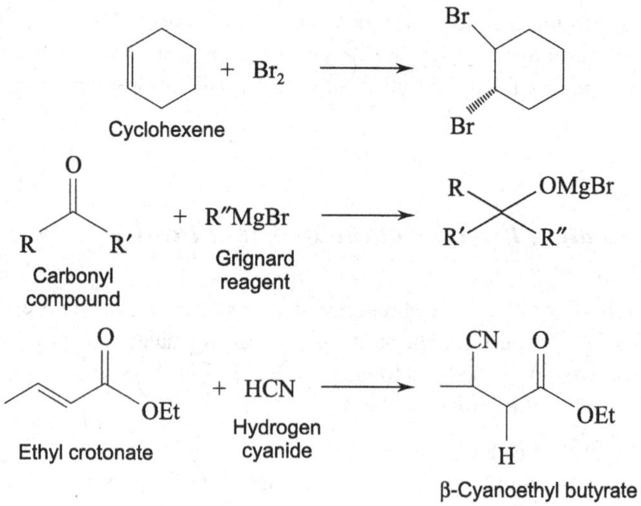

Scheme 1.20 Examples of addition reactions

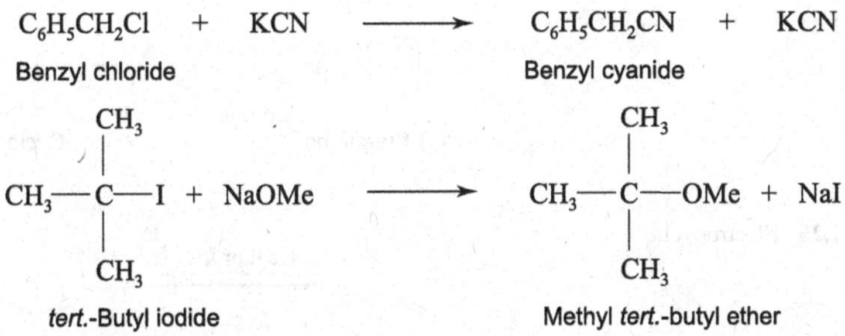

$$C_6H_5CH_2Cl + KCN \longrightarrow C_6H_5CH_2CN + KCN$$

Benzyl chloride   Benzyl cyanide

$$CH_3\text{—}\overset{\overset{\displaystyle CH_3}{|}}{\underset{\underset{\displaystyle CH_3}{|}}{C}}\text{—}I + NaOMe \longrightarrow CH_3\text{—}\overset{\overset{\displaystyle CH_3}{|}}{\underset{\underset{\displaystyle CH_3}{|}}{C}}\text{—}OMe + NaI$$

tert.-Butyl iodide   Methyl *tert.*-butyl ether

Scheme 1.21 Examples of substitution reaction

$$C_6H_5COOCH_3 + KI \longrightarrow C_6H_5COOK + CH_3I$$

Methyl benzoate   Pot-benzoate   Methyl iodide

Scheme 1.22 Synthesis of methyl iodide

alcohol to generate an olefin and loss of an alcohol from a hemiacetal to give an aldehyde (Scheme 1.23).

As in the case of substitution reactions, the environmental implications of the leaving group should be examined, evaluated and controlled.

(e) **Pericyclic Reactions**

These are concerted reactions and are characterized by the making or breaking of bonds in a single concerted step through a cyclic transition state involving π or σ electrons. Energy of activation for pericyclic reaction is supplied by heat in a thermally induced reaction and by ultraviolet light in a photo-induced reaction. Pericyclic reactions are highly stereospecific, and often the thermal and photochemical processes yield products with different but specific stereochemistry. Since pericyclic reactions do not involve ionic or free radical intermediates, solvents and nucleophiles or electrophilic reagents have no effect on the course of the reaction. We normally come across three types of pericyclic reactions.

(i) **Cycloaddition reactions**: In these reactions, two molecules combine to form a ring; two π bonds being converted to two signal bonds in the

Scheme 1.23 Some examples of elimination reaction

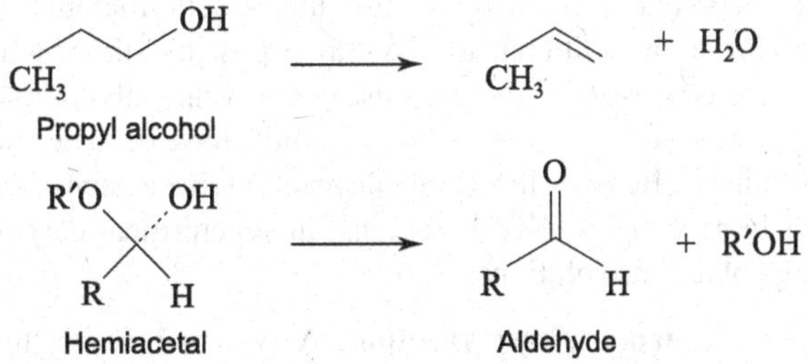

Propyl alcohol $\longrightarrow$ + H_2O

Hemiacetal $\longrightarrow$ Aldehyde   + R'OH

Scheme 1.24 Cycloaddition reaction

1,3-Butadiene   Ethene   Cyclohexene

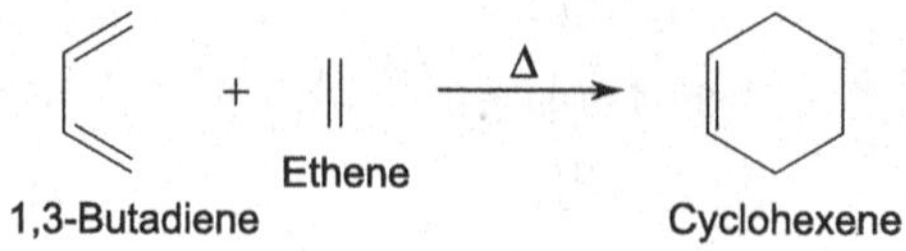

Scheme 1.25 Electrocyclic reaction

1,3,5-Hexatriene   Heat or $h\nu$ 1,3-Cyclohexadiene

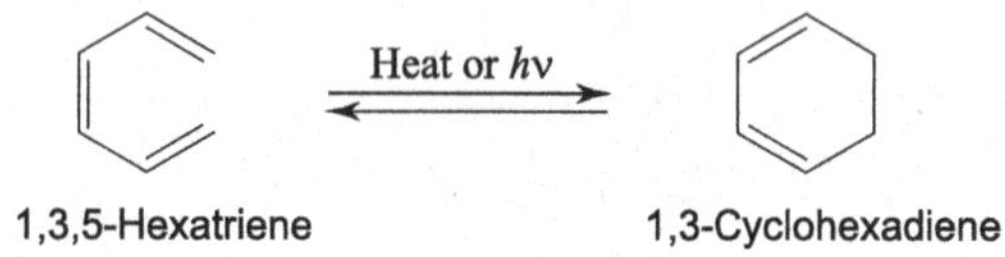

Scheme 1.26 Sigmatropic rearrangement

1,3-Pentadiene

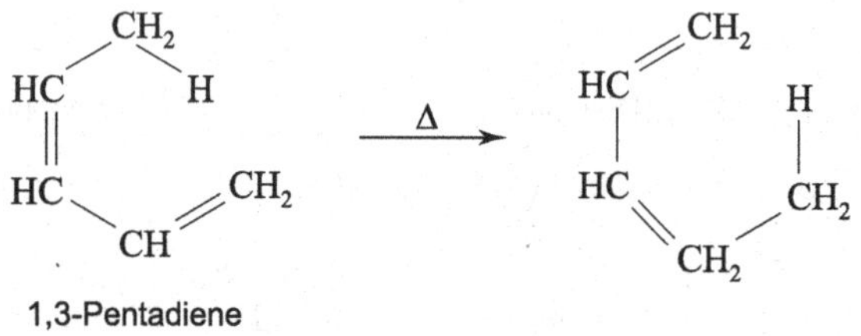

process. The most common example of a cycloaddition reaction is the Diels–Alder reaction (Scheme 1.24).

(ii) **Electrocyclic reactions**: These are reversible reactions in which a compound with two π electrons are used to form a sigma bond (Scheme 1.25).

(iii) **Sigmatropic rearrangements**: These are concerted intramolecular rearrangements in which an atom or a group of atoms shift from one position to another (Scheme 1.26).

Pericyclic reactions are very convenient from the point of view of environmental problems, since there is no by-product obtained in these reactions.

1.7.3   Selection of Appropriate Solvent

The solvent selected for a particular reaction should not have any environmental pollution or health hazards. As far as possible, the reactions should be performed in aqueous phase or without the use of a solvent (solventless reaction). Recently, a novel type of solvents known as 'ionic liquids' have been developed and are used in various synthesis. Besides, these organic reactions have been carried out in supercritical water or in near water (NCW) regions, in supercritical carbon dioxide and polyethylene glycol and its solutions.

(i) **Aqueous phase reaction**: A typical reaction that has been carried out in aqueous phase is the Diels–Alder reaction (Scheme 1.27).

Besides this, a number of other reactions have also been performed in aqueous phase; these will be discussed subsequently.

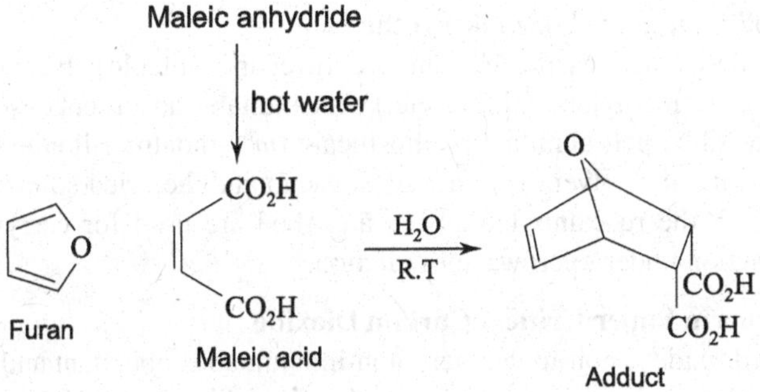

Maleic anhydride

hot water

Furan   Maleic acid   Adduct

Scheme 1.27 Aqueous Diels–Alder reaction

(ii) **Reactions in ionic liquid**: Ionic liquids are made up of at least two components in which the cation and anion can be varied. In these cases, the properties, such as melting point, viscosity, density and hydrophobicity, can be varied by simple changes to the structure of ions. The ionic liquids are immiscible in water. By choosing the correct ionic liquids, higher product yield can be obtained and a reduced amount of waste is produced in a given reaction.

Ionic liquids are good solvents for a wide range of both inorganic and organic materials. They are also immiscible with a number of organic solvents and provide a non-aqueous, polar alternative for two-phase system.

Ionic liquid find its applications in alkylations [4], allylations [5], hydroformy-lations [6], epoxidations [7], synthesis of ethers [8], Friedel–Crafts reaction [9], Diels–Alder reaction [10], Knoevengal condensation [11] and Wittig reaction [12].

For more details about the applications of ionic liquids, see Ref. [13].

(iii) **Organic synthesis in solid state**: The organic syntheses in solid state (viz., solvent-free organic synthesis and transformations) are mostly green reactions. These are of two types:

(a) **Solid phase organic synthesis without any solvent**
     The earliest record of an organic reaction in dry state is the Claisen rearrangement of allyl phenyl ether to *o*-allyl phenol (Scheme 1.28).
     A large number of reactions have now been performed without any solvent. These will be discussed subsequently.

Scheme 1.28 Claisen rearrangement

Allyl phenyl ether   *o*-Allyl phenyl

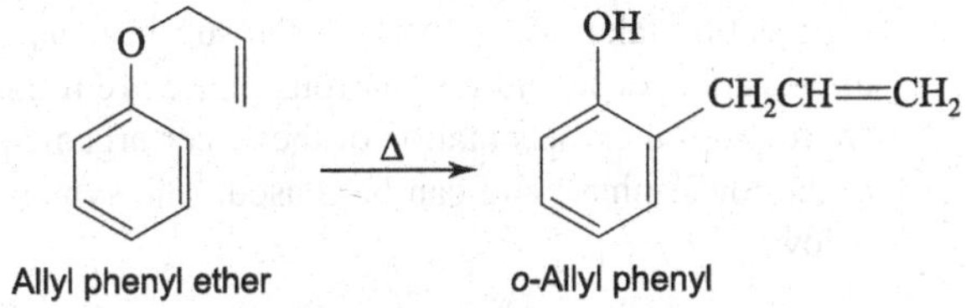

(b) **Solid-supported organic synthesis**

In these reactions, the reactants are stirred in a suitable solvent (e.g., water, alcohol, methylene chloride etc.) with a suitable adsorbent or solid support like silica gel, alumina, phyllosilicate (m^{n+}-montmorillonite etc.). After stirring, the solvent is removed in vacuo and the dried solid supports on which the reactants have been adsorbed are used for carrying out the reaction under microwave irradiation.

(iv) **Reactions in Super Critical Carbon Dioxide**

Carbon dioxide is non-toxic, inert, non-inflammable, abundant and is cheaper. As a solvent, it is used as super critical carbon dioxide, that is, above its critical temperature of 31.1 °C and critical pressure of 7.38 MPa. In this region, the density of SC-CO_2 and its solvent properties can be varied by change of pressure and temperature. Being a green solvent, it is used for a number of transformations in organic synthesis. An advantage associated with the use of SC-CO_2 is that after the reaction is complete the final product can be easily isolated by reducing the pressure or by removal of CO_2.

(v) **Reactions in polyethylene glycol and its solutions**

Polyethylene glycol (PEG, HO-$CH_2CH_2O)_n$ is available in a number of molecular weights ranging from 200 to tens of thousands. At room temperature, PEG-H_2O solutions are hydroscopic polymers, a colourless liquid of molecular weight less than 600 and a waxy white solid of molecular weight more than 800.

Polyethylene glycol has been approved by FDA for internal consumption. The aqueous solutions of PEG are biocompatible and are used in tissue culture and for the preservation of organs. PEG has been used as an alternative solvent to VOC. Also, PEG has low flammability and is biodegradable. PEG has been used in a number of organic transformations involving substitution reactions, oxidations, reductions, Williamson ether synthesis and as PTC.

(vi) **Reactions in Fluorous Solvents**

Fluorous solvents are mostly perfluorohydrocarbons, like perfluorohexane, perfluoroheptane, perfluorocyclohexane and so on. These are useful as solvent in synthetic organic chemistry. The fluorous solvents are virtually insoluble in common organic solvents at ambient temperature and form biphasic mixture in the presence of organic solvents like toluene or dichloro methylene. The miscibility increases with rise of temperature. The organic substrate is dissolved in an organic phase and perfluoro-tagged catalyst or reagent in the fluorous solvent added. At higher temperature, the biphasic mixture becomes monophasic and the reaction takes place between the substrate and the reagent. After cooling, the organic phase and the fluorous phase are reformed. The formed products are recovered by decantation of the upper organic phase. The catalyst remains in the lower phase and can be reused. The synthesis is represented as shown below:

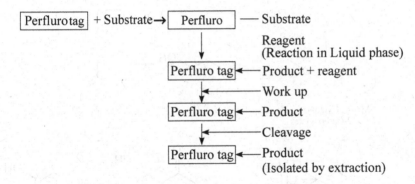

A number of organic transformations can be carried out using fluorous solvents.

1.7.4 Reagents

If possible, a green reagent is used. A reagent is considered to be green if it has no adverse effect on the environment. Some of such reagents include oxygen, singlet oxygen, ozone, hydrogen peroxide, dioxiranes, oxone ($2KHSO_4 \cdot KHSO_4 \cdot K_2SO_4$), peroxy acids, dimethyl carbonate and polymer-supported reagents.

It is best to use polymer-supported reagents. These are obtained by linking an organic reagent to a polymeric material (like polystyrene). The organic substrate is then added to the reaction mixture containing polymer-supported reagent. On completion of the reaction, the spent reagent is removed from the solution and the desired product is isolated. Special advantage of such reagents is the ease of separation of the product from the spent reagent. Various steps involved in organic synthesis are given below.

Polymer (P) + Reagent (R) ⟶ Polymer Supported reagent [P – R]

               ↓ Substrate

    Formed product + Spent Polymer + unused polymer reagent

            | Filtration

Filtrate	Residue
Formed product	Spent polymer
	+
	unused polymeric reagent

Some of the commonly used polymer-supported reagents include polymer-supported chromic and, polymeric Wittig reagent, polymer-supported aluminium chloride and polymer-supported trisubstituted phosphine dichloride.

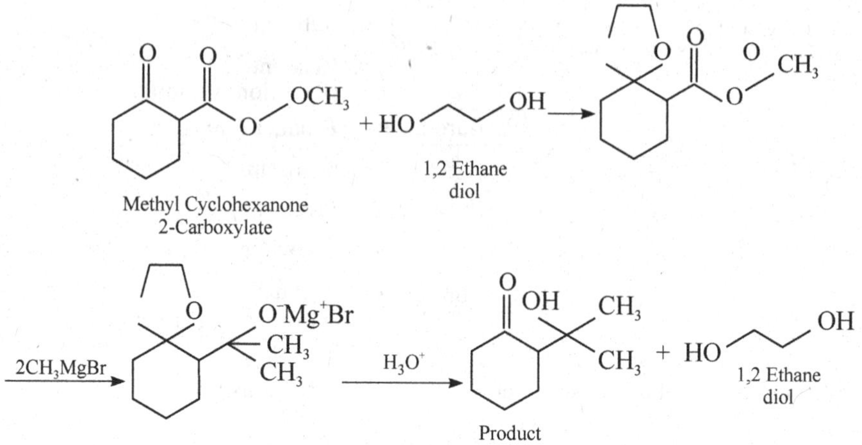

Scheme 1.29 Use of protecting group

1.7.5   Use of Protecting Groups

The use of protecting groups should be avoided as far as possible. These generate wastes. However, protecting groups are sometimes needed in order to solve a chemoselectivity problem. These groups should be added only in stoichiometric amounts and removed after the reaction is complete. Since these protecting groups are not incorporated into the final product, their use makes the reaction less atom-economical. A typical example of such a protecting group is 1,2 ethanediol, which is used to protect a keto group from reacting with a Grignard reagent as shown below (Scheme 1.29).

1.7.6   Use of Catalysts

It is well known that catalysts facilitate organic transformations. These can be affected in short duration of time and consume less energy, giving good yields. The catalysts are not consumed in a reaction and can be reused. Some of the reactions which do not take place normally can be made to go in the presence of catalyst. An example of such a reaction is given below (Scheme 1.30).

The catalysts are useful for the selective reduction of a triple bond to a double bond (Scheme 1.31).

The commonly useful catalysts include phase transfer catalyst and biocatalyst.

Scheme 1.30 Use of a catalyst

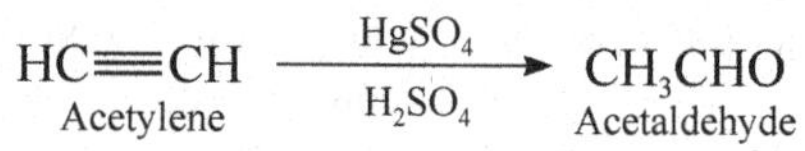

$$HC\equiv CH \xrightarrow[\text{H}_2\text{SO}_4]{\text{HgSO}_4} CH_3CHO$$

Acetylene   Acetaldehyde

Scheme 1.31 Reduction of
a triple bond to a double
bond

$$H_3CC \equiv CH + H_2 \xrightarrow{Pd - BaSO_4} H_3C\ CH = CH_2$$

Propyne   Propene

The phase transfer catalysts (PTC) are ionic substances, usually quaternary ammonium salts, in which the size of the hydrocarbon group in the cation is large enough to confer good solubility of its salt in organic solvent. The PTC reaction is, in fact, a methodology for accelerating a reaction between water-insoluble organic compound and water-soluble reagents. The main function of a catalyst (PTC) is to transfer the anion from aqueous phase to organic phase. As an example, heating and stirring a mixture of 1-chlorooctane with sodium cyanide for several days give practically no yield of 1-cyanooctane. However, addition of a small amount of a PTC makes the reaction to go on completion in about 2 h in excellent yield (C. M. Starks, J. Am. Chem. Soc., 1971, 93, 195) (Scheme 1.32).

The PTC methodology has been used for a number of organic reactions. Some of these include Hofmann catalysed carbylamine reaction, esterification, generation of dichlorocarbene, diazomethane, saponification, Darzens reaction, Williamsons ether synthesis, Wittig reactions and so on.

The biocatalysts are well-known enzymes and catalyse organic transformations. These can prevent generation of wastes by performing the process with high stereo- and regioselectivity and use water as the green solvent. In fact, the biocatalysts have become one of the greenest technologies. Some of the reactions which can be performed using biocatalysts include oxidations and reductions and so on.

Besides the above, another type of catalysts is **polymer-supported catalysts**. These are prepared by linking a polymer with a catalyst. The advantage is that such catalysts after the reaction can be recovered (by filtration) and can be reused. Some of the polymer-supported catalysts include polystyrene-aluminium chloride, polystyrene-based super acid catalyst, polymeric electrocyclic catalyst, polymer-supported PTC, polymeric triphase catalyst and polymeric photosensitizer.

1.7.7 Energy Requirement

Energy is essential for a chemical reaction. Its requirements, however, should be kept to a minimum. As an example, in case the starting materials are soluble in a particular solvent the reaction mixture has to heated for completion of the reaction. The time

$$CH_3(CH_2)_6\ CH_2Cl + NaCN + H_2O \xrightarrow[\substack{PTC \\ (CH_3CH_2)_{15}\ P^+(nBu)_3 \\ \Delta 105"c,\ 2Hr}]{decane} CH_3(CH_2)_6\ CH_2CN$$

1–chlorooctane   1–cyanooctane 95%

Scheme 1.32 Use of phase transfer reagent

necessary for heating for the completion of the reaction should be minimum so that minimum amount of energy is needed. It is well known that use of a catalyst lowers the requirement of energy for a reaction. The requirement for supplying energy can be best affected by microwave heating, sonication or photochemical activation.

Microwave-assisted organic reactions can be performed using water or organic solvents. The reactions can also be performed in solid state.

A number of organic reactions can be performed using microwave heating. These include Hofmann elimination, saponification, oxidation, esterification, trans-esterification, cycloaddition reactions, Diels–Alder reaction, Claisen rearrangement, reduction and so on.

Ultrasound has been used to accelerate a number of organic reactions. Some examples of reactions assisted by ultrasound include esterification, saponification, cyclopropanation, oxidation, reduction, hydroborations, coupling reactions, generation of dichlorocarbene, Strecker synthesis, Bouveault reaction, Reformatsky reaction, Grignard reagents, Cannizzaro reaction, Curtius rearrangement, Dieckmann cyclisation and so on.

Besides microwaves and sonication, energy for a chemical reaction can also be supplied photochemically, that is, carrying out the reaction in the presence of sunlight or UV lamp. A number of reactions can be carried out. These include synthesis of benzopinacol, isomerization of olefins, cycloaddition reactions, photo-oxygenation, Barton reaction, Friedel–Crafts reaction and so on.

1.8   Some Examples of Green Synthesis

1.8.1   Adipic Acid

On a global scale, approximately 1.9 billion kg of adipic acid per annum are required for the manufacture of nylon-66, polyurethane, lubricants, plasticizers and so on. Adipic acid was earlier synthesized from benzene, a non-renewable resource. Also, inhalation of benzene can cause cancer and leukaemia.

Various steps involved for the conversion of benzene into adipic acid are given in Scheme 1.33.

In the last step of the above synthesis, considerable amount of nitrous oxide gas (N_2O, also known as laughing gas) is produced as by-product, which finds its way into the atmosphere. The N_2O released rises to the stratosphere region where it causes destruction of the ozone layer. Due to this the UV radiation from the sun reaches the earth surface, which in turn result in increased risk of skin cancer and cataract.

In view of the above, attempts were made to manufacture adipic acid by a green process. This was achieved by John W. Frost and Karan M. Draths from Michigan State University. In this process glucose is converted into adipic acid using *Escherichia coli* bacteria, which is genetically engineered using recombinant DNA technology. In this process, a segment of DNA responsible for the desired quality is

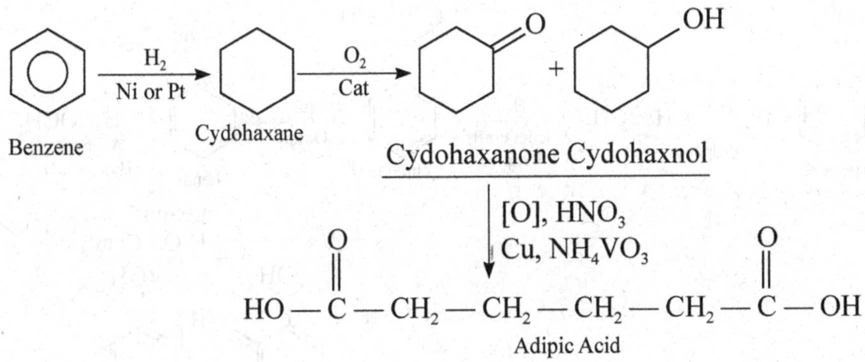

Scheme 1.33 Synthesis of adipic acid from benzene

spliced out of one organism and subsequently added to the DNA of another segment, giving a second organism of desired quality that was present in the first. This technology alters the metabolic pathway of glucose in *E. coli*. Various steps involved in the conversion of glucose into adipic acid are given in Scheme 1.34.

As observed, the above green synthesis of adipic acid uses glucose, a non-toxic renewable resource (glucose can be obtained from biomass). The biocatalytic process has been used for the conversion of starch into glucose. However, the use of cellulose feed stock (derived from agricultural waste) could make the biocatalytic synthesis of adipic acid more commercially viable.

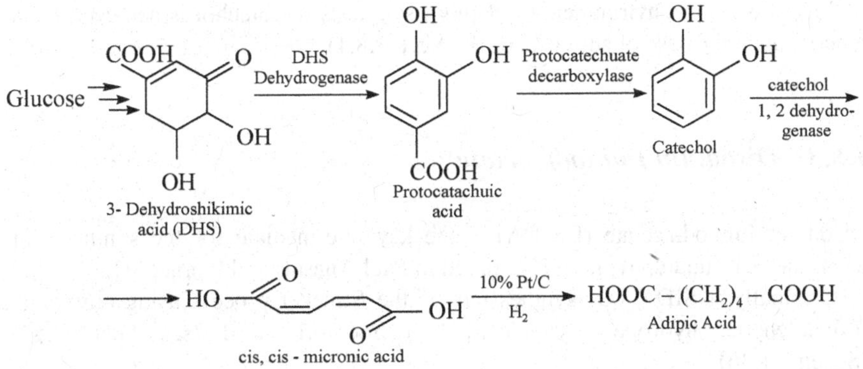

Scheme 1.34 Green synthesis of adipic acid from glucose

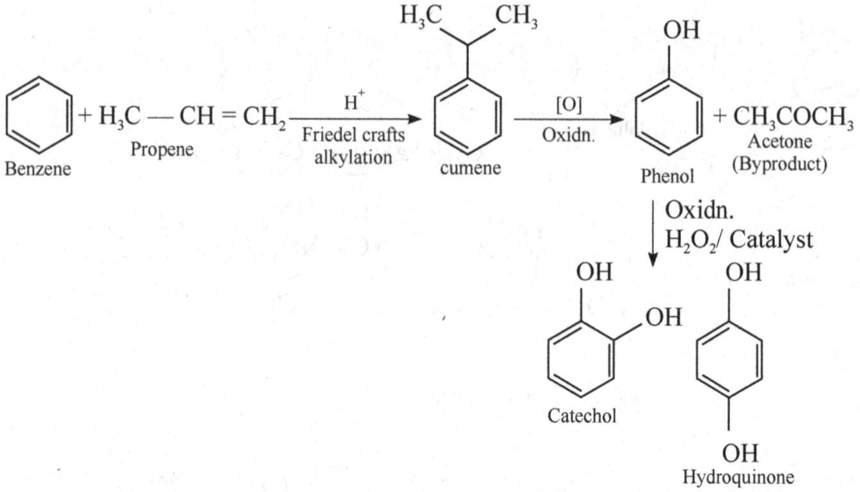

Scheme 1.35 Synthesis of catechol from benzene

1.8.2 Catechol

Catechol like adipic acid is also obtained from benzene (J. W. Frost, K. M. Draths, Chemistry in Britain, 1995, 31(3), 206–210). Various steps involved in the synthesis of catechol from benzene are given in Scheme 1.35.

The problems associated with the use of benzene have already been discussed in Sect. 1.8.1. An environmentally benign synthesis of catechol is achieved from glucose as in the case of adipic acid (see Sect. 1.8.1).

1.8.3 Disodium Iminodiacetate

Disodium iminodiacetate (DSIDA) is the key intermediate for the synthesis of Monsanto's Round up (r) herbicide (cited in Paul Anastas and Warner [1]).

Originally, DSIDA was synthesized by the Strecker process using ammonia, formaldehyde, hydrogen cyanide and hydrochloric acid (see Sect. 2.52) (Scheme 1.36).

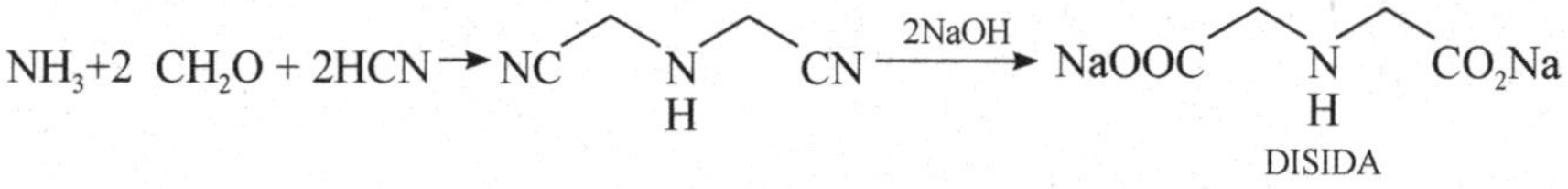

Scheme 1.36 Manufacturing of DSIDA using Strecker process

In the above process, HCN, a deadly substance, is used. Additional waste containing traces of CN^- and HCHO are also generated to an extent of about 1 kg per 7 kg of the product. This waste must be treated before disposal.

An alternative green process for the manufacture of DSIDA has been developed by Monsanto. It involves copper-catalysed dehydrogenation of diethanol amine (DEA). This procedure does not involve the use of HCN and HCHO and is safer giving better yield.

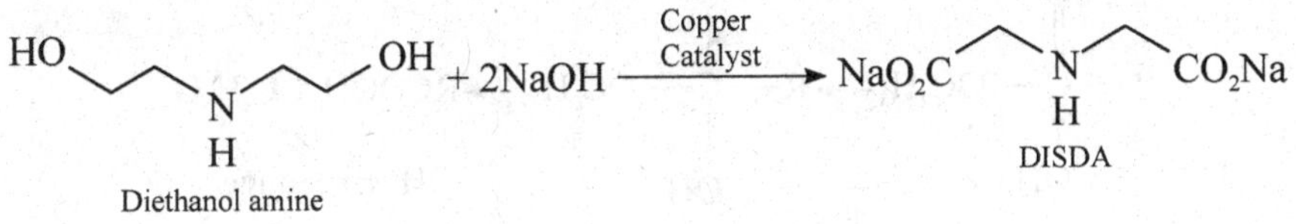

Alternative synthesis of DSIDA

1.8.4 Hofmann Elimination

For details see Sect. 2.33.

1.8.5 Benzoic Acid from Methyl Benzoate

Methyl benzoate on heating with aqueous KOH followed by acidification gives benzoic acid. The process is known as saponification (Scheme 1.37).

In the above process, better yields are obtained by heating the aqueous alkali under microwave irradiation for 2.5 min to give the corresponding acid (Scheme 1.38).

(Conversion of methyl benzoate, into benzoic acid with aq. alkali using microwave irradiation).

Even hindered esters can be saponified by the above process.

The mechanism of saponification is given in Scheme 1.39:

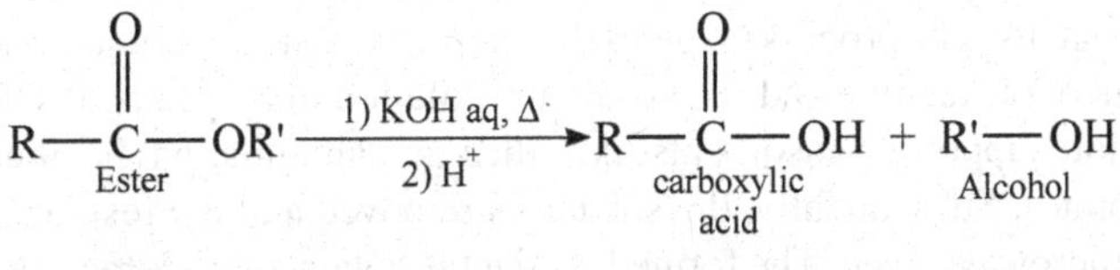

Scheme 1.37 Saponification of esters

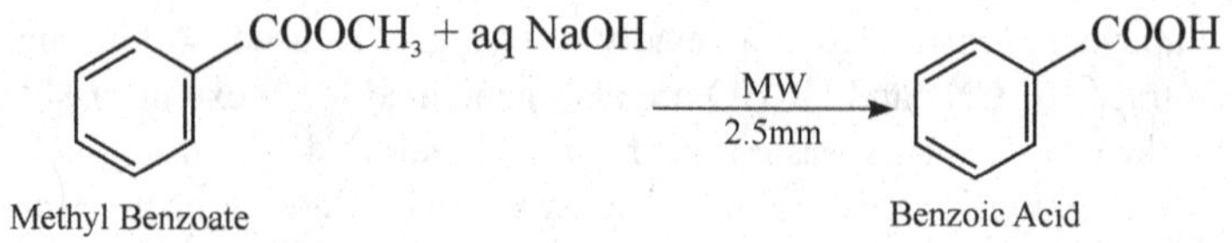

COOCH$_3$ + aq NaOH

$\xrightarrow[\text{2.5mm}]{\text{MW}}$

COOH

Methyl Benzoate   Benzoic Acid

Scheme 1.38 Saponification of esters using microwaves

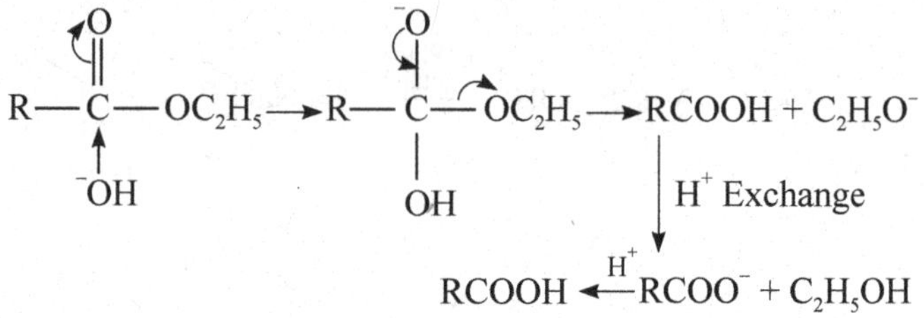

$$R—\overset{\overset{\displaystyle O}{\|}}{\underset{\underset{\displaystyle ^-OH}{\uparrow}}{C}}—OC_2H_5 \longrightarrow R—\overset{\overset{\displaystyle ^-O}{\|}}{\underset{\underset{\displaystyle OH}{|}}{C}}—OC_2H_5 \longrightarrow RCOOH + C_2H_5O^-$$

$$\Big\downarrow H^+ \text{ Exchange}$$

$$RCOOH \xleftarrow{H^+} RCOO^- + C_2H_5OH$$

Scheme 1.39 Mechanism of saponification

Scheme 1.40 Oxidation of toluene to benzoic acid

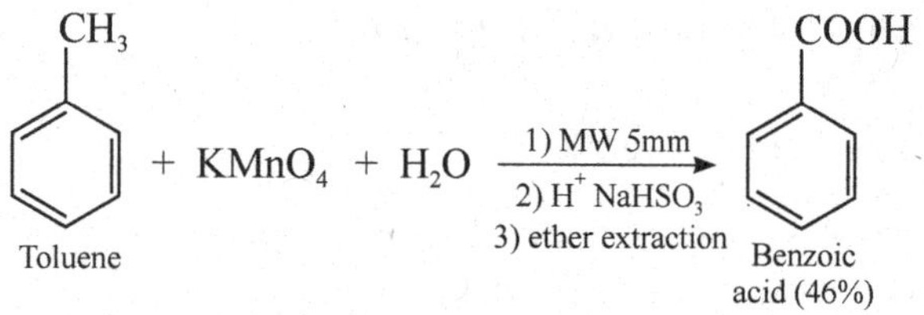

CH$_3$

$+$ KMnO$_4$ $+$ H$_2$O $\xrightarrow[\substack{\text{2) H}^+\text{ NaHSO}_3 \\ \text{3) ether extraction}}]{\text{1) MW 5mm}}$

COOH

Toluene   Benzoic
acid (46%)

1.8.6 Benzoic Acid by Oxidation of Toluene

Toluene can be oxidized to benzoic acid by heating with KMnO$_4$ solution and KOH for 5 min in a microwave oven (Scheme 1.40).

1.8.7 Oxidation of Alcohols to Carbonyl Compounds

Alcohols could be converted to carbonyl compounds by heating in solid state using supported reagents. The process involves dissolving the reactants in suitable solvents, like water, alcohol, acetone and so on, and the solution thus obtained is stirred with a suitable solid support or adsorbents like silica gel, alumina, phyllosilicate (M^{n+1}-montmorillonite). Subsequently, the solvent is removed and the residual product is heated in a microwave oven. The formed product is obtained by extraction with ester.

Some of the supporting agents used are Clayfen (clay-supported iron (III) nitrate), Clay-top-H$_2$O$_2$, manganese dioxide-silica, Wet CrO$_3$ Al$_2$O$_3$, copper sulphate-alumina and so on (R. S. Varma, Green Chemistry, 1999, 417) (Scheme 1.41).

Scheme 1.41 Oxidation of alcohols to carbonyl compounds

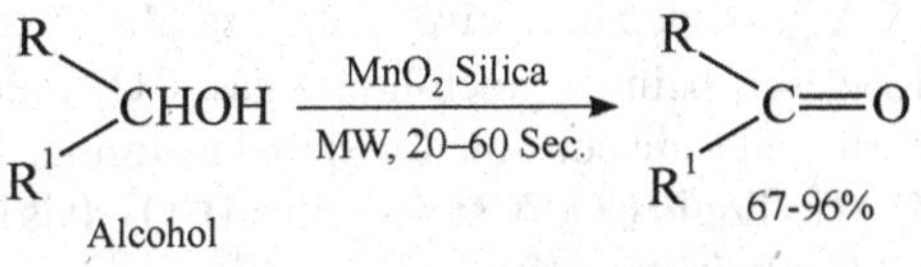

Scheme 1.42 Decarboxylation

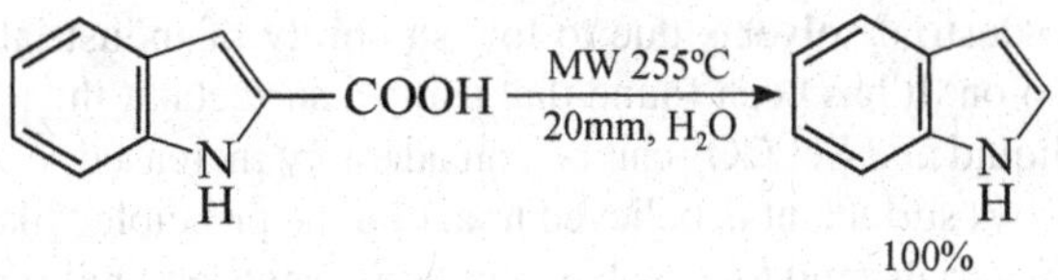

1.8.8 Diels Alder Reaction

For details see Sect. 2.22.

1.8.9 Decarboxylation Reactions

Carboxylic acids can be decarboxylated by heating in quinoline in a microwave oven. A Typical example is given in Scheme 1.42.

1.8.10 Sonochemical Simmons–Smith Reaction

For details, see Sect. 2.4.8.1.

1.8.11 Surfactants for Carbon Dioxide

Volatile organic solvents are known to be used to an extent of about 30 million pounds per annum for various purposes. The use of VOCs is responsible for a number of environmental problems.

A possible alternative to the use of VOCs and CFCs is to use CO_2 as a solvent. However, being a green-house gas, CO_2 reflects back the IR radiations toward the earth and is responsible for global warming. The problem can be addressed by using CO_2 as liquid CO_2 or super critical CO_2. It is known that SC-CO_2 is used as a solvent to decaffeinate coffee.

CO_2 in liquid or SC-CO_2 form has so far not been used as a replacement of industrial solvents due to low solubility of industrial materials, like polymers and so on. It has been found that using a surfactant, the solubility of such substances in liquid and SC-CO_2 can be considerably increased.

A surfactant is believed to act on the principle "like dissolving like". Thus, polar materials tend to dissolve in polar solvents and non-polar solutes tend to dissolve in non-polar solvents. A surfactant has one end having polarity similar to the polarity of the particle to be emulsified and the other end having polarity similar to the polarity of the solvent. The surfactant molecules, in fact, assemble themselves in a spherical structure called a micelle. Thus, a surfactant acts to stabilize the non-polar particle in the polar solvent, resulting in the dissolution of a non-polar solute in a polar solvent. In this way the soaps and detergents act as surfactants to dissolve grease in water.

It was found by Joseph M. Desimone of the University of North Carolina and North Carolina State University that polymers having a carbon back-bone chain with most of the carbons attached to fluorine atoms are soluble in liquid and SC-CO_2. Such polymers are called fluoropolymers. In fact, Desimone synthesized such fluoropolymers having polystyrene blocks which are insoluble in CO_2 and poly 1,1-dihydroperfluorooctyl acrylate block and a graft segment which is soluble in CO_2. A typical fluoropolymer is given in Scheme 1.43:

The waxes, grease and oil present in the fabric get trapped in the micelle structure (of the fluoropolymer) and get carried away in liquid or super critical CO_2 which is used as a solvent.

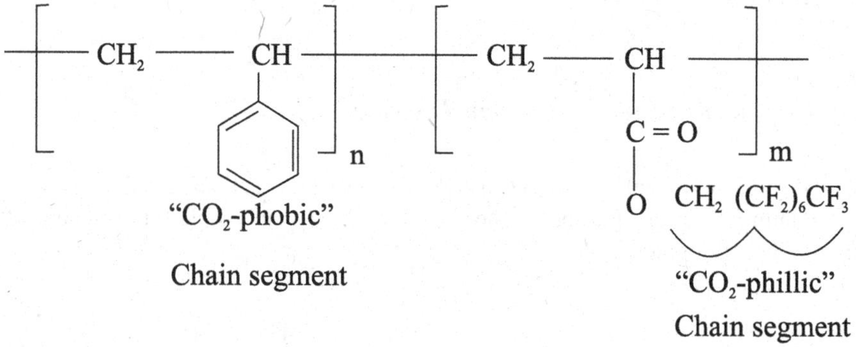

Scheme 1.43 A typical fluoropolymer

Desimone along with T. Romack and J. MacClain produced dry cleaning machines which use liquid CO_2 and a surfactant for dry cleaning of fabrics. The net result is replacement of environmentally harmful PERC which is a suspected human carcinogen.

Source: Real-world cases in Green Chemistry, American Chemical Society, 2000 Page 13.

1.8.12   A Safe Marine Antifoulant

Fouling involves growth of barnacles, algae, plants and diatoms on the hulls of boats resulting in hydrodynamic drug on the boat. This increases the cost of additional fuel consumption. Normally, tributyltin compounds, for example, tributyltin oxide (TBTO) were used to reduce or prevent fouling. TBTO is used by mixing with paint, which is used to paint the boat (Scheme 1.44).

The main drawbacks in the use of TBTO are the tendency of bioaccumulation in marine animals, toxicity to many organisms and being persistent in the environment.

An active foul-inhibiting agent, 4, 5-dichloro-2-n-octyl-4-isothiazoline-3-one (DCOI) has been developed by Rohm and Haas (Scheme 1.45).

The DCOI is far less persistent in marine environments. After being leaked from hull coatings, DCOI get bound to soil particles in the sediments and is less available and so does not pose threat to non-target organisms. In the organisms DCOI gets broken quickly.

Source: Real-world cases in Green Chemistry, American Chemical Society, 2000, Page 37–41.

Scheme 1.44   TBTO

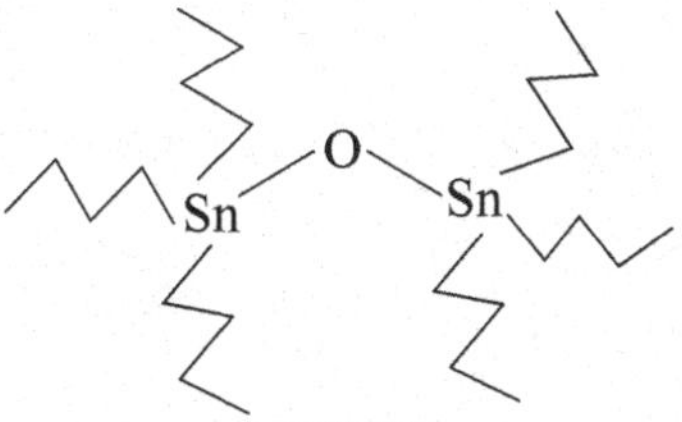

(TBTO)

Scheme 1.45   DCOI

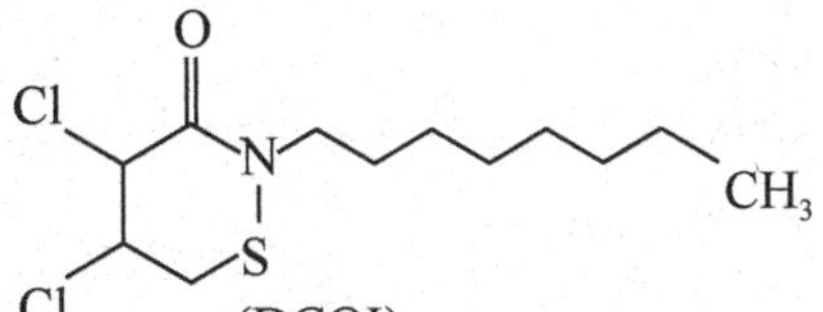

(DCOI)

References

1. Paul T. Anastas and John C. Warner, Green Chemistry Theory and Practice, Oxford University Press, New York 1998.
2. Barry M. Trost, Science, 1991, *254*, 1471.
3. Roser A. Sheldon, Chem. Ind. (London), 1992, 903.
4. Chauvin, Y., Hirschauer, H., Olivier, H., J. Mol. Catal., 1994, *92*, 155.
5. Chen, W., Xu L., Chatterton, C., Xiao, Chem. Commun., 1999, 1247.
6. Fuller, A., Breda, A.C., Carlin, R.T., J. Electrochem. Soc., 1997, *144*, 67; Favrc, F., Olivier, H. *et al.*, Chem. Commun., 2001, 1360.
7. Song, C.E., Roh, E.J., Chem. Commun., 2000, 837.
8. Fraga-Dubreeuil, J., Bourahla, K., Rahmouni, M., Bazureau, J.P., Hamelin, J., Catal. Commun., 2002, *3*, 185.
9. Boon, J.A., Levisky, J.A., Pflug, J.I., Wilkes, J.S., J. Org. Chem., 1986, *51*, 480.
10. Earle, M.S., McMormae, P.B., Sedden, K.R., Green Chem., 1999, *1*, 23.
11. Harjani, J.R., Orara, S.J., Salunkhe, M.M., Tetrahedron Lett., 2002, *43*, 1127.
12. Wheeler, C., West, K.N., Liottda, C.L. Eckert, C.A., Chem. Commun., 2001, *88*.
13. Ahluwalia V.K., Aggarwal, R. Organic Synthesis, Special Techniques, Narosa Publishing House, 2005, 195–210.
14. V.K. Ahluwalia and R.S. Verma, Green Solvents for Organic Synthesis, Narosa Publishing House, 2009.

Chapter 2
Green Reactions

Introduction

A majority of organic reactions require the use of volatile solvents and dry conditions, and produce a number of by-products which are harmful to the environment. With the advancement of knowledge and development of new and better techniques, a large number of organic reactions could be carried out in eco-friendly conditions. It has now been possible to carry out those reactions which earlier needed anhydrous conditions, in aqueous phase and in supercritical water or in near critical water (NCW) regions. Besides water, reactions have also been conducted using supercritical carbon dioxide, ionic liquids, polyethylene glycol and its solutions. Also, the yields in a number of reactions have been increased by the use of microwaves, sonication and fluorous phase techniques, and catalysts including biocatalysts. Such reactions are now designated as green reactions. For the sake of understanding, the reaction conditions used earlier and their mechanism as well as the new conditions which make them green are as follows.

2.1 Acyloin Condensation [1]

The process consists of treating carboxylic ester with metallic sodium in large volume of benzene or toluene followed by protic solvents. The product obtained is α-hydroxy ketone, called acyloin. This condensation is called acyloin condensation. In this reaction, large dilution is required to ensure intramolecular condensation as against intermolecular reaction. A typical acyloin condensation is given in Scheme 2.1.

Mechanism

A radical mechanism is believed to occur. The metallic sodium donates its electron to the carbonyl carbon to give the species (1), which dimerizes to give (2). Subsequent loss of alkoxy group produces 1,2-diketone (3). Further reduction gives sodium salt

© The Author(s) 2021
V. K. Ahluwalia, *Green Chemistry*,
https://doi.org/10.1007/978-3-030-58513-6_2

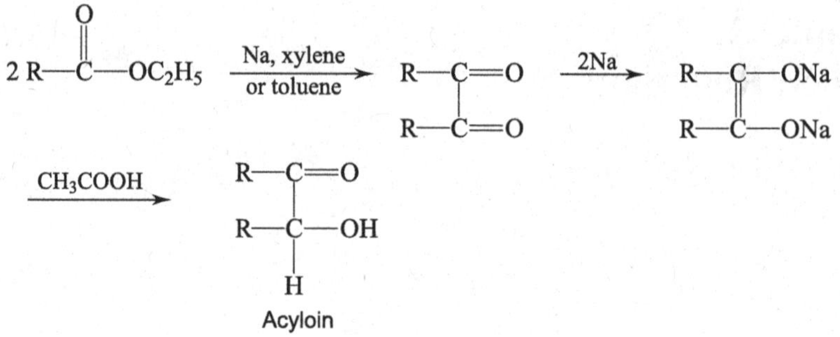

Scheme 2.1 Acyloin condensation

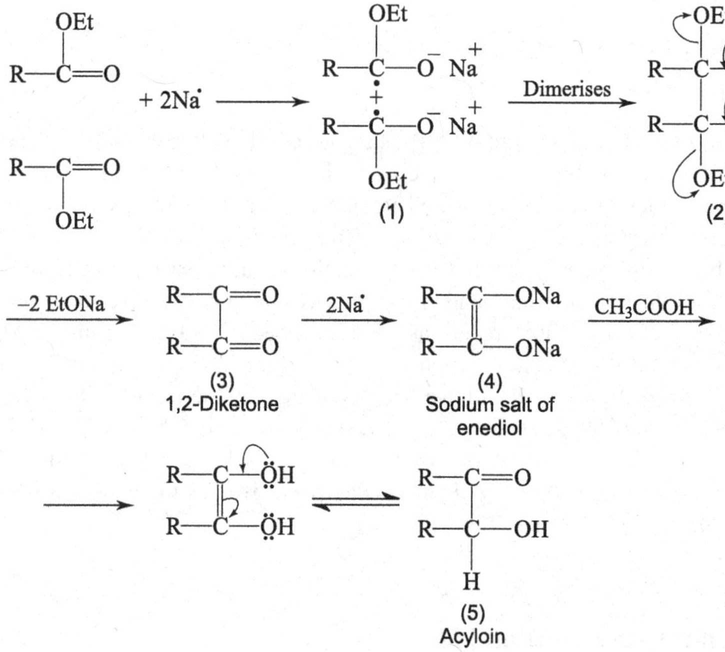

Scheme 2.2 Mechanism of acyloin condensation

of enediol (4). Finally, addition of acids yields 1,2-diol which tautomerizes to acyloin (5) (Scheme 2.2).

Large ring compounds (cyclic acyloins) are obtained by using long-chain dicarboxylic esters. The method is best suited for closing rings of ten members or more. In cyclic acyloins, dilution technique is not required. The yield is 60–95% for 10–20-membered ring (Scheme 2.3).

In case of cyclic acyloins, the mechanism suggested is given in Scheme 2.4.

In acyloin condensation, much better results are obtained if the reactions are conducted in trimethyl silane; even four-membered rings can be prepared easily under these conditions. The chlorotrimethyl silane acts as a scavenger for the alkoxide ion

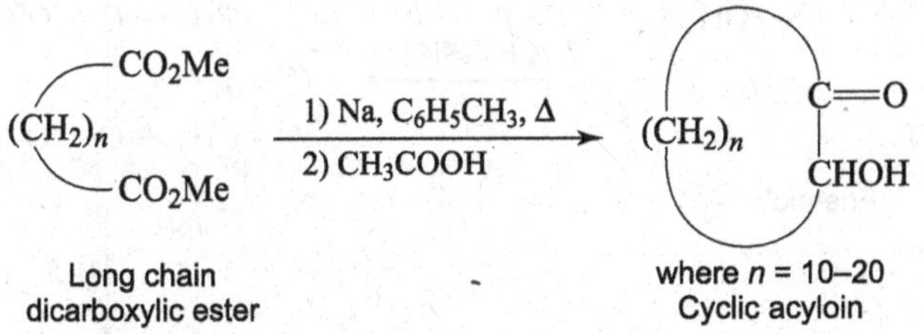

Scheme 2.3 Synthesis of large ring compounds

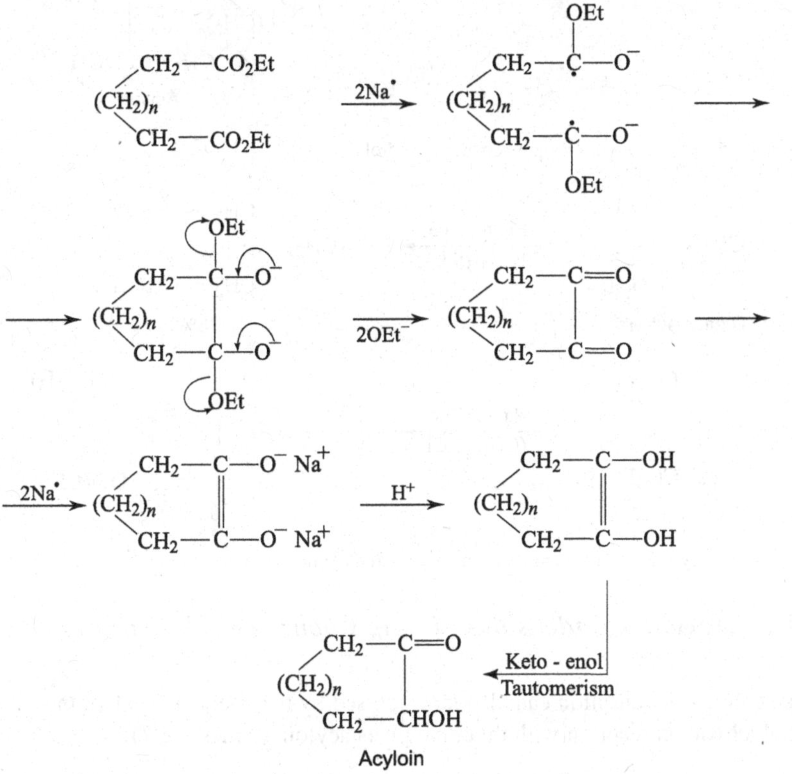

Scheme 2.4 Mechanism in case of cyclic acyloins

liberated during the reaction (*see* Scheme 2.4) so that the reaction medium is kept neutral and the wasteful base-catalysed side reactions, such as β-elimination and Claisen or Dieckmann condensations, are avoided. Also, the oxygen-sensitive ene-diol is protected as the bis-trimethyl silyl ether, which may be isolated and purified before hydrolysis to acyloin (Scheme 2.5).

Using the above methodology, diethyl suberate was converted into seven-membered cyclic acyloin with 75% yield and the four-membered ring compound with 90% yield (Scheme 2.6).

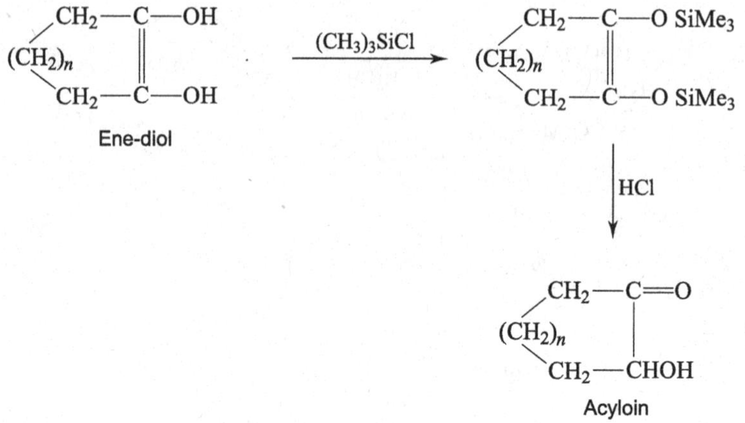

Scheme 2.5 Protection of oxygen-sensitive enediol

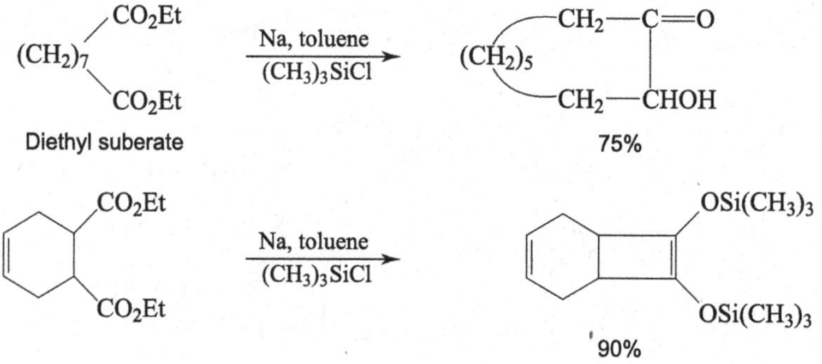

Diethyl suberate   75%

90%

Scheme 2.6 Synthesis of seven- and four-membered cyclic acyloin

2.1.1 Acyloin Condensation Using Coenzyme, Thiamine

The acyloin condensation can also be catalysed by the coenzyme 'thiamine'. Thus, acetaldehyde on reaction with thiamine gives acyloin (Scheme 2.7).

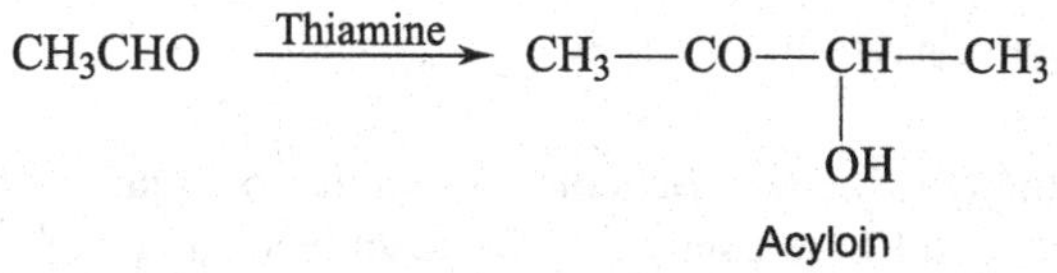

$$CH_3CHO \xrightarrow{\text{Thiamine}} CH_3—CO—\underset{\underset{OH}{|}}{CH}—CH_3$$

Acyloin

The mechanism of the reaction is given below.

The above reaction is similar to benzoin condensation.

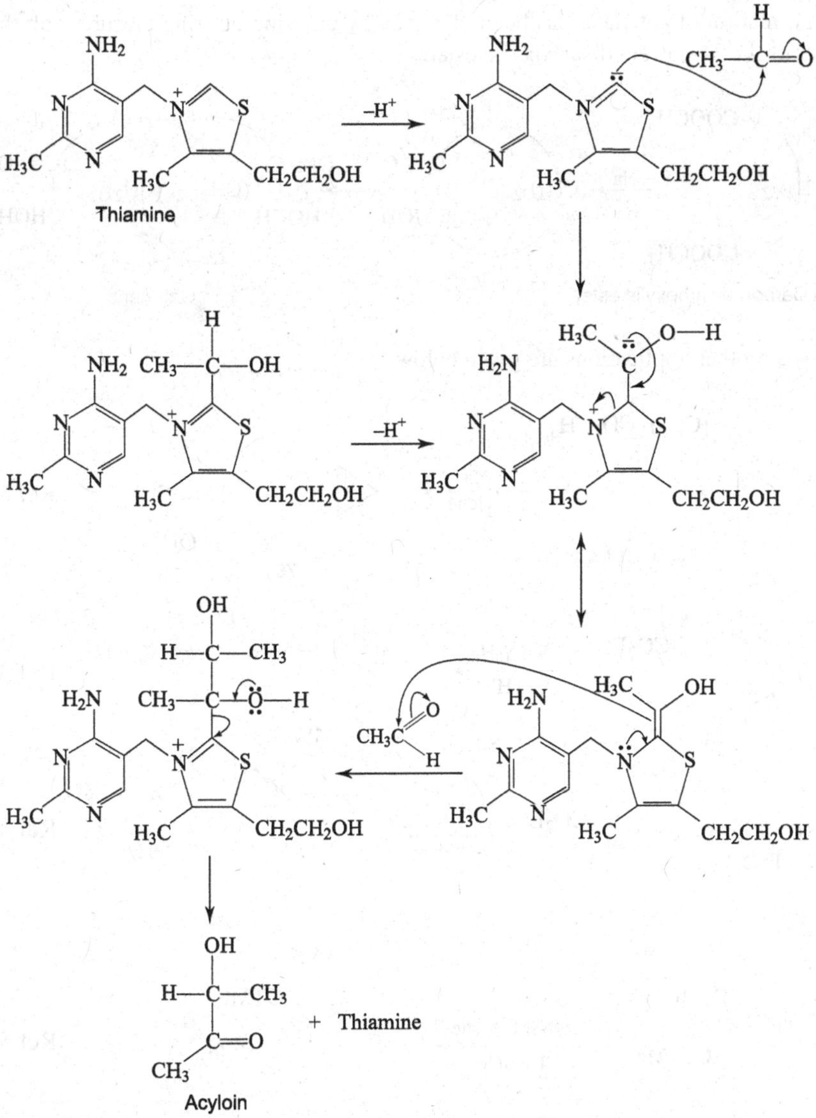

Scheme 2.7 Acyloin condensation using coenzyme thiamine

2.1.2 Applications

Acyloin condensation is a very useful synthetic reaction.

1. Cyclic acyloins can be conveniently prepared (*see* Schemes 2.3, 2.4, 2.5 and 2.6).
2. Preparation of catenane.

Formation of catenane has been observed by carrying out ring closure with the ester of the 34-carbon dicarboxylic ester.

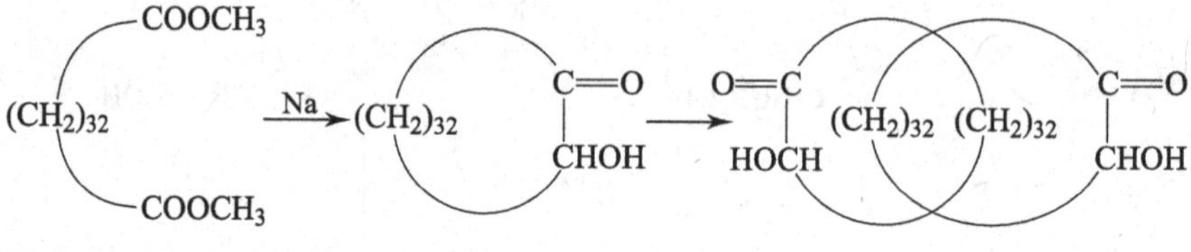

34-Carbon dicarboxylic ester   Catenane

Some other applications are given below.

(i) $\xrightarrow[\text{xylene}]{\text{Na}}$ 76%   Ref 1

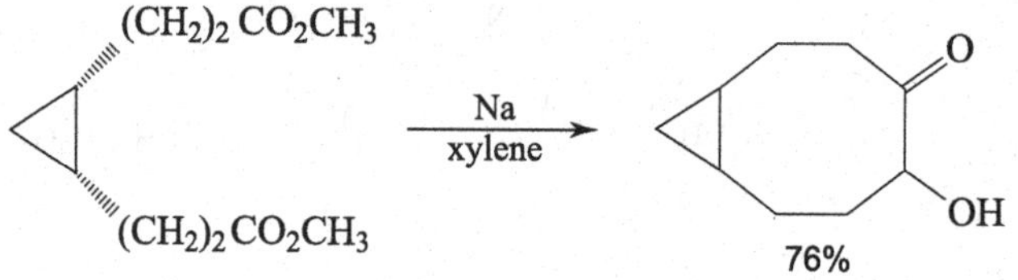

(ii) $\xrightarrow[\text{THF}]{\text{NaC}_{10}\text{H}_8}$ 11%   Ref 3

(iii) $\xrightarrow[\text{xylene}]{\text{Na}}$ 70%   Ref 4

(iv) $\xrightarrow[\text{2) TMSCl}]{\text{1)Na/Toluene}}$ 78%   Ref 5

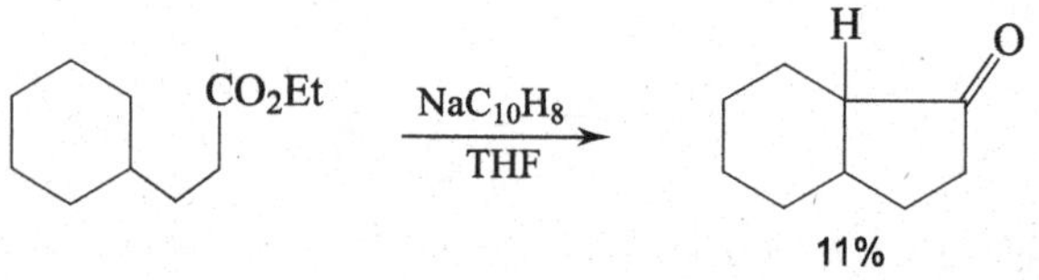

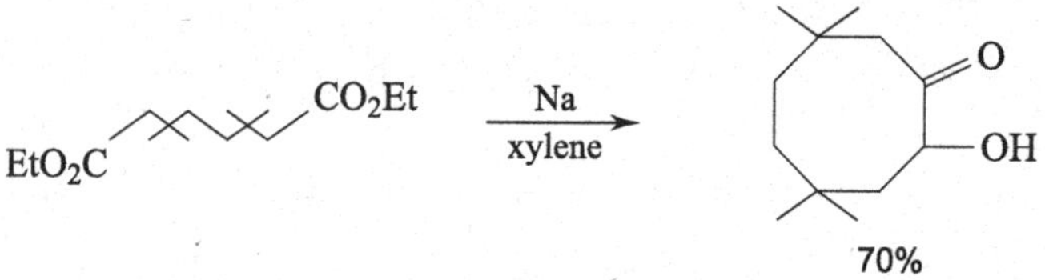

$\downarrow$ SiO$_2$

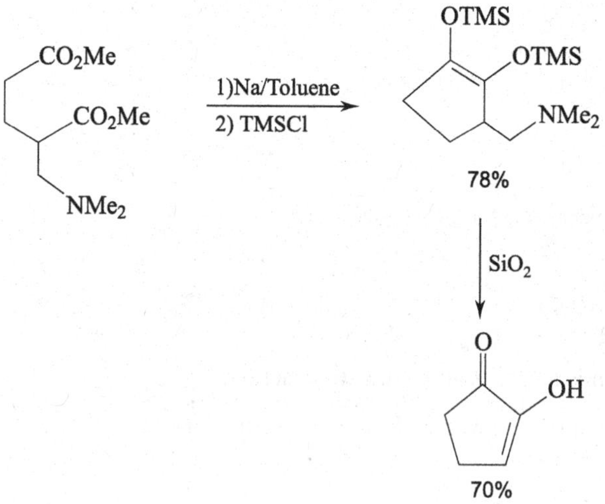

70%

$$\underset{\text{Acetaldehyde}}{CH_3CHO} + CH_3CHO \underset{}{\overset{^-OH}{\rightleftharpoons}} \underset{\substack{\beta\text{-Hydroxy butyraldehyde} \\ \text{(aldol)}}}{CH_3\overset{\overset{\displaystyle OH}{|}}{CH}CH_2CHO}$$

Scheme 2.8 Aldol condensation

2.2 Aldol Condensation [6]

The aldol condensation is one of the most important carbon–carbon bond-forming reactions in organic synthesis. Self-condensation of aldehydes (having α-hydrogen atom) on warming with dilute alkali to give β-hydroxyaldehydes (known as aldols) is called aldol condensation. A typical example is the reaction of acetaldehyde with dilute alkali (Scheme 2.8).

The aldol reaction can take place between two identical or different aldehydes or ketones, and an aldehyde and a ketone.

Mechanism

The first step is the removal of a proton from the α-carbon of one molecule of acetaldehyde by a base (hydroxide ion) to give an enolate ion, which is resonance stabilized. The formed enolate ion then acts as a nucleophile—as a carbanion—and attacks the carbonyl carbon of a second molecule of acetaldehyde, producing an alkoxide anion. Finally, the alkoxide anion removes a proton from a molecule of water to form the aldol. The various steps are shown in Scheme 2.9.

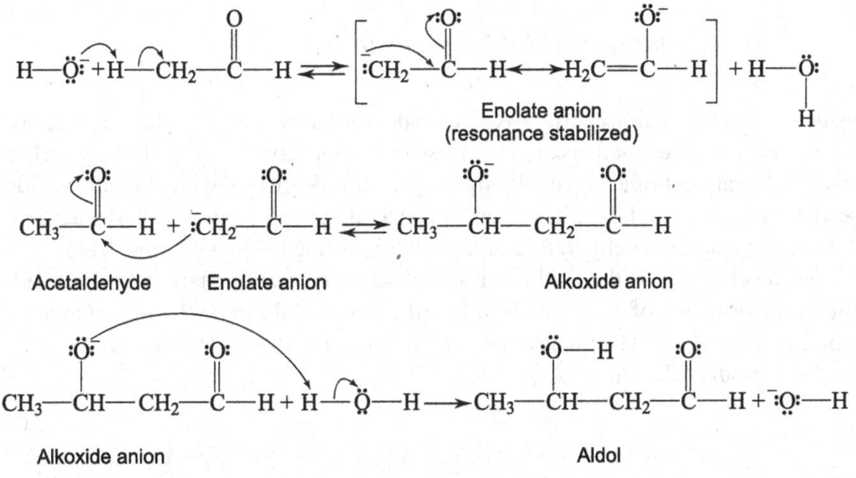

Scheme 2.9 Mechanism of aldol condensation

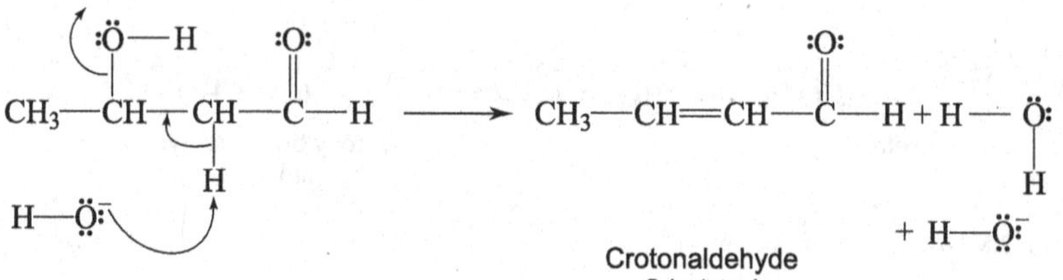

$$CH_3-CH-CH-C-H \longrightarrow CH_3-CH=CH-C-H + H-\ddot{O} + H-\ddot{O}$$

Crotonaldehyde
2-butenal

Scheme 2.10 Dehydration of aldols under basic conditions

$$2CH_3CCH_3 \xrightarrow{\text{HCl}} H_3C-C-CH=C-CH_3 + H_2O$$

Acetone   4-Methyl-3-penten-2-one

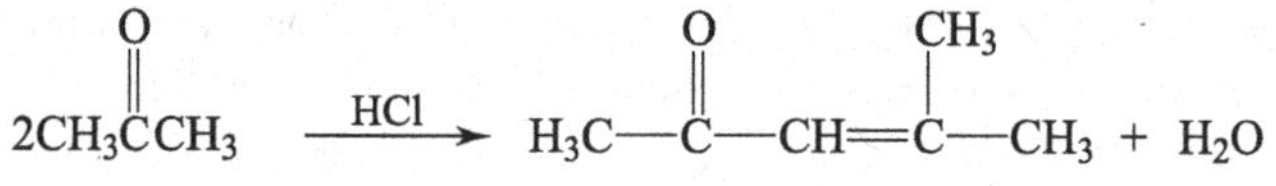

Scheme 2.11 Acid-catalysed dehydration of aldols

The Aldol

The aldol obtained under basic conditions (in the example cited above) when on heating, dehydration takes place to give crotonaldehyde (2-butenal). The dehydration is prompted due to acidity of the remaining α-hydrogen and also because the product is stabilized by having conjugated double bonds (Scheme 2.10).

In some cases, the dehydration in the aldol occurs readily and it is not possible to isolate the aldol.

2.2.1 Acid-Catalysed Aldol Condensation

In the aldol condensation cited above, the condensation takes place in the presence of a base. However, aldol condensations can also be brought about with acid catalysts. For example, treatment of acetone with hydrogen chloride gives the aldol condensation product, viz., 4-methyl-3-penten-2-one. In general, in acid-catalysed aldol reactions, there is simultaneous dehydration of the initially formed aldol (Scheme 2.11).

The mechanism of the acid-catalysed aldol condensation starts with the acid-catalysed formation of enol, which adds to the protonated carbonyl group of another molecule of acetone. The final step is proton transfer and dehydration leading to the final end product (Scheme 2.12).

2.2.2 Crossed Aldol Condensation

An aldol condensation that uses two different carbonyl compounds is called a crossed aldol condensation. In such a situation, the following three situations may occur:

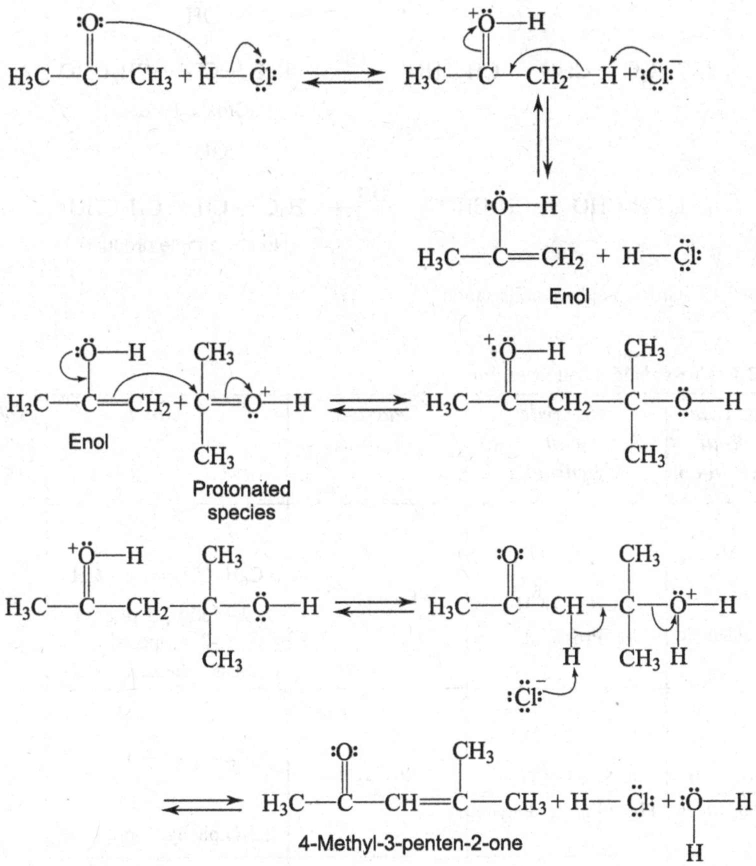

Scheme 2.12 Mechanism of acid-catalysed aldol condensation

(i) Crossed aldol condensation between two different aldehydes

In case both the aldehydes have α-hydrogen(s), then both can form carbanions, and so a mixture of four products are formed. Such a reaction has no synthetic utility. If, on the other hand, one of the aldehydes has no α-hydrogen, then in such a case two products are formed as shown in Scheme 2.13.

The formation of the crossed product can be achieved (and the formation of normal simple product can be avoided) by placing the aldehyde which has no α-hydrogen along with sodium hydroxide in a flask and then slowly adding (dropwise) the aldehyde with an α-hydrogen to the mixture. Under these conditions, the concentration of the reactant with an α-hydrogen is always low and much of the reactant is present as an enolate anion. So, the main reaction that takes place is between this enolate anion and the component that has no α-hydrogen. In other words, major amount of crossed aldol condensation product will be obtained. Table 2.1 gives some of the typical crossed aldol condensation.

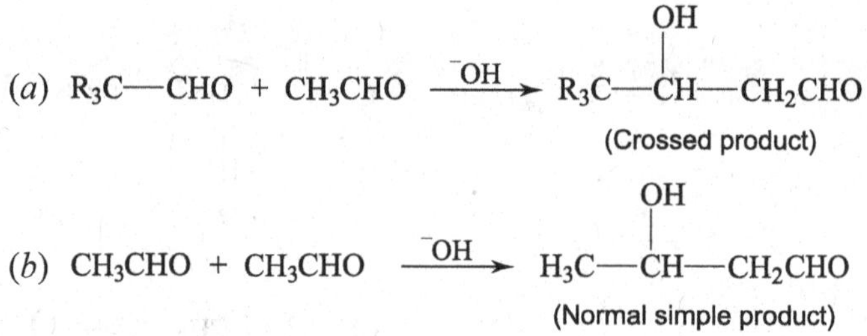

$$(a) \quad R_3C\text{—}CHO + CH_3CHO \xrightarrow{\ ^-OH\ } R_3C\overset{\overset{\displaystyle OH}{|}}{\text{—}CH}\text{—}CH_2CHO$$

(Crossed product)

$$(b) \quad CH_3CHO + CH_3CHO \xrightarrow{\ ^-OH\ } H_3C\overset{\overset{\displaystyle OH}{|}}{\text{—}CH}\text{—}CH_2CHO$$

(Normal simple product)

Scheme 2.13 Crossed aldol condensation

Table 2.1 Crossed aldol condensation

Aldehyde without α-hydrogen	Aldehyde with α-hydrogen	Reaction conditions	Product	Yield %
$\overset{O}{\overset{\|\|}{C_6H_5CH}}$ Benzaldehyde	$\overset{O}{\overset{\|\|}{CH_3CH_2CH}}$ Propanal	$-OH/10°C$	$C_6H_5\,CH\!\!=\!\!\overset{\overset{\displaystyle CH_3}{\|}}{C}\!\!-\!\!\overset{\overset{\displaystyle O}{\|\|}}{CH}$ 2-Methyl-3-phenyl-2-propenal (α-Methylcinnamaldehyde)	68
$\overset{O}{\overset{\|\|}{C_6H_5C}}\text{—}H$ Benzaldehyde	$\overset{O}{\overset{\|\|}{C_6H_5CH_2CH}}$ Phenyl acetaldehyde	$-OH/20°C$	$C_6H_5CH\!\!=\!\!\overset{\overset{\displaystyle O}{\|\|}}{C}\!\!\underset{\underset{\displaystyle C_6H_5}{\|}}{CH}$ 2,3-Diphenyl-2-propenal	65
$H\text{—}\overset{O}{\overset{\|\|}{C}}\text{—}H$ Formaldehyde	$CH_3\overset{\overset{\displaystyle CH_3}{\|}}{CH}\text{—}\overset{\overset{\displaystyle O}{\|\|}}{CH}$ 2-Methyl propanal	$40°C\ dil.Na_2CO_3$	$CH_3\text{—}\overset{\overset{\displaystyle CH_3}{\|}}{\underset{\underset{\displaystyle CH_2OH}{\|}}{C}}\text{—}\overset{\overset{\displaystyle O}{\|\|}}{CH}$ 2-Hydroxy methyl-2-Methyl propanal	> 64%

(*ii*) Crossed aldol condensation between two different ketones

In such cases, poor yield is obtained and so it is not much useful. The poor reactivity of carbonyl carbons of ketones is responsible for low yields.

(*iii*) Crossed aldol condensation between an aldehyde and a ketone

(*a*) When an aldehyde and a ketone, both having α-hydrogens, are condensed, only two products are obtained (Scheme 2.14). This is because ketones are poor carbanion acceptors and do not undergo self-condensation.

Normally, the crossed product predominates. The formation of aldol can be minimized by slow addition of the aldehyde to the mixture of ketone and alkali.

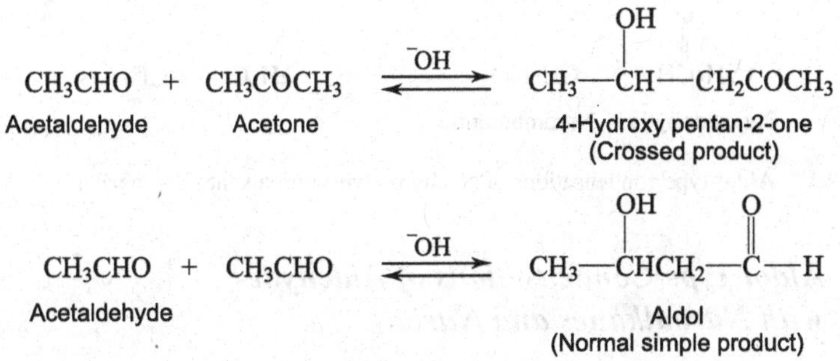

$$CH_3CHO + CH_3COCH_3 \xrightarrow{\;^-OH\;} CH_3-\overset{\overset{\displaystyle OH}{|}}{CH}-CH_2COCH_3$$

Acetaldehyde   Acetone   4-Hydroxy pentan-2-one
(Crossed product)

$$CH_3CHO + CH_3CHO \xrightarrow{\;^-OH\;} CH_3-\overset{\overset{\displaystyle OH}{|}}{C}HCH_2-\overset{\overset{\displaystyle O}{\|}}{C}-H$$

Acetaldehyde   Aldol
(Normal simple product)

Scheme 2.14 Crossed aldol condensation between an aldehyde and a ketone

$$CH_3\overset{\overset{\displaystyle O}{\|}}{C}CH_3 + CH_2{=}O \xrightarrow{\;^-OH\;} CH_3-\overset{\overset{\displaystyle O}{\|}}{C}-CH_2CH_2OH$$

Acetone   Formaldehyde   3-Ketobutanol

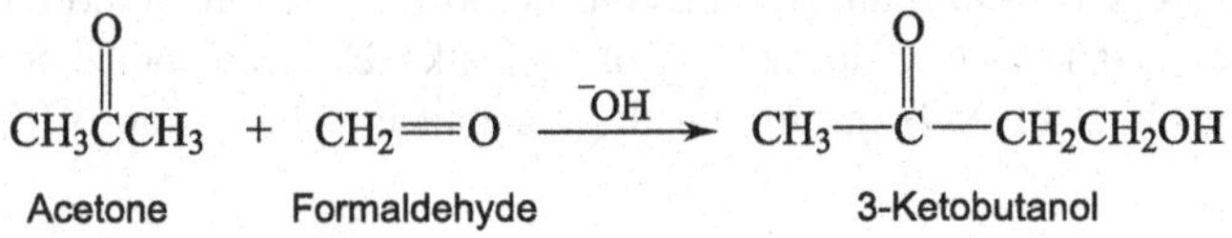

Scheme 2.15 Mixed aldol condensation

$$CH_3\overset{\overset{\displaystyle O}{\|}}{C}-CH_3 + 6\,CH_2{=}O \xrightarrow{\;^-OH\;} (HOH_2C)_3C-\overset{\overset{\displaystyle O}{\|}}{C}-C(CH_2OH)_3$$

Acetone   Formaldehyde

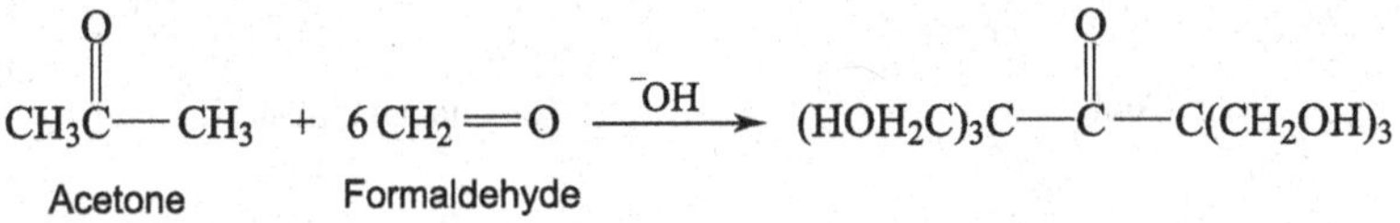

Scheme 2.16 In mixed aldol condensation all alpha hydrogens are replaced

(b) When the reaction is between a ketone and an aldehyde with no α-hydrogen, only one product is obtained. Such a condensation is called **Claisen–Schmidt** reaction.

A reaction worth mentioning as a crossed aldol condensation between an aldehyde and a ketone is the mixed aldol condensation of acetone with formaldehyde. In this case, formaldehyde cannot form an enolate since it lacks α-hydrogen. However, it is a good electron-pair acceptor because of freedom from steric hinderance and because it has a usually weak carbonyl bond. Acetone forms an enolate easily, but it is a relatively poor acceptor. So, the following reaction occurs (Scheme 2.15):

The reaction does not stop as indicated in Scheme 2.15. In fact, all six α-hydrogens can be replaced by –CH₂OH groups (Scheme 2.16).

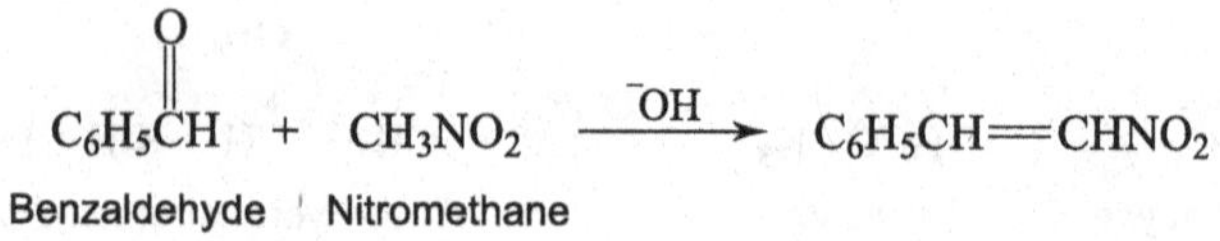

$$C_6H_5\overset{\overset{\displaystyle O}{\|}}{C}H \ + \ CH_3NO_2 \ \xrightarrow{\ ^-OH\ } \ C_6H_5CH\!\!=\!\!CHNO_2$$

Benzaldehyde ¦ Nitromethane

Scheme 2.17 Aldol-type condensations of aldehydes with nitroalkanes and nitriles

2.2.3 Aldol Type Condensations of Aldehydes with Nitroalkanes and Nitriles

(i)   Condensation with nitroalkanes

The α-hydrogens of nitroalkanes are considerably acidic, much more than those of aldehydes and ketones. The acidity of nitroalkanes is attributed to electron-withdrawing effect of NO_2 group and the resonance stabilization of the formed anion.

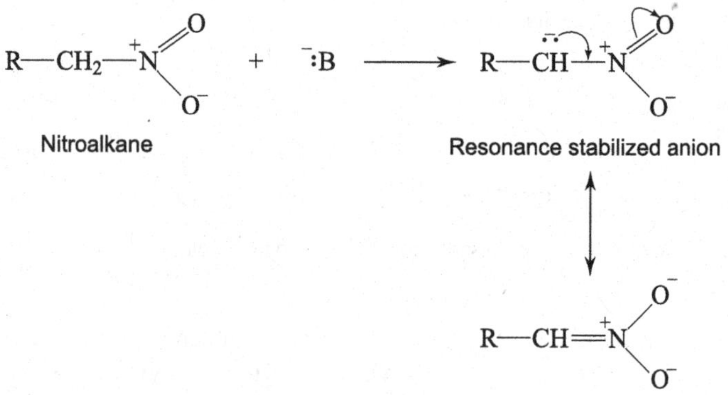

Nitroalkane   Resonance stabilized anion

Nitroalkanes having α-hydrogens undergo base-catalysed aldol-type condensations with aldehydes and ketones. An example is the condensation of benzaldehyde and nitromethane (Scheme 2.17).

The rate of the above condensation is increased by using basic alumina catalyst and **sonication**.

The reaction of a nitroalkane with an aldehyde in the presence of base is called **Henry reaction**. This also forms the subject-matter of subsequent Sect. 2.31.

(ii) Condensation with nitriles

Like nitroalkanes, the α-hydrogens of nitriles are also acidic (but less than those of aldehydes and ketones) and so undergo aldol-type condensations. One such example is the condensation of benzaldehyde with phenylacetonitrile (Scheme 2.18).

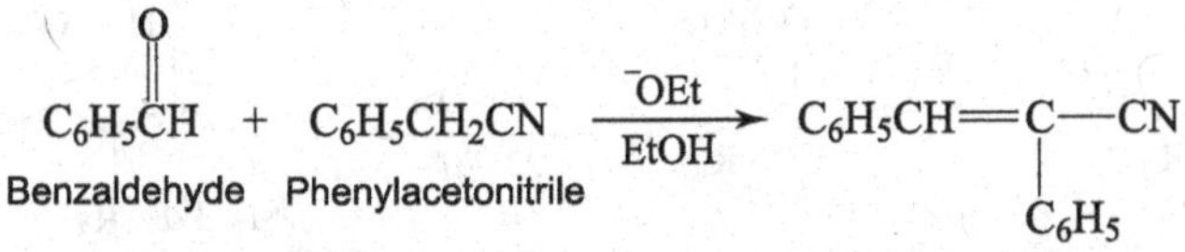

Scheme 2.18 Condensation of benzaldehyde with phenylacetonitrile

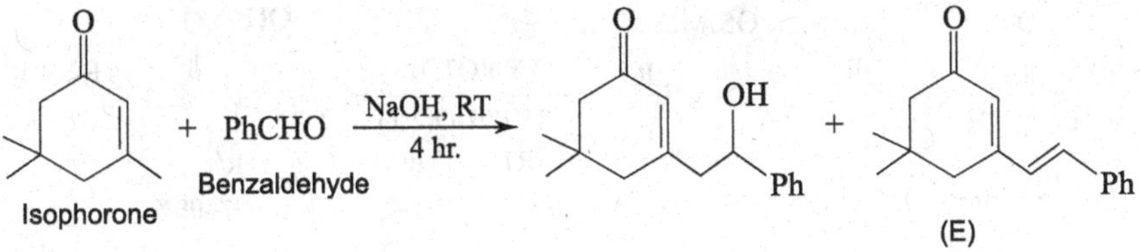

	Water	CTACl	TBACl
(addition)	24%	–	27%
(E)	–	80%	58%

Scheme 2.19 Vinylogous aldol reaction

2.2.4 Vinylogous Aldol Reaction

The γ-hydrogen of an α,β-unsaturated ketones, nitriles and esters is 'active' hydrogen and so electrophilic addition takes place at γ-position. This is known as **vinylogous aldol addition**, when the electrophile is an aldehyde. Thus, the reaction of isophorone with benzaldehyde gives only vinylogous aldol addition in low yields. However, in the presence of CTACl, the condensation product, (E) benzylidene isophorone is obtained in excellent yield. By using tetrabutylammonium chloride (TBACl), a mixture of addition and condensation product is obtained [8] (Scheme 2.19).

2.2.5 Aldol Condensation of Silyl Enol Ethers in Aqueous Media

The aldol condensation of silyl enol ethers with benzaldehydes catalysed by titanium tetrachloride was first reported [9] in 1973. However, these reactions are carried out in anhydrous solvents [10]. It has now been possible to perform aldol condensation of silyl enol ethers with aldehydes in aqueous phase [11] (Scheme 2.20).

The aqueous phase reaction (Scheme 2.20) was carried out without any catalyst, but it took several days for completion, since water serves as a weak Lewis acid. The addition of stronger Lewis acid (e.g., ytterbium triflate) greatly improved [12] the yield and the rate (Scheme 2.21).

The reaction of silyl enol ether of propiophenone with commercial formaldehyde in the presence of ytterbium triflate gave the adduct (Scheme 2.22).

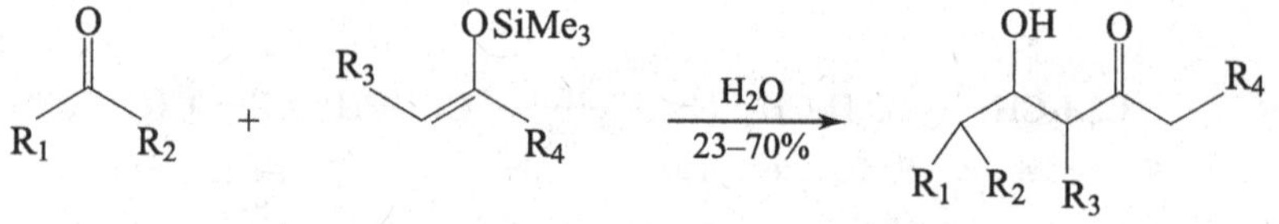

Scheme 2.20 Aldol condensation of silyl enol ethers in aqueous solution

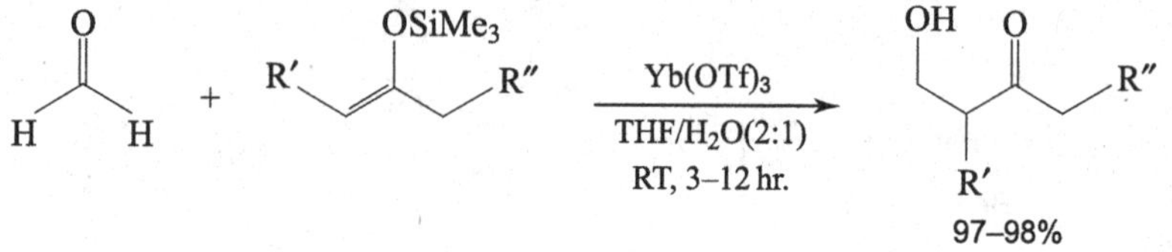

Scheme 2.21 In aqueous phase reaction addition of stronger base improves the yield

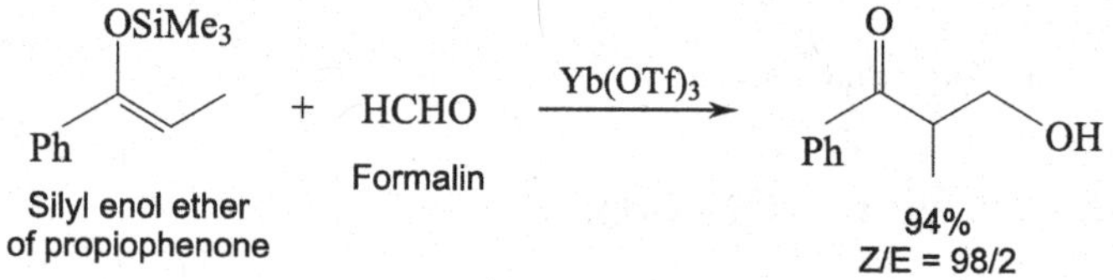

Scheme 2.22 Silyl enol ether of propionaldehyde on reaction with HCHO in the presence of Yb(OTf)$_2$ gave the adduct

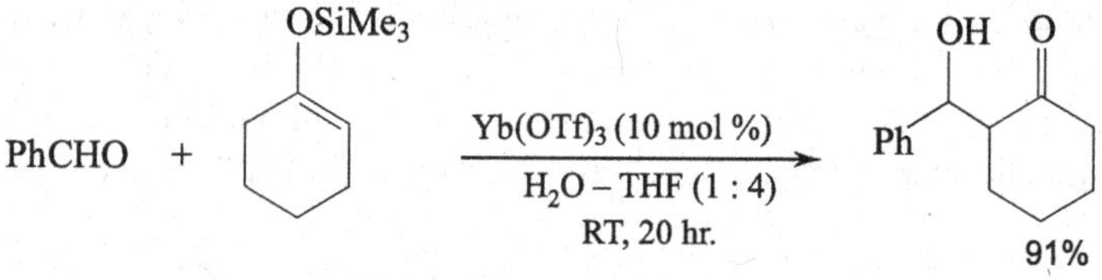

Scheme 2.23 The reaction of 1-trimethylsilyloxycyclohexene in the presence of Yb(OTf)$_3$ gave a mixture of adducts

Aldehydes other than formaldehydes can also be used. Thus, the reaction of 1-trimethylsilyloxycyclohexene with benzaldehyde in the presence of catalytic amount of Yb(OTf)$_3$ (10 mol%) in H$_2$O-THF (1:4) give 91% yield of the adduczt [12, 14] (Scheme 2.23).

2.2.6 Aldol Condensation in Solid Phase

The aldol condensation of the lithium enolate of methyl 3,3-dimethylbutanoate with aromatic aldehydes gave [15] 8:92 mixture of the *syn* and *anti* products with 70% yield (Scheme 2.24).

The above reaction (Scheme 2.24) is carried out by mixing freshly ground mixture of the starting materials in vacuum for 3 days at room temperature.

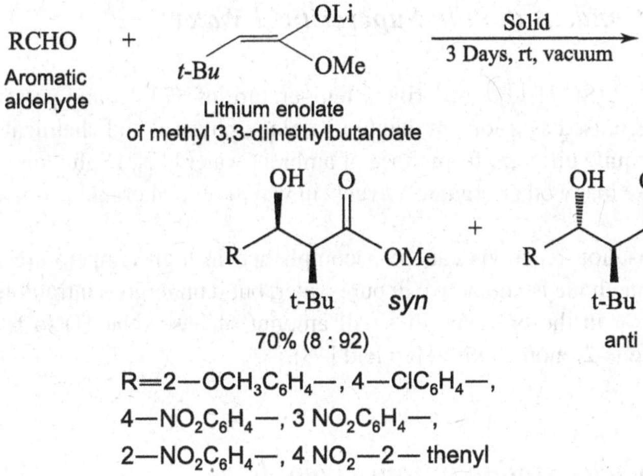

Scheme 2.24 An aldol condensation in solid phase

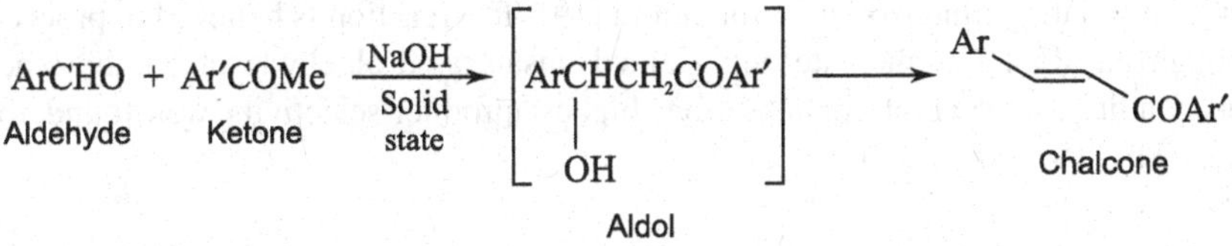

Ar	Ar'	Reaction time	Yield aldol	Yield Chalcone
Ph	Ph	30	10	–
p-Me C6H4–	• Ph	5	–	97
p-Me C6H4–	p-Me C6H4–	5	–	99
p-Cl C6H4–	Ph	5	–	98
p-Cl C6H4	p-MeOC6H4–	10	–	79

Scheme 2.25 Solid-state reaction of ArCHO and Ar'COMe in the presence of NaOH forms aldol

In the absence of any solvent, some aldol condensations proceed [16] more efficiently and stereoselectively. In this procedure, an appropriate aldehyde and ketone, and NaOH are grounded in a pestle and mortar at room temperature for 5 min, and the product obtained is the corresponding chalcone. In this case, the initially formed aldol dehydrates easily (Scheme 2.25).

Use of alcohol as solvent in the above procedure using conventional procedure gave only aldol with poor yield (10–25%).

2.2.7 Aldol Condensation in Supercritical Water

Supercritical water (SC-H$_2$O) with critical temperature of 374 °C and pressure of 22.1 MPa has been used as a solvent due to its unique physical and chemical properties, which are quite different from those of ambient water [17]. High temperature water behaves like many other organic solvents in which several organic compounds are soluble.

Aldol condensation reactions can be accomplished in high temperature water. Though 2,5-hexanedione is unreactive in pure water, but it undergoes intramolecular aldol condensation in the presence of small amount of base (NaOH) to form 3-methylcyclopentene-2-enone with 81% yield [18].

2.2.8 Aldol Condensation in Ionic Liquids ·

Aldol condensation of propanol to form 2-methylpent-2-enal has been carried out in non-coordinating imidazolium ionic liquid [19]. The reaction is believed to proceed through an aldol intermediate and yielded unsaturated aldehyde under the reaction conditions. In aldol condensation, highest product selectivity was found for [b$_{min}$][PF$_6$] (Scheme 2.26).

2.2.9 Asymmetric Aldol Condensations

Asymmetric version of aldol condensation has been utilized for enantioselective C-bond forming.

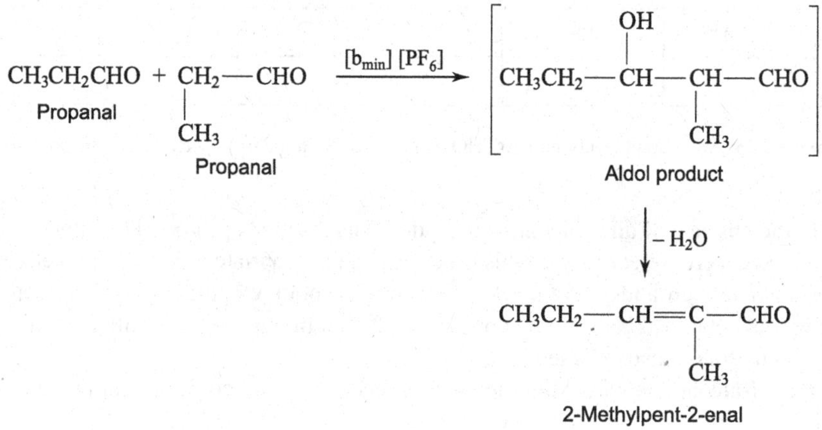

Scheme 2.26 Aldol condensation in ionic liquids

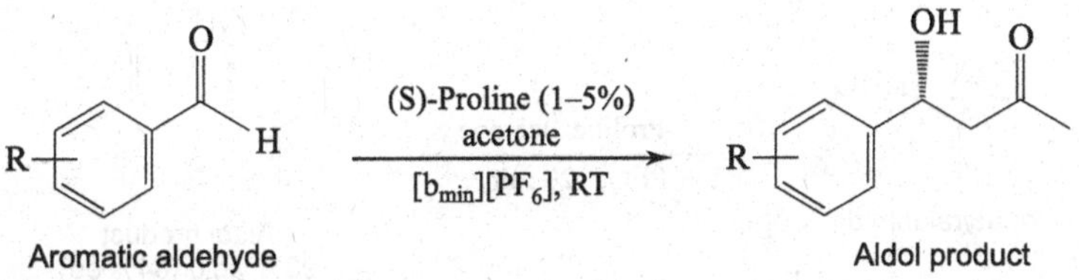

Scheme 2.27 Asymmetric aldol condensation

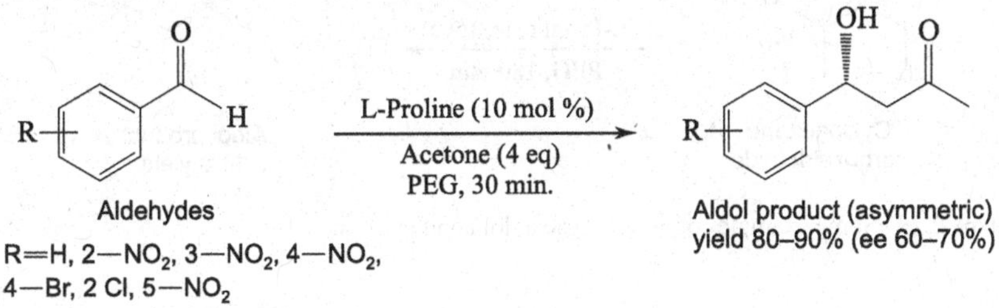

R=H, 2—NO$_2$, 3—NO$_2$, 4—NO$_2$,
4—Br, 2 Cl, 5—NO$_2$

Scheme 2.28 Asymmetric condensation between aldehydes in the presence of acetone and PEG gave asymmetric aldol products

The proline-catalysed asymmetric direct aldol reaction of different aromatic alde-hydes with acetone and other ketones in the ionic liquid [b$_{min}$][PF$_6$] gave good yield of the aldol product with reasonable enantioselectivities (Scheme 2.27) [20].

L-proline-catalysed direct asymmetric aldol reaction of acetone with various aromatic and aliphatic aldehydes in polyethylene glycol (PEG-400) has also been reported [21] to give asymmetric aldol products (Scheme 2.28).

In a similar way, aliphatic aldehydes isobutyraldehyde and cyclohexane carbox-aldehyde have been used as substrates [21] and the asymmetric aldol products are obtained with 90 and 65 yield, respectively (Scheme 2.29).

In the above condensations (Schemes 2.28 and 2.29), polyethylene glycol (PEG) HO–(CH$_2$CH$_2$O)$_n$–H used as a solvent is a biologically compatible product, a green solvent, is commercially available and is reusable.

2.2.10 Applications

A number of synthetic procedures based on aldol condensation of aldehydes and ketones are as follows:

(*i*) β-Ionone required for the synthesis of vitamin A is prepared by the condensa-tion of citral with acetone followed by subsequent treatment with boron trifluoride (Scheme 2.30).

(*ii*) A commercially important mixed condensation involves the reaction of acetaldehyde and excess formaldehyde in the presence of calcium hydroxide to

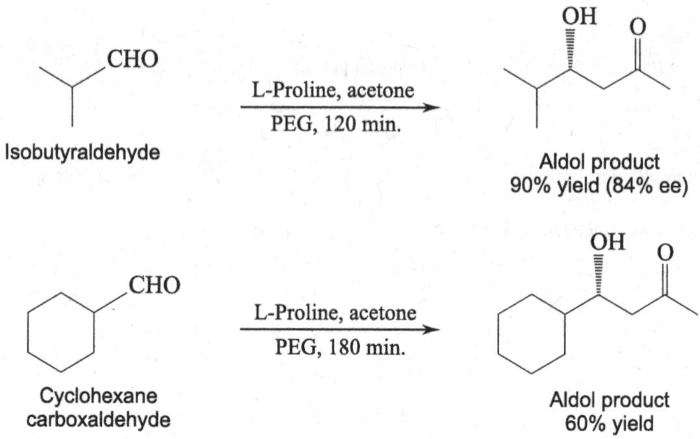

Isobutyraldehyde + CHO →(L-Proline, acetone / PEG, 120 min.)→ **Aldol product** 90% yield (84% ee)

Cyclohexane carboxaldehyde →(L-Proline, acetone / PEG, 180 min.)→ **Aldol product** 60% yield

Scheme 2.29 Some examples of asymmetric aldol condensation

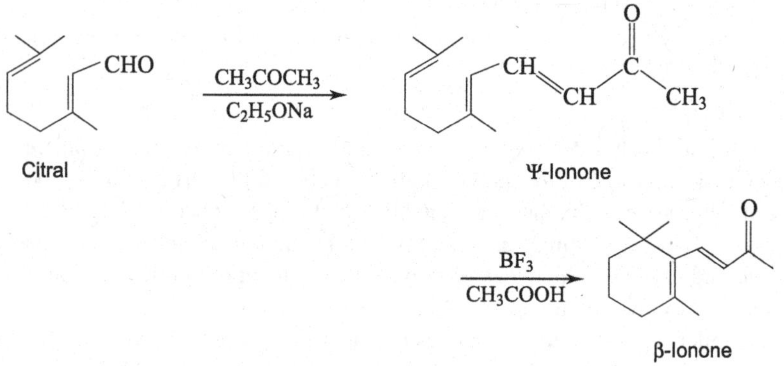

Citral →(CH$_3$COCH$_3$ / C$_2$H$_5$ONa)→ **Ψ-Ionone** →(BF$_3$ / CH$_3$COOH)→ **β-Ionone**

Scheme 2.30 Synthesis of β-ionone from citral

give trihydroxymethyleneacetaldehyde (which has no α-hydrogen) and undergoes **crossed Cannizzaro reaction** with formaldehyde to give a tetrahydroxy alcohol, known as pentaerythritol (Scheme 2.31).

(*iii*) Aldol condensation of acetone in the presence of acids (dry hydrogen chloride gas) gives mesityl oxide and phorone (Scheme 2.32).

(*iv*) An important class of cyanine dyes (photographic sensitizers) is obtained by the aldol condensation of pyridine and quinoline (having methyl groups at positions 2 and 4) with aldehydes (Scheme 2.33).

(*v*) Acetylenic alcohol, viz., 2-butyne-1,4-diol, a valuable commercial product is obtained by aldol-type condensation of acetylene (compounds having acidic C—H bonds) with formaldehyde in the presence of Cu$_2$C$_2$ (Scheme 2.34).

The above process is called **ethynylation**.

(*vi*) Primary and secondary nitro compounds undergo aldol-type additions (Scheme 2.35).

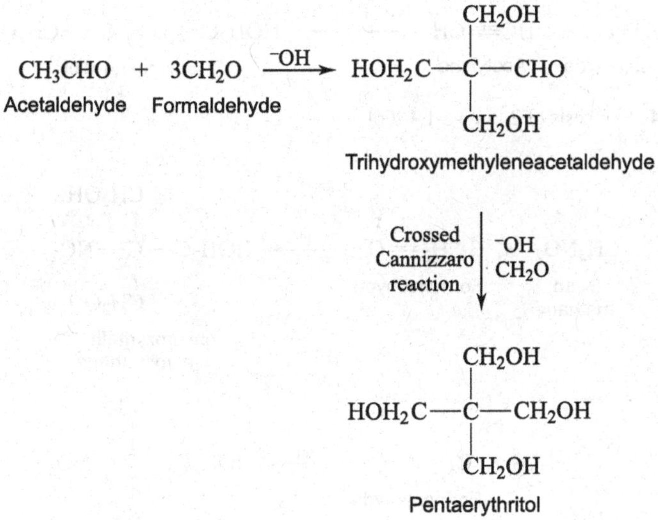

$$CH_3CHO + 3CH_2O \xrightarrow{^-OH} HOH_2C\underset{\underset{CH_2OH}{|}}{\overset{\overset{CH_2OH}{|}}{C}}CHO$$

Acetaldehyde   Formaldehyde

Trihydroxymethyleneacetaldehyde

Crossed Cannizzaro reaction $\Big\downarrow$ ^-OH CH_2O

$$HOH_2C\underset{\underset{CH_2OH}{|}}{\overset{\overset{CH_2OH}{|}}{C}}CH_2OH$$

Pentaerythritol

Scheme 2.31 Synthesis of pentaerythritol

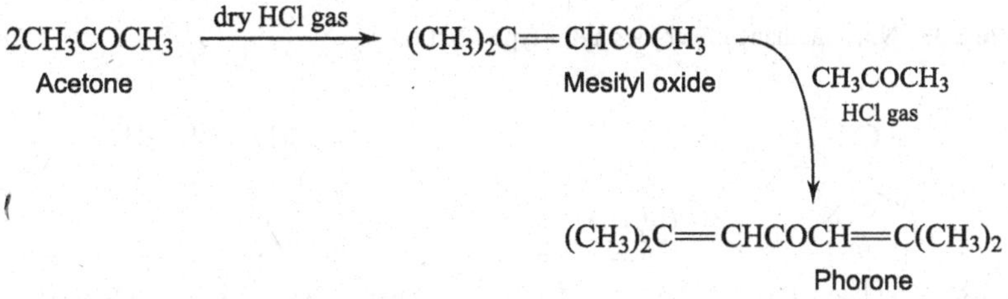

$$2CH_3COCH_3 \xrightarrow{\text{dry HCl gas}} (CH_3)_2C{=}CHCOCH_3 \quad\overset{CH_3COCH_3}{\underset{\text{HCl gas}}{\diagdown}}$$

Acetone   Mesityl oxide

$$(CH_3)_2C{=}CHCOCH{=}C(CH_3)_2$$

Phorone

Scheme 2.32 Conversion of CH_3COCH_3 into mesityl oxide and phorone

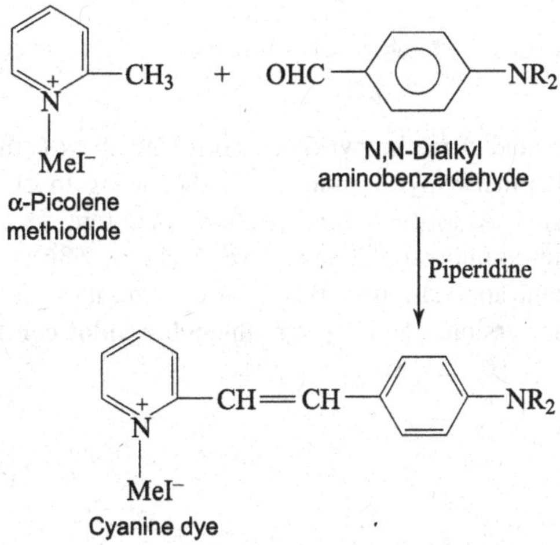

α-Picolene methiodide   N,N-Dialkyl aminobenzaldehyde

Piperidine

Cyanine dye

Scheme 2.33 Synthesis of cyanine dyes

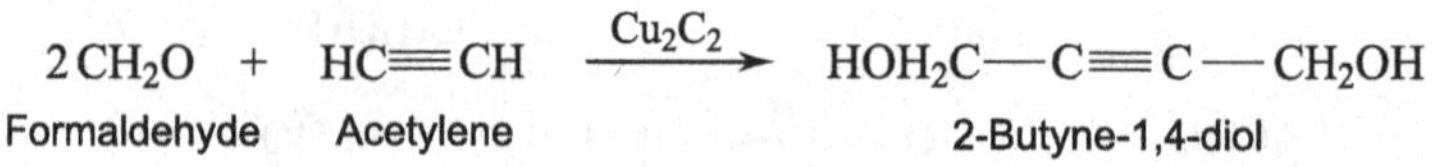

$$2\,CH_2O \;+\; HC\!\equiv\!CH \;\xrightarrow{Cu_2C_2}\; HOH_2C\!-\!C\!\equiv\!C\!-\!CH_2OH$$

Formaldehyde Acetylene 2-Butyne-1,4-diol

Scheme 2.34 Synthesis of 2-butyne-1,4-diol

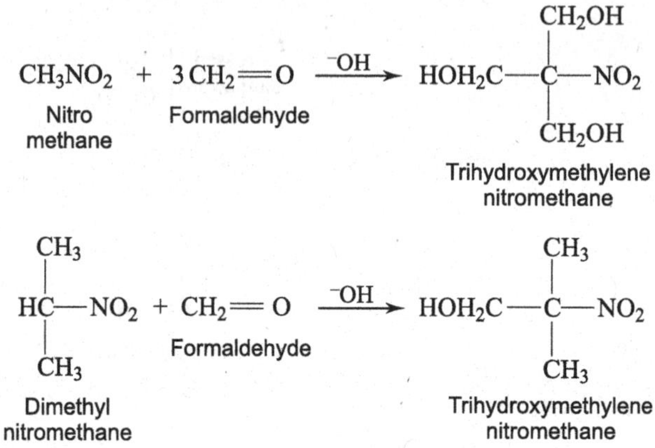

$$CH_3NO_2 \;+\; 3\,CH_2\!=\!O \;\xrightarrow{-OH}\; HOH_2C\!-\!\underset{\underset{CH_2OH}{|}}{\overset{\overset{CH_2OH}{|}}{C}}\!-\!NO_2$$

Nitro methane   Formaldehyde

Trihydroxymethylene nitromethane

$$\underset{\underset{CH_3}{|}}{\overset{\overset{CH_3}{|}}{HC}}\!-\!NO_2 \;+\; CH_2\!=\!O \;\xrightarrow{-OH}\; HOH_2C\!-\!\underset{\underset{CH_3}{|}}{\overset{\overset{CH_3}{|}}{C}}\!-\!NO_2$$

Dimethyl nitromethane   Formaldehyde   Trihydroxymethylene nitromethane

Scheme 2.35 Nitro methane undergoes aldol-type reaction

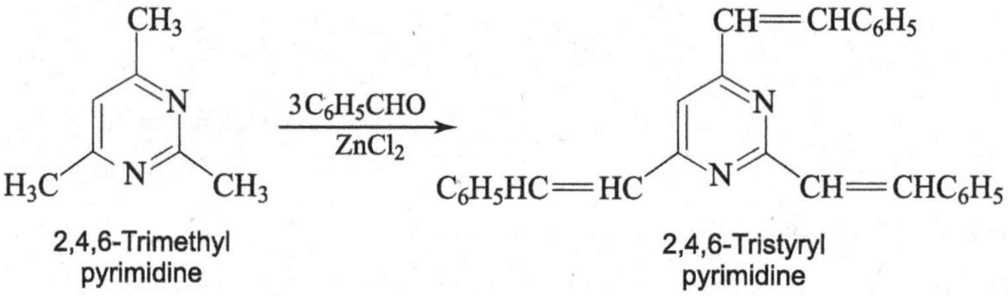

2,4,6-Trimethyl pyrimidine $3\,C_6H_5CHO$ / $ZnCl_2$ 2,4,6-Tristyryl pyrimidine

Scheme 2.36 Another example of aldol-type condensation

(*vii*) As with 2- and 4-methylpyridines (see (*iv*) above), the methyl hydrogens of 2-, 4- and 6-methylpyrimidine are acidic owing to electron withdrawal by the nitrogen atoms. Aldol-type condensations therefore occur with aldehydes (5-methylpyrimidines are not similarly reactive) (Scheme 2.36).

(*viii*) An important application of the aldol condensation in organic synthesis is its intramolecular version, called '**intramolecular aldol condensation**'. Some examples are given in [16, 22] (Scheme 2.37).

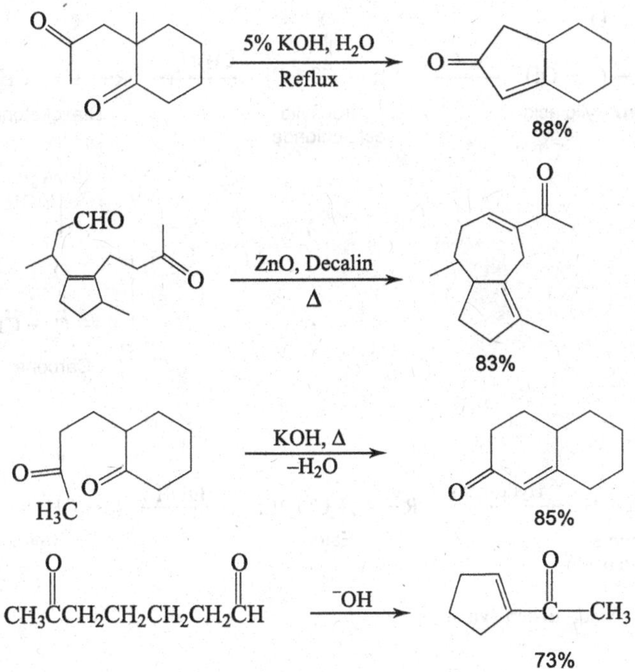

Scheme 2.37 Some examples of intramolecular aldol condensation

2.3 Arndt–Eistert Synthesis [23]

It is a convenient method of converting an acid (RCOOH) to the next homologous acid (RCH₂COOH). In this procedure, the acid is first converted to its acid chloride, which is then treated with diazomethane (CH_2N_2) resulting in the formation of diazoketone ($RCOCHN_2$). The diazoketone on treatment with silver oxide gets converted to a ketene ($RCH == C == O$), which gets esterified under the conditions of the reaction. Subsequent hydrolysis of the ester provides the homologous acid (RCH_2COOH). Various steps are shown in Scheme 2.38.

As seen in Scheme 2.38, the conversion of the diazoketone into the ketene involves a rearrangement known as **Wolff rearrangement** under the catalytic influence of silver oxide (*see* mechanism given ahead).

Mechanism

The various steps involved in the mechanism are:

(i)   Nucleophilic attack of diazomethane on the carbonyl carbon of the acid chloride to give the diazoketone.
(ii) Diazoketone eliminates a molecule of nitrogen to form a carbene.
(iii) Rearrangement of the carbene to the ketene (**Wolff's rearrangement**).

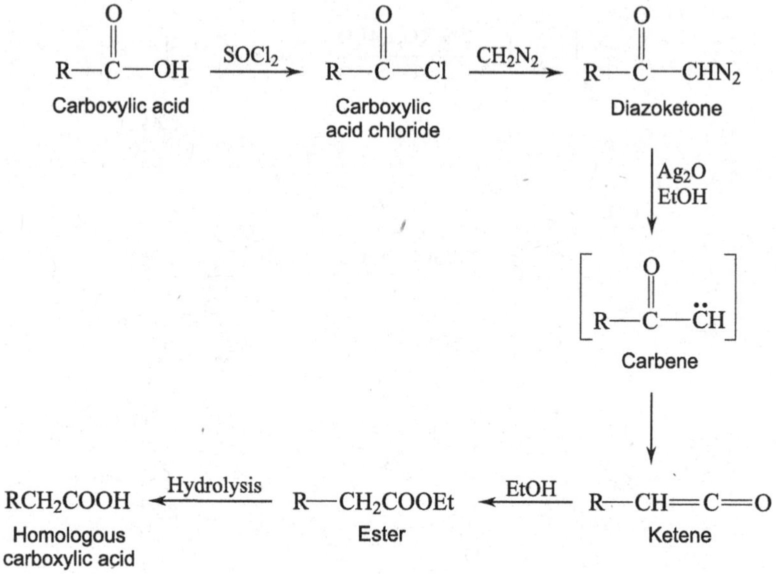

$$R-\overset{\overset{\displaystyle O}{\|}}{C}-OH \xrightarrow{SOCl_2} R-\overset{\overset{\displaystyle O}{\|}}{C}-Cl \xrightarrow{CH_2N_2} R-\overset{\overset{\displaystyle O}{\|}}{C}-CHN_2$$

Carboxylic acid   Carboxylic   Diazoketone
                  acid chloride

$$\Big\downarrow \begin{matrix} Ag_2O \\ EtOH \end{matrix}$$

$$\left[R-\overset{\overset{\displaystyle O}{\|}}{C}-\ddot{C}H \right]$$

Carbene

$$RCH_2COOH \xleftarrow{Hydrolysis} R-CH_2COOEt \xleftarrow{EtOH} R-CH=C=O$$

Homologous   Ester   Ketene
carboxylic acid

Scheme 2.38 Arndt–Eistert synthesis

(iv)   The reactive ketene reacts with the nucleophile present (H_2O) to form the higher homologue of the acid.

The various steps of mechanism are depicted in Scheme 2.39.

The above mechanism is supported by the fact that the formed intermediate ketene can be trapped. Also, the isotopic labelling experiment has shown that the carbonyl of the acid chloride or the diazoketone is present in the resulting acid as the carbonyl carbon (Scheme 2.40).

2.3.1   Applications (Scheme 2.41)

(i) Synthesis of higher homologues of carboxylic acids, amides and esters.
The ketene obtained by the Wolff's rearrangement of the carbene (obtained as an intermediate during Arndt-Eistert synthesis) on reaction with H_2O, NH_3 or R^1OH gives the corresponding carboxylic acid, amide or ester, respectively, all of which have one carbon atom more than the starting carboxylic acid.

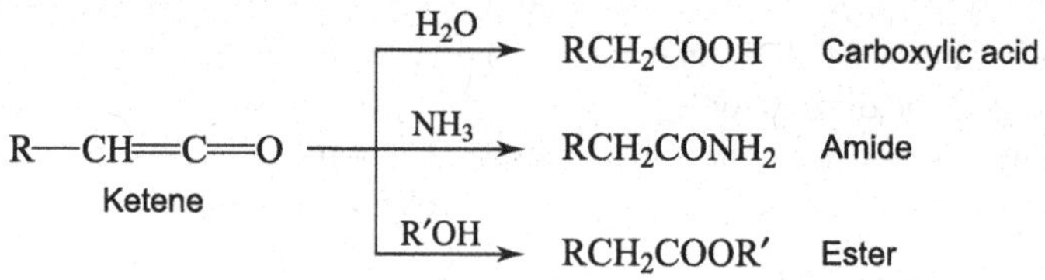

$$R-CH=C=O \begin{cases} \xrightarrow{H_2O} RCH_2COOH & \text{Carboxylic acid} \\ \xrightarrow{NH_3} RCH_2CONH_2 & \text{Amide} \\ \xrightarrow{R'OH} RCH_2COOR' & \text{Ester} \end{cases}$$

Ketene

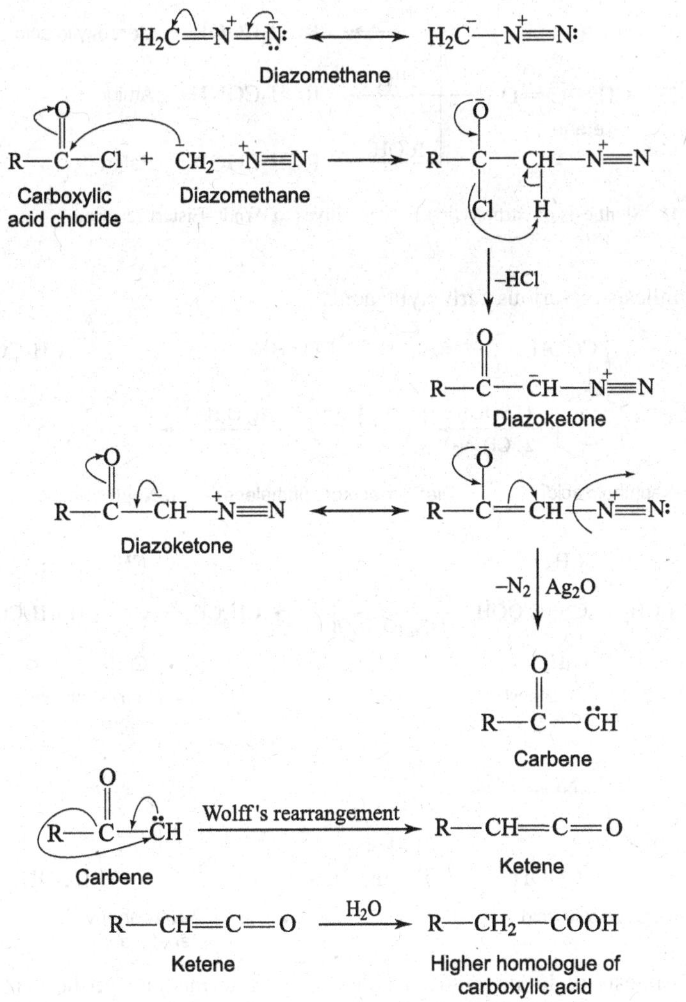

$$H_2\overset{..}{C}=\overset{+}{N}=\overset{..}{N}: \quad \longleftrightarrow \quad H_2\overset{-}{C}-\overset{+}{N}\equiv N:$$

Diazomethane

$$R-\overset{O}{\underset{\|}{C}}-Cl \;+\; CH_2-\overset{+}{N}\equiv N \quad \longrightarrow \quad R-\overset{\overset{O}{|}}{\underset{\underset{Cl}{|}}{C}}-\overset{+}{\underset{\underset{H}{|}}{CH}}-\overset{+}{N}\equiv N$$

Carboxylic   Diazomethane
acid chloride

$$\downarrow -HCl$$

$$R-\overset{O}{\underset{\|}{C}}-\overset{-}{C}H-\overset{+}{N}\equiv N$$

Diazoketone

$$R-\overset{O}{\underset{\|}{C}}-\overset{-}{C}H-\overset{+}{N}\equiv N \quad \longleftrightarrow \quad R-\overset{\overset{O}{|}}{C}=CH-\overset{+}{N}\equiv \overset{..}{N}:$$

Diazoketone

$$\downarrow -N_2 \quad Ag_2O$$

$$R-\overset{O}{\underset{\|}{C}}-\overset{..}{C}H$$

Carbene

$$R-\overset{O}{\underset{\|}{C}}-\overset{..}{C}H \quad \xrightarrow{\text{Wolff's rearrangement}} \quad R-CH=C=O$$

Carbene   Ketene

$$R-CH=C=O \quad \xrightarrow{H_2O} \quad R-CH_2-COOH$$

Ketene   Higher homologue of
                                    carboxylic acid

Scheme 2.39 Mechanism of Arndt–Eistert reaction

$$R-\overset{O}{\underset{\|}{\overset{13}{C}}}-Cl \quad \longrightarrow \quad R-\overset{O}{\underset{\|}{\overset{13}{C}}}-CHN_2 \quad \xrightarrow[H_2O]{Ag_2O} \quad RCH_2\overset{13}{C}OOH$$

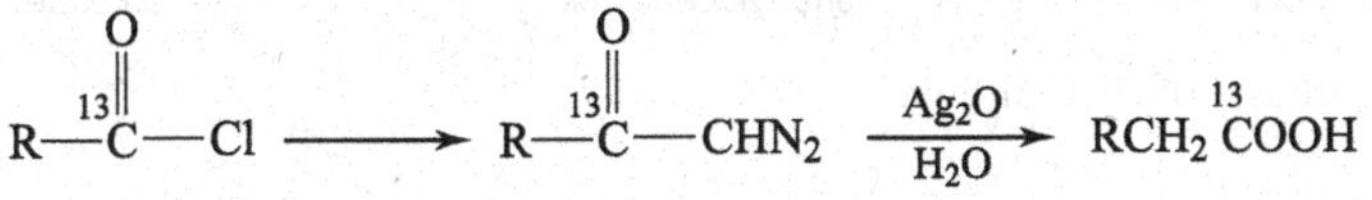

Scheme 2.40 Mechanism of Arndt–Eistert reactions

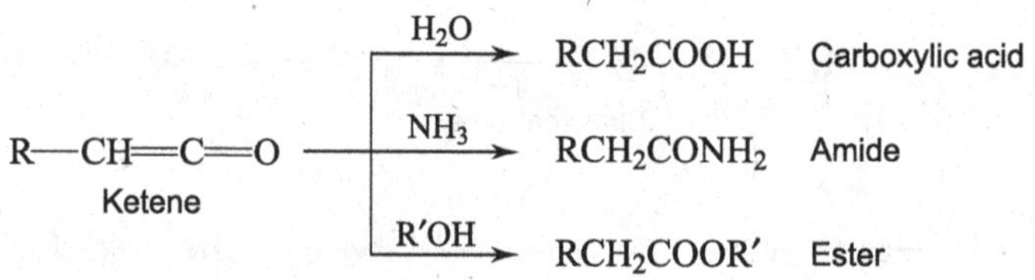

$$R-CH=C=O \quad \text{Ketene}$$

Ketene reacts:
- with H_2O → RCH_2COOH Carboxylic acid
- with NH_3 → RCH_2CONH_2 Amide
- with $R'OH$ → RCH_2COOR' Ester

Scheme 2.41 Synthesis of carboxylic acid derivatives to Arndt–Eistert reaction

(*ii*) Synthesis of various carboxylic acids

(*a*)

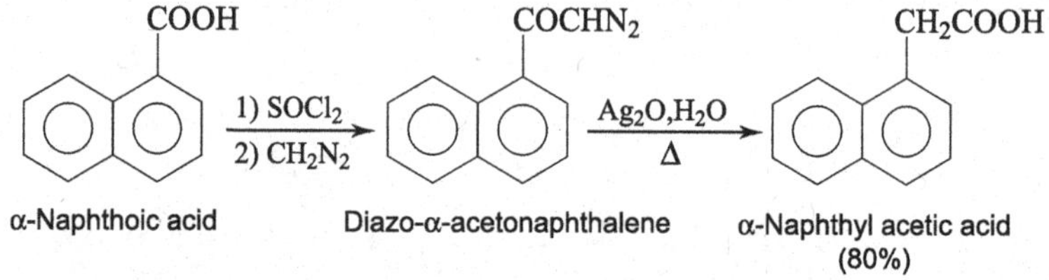

α-Naphthoic acid $\xrightarrow[\text{2) CH}_2N_2]{\text{1) SOCl}_2}$ Diazo-α-acetonaphthalene $\xrightarrow[\Delta]{\text{Ag}_2O, H_2O}$ α-Naphthyl acetic acid (80%)

(*b*)

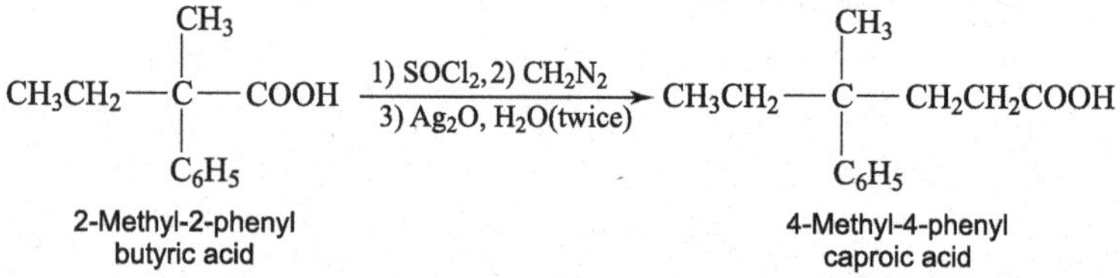

$$CH_3CH_2-\underset{\underset{C_6H_5}{|}}{\overset{\overset{CH_3}{|}}{C}}-COOH \xrightarrow[\text{3) Ag}_2O, H_2O(\text{twice})]{\text{1) SOCl}_2, \text{2) CH}_2N_2} CH_3CH_2-\underset{\underset{C_6H_5}{|}}{\overset{\overset{CH_3}{|}}{C}}-CH_2CH_2COOH$$

2-Methyl-2-phenyl butyric acid

4-Methyl-4-phenyl caproic acid

(*c*)

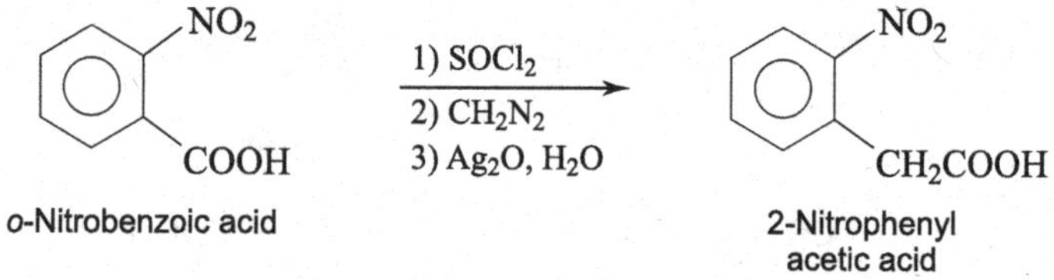

o-Nitrobenzoic acid $\xrightarrow[\text{3) Ag}_2O, H_2O]{\text{1) SOCl}_2 \quad \text{2) CH}_2N_2}$ 2-Nitrophenyl acetic acid

(*iii*) Synthesis of homoveratroyl chloride, an intermediate in the synthesis of papaverine

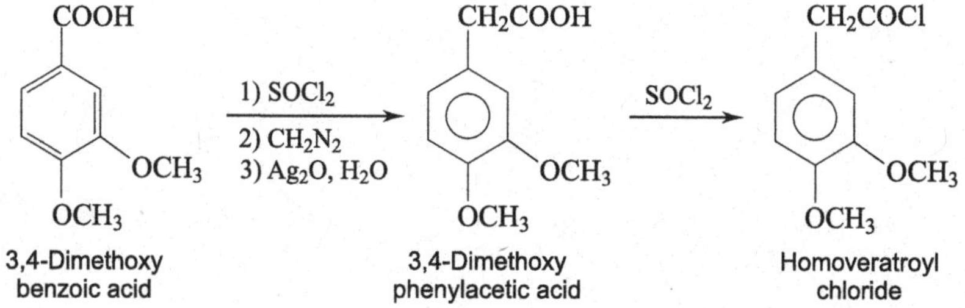

3,4-Dimethoxy benzoic acid $\xrightarrow[\text{3) Ag}_2O, H_2O]{\text{1) SOCl}_2 \quad \text{2) CH}_2N_2}$ 3,4-Dimethoxy phenylacetic acid $\xrightarrow{\text{SOCl}_2}$ Homoveratroyl chloride

(*iv*) Synthesis of mescaline

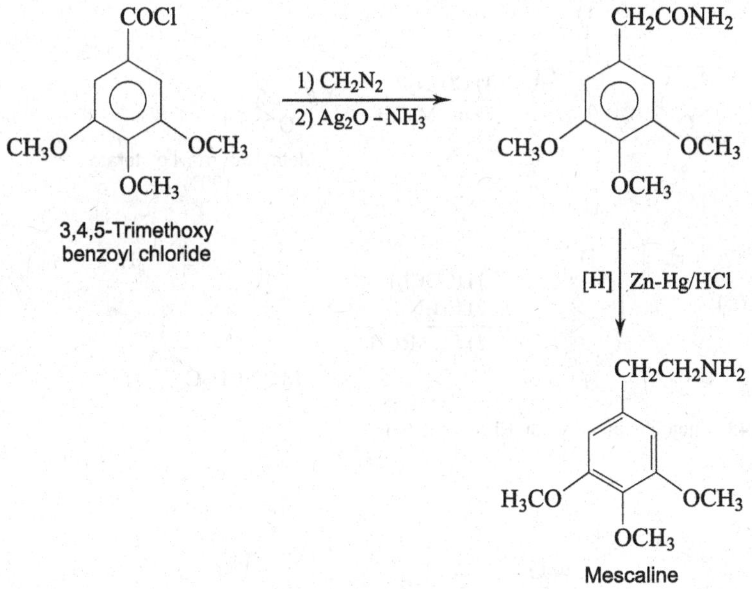

3,4,5-Trimethoxy benzoyl chloride

1) CH_2N_2
2) $Ag_2O - NH_3$

[H] | Zn-Hg/HCl

Mescaline

(v) Synthesis of the ω-hydroxy ketones (ketoalcohols)
Diazoketones on treatment with aqueous formic acid give ω-hydroxy ketones. In the absence of catalyst, the diazoketone is hydrolysed to a carboxylic acid.

$$RCOCHN_2 \xrightarrow{HCOOH + H_2O} RCOCH_2OH$$

(vi) Trimethylsilyl diazomethane can also be used [24] for homologation in Arndt–Eistert synthesis.

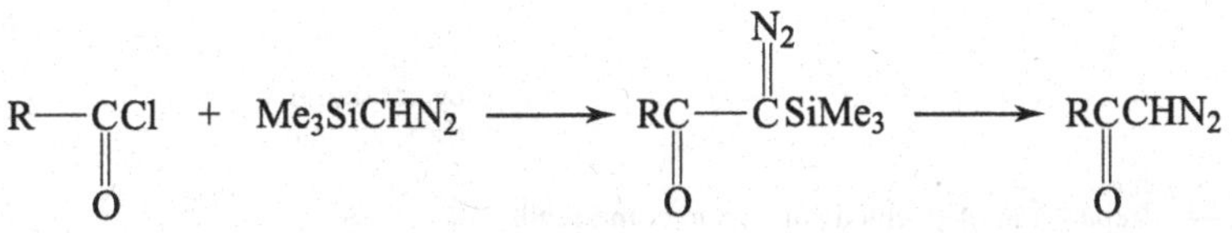

$$R-\underset{O}{\overset{}{C}}Cl + Me_3SiCHN_2 \longrightarrow RC\overset{N_2}{\underset{O}{-C}}SiMe_3 \longrightarrow R\underset{O}{C}CHN_2$$

(vii) A **photochemical Arndt–Eistert** reaction can also be performed. Some examples are:
(a) Synthesis of methyl γ-cyclohexyl butyrate [25] (Scheme 2.42).

(b) Synthesis of methyl δ-furanyl glutarate [26, 27] (Scheme 2.43).

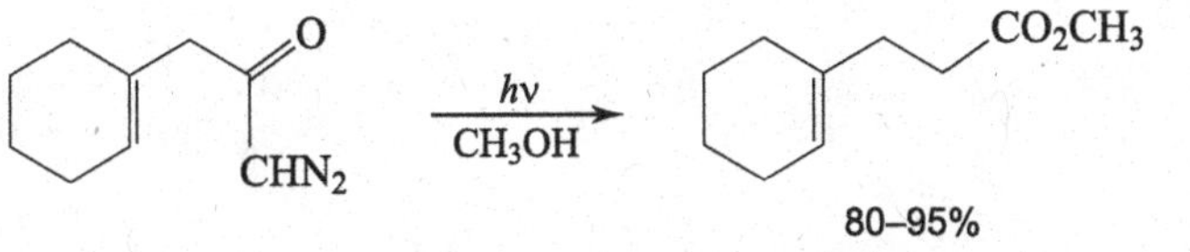

$\xrightarrow[CH_3OH]{h\nu}$

80–95%

Scheme 2.42

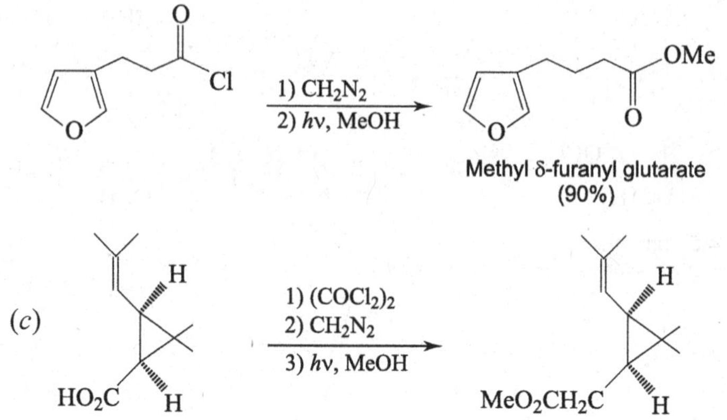

Scheme 2.43 Photochemical Arndt–Eistert reaction

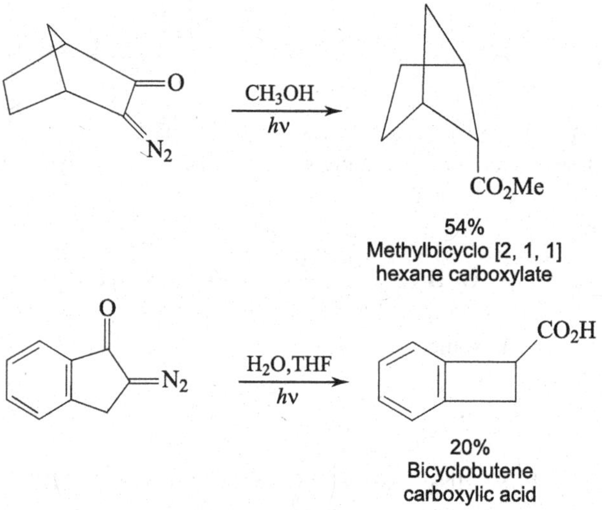

Scheme 2.44 Preparation of strained small ring compounds

(*viii*) A variation of Arndt–Eistert reaction is that in cyclic diazoketones the rearrangement leads to ring contraction. This reaction has been widely used [28] for the preparation of strained small ring compounds, such as bicyclo [2,1,1]-hexane and benzocyclobutene (Scheme 2.44).

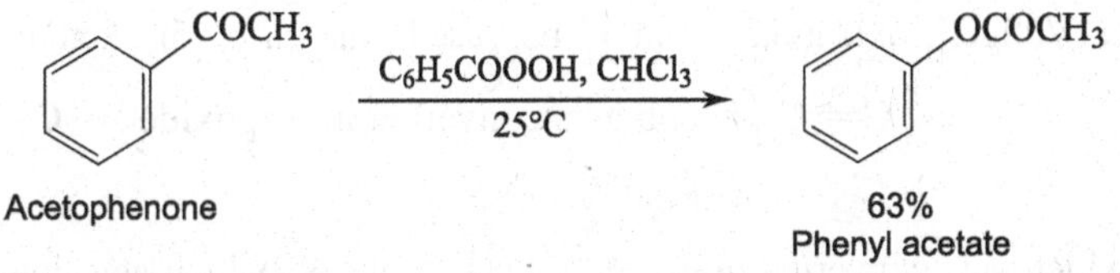

Scheme 2.45   Baeyer–Villiger oxidation

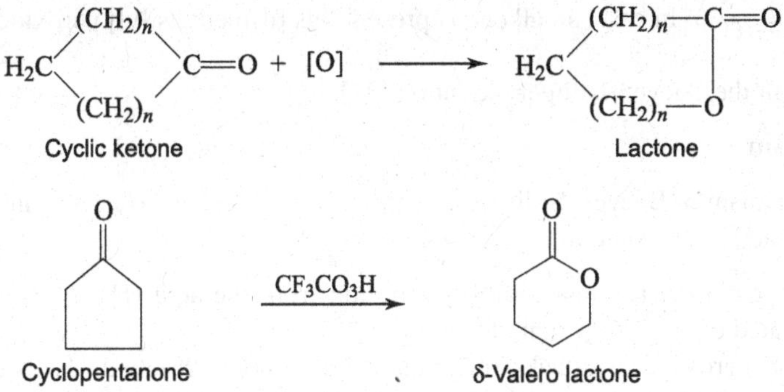

Scheme 2.46   Baeyer–Villiger oxidation

2.4   Baeyer–Villiger Oxidation

The oxidation of ketones to esters with hydrogen peroxide or with peracids (RCO_3H) is known as Baeyer–Villiger oxidation [29]. The reaction can be brought about conveniently by hydrogen peroxide in weak basic solution, peroxy sulfuric acid (Caro's acid) or per acids like trifluoroacetic acid, per benzoic acid, performic acid and m-chloroperbenzoic acid. With Caro's acid, the rearrangement step is much faster than with peracetic acid because sulphate is a better leaving group than acetate. The most efficient reagent is trifluoroacetic acid [30]. A typical example of Baeyer–Villiger oxidation is the reaction of acetophenone with perbenzoic acid at room temperature to give phenyl acetate with 63% yield (Scheme 2.45).

Baeyer–Villiger oxidation converts cyclic ketones to lactones with ring expansion (Scheme 2.46).

This is a convenient method for the synthesis of lactones. The overall reaction is an insertion of oxygen atom between the carbonyl carbon and the adjacent carbon (in ketone). Organic solvents which are inert under the conditions of the reaction may be used. Of course, the choice of the solvent depends on the solubility of the reactants. Solvents like acetic acid and chloroform are commonly used. An important modification of the Baeyer–Villiger oxidation is **Dakins reaction**.

Baeyer–Villiger oxidation cannot be used in case of ketone which

contain $\quad$ C=C $\left(\text{which gets converted into epoxide} -\overset{\displaystyle \text{C}-\text{C}}{\underset{\text{O}}{\diagdown\diagup}}-\right)$,

—S— (which gets converted into $-\underset{\underset{O}{\|}}{S}$—) or $\diagdown$N—R (which gets converted

into $-\underset{\overset{|}{R}}{N}\rightarrow$O).). In case, an alkene is present, bis [trimethylsilyl] peroxide is used

to carry out the Baeyer–Villiger oxidation [31].

Mechanism

The mechanism of Baeyer–Villiger oxidation is not clear. However, it is understood that the reaction takes the following course:

(*i*) The carbonyl reactant removes a proton from the acid (H—A) to give the protonated carbonyl reactant (1).

(*ii*) The peroxy acid attacks the protonated carbonyl reactant (1) to give the oxonium ion (2).

(*iii*) A proton is removed from the oxonium ion (2) to give the species (3).

(*iv*) The species (3) abstracts a proton from the acid (H—A) to give the species (4).

(*v*) The phenyl group migration with an electron pair takes place (from the species 4) to the adjacent oxygen, simultaneously with the departure of RCO$_2$H as a leaving group to give 5.

(*vi*) Final step is the removal of a proton which results in the formation of ester (Scheme 2.47).

The products of Baeyer–Villiger oxidation show that a phenyl group has a greater tendency to migrate than a methyl group. Had this not been the case, the product would have been C$_6$H$_5$COOCH$_3$, and not CH$_3$COOC$_6$H$_5$. The above mechanism is supported by the observation that the labelled carbonyl oxygen atom of the ketone becomes the carbonyl oxygen atom of the ester (the ester has the same ^{18}O content as the ketone) (Scheme 2.48).

Migratory Aptitude

The tendency of a group to migrate is called its migratory aptitude. Studies have shown that in the Baeyer–Villiger oxidation (and also other reactions), the migratory aptitude of the groups is H > phenyl > 3° alkyl > 2° alkyl > 1° alkyl > methyl. In all cases, this order is for groups migrating with their electron pairs, i.e., as anions. The aryl group migrates in preference to methyl and primary alkyl groups. In the aryl series, migration is facilitated by electron-releasing para substituents. Thus, migratory aptitude among aryl group is *p*-CH$_3$OC$_6$H$_4$ > C$_6$H$_5$ > *p*-O$_2$NC$_6$H$_4$. For

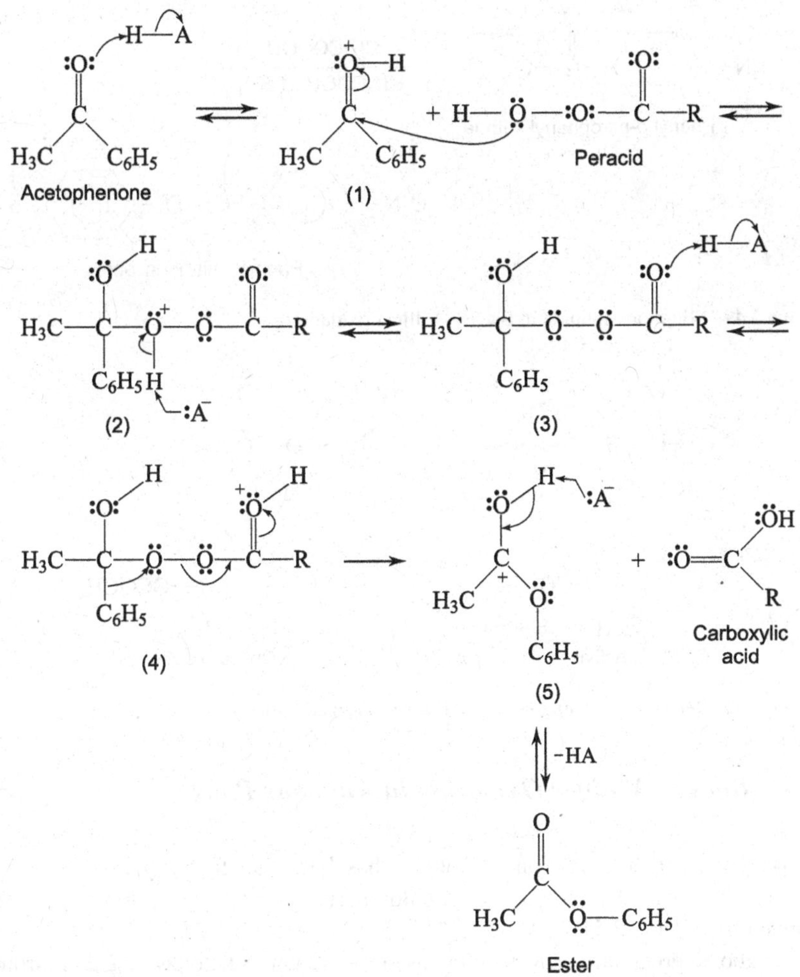

Scheme 2.47 Mechanism of Baeyer–Villiger reaction

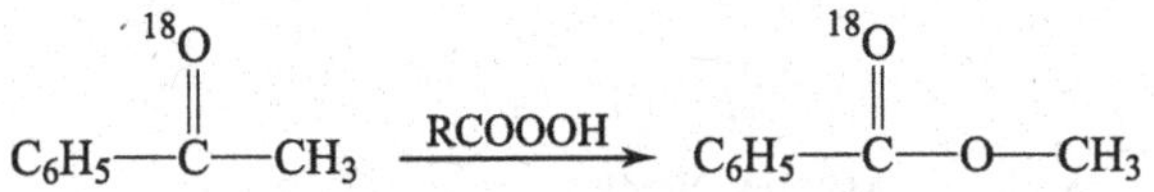

Scheme 2.48 The Baeyer–Villiger oxidation shows that a phenyl group has a greater tendency to migrate than a methyl group

example, phenyl p-nitrophenyl ketone yields only phenyl p-nitrobenzoate by the migration of phenyl group (Scheme 2.49).

The Baeyer–Villiger oxidation takes place with retention of configuration of the migrating group. Two such examples are given in Scheme 2.50.

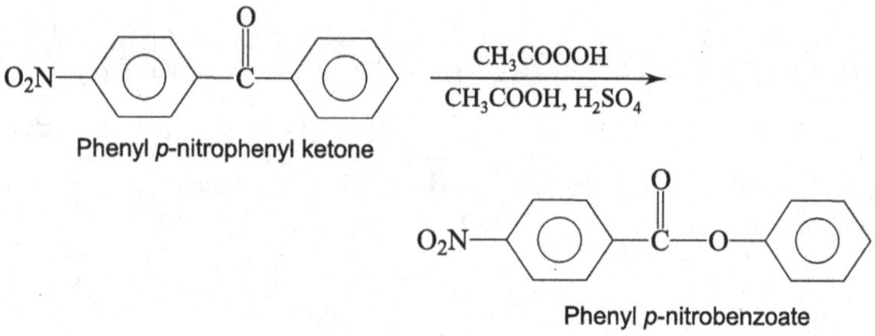

Scheme 2.49 Migratory aptitude in Baeyer–Villiger oxidation

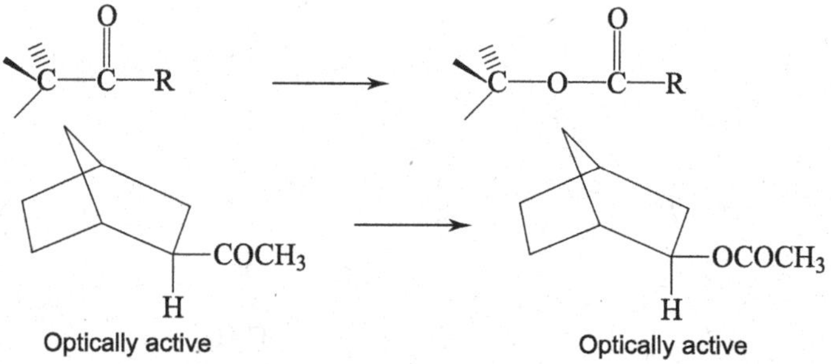

Scheme 2.50 Retention of configuration in Baeyer–Villiger oxidation

2.4.1 Baeyer–Villiger Oxidation in Aqueous Phase

The Baeyer–Villiger oxidation of ketones has been satisfactorily carried out in aqueous heterogenous medium with m-chloroperbenzoic acid [32]. Some examples are given in Scheme 2.51.

The above procedure can also be used for reactive ketones (e.g., anthrone, which usually gives anthraquinone) and ketones which are unreactive or give expected lactones in organic solvents with difficulty [33] that can also be oxidized (Scheme 2.52).

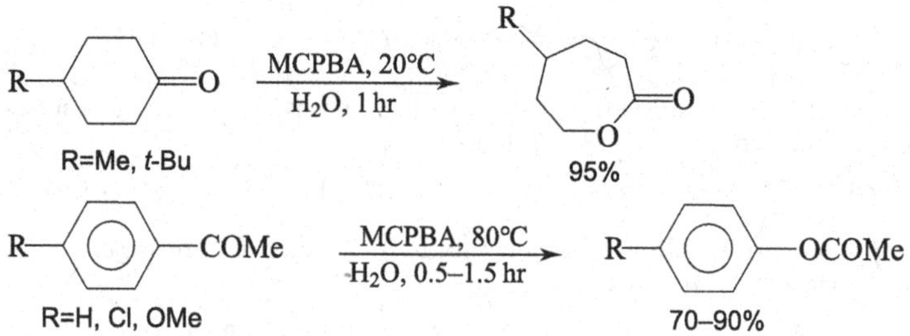

Scheme 2.51 Baeyer–Villiger oxidation in aqueous phase

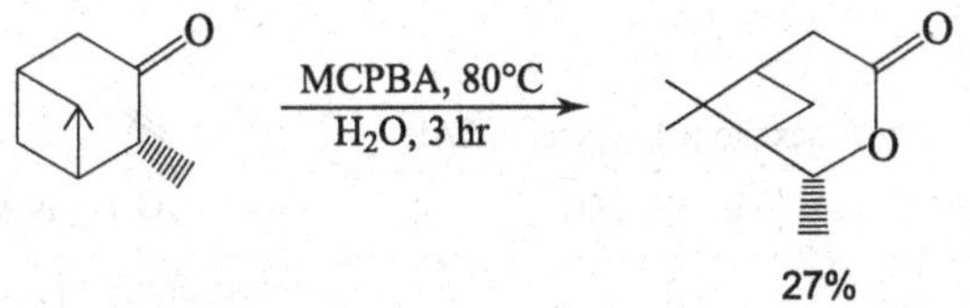

27%

Scheme 2.52 Unreactive ketones can also be oxidized in aqueous phase

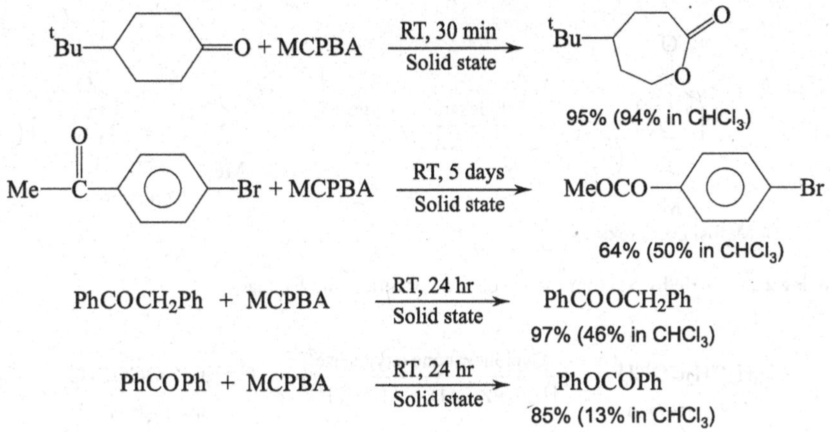

95% (94% in CHCl$_3$)

64% (50% in CHCl$_3$)

PhCOCH$_2$Ph + MCPBA $\xrightarrow[\text{Solid state}]{\text{RT, 24 hr}}$ PhCOOCH$_2$Ph

97% (46% in CHCl$_3$)

PhCOPh + MCPBA $\xrightarrow[\text{Solid state}]{\text{RT, 24 hr}}$ PhOCOPh

85% (13% in CHCl$_3$)

Scheme 2.53 Examples of Baeyer–Villiger oxidation in solid state

2.4.2 Baeyer–Villiger Oxidation in Solid State

Some Baeyer–Villiger oxidation of ketones with *m*-chloroperbenzoic acid proceed much faster in the solid state than in solution. In this procedure, a mixture of powdered ketone and 2 mol equivalent of *m*-chloroperbenzoic acid is kept at room temperature to give the product [34]. Some examples are given in Scheme 2.53 (the yield obtained using CHCl$_3$ is also included for the sake of comparison).

2.4.3 Enzymatic Baeyer–Villiger Oxidation

A number of Baeyer–Villiger oxidations have been carried out with enzymes. A typical transformation is the enzymatic Baeyer–Villiger oxidation, which gives lactone from cyclohexanone using [35, 36] a purified cyclohexanone oxygenase enzyme (Scheme 2.54).

Similarly, 4-methyl cyclohexanone can be converted into the corresponding lactone in 80% yield [37] with > 98% cyclohexanone oxygenase obtained from Acinetobacter (Scheme 2.55).

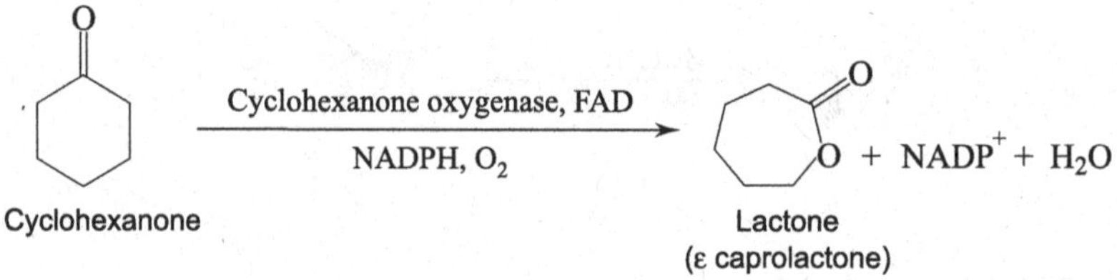

Scheme 2.54 Enzymatic Baeyer–Villiger oxidation

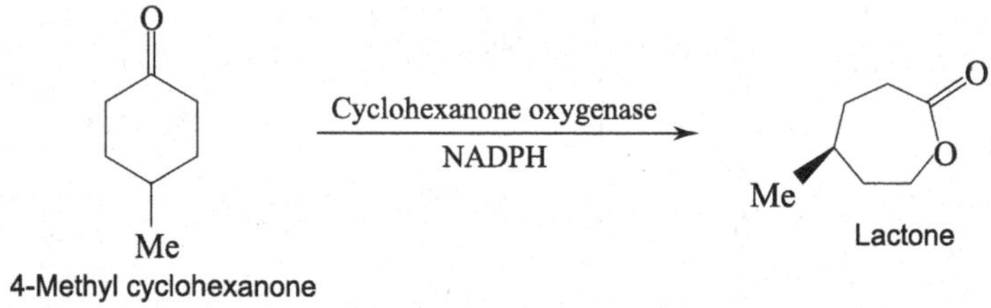

Scheme 2.55 4-Methyl cyclohexanone can be converted into lactone

$$C_6H_5CH_2COCH_3 \xrightarrow[\text{O}_2,\ \text{ENZ-FAD, NADPH, H}^+]{\text{Cyclohexanone oxygenase}} C_6H_5CH_2OCOCH_3$$

Phenylacetone   Benzylacetate

Scheme 2.56 Synthesis of benzylacetate

Cyclohexanone oxygenase in the presence of NADPH (reduced nicotinamide adenine dinucleotide phosphate) converts [37] phenyl acetone into benzylacetate (Scheme 2.56).

A number of enzymatic Baeyer–Villiger oxidations have been reported in steroids (see Sect. 2.4.4 "Applications").

2.4.4 Applications

Baeyer–Villiger oxidation has great synthetic utility. Some important applications are:

(i) Transformation of ketones into esters. An oxygen atom is introduced between the carbon of the carbonyl and the adjacent carbon. This reaction is applicable to both acyclic and cyclic ketones. Oxidation of cyclic ketones results in ring expansion and forms lactones as illustrated by the conversion of cyclopentanone to δ-valerolactone (Scheme 2.57). Similarly, camphor on Baeyer–Villiger oxidation gives α-compholide with 30% yield using Caro's acid, and 2,3-dimethylcyclohexanone is converted [38] into lactone by this method.

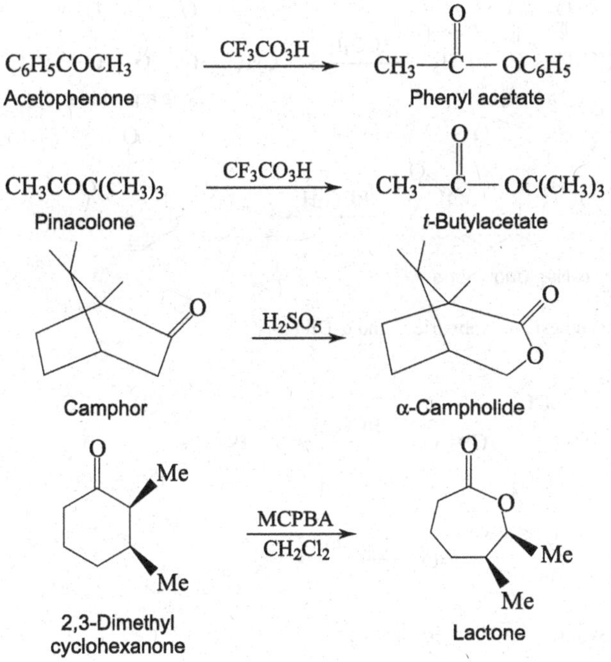

$$C_6H_5COCH_3 \xrightarrow{CF_3CO_3H} CH_3-\overset{\overset{\displaystyle O}{\|}}{C}-OC_6H_5$$

Acetophenone

Phenyl acetate

$$CH_3COC(CH_3)_3 \xrightarrow{CF_3CO_3H} CH_3-\overset{\overset{\displaystyle O}{\|}}{C}-OC(CH_3)_3$$

Pinacolone

t-Butylacetate

Camphor

$\xrightarrow{H_2SO_5}$

α-Campholide

2,3-Dimethyl cyclohexanone

$\xrightarrow[CH_2Cl_2]{MCPBA}$

Lactone

Scheme 2.57 Transformation of ketones into esters

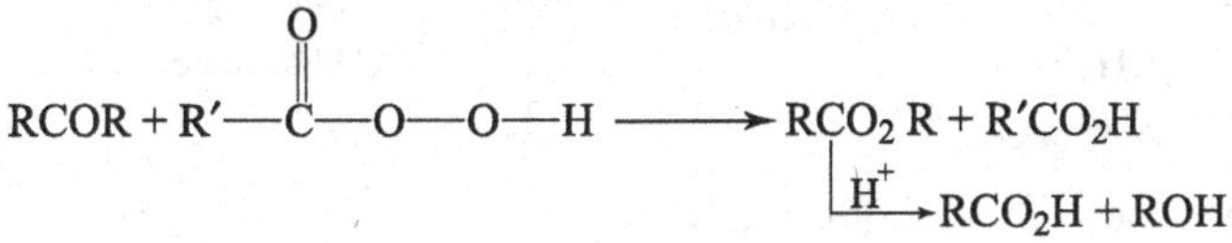

$$RCOR + R'-\overset{\overset{\displaystyle O}{\|}}{C}-O-O-H \longrightarrow RCO_2R + R'CO_2H$$

$$\underset{H^+}{\big\lfloor} \rightarrow RCO_2H + ROH$$

Scheme 2.58 Synthesis of carboxylic acids

(ii) Synthesis of carboxylic acids from ketones or aldehydes. Baeyer–Villiger oxidation of aldehydes or ketones gives esters, which on saponification gives carboxylic acids (Scheme 2.58).

(iii) Synthesis of anhydrides. α-Diketones on oxidation under Baeyer–Villiger conditions give anhydrides (Scheme 2.59).

(iv) Synthesis of large ring lactones (which are difficult to prepare by other methods) (Scheme 2.60).

(v) Synthesis of long-chain hydroxyesters. The long-chain lactones obtained by Baeyer–Villiger oxidation of ketone (*see* (*iv*) above) on treatment with conc. H_2SO_4/C_2H_5OH give long-chain α,ω-hydroxyesters (Scheme 2.61).

(vi) Biochemical Baeyer–Villiger oxidations of steroids.

19-Nortestosterone on treatment with *Aspergillus tamarii* gives 70% yield of 19-nortestololactone [39]. Progesterone and testosterone are converted into Δ^1-dehydrotestololactone by fermentation with *Cylindrocarpon radicicola* [40].

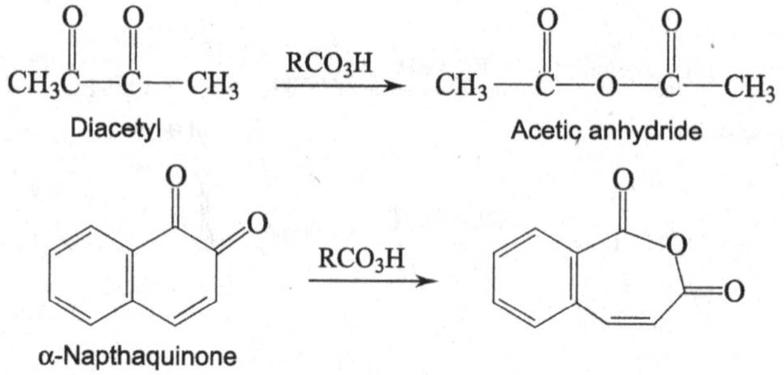

Scheme 2.59 Synthesis of Anhydrides and α-Diketones

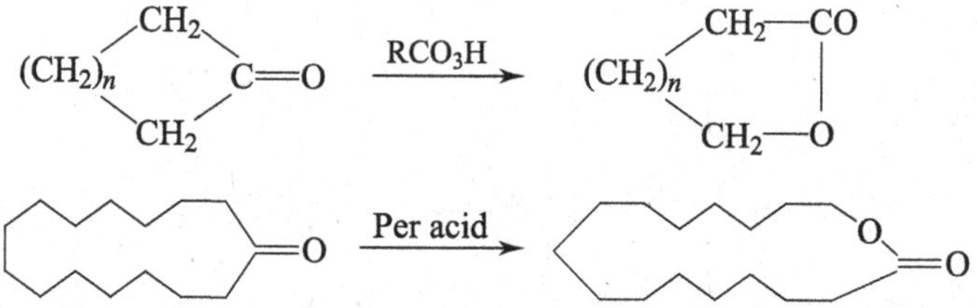

Scheme 2.60 Synthesis of large ring lactones

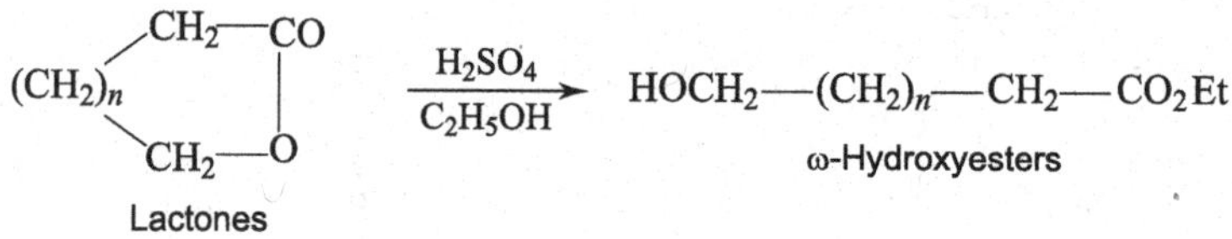

Scheme 2.61 Synthesis of long-chain hydroxyesters

Testololactone is obtained from progesterone by oxidation with *Penicillium chrysogenum* and from 4-androstene-3,17-dione by treatment with *Penicillium lilacinum* [41] (Scheme 2.62).

(vii) Synthesis of bicyclic lactone with retention of configuration [42] (Scheme 2.63).

(viii) Synthesis of phenols

The ester formed in Baeyer–Villiger oxidation can be hydrolysed to the corresponding phenol. Thus, it provides a route to transform aldehydes or ketones to phenols. For example, veratraldehyde is converted into 3,4-dimethoxyphenol (*see* also **Dakins reaction**) (Scheme 2.64).

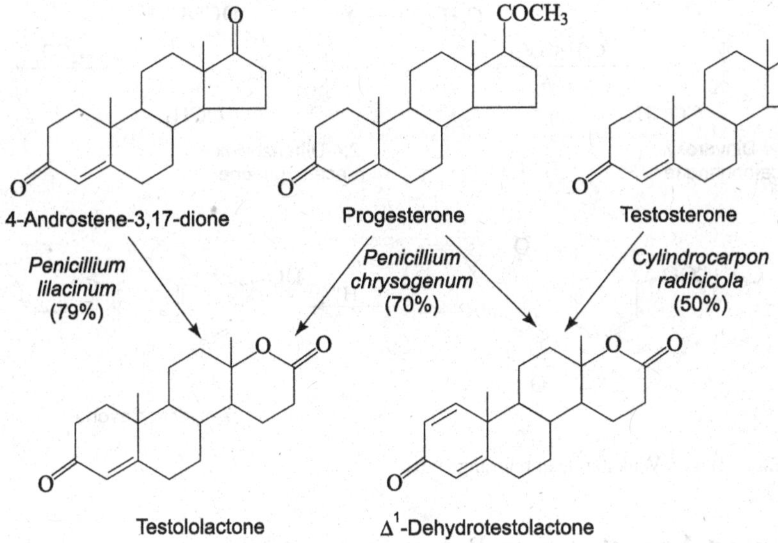

Scheme 2.62 Biochemical Baeyer–Villiger oxidations of steroids

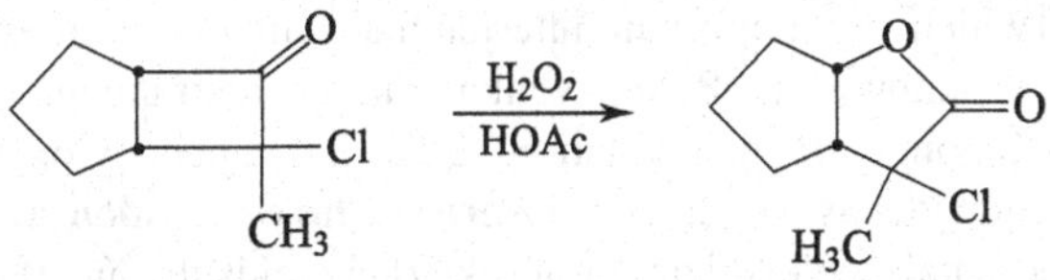

Scheme 2.63 Synthesis of bicyclic lactone

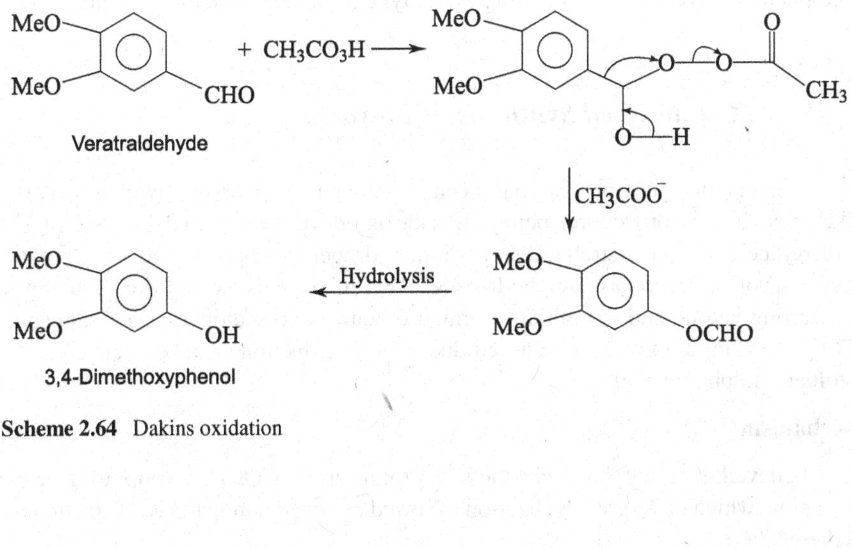

Scheme 2.64 Dakins oxidation

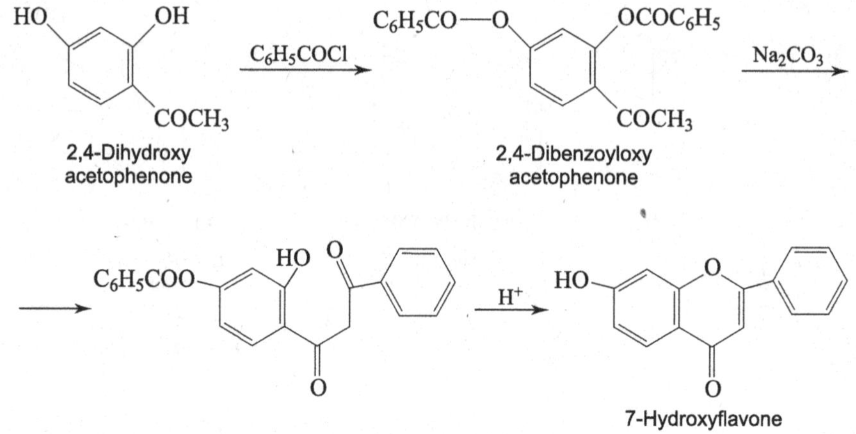

Scheme 2.65 Baker–Venkataraman rearrangement

2.5 Baker–Venkataraman Rearrangement [43]

The base-catalysed rearrangement of o-acyloxy (preferably o-benzoyloxy) ketones to β-diketones (which are important intermediates in the synthesis of flavones or chromones) is known as Baker–Venkataraman rearrangement. Thus, 2,4-dihydroxyacetophenone on benzoylation give 2,4-dibenzoyloxyacetophenone, which on treatment with base (undergoes **internal Claisen condensation**) to give the diketone; the acid-catalysed cyclization of diketone yields flavone (Scheme 2.65).

The whole procedure (Scheme 1) is known as Baker–Venkataraman flavone synthesis. Use of 2,4-diacetoxy acetophenone in place of 2,4 dibenzyloxy-acetophenone give the corresponding 2-methyl chromone as the final product.

2.5.1 PTC-Catalysed Synthesis of Flavones

A convenient one-step process has been developed [44] for the synthesis of β-diketones. In this procedure, benzylchloride is added to a stirred mixture of 2-hydroxyacetophenone, tetrabutylammonium hydrogen sulphate and aqueous potassium carbonate or potassium hydroxide. The mixture is stirred for 2 h until the starting ketone and the initially formed o-benzoyloxyacetophenone disappears (TLC). Working up gives the required diketone. The diketone can be cyclized with p-toluene sulphonic acid.

Mechanism

It is believed that the base abstracts a proton from COCH$_3$ group to give a carbanion, which undergoes cyclization followed by ring opening to give β-diketones (Scheme 2.66).

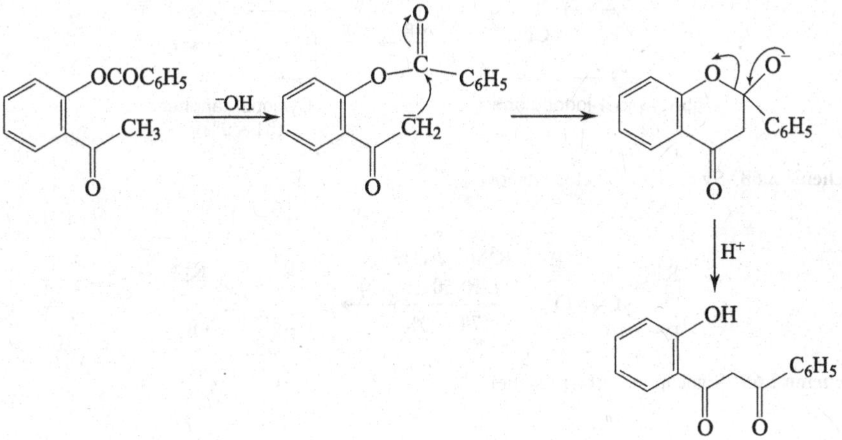

Scheme 2.66 Mechanism of Baker–Venkataraman rearrangement

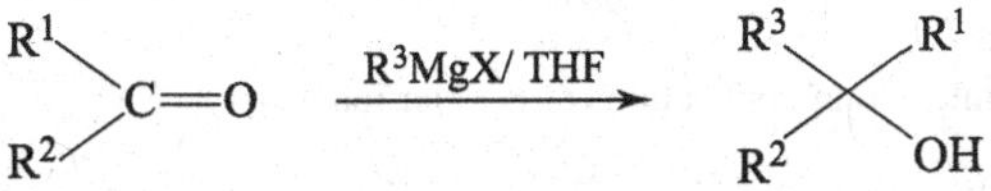

Scheme 2.67 Barbier reaction

2.5.2 *Application*

Baker–Venkataraman procedure is used for the synthesis of flavones and chromones.

2.6 Barbier Reaction [45]

The reaction of a ketone with an organometallic reagent to give the corresponding alcohol is known as Barbier reaction (Scheme 2.67).

The original process discovered by Barbier offers distinct advantages, but has a number of drawbacks. It is very important to prepare the organometallic reagent in situ (in the presence of substrate). As the organometallic intermediate is generally very reactive with water, Barbier-type reactions were earlier carried out in anhydrous solvents. Barbier reaction can only be performed with reactive alkyl halides.

An important Barbier reaction [46] is useful in the synthesis of cyclopentanone (Scheme 2.68).

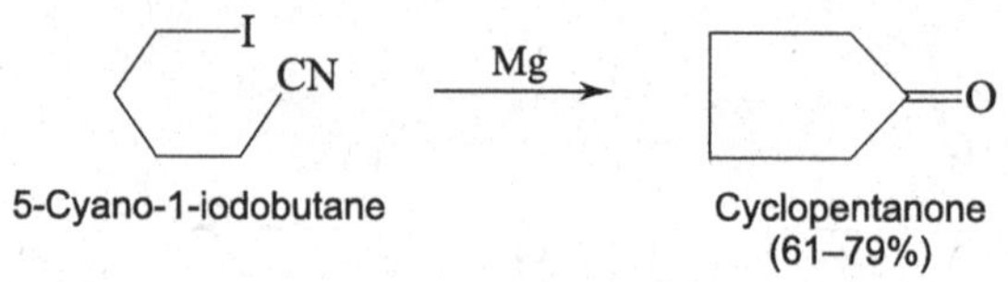

5-Cyano-1-iodobutane   Cyclopentanone
                                                 (61–79%)

Scheme 2.68 Synthesis of cyclopentanone

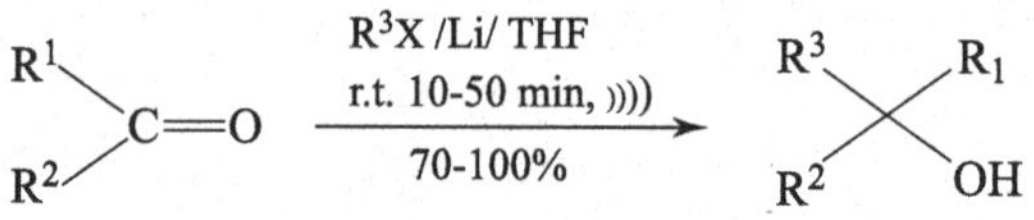

Scheme 2.69 Sonication Barbier reaction

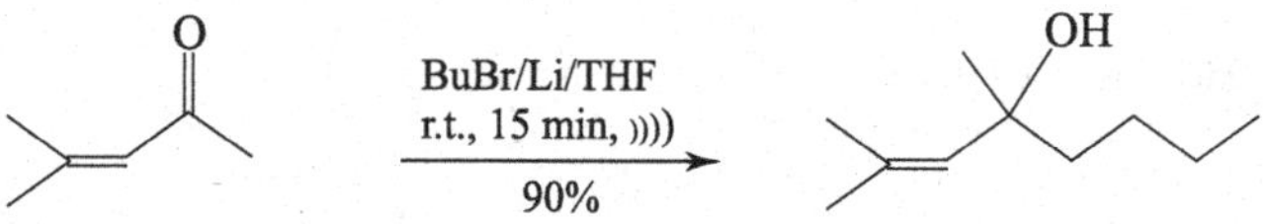

Scheme 2.70 Some interesting applications of Barbier reactions

2.6.1 Barbier Reaction Under Sonication

The first improvement was obtained by the replacement of magnesium by lithium [47], but the most crucial step was made by **sonication** [48]. There are significant advantages in this method, as the reactions can be carried out in commercial THF and are largely free from side reactions, such as reduction and enolization, which are common in conventional procedure. Even allyl or benzyl halides give much better yields (>95%), and very little Wurtz coupling that predominates in non-ultrasonic conventional method has been observed.

The modified Barbier reaction using Li/THF/sonication is represented as in Scheme 2.69.

2.6.2 Applications (Scheme 2.70)

(*i*) In Barbier reaction, α,β-unsaturated ketones can be used with good results.
(*ii*) An intermediate required for the synthesis of sesquicarene can be obtained by a sonochemical Barbier step [49].

(iii) Synthesis of pentalenic acid [50] starting from dimethylcyclopentanone.

Diethylcyclopentenone   Pentalenic acid

(iv) Examples of the sonochemical Barbier reaction is described using benzylic halides [51, 52]. In these reactions, there is no Wurtz coupling.

(v) An allylic phosphate can be used in place of the corresponding halide [53].

(vi) An intramolecular reaction from the following substrate is the key step in the synthesis of trichodiene [54].

(vii) Barbier procedure can be applied to amides. The usual reaction of organometallic compounds with amides gives carbonyl compounds along with the formation of many by-products. However, under Barbier sonochemical procedure, several improvements are observed and good yields are obtained [55, 56].

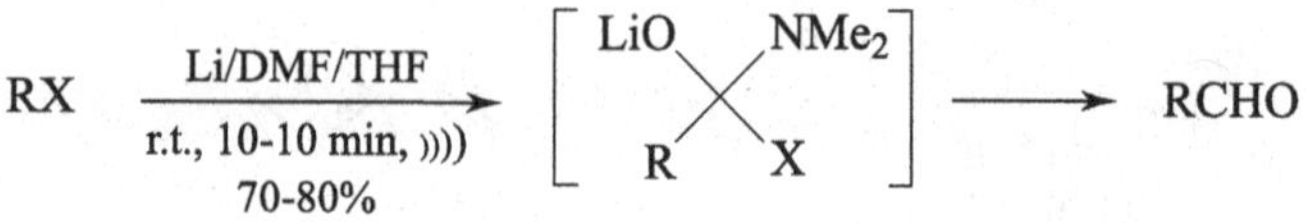

$$RX \xrightarrow[\substack{r.t., 10\text{-}10 \text{ min},)))\\ 70\text{-}80\%}]{Li/DMF/THF} \left[\begin{array}{c} LiO \quad NMe_2 \\ \diagdown\!\!\diagup\!\!\diagdown \\ R \quad X \end{array} \right] \longrightarrow RCHO$$

R = aryl, aryl, benzyl

(*viii*) Barbier procedure can be applied to isocyanates. Sonochemical Barbier procedure for the reaction with isocyanates gives much better result. However, in this case, better results are obtained [57] with sodium or magnesium. A comparison of the result with different metals is given below.

$$tBuN=C=O \xrightarrow[r.t]{Metal \,|PhBr|\, THF} PhCONHBu(t)$$

Metal	Conditions	Time (h)	% Yield
Na	 ⌒→	48 h	53
Li	))))	15 min.	51
Na	))))	45 min.	78
Mg	))))	15 min.	91

Besides what have been stated above, a number of other applications of sonochemical Barbier procedure have been recorded [58].

2.7 Barton Reaction

In Barton reaction [59], a methyl group in the δ-position to an OH group is converted into an oxime group, which can be oxidized to a CHO group. In this procedure, the alcohol is first converted to the nitrite ester. Photolysis of the nitrite results in the conversion of the nitrite group to the OH group and nitrosation of the methyl group. Hydrolysis of the oxime tautomer gives the aldehyde. The overall reaction is shown in Scheme 2.71.

Mechanism

The alcohol on reaction with nitrosyl chloride (NOCl) gives the nitrite, which on photolysis undergoes homolytic cleavage to give alkoxy radical; subsequently, a

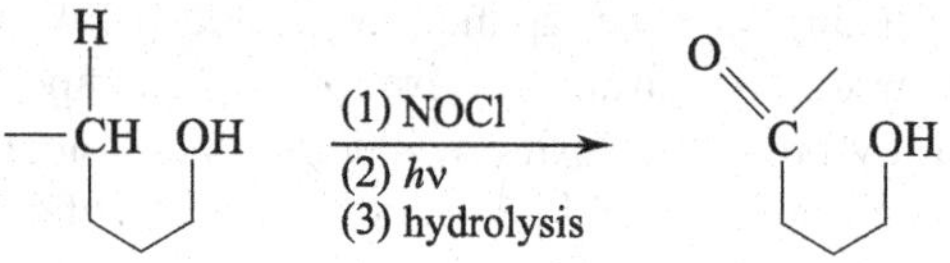

$$\begin{array}{c} H \\ | \\ \text{—CH} \quad OH \end{array} \xrightarrow[\substack{(2)\ h\nu \\ (3)\ hydrolysis}]{(1)\ NOCl} \begin{array}{c} O \\ \diagdown \\ C \quad OH \end{array}$$

Scheme 2.71 Barton reaction

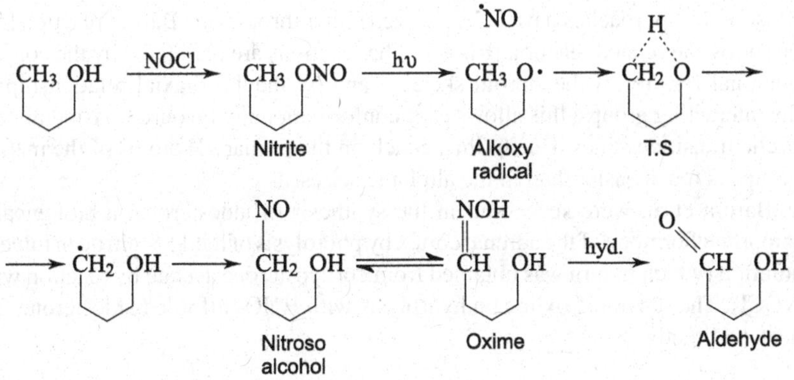

Scheme 2.72 Mechanism of Barton reaction

hydrogen atom is abstracted from a carbon atom in a δ-position to the original hydroxyl group to give nitroso alcohol, which tautomerizes to the oxime. The transfer of hydrogen to alkoxy-free radical takes place via a six-member transition state [60]. Finally, the oxime group can be hydrolysed to the aldehyde group. Various steps involved are shown in Scheme 2.72.

Barton reaction provides a procedure to oxidize a carbon atom separated from an OH group by three other carbon atoms.

2.7.1 Applications (Scheme 2.73)

1. The most important synthetic application of the Barton reaction has been in the steroid series, particularly in the functionalization of the ten non-activated C-18 and C-19 angular methyl group by photolysis of the nitrites of suitably disposed hydroxyl group. In principle (*see* structure below), C-18 methyl group can be attacked by an alkoxy radical at C-8, C-11, C-15 or C-20. The C-19 methyl group can be attacked by an alkoxy radical at C-2, C-4, C-6 and C-11.

Attack of C-18 and C-19 methyl groups by various alkoxy radicals.

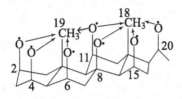

Scheme 2.73 Some important applications of Barton reaction

Most of the approaches have been realized either through the Barton reaction [59, 60] or by the related reaction [61–63]. The reactions are facilitated by the conformational rigidity of the steroid skeleton and by the 1,3-diaxial relationship of the interacting groups; this allows easy conformationally favoured six-membered cyclic transition states. Due to this, attack on the primary H atoms of the methyl groups is much easier than in the aliphatic series.

2. Barton et al. were successful in the synthesis of aldosterone, a biologically important hormone of the adrenal cortex by photolysis of the 11 β-nitrite in toluene solution (which in turn was obtained from corticosterone acetate by reaction with NOCl). The separated oxime on hydrolysis with HNO_2 afforded aldosterone-21-acetate directly.

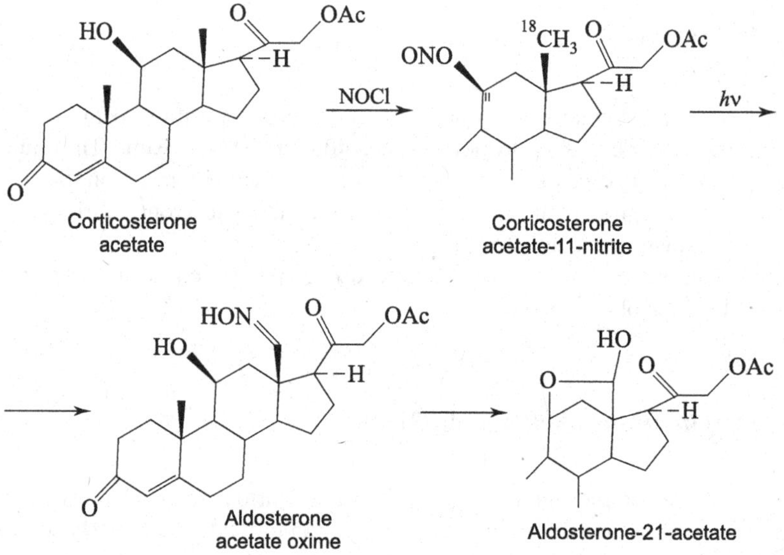

Corticosterone
acetate

Corticosterone
acetate-11-nitrite

Aldosterone
acetate oxime

Aldosterone-21-acetate

3. Synthesis of perhydrohistrionicotoxin was effected [64] by photolysis of the appropriate nitrite; the formed oxime on Beckmann rearrangement yielded the bicyclic lactam.

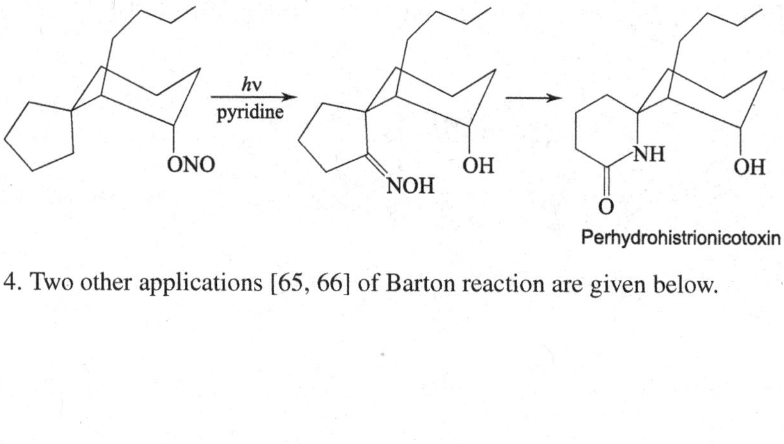

Perhydrohistrionicotoxin

4. Two other applications [65, 66] of Barton reaction are given below.

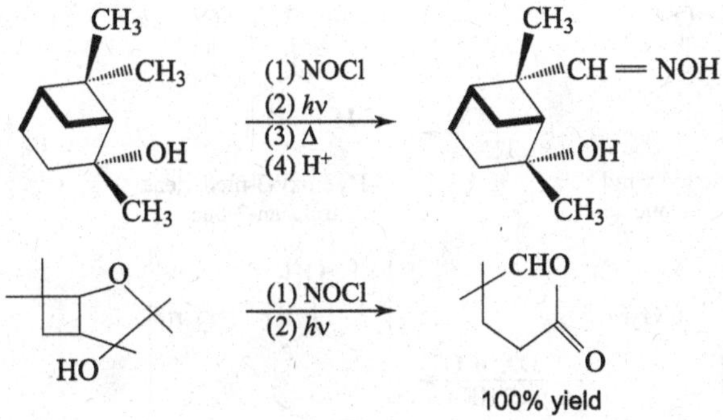

5. Barton reaction is useful for the production of tritium-labelled aldosterone (1), which finds its use in a medical diagnostic aid. The reaction involves photochemical reaction of the nitrite of pregnane derivative (2) to give the C-18 oxime (3), which on heating gives the nitrone [67] (4). The nitrone can be converted into a series of conventional steps into 1,2-didehydroaldosterone acetate (5). Final catalytic tritiation of (5) and subsequent hydrolysis [68] gave the radio-labelled aldosterone (1).

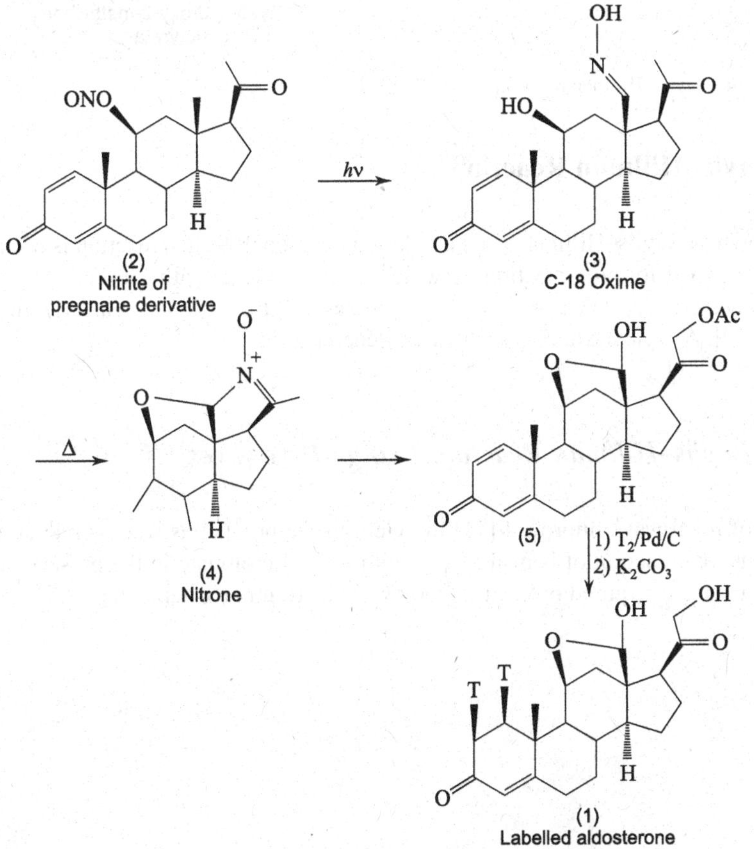

(2)
Nitrite of
pregnane derivative

(3)
C-18 Oxime

(4)
Nitrone

(5)
1) T$_2$/Pd/C
2) K$_2$CO$_3$

(1)
Labelled aldosterone

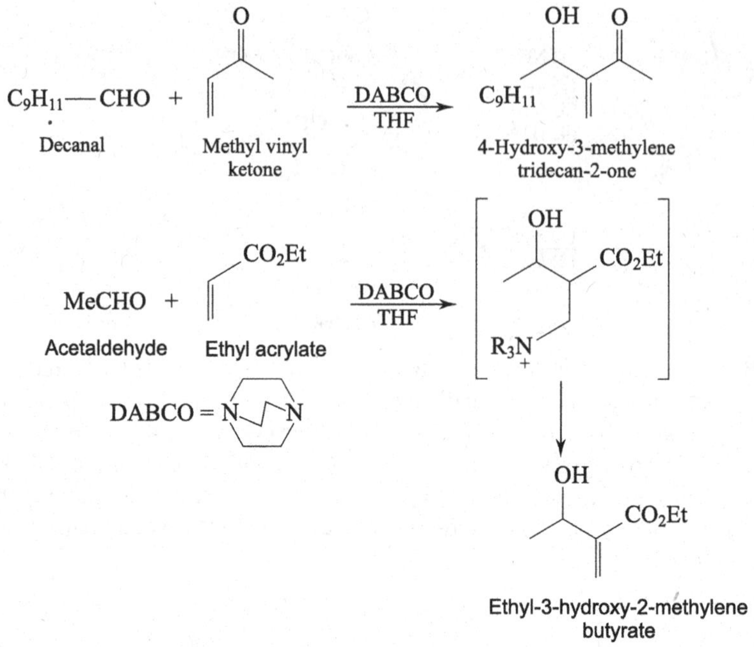

Scheme 2.74 Baylis–Hillman reaction

2.8 Baylis-Hillman Reaction

Also known as Baylis–Hillman vinyl alkylation reaction, [69] this reaction is a very useful C–C bond-forming reaction, in which aldehydes react with acrylates to give α-(hydroxyalkyl) acrylates or with vinyl ketones to give α-(hydroxylalkyl) vinyl ketones [70]. A typical example is given in Scheme 2.74.

2.8.1 Baylis–Hillman Reaction Using Microwaves

The use of microwave improved [71] the yield of the product as well as reduction time. Thus, the reaction of benzaldehyde with methyl crotonate in the presence of DABCO gave the required product in good yield in 10 min (Scheme 2.75).

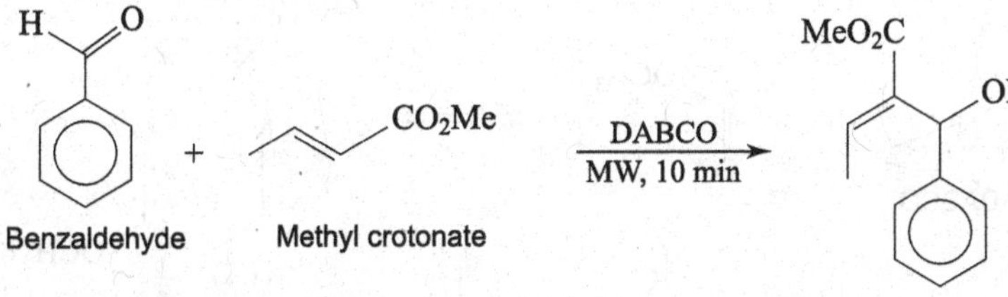

Scheme 2.75 Some chemical Baylis–Hillman reaction

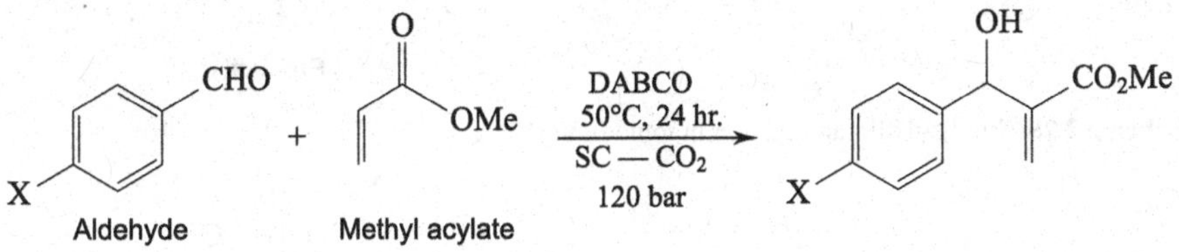

Scheme 2.76 Baylis–Hillman reaction in SC–CO$_2$

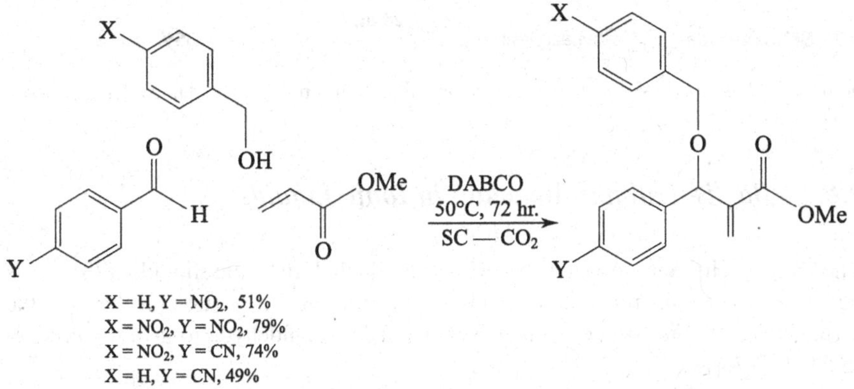

X = H, Y = NO$_2$, 51%
X = NO$_2$, Y = NO$_2$, 79%
X = NO$_2$, Y = CN, 74%
X = H, Y = CN, 49%

Scheme 2.77 In three-component coupling in the presence of SC CO$_2$ and alcohol. The major product is an ether

2.8.2 Baylis–Hillman Reaction in Supercritical Carbon Dioxide

The Baylis–Hillman reaction has also been carried out in SC–CO$_2$ and gives better conversion and reaction rates [72] as compared to solution-phase reactions (Scheme 2.76).

If the above reaction is carried out in the presence of an alcohol, the major product is an ether from a three-component coupling reaction, which occurs only in the presence of SC–CO$_2$ (Scheme 2.77).

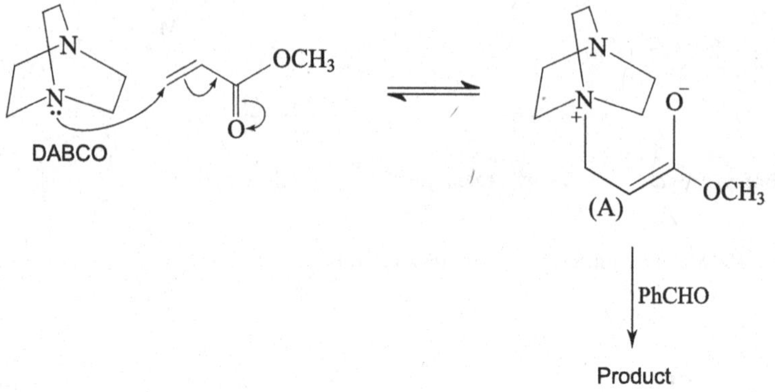

Scheme 2.78 Baylis–Hillman reaction in ionic liquids

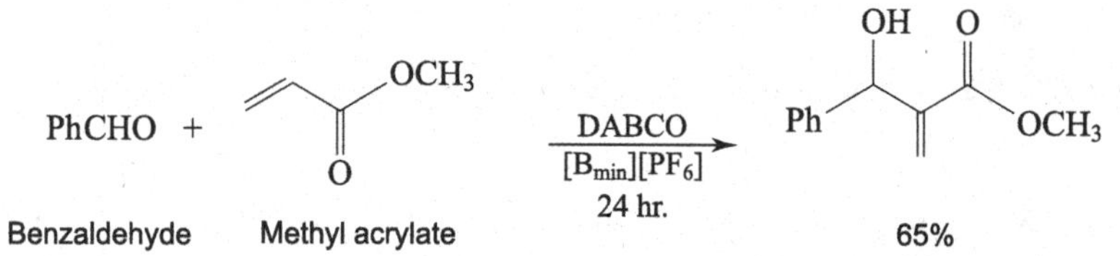

Scheme 2.79 In Baylis–Hillman reaction in ionic liquids best results are obtained in the presence of DABCO

2.8.3 Baylis–Hillman Reaction in Ionic Liquids

The Baylis–Hillman reaction has been conducted in ionic liquids [73]. It is believed [73] that the reaction proceeds via an addition–elimination mechanism, the formed zwitterionic speeches, such as (A) attacks the aldehyde to give the product (Scheme 2.78).

Though a number of tertiary amines have been used for the reaction, the base of choice is (diazabicyclo[2.2.2]octane (DABCO). The Baylis–Hillman reaction between benzaldehyde and methyl acylate in the ionic liquid [b$_{min}$][PF$_6$] was found to be 33 times faster than the reaction in CH$_3$CN, although only moderate yield to the desired product was obtained [74] (Scheme 2.79).

It was found [74] that under basic reaction condition, the aldehyde was being consumed with imidazolium cation. This accounts for the low yields and also demonstrated that the ionic liquids are not always inert solvents [74]. Thus, it was shown that the acidic nature of the C(2) hydrogen of the imidazolium cation was responsible for the side reaction (Scheme 2.80).

In each successive recycle, more and more of the addition product is formed between the deprotonated imidazolium salts, and the aldehyde was accumulated. Thus, less ionic liquid was left for reacting with the aldehyde to give the side reaction product and so more of the aldehyde was available for the normal Baylis–Hillman

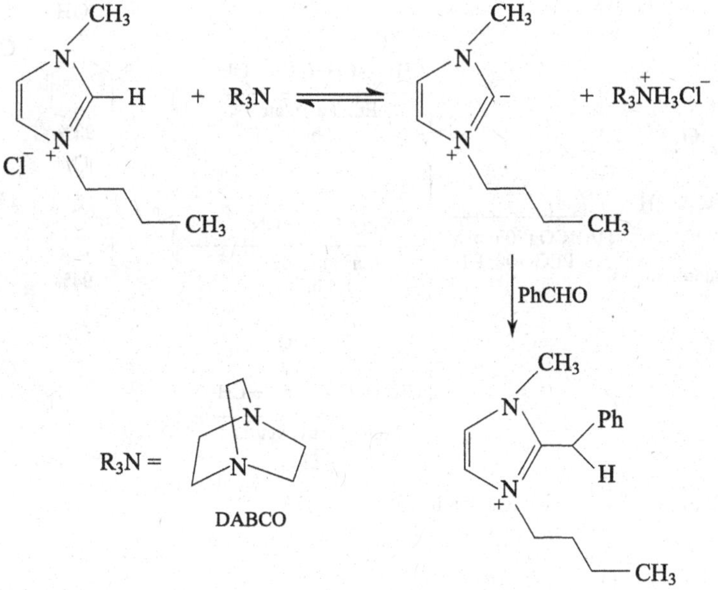

Scheme 2.80 A side reaction in Baylis–Hillman reaction in ionic liquid

reaction. Thus, for subsequent recycling the yield improved. On the basis of the results obtained, it was concluded that caution must be exercised when using ionic liquids from one reaction to another; in such cases, a mixture of products was obtained.

In order to overcome the problem of low yields due to the formation of side reaction product (because of the acidity of C(2) imidazolium cation), ionic liquids substituted at the 2-position were used [75]. It was found [75] that the Baylis–Hillman reaction between a variety of aldehydes and methyl acrylate proceeded smoothly in the ionic liquid [b_{mmin}] [PF_6], in contrast to the results obtained with [b_{min}] [PF_6].

The Baylis–Hillman reaction has been successfully carried out in the presence of imidazolinium-based ionic liquids containing a phenyl group of C(2) position [cation = (m Ph_{min})] (A), which are stable to a variety of strongly basic conditions [76].

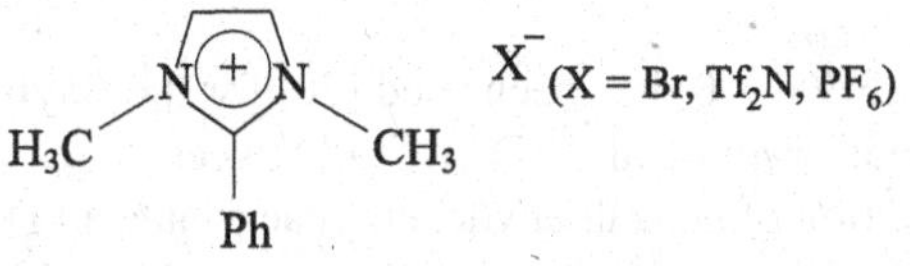

The Baylis–Hillman reaction has been performed in the presence of chloroaluminate ionic liquids [77]. It has been found that 1-ethyl-3-methylimidazolium chloride (EMIC + AlCl$_3$) is a more efficient chloroaluminate ionic liquid.

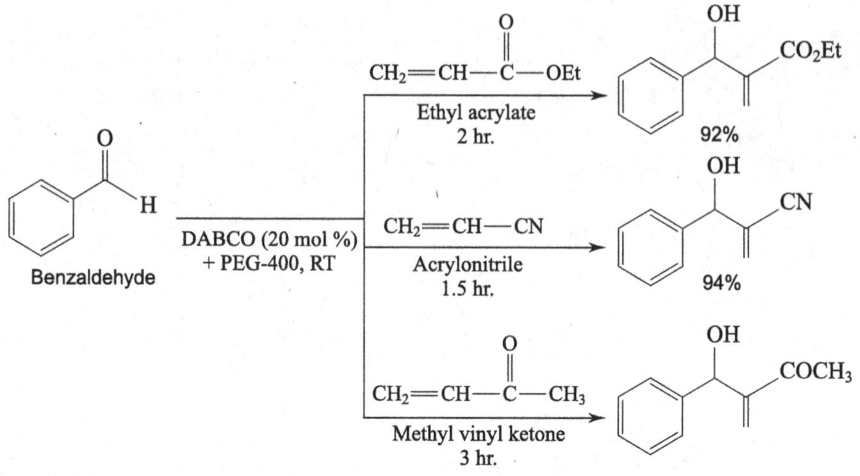

Scheme 2.81 Baylis–Hillman reaction in PEG

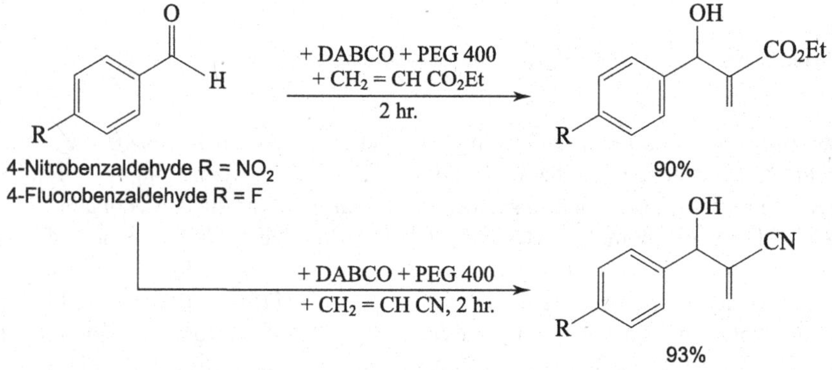

Scheme 2.82 Some examples of Baylis–Hillman reaction in PEG

2.8.4 Baylis–Hillman Reaction in Polyethylene Glycol (PEG)

Polyethylene glycol (PEG-400) has been used [78] for the Baylis–Hillman reaction using the conventional basic catalyst DABCO between unreactive aldehydes and olefins. Thus, the reaction of benzaldehyde, ethyl acrylate and DABCO in PEG-400 at room temperature for 2 h gave the Baylis–Hillman product with 92% yield. Similar products were obtained by the reaction of benzaldehyde with acrylonitrile or methyl vinyl ketone (Scheme 2.81).

4-Nitrobenzaldehyde reacted with ethyl acrylate and acrylonitrile giving the expected product in 90 and 93%, respectively, after 2 h. 4-Fluorobenzaldehyde reacted much faster forming the expected product in 4 h, while the same reaction took over 60 h using triethylamine (Scheme 2.82).

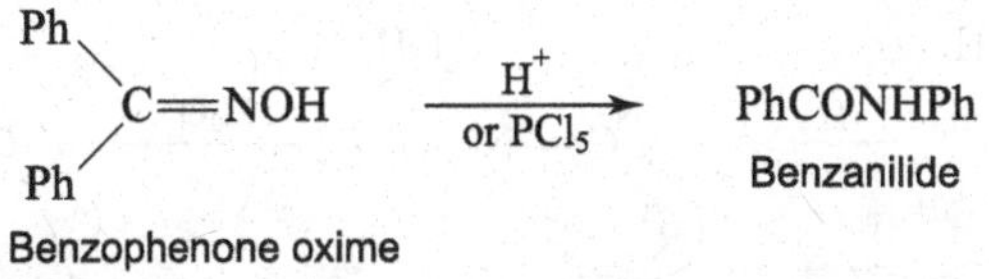

Ph
$\quad$C=NOH $\quad\xrightarrow[\text{or PCl}_5]{\text{H}^+}\quad$ PhCONHPh
Ph
$\qquad\qquad\qquad\qquad\qquad\qquad$ Benzanilide

Benzophenone oxime

Scheme 2.83 Beckmann rearrangement

Some other aldehydes like 2-chloro-5-nitrobenzaldehyde, 2-furaldehyde and 2-thiophene carboxaldehyde showed similar results. Aliphatic aldehydes like 3-phenyl propanol ($C_6H_5CH_2CH_2CHO$), isobutyraldehyde [$(CH_3)_2$ CHCHO] and hexanal

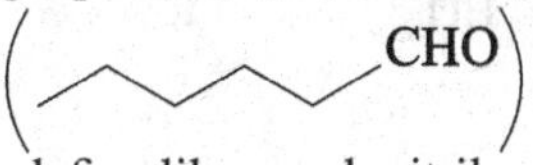

$\left(\text{\Large\char`\~\char`\~\char`\~CHO}\right)$ also underwent Baylis–Hillman reaction with activated

olefins like acrylonitrile and ethyl acrylate in 75%, 86% and 80% yields, respectively. Also, formaldehyde and *trans* cinnamaldehyde reacted with acrylates in PEG to provide good yields of the expected products.

Polyethylene glycol is a rapid and recyclable medium for the Baylis–Hillman reaction, which is one of the very few reactions where there is 100% atom economy.

The Baylis–Hillman reaction is also known as **Morita–Baylis–Hillman** reaction (MBH reaction) [72, 79] and has been carried out in supercritical carbon dioxide.

2.9 Beckmann Rearrangement

The rearrangement of oximes under the influence of a variety of acidic reagents (e.g., PCl_5, P_2O_5, $SOCl_2$, $POCl_3$, H_2PO_4 and H_3SO_4) to N-substituted amides is known as Beckmann rearrangement. A typical example is the rearrangement of benzophenone oxime to benzanilide (Scheme 2.83).

The reaction is highly stereospecific where the migrating group is anti to the oxime hydroxyl group. The mechanism of the reaction is shown in Scheme 2.84.

2.9.1 Beckmann Rearrangement Under Microwave Irradiation

A solid-state microwave-assisted Beckmann rearrangement has been carried out [81, 82]. In this procedure, the oxime of a ketone is mixed with montmorillonite K 10 clay in dry media and the mixture was irradiated for 7 min in a microwave oven to give the corresponding anilide in 91% yield (Scheme 2.85).

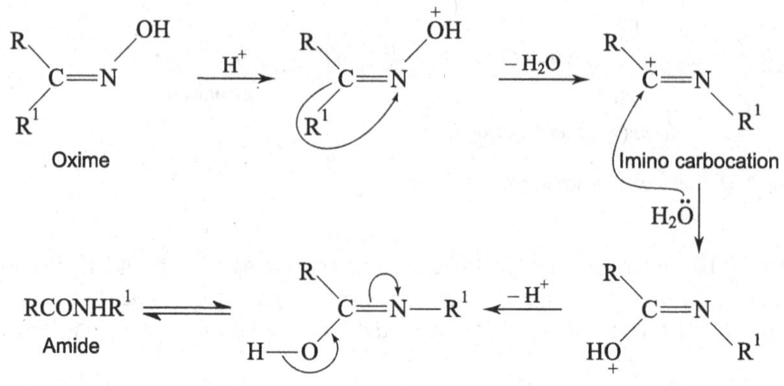

Scheme 2.84 Mechanism of Beckmann rearrangement

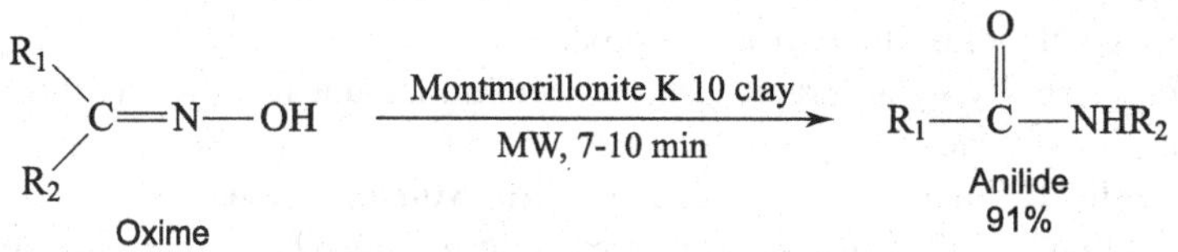

Scheme 2.85 Microwave-assisted Beckmann rearrangement

2.9.2   Beckmann Rearrangement in Ionic Liquids

The Beckmann rearrangement of cyclohexanone oxime to caprolactam has been carried out in ionic liquids. The ionic liquid to be used must serve the dual role of catalyst and the reaction medium [83] (Scheme 2.86).

On an industrial scale, Beckmann rearrangement is carried out with corrosive oleum, which is then neutralized using ammonium hydroxide. The result is generation of large amount of ammonium sulphate as by-product [84]. The main difficulty in using acidic ionic liquids is that the caprolactam product, due to the basic nature, combines with the ionic liquid making product separation impossible. The use of caprolactam-based ionic liquid coupled with a dynamic exchange between the caprolactam product and the ionic liquid allows facile product isolation. However, better yields are obtained when the product was isolated by chromatography [85].

2.10   Benzil-Benzilic Rearrangement

The α-diketones (benzils) undergo a base-catalysed reaction called benzil-benzilic acid rearrangement [86]. Initially, the salts of α-hydroxy carboxylic acids are formed, which on acidification yield the hydroxy carboxylic acid. Thus, benzil on treatment with KOH followed by acidification yields benzilic acid (Scheme 2.87).

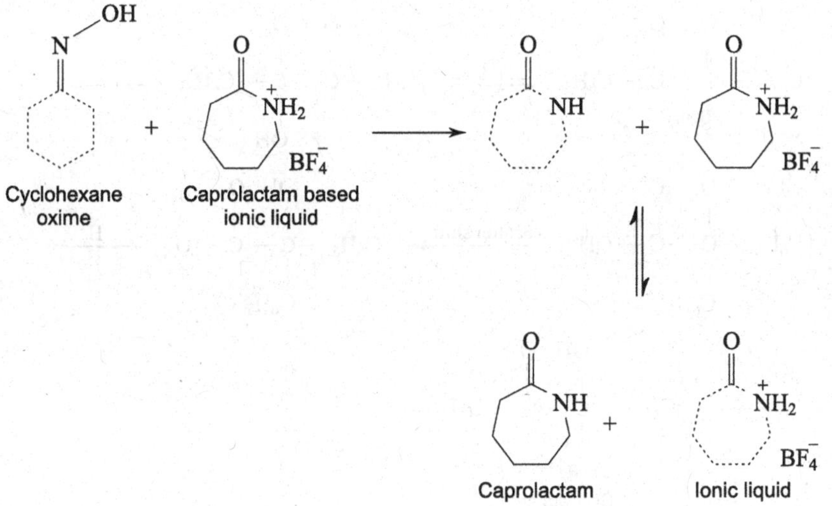

Scheme 2.86 Beckmann rearrangement in ionic liquids

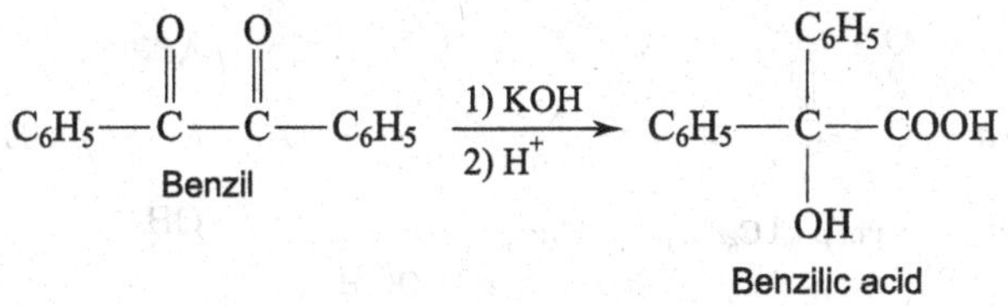

Scheme 2.87 Benzil-benzilic rearrangement

The mechanism of benzil-benzilic acid rearrangement is given in Scheme 2.88.

2.10.1 Benzil-Benzilic Acid Rearrangement under Microwave Irradiation

It is found that benzil-benzilic acid rearrangement proceeds more efficiently and faster in solid state, [87] and it takes 0.1–6 h for completion and the yields are 70–93%. The benzil-benzilic acid rearrangement could also be conducted in solid state by MW irradiation [88] reducing considerably the time of the reaction and result in good yield (Scheme 2.89).

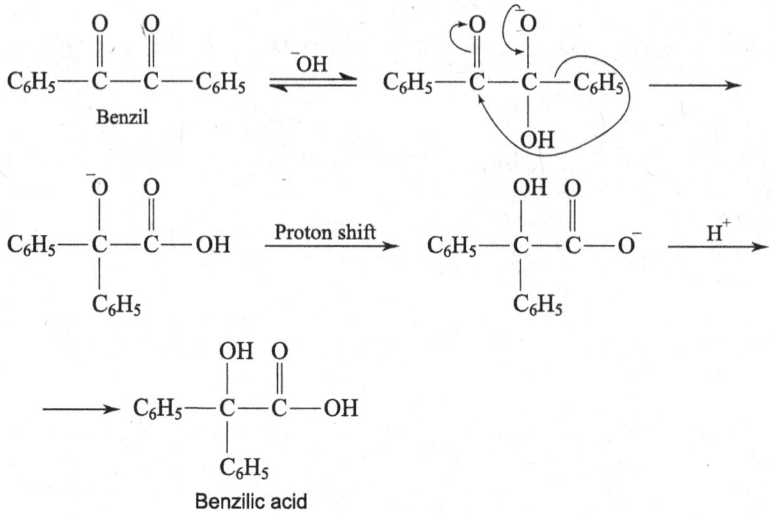

Scheme 2.88 Mechanism of benzil-benzilic rearrangement

Ar = Ar′ = Ph, p-Cl C$_6$H$_4$, p-NO$_2$C$_6$H$_4$,
Ar = Ph, Ar′ = p-Cl C$_6$H$_4$, p-NO$_2$C$_6$H$_4$, p-MeOC$_6$H$_4$

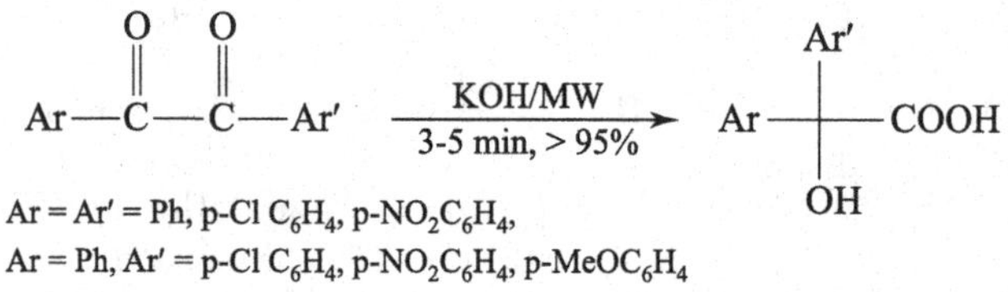

Scheme 2.89 Microwave-assisted benzil-benzilic rearrangement

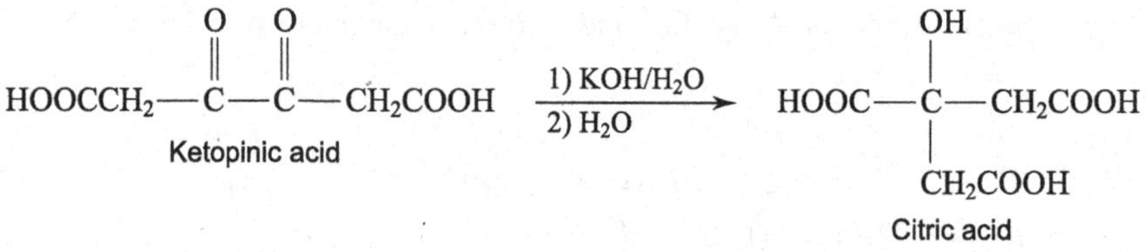

Scheme 2.90 Synthesis of citric acid

2.10.2 Applications

The benzil-benzilic acid rearrangement is useful for the synthesis of citric acid (Scheme 2.90).

The benzil-benzilic acid rearrangement of furil, 9,10-phenanthraquinone [89] and cyclohexane-1,2-dione is described in Scheme 2.91.

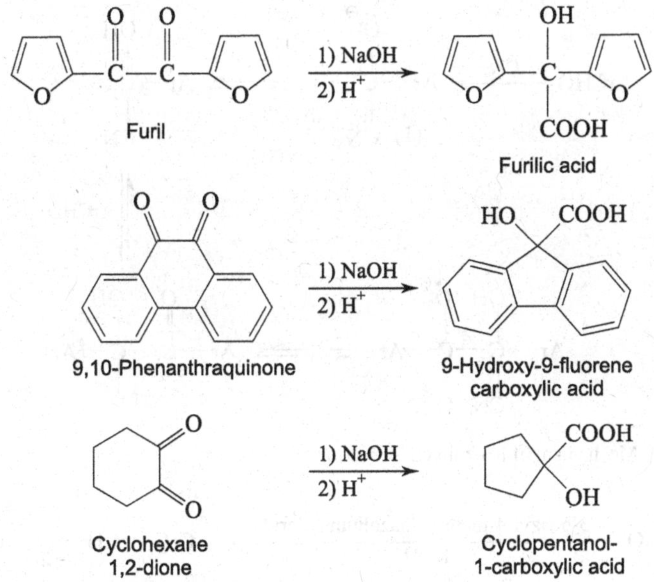

Scheme 2.91 Some examples of benzil-benzilic acid rearrangement

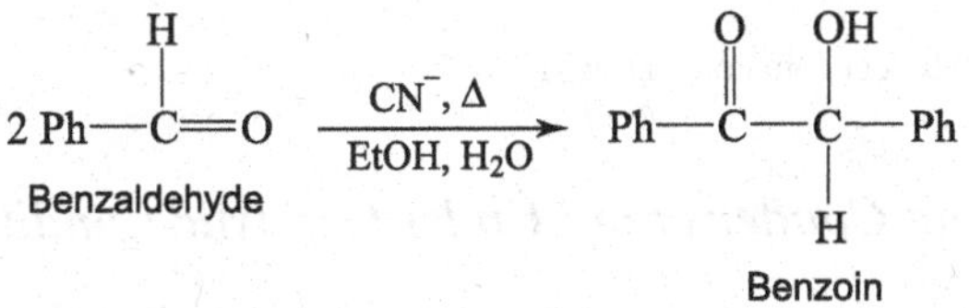

Scheme 2.92 Benzoin condensation

2.11 Benzoin Condensation [90]

Aromatic aldehydes (having no α-hydrogen) on treatment with sodium or potassium cyanide undergo self-condensation to give α-hydroxy ketones (benzoin). This is known as benzoin condensation (Scheme 2.92).

Benzoin condensation does not take place with aliphatic aldehydes under these conditions.

Mechanism

Benzoin condensation is believed to occur through a **Knoevenagel type addition**. The cyanide ion attacks the carbonyl group of the aldehyde to give carbanion (1), which reacts with a second molecule of the aldehyde (Scheme 2.93).

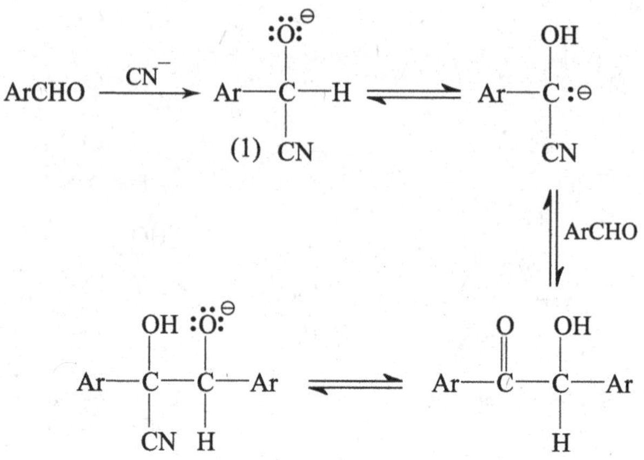

Scheme 2.93 Mechanism of benzoin condensation

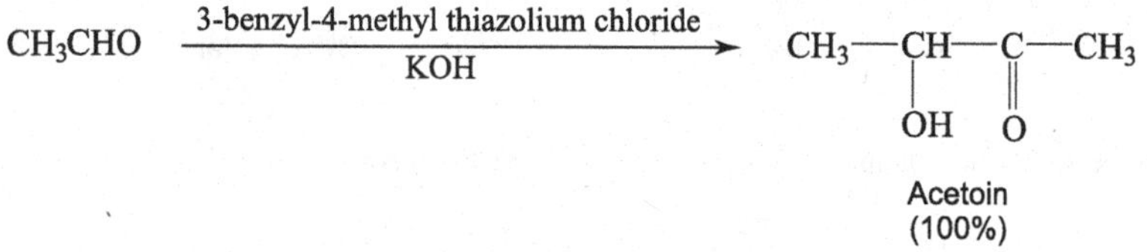

Acetoin
(100%)

Scheme 2.94 Catalytic benzoin condensation

2.11.1 Benzoin Condensation Under Catalytic Conditions

It has been found [92] that even aliphatic aldehydes like acetaldehyde undergo benzoin condensation with solid potassium hydroxide and using 3-benzyl-4-methylthiazolium chloride as a catalyst. Thus, acetaldehyde yields acetoin in quantitative yield (Scheme 2.94).

It is found that aromatic aldehydes reacted for few minutes under the above conditions, whereas aliphatic aldehydes required 5–10 h.

Benzoin condensations of aromatic aldehydes with aqueous sodium cyanide are catalysed [93] by quaternary ammonium salts (Scheme 2.95).

In a similar way, acyloin condensations with aromatic or aliphatic aldehydes proceed remarkably well using N-laurylthiazolium bromide as catalyst with an aqueous phosphate solution [94] (Scheme 2.96).

It was found [95] by Breslow that the benzoin condensation in aqueous media using inorganic salts (e.g., LiCl) is about 200 times faster than in ethanol without any salt. The benzoin condensation was also accelerated by the addition of γ-cyclodextrin, whereas addition of β-cyclodextrin inhibited the condensation.

Unsymmetrical or mixed benzoins may often be obtained in good yield from two different aldehydes (Scheme 2.97).

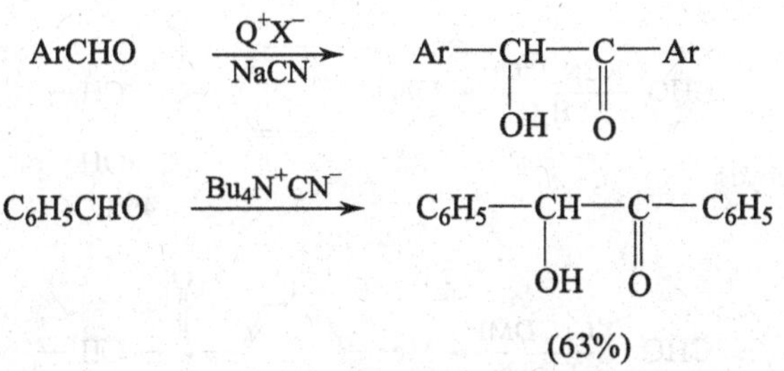

$$ArCHO \xrightarrow[NaCN]{Q^+X^-} Ar-\underset{\underset{OH}{|}}{CH}-\underset{\underset{O}{\parallel}}{C}-Ar$$

$$C_6H_5CHO \xrightarrow{Bu_4N^+CN^-} C_6H_5-\underset{\underset{OH}{|}}{CH}-\underset{\underset{O}{\parallel}}{C}-C_6H_5$$

(63%)

Scheme 2.95 Some benzoin condensations are catalysed by quaternary ammonium salts

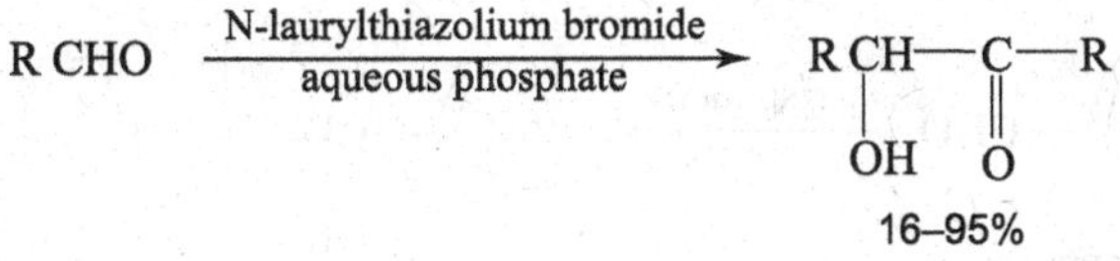

$$R\,CHO \xrightarrow[aqueous\ phosphate]{N\text{-laurylthiazolium bromide}} R\,\underset{\underset{OH}{|}}{CH}-\underset{\underset{O}{\parallel}}{C}-R$$

16–95%

Scheme 2.96 Acyloin condensation

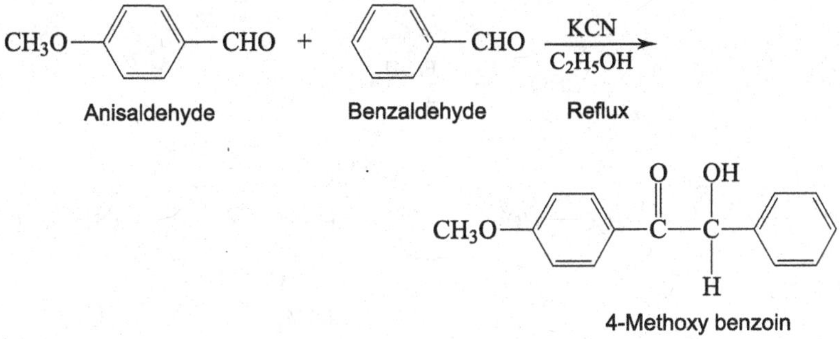

Anisaldehyde   Benzaldehyde   Reflux

$$CH_3O-\text{(ring)}-\underset{\underset{O}{\parallel}}{C}-\underset{\underset{H}{|}}{\overset{OH}{C}}-\text{(ring)}$$

4-Methoxy benzoin

Scheme 2.97 Synthesis of unsymmetrical benzoins

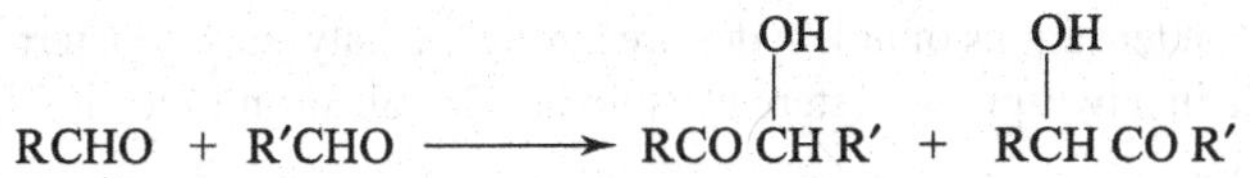

$$RCHO + R'CHO \longrightarrow RCO\,\underset{\underset{}{}}{\overset{OH}{CH}}\,R' + R\,\underset{}{\overset{OH}{CH}}\,CO\,R'$$

A typical example is

The above mixed benzoin condensation is catalysed by thiazolium salts [96].

Benzoin condensation can also be brought about by coenzyme 'Thiamine' (R. Breslow, J. Am. Chem. Soc., 1958, *80*, 3719).

For mechanism, *see* acyloin condensation (Sect. 2.1).

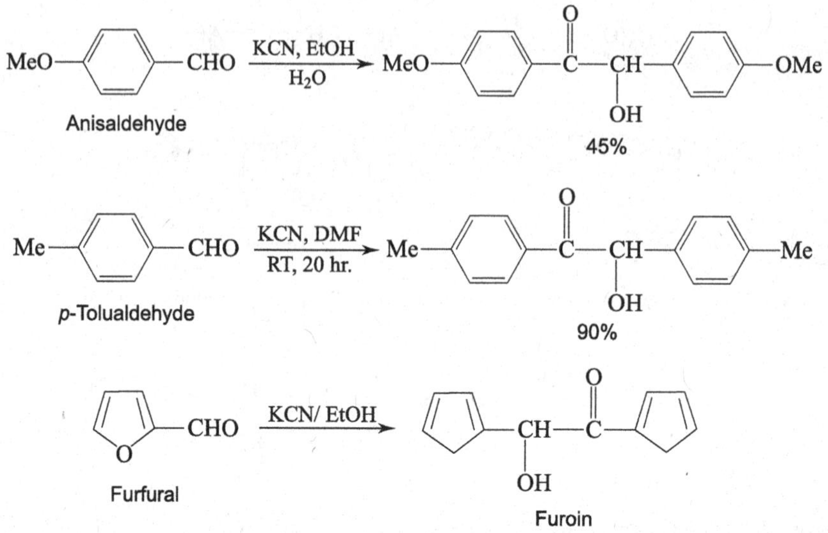

Scheme 2.98 Some applications of benzoin condensation

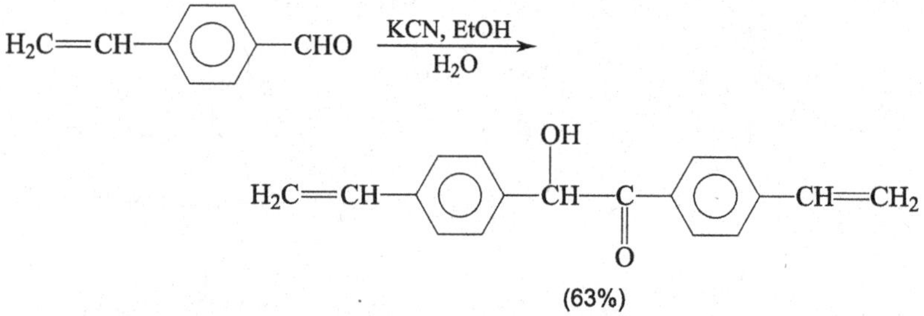

Scheme 2.99 Another application of benzoin condensation

Benzoin condensations of aldehydes are strongly catalysed by quaternary ammonium cyanide in a two-phase system (J. Solodav, Tetrahedron Lett., 1971, 287).

2.11.2 Applications

Anisaldehyde and p-tolualdehyde give the corresponding α-hydroxyketones [97, 98]. Similarly, furfural gives furoin (Scheme 2.98).

Another application is benzoin condensation of p-formyl styrene gives [99] the corresponding product (Scheme 2.99).

Benzoins are useful intermediates for the synthesis of other compounds, since they can be oxidized to α-diketones and reduced under different conditions to give various products. These reactions are summarized in Scheme 2.100.

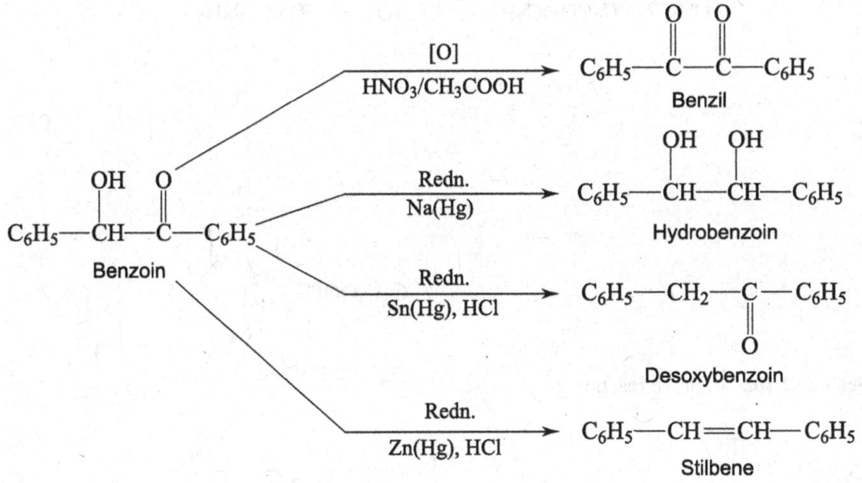

Scheme 2.100 Some useful products obtained from benzoins

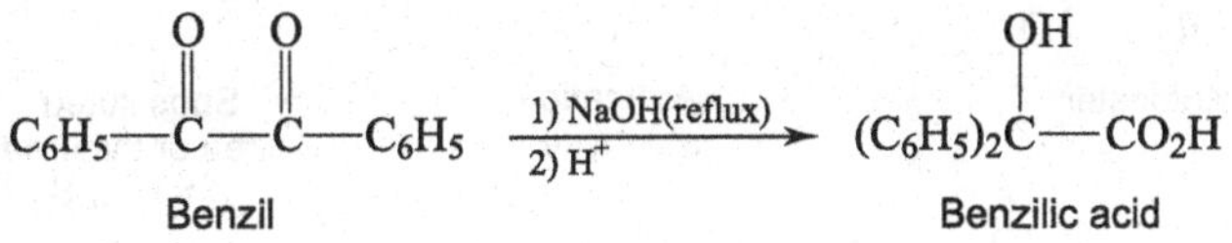

Scheme 2.101 Synthesis of benzilic acid

The oxidation product, viz., benzil (a α-diketone) obtained by nitric acid oxidation of benzoin, undergoes an interesting base-catalysed rearrangement to the α-hydroxy acid, benzilic acid (see Sect. 2.10) (Scheme 2.101).

2.12 Biginelli Reaction [100]

It involves the acid-catalysed condensation of an aldehyde, a β-ketoester and urea to yield tetrahydropyrimidinones (Scheme 2.102).

2.12.1 Biginelli Reaction Under Microwave Irradiation

In this procedure, a neat mixture of β-ketoesters, aryl aldehyde and urea or thiourea in the presence of polyphosphate ester (PPE) is subjected to irradiation in a domestic microwave oven for 1–5 min and the formed dihydropyrimidines were obtained in 61–95% yield after aqueous work-up (Scheme 2.103) [101].

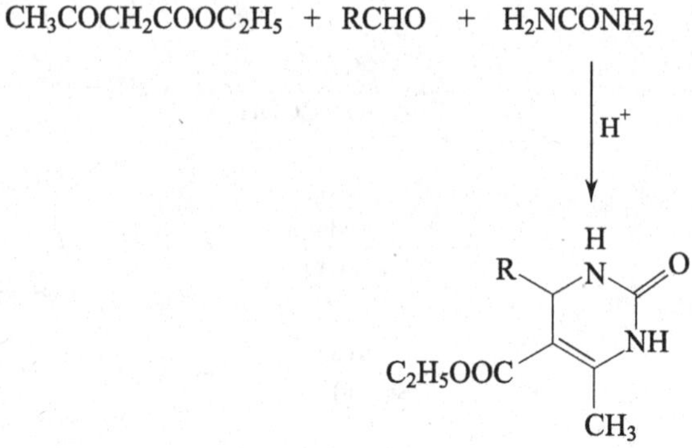

$$CH_3COCH_2COOC_2H_5 \ + \ RCHO \ + \ H_2NCONH_2$$

Scheme 2.102 Biginelli reaction

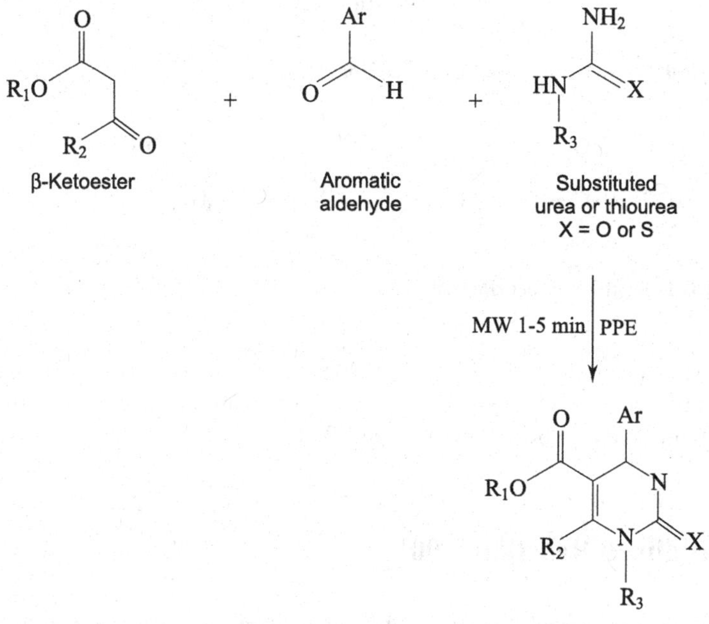

β-Ketoester   Aromatic   Substituted
                     aldehyde   urea or thiourea
                                       X = O or S

MW 1-5 min | PPE

Scheme 2.103 Microwave-assisted Biginelli reaction

2.12.2 Biginelli Reaction in Ionic Liquids

Ionic liquids like [b$_{min}$][PF$_4$] and [b$_{min}$] [PF$_6$] have been used as catalyst for the Biginelli reaction under solvent-free conditions (Scheme 2.104).

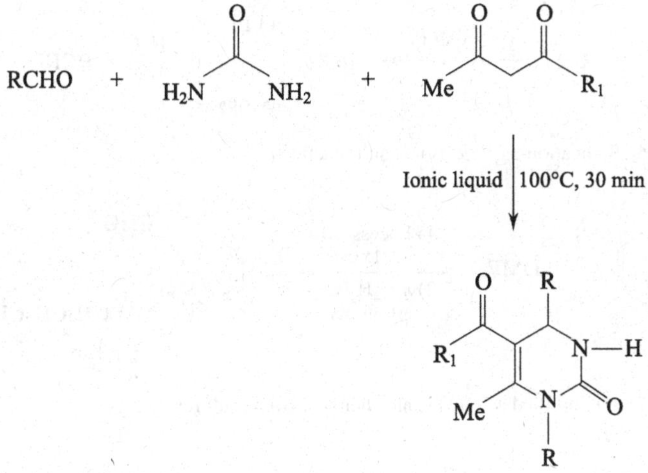

Scheme 2.104 Biginelli reaction in ionic liquids

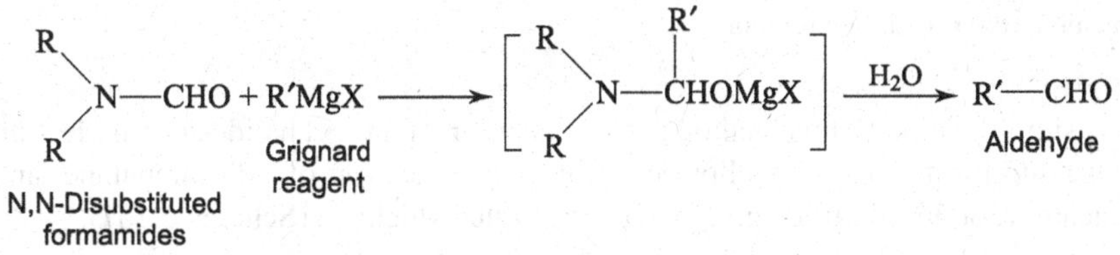

Scheme 2.105 Bouveault reaction

2.13 Bouveault Reaction [103]

The action of Grignard reagents on N,N-disubstituted formamides yield aldehyde. This is known as Bouveault reaction or **Bouveault aldehyde synthesis** (Scheme 2.105).

2.13.1 Bouveault Reactions Under Sonication

In place of Grignard reagents, organolithium reagents give better yields [104]. The organolithium reagents are obtained by sonication of aryl halides with lithium using low-intensity ultrasonic. These reagents are used in Bouveault reaction and give higher yields than the traditional methods [55] (Scheme 2.106).

In non-ultrasonic Bouveault reaction, which suffers from numerous side reactions, the method is improved when DMF is replaced by more elaborate and expensive formamide, $Me_2NCH_2CH_2N(Me)CHO$.

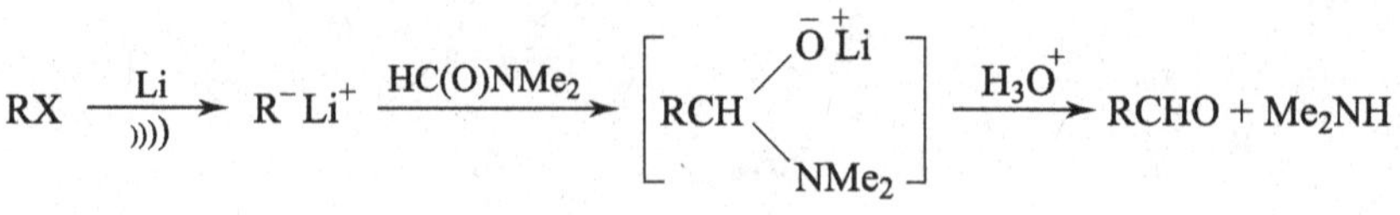

$$RX \xrightarrow[))))]{Li} R^-Li^+ \xrightarrow{HC(O)NMe_2} \left[RCH \begin{array}{c} \overset{-}{O} \overset{+}{Li} \\ \diagup \\ \diagdown \\ NMe_2 \end{array} \right] \xrightarrow{H_3O^+} RCHO + Me_2NH$$

Scheme 2.106 Sonication-assisted Bouveault reactions

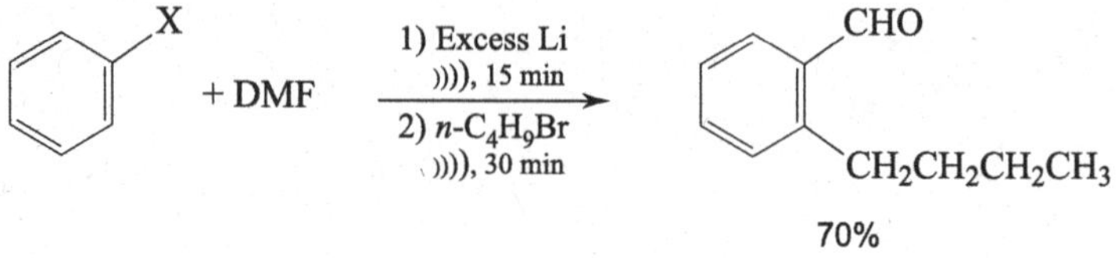

Scheme 2.107 A simplified assisted sonic chemical Bouveault reaction

$$\underset{\text{Benzaldehyde}}{C_6H_5CHO} + C_6H_5CHO \xrightarrow{KOH} \underset{\text{Benzyl alcohol}}{C_6H_5CH_2OH} + \underset{\text{Pot. benzoate}}{C_6H_5COO-K^+}$$

Scheme 2.108 Cannizzaro reaction

A simplification of this method is the sonication of an aryl halide and amide with excess lithium for 15 min followed by dropwise addition of 1-bromobutane, and sonication for 30 min more gives the o-substituted aldehyde (Scheme 2.107).

Use of iodomethane in place of n-butyl bromide in the above reaction gives o-tolualdehyde [105].

2.14 Cannizzaro Reaction [106]

Aldehydes without α-hydrogen(s) on treatment with concentrated aqueous alkali undergo self-oxidation and reduction to give an alcohol, and the salt of the corresponding carboxylic acid (Scheme 2.108).

This disproportionation or self-oxidation and reduction of aromatic aldehydes, devoid of α-hydrogen is known as Cannizzaro reaction [106].

Cannizzaro reaction best proceeds with aromatic aldehydes without α-hydrogen, and with some aliphatic aldehydes like formaldehyde and dimethyl acetaldehyde (which do not have α-hydrogen also undergo Cannizzaro reaction (Scheme 2.109).

Mechanism

Cannizzaro reaction involves transfer of a hydrogen atom from one molecule of the aldehyde to another. This has been established by using deuterated benzaldehyde instead of benzaldehyde (Scheme 2.110).

$$HCHO + NaOH \xrightarrow{\Delta} CH_3OH + HCOONa$$

Formaldehyde   Methyl alcohol   Sod. formate

$$2(CH_3)_2CHCHO + NaOH \xrightarrow{\Delta} (CH_3)_2CHCH_2OH + (CH_3)_2CHCOONa$$

Dimethyl acetaldehyde   2-Methyl-1-propanol   Sod. 2-methyl-1-
propionate

Scheme 2.109 Cannizzaro reaction at aldehydes which do not have α-hydrogens

$$2C_6H_5CDO + OH^- \xrightarrow{H_2O} C_6H_5CD_2OH + C_6H_5\overset{\overset{O}{\|}}{C}-O^-$$

Scheme 2.110 Mechanism of Cannizzaro reaction

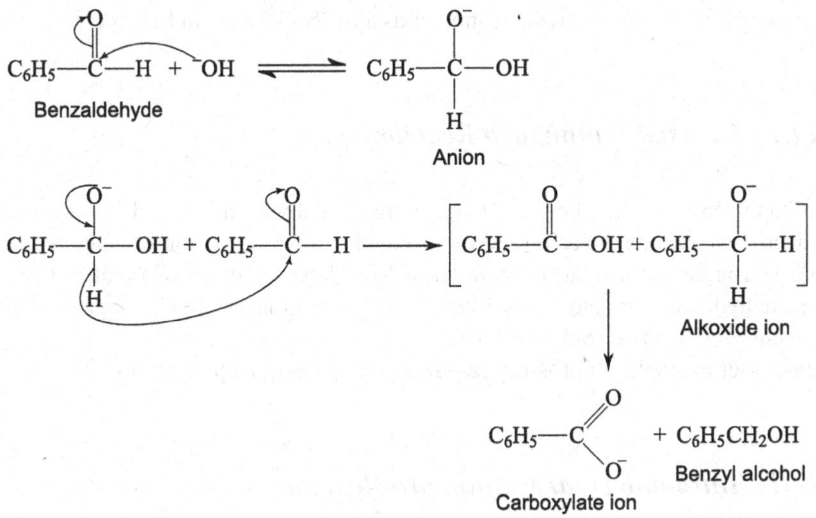

Scheme 2.111 Mechanism of Cannizzaro reaction

It is found that the benzyl alcohol formed is exclusively deuterated. So, the possibility of exchange of hydrogen with hydrogen atoms in the solvent is ruled out. Thus, a transfer of hydrogen (or deuterium) takes place between the carbonyl atoms of the two aldehyde molecules.

Cannizzaro reaction proceeds by formation of an anion (by reaction with base) which may transfer a hydride ion intermolecularly to the carbonyl of another aldehyde molecule forming the carboxylic acid and the alkoxide ion. Final step is the shifting of a proton from the acid to the alcohol (Scheme 2.111).

$$R-CHO + CH_2O \xrightarrow{\ ^-OH\ } RCH_2OH + HCOO^-$$

Scheme 2.112 Crossed Cannizzaro reaction

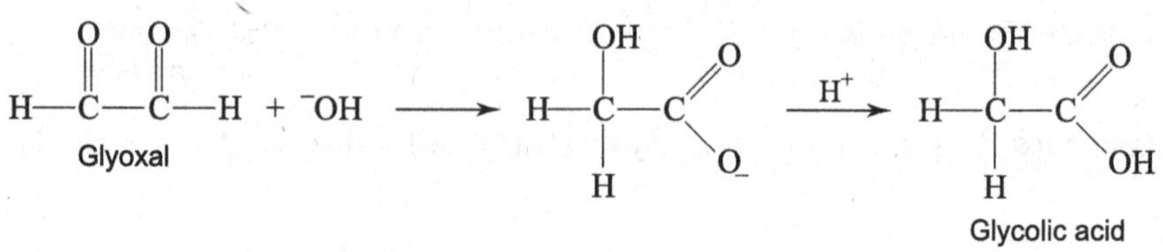

Scheme 2.113 An intramolecular Cannizzaro reaction

If the Cannizzaro reaction is performed in D_2O, the alcohol formed has no deuterium establishing that the mechanism involves a direct transfer of hydrogen from one molecule of aldehyde to another as already depicted in Scheme 3.

2.14.1 Crossed Cannizzaro Reaction

The Cannizzaro reaction between two different aldehydes may yield four different products (two carboxylic acids and two alcohols). Such a reaction is called crossed Cannizzaro reaction and has no synthetic value. However, if one of the aldehydes is formaldehyde, the formate ion and the alcohol corresponding to the other aldehyde are exclusively formed (Scheme 2.112).

For other examples of crossed Cannizzaro reaction, *see* applications.

2.14.2 Intramolecular Cannizzaro Reaction

Certain compounds that contain two carbonyl groups undergo **internal Cannizzaro reaction**. For example, glyoxal on treatment with base gives glycolic acid (Scheme 2.113).

In a similar way, phenyl glyoxal undergoes **intramolecular Cannizzaro reaction** to give mandelic acid (Scheme 2.114).

An analogous reaction occurs with benzil, which results in carbon skeleton rearrangement and is known as **benzil-benzilic acid rearrangement** (Scheme 2.115).

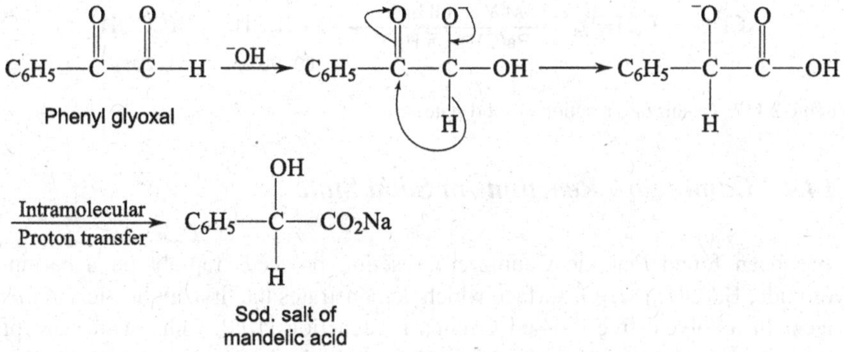

Scheme 2.114 Another example of intramolecular Cannizzaro reaction

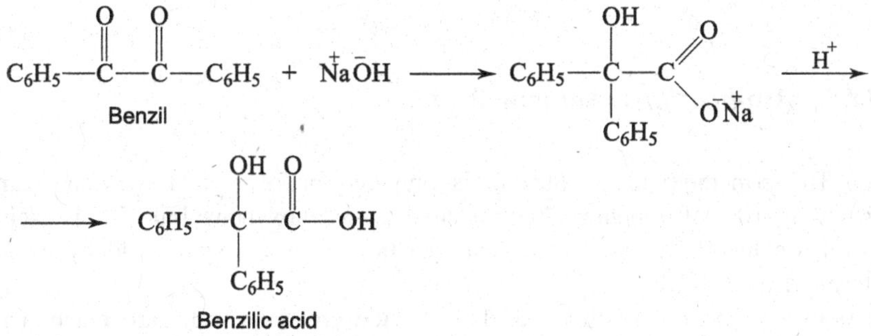

Scheme 2.115 Synthesis of benzilic acid

$$C_6H_5CHO \xrightarrow[\text{)))), 10 min}]{\text{Ba(OH)}_2\text{EtOH}} C_6H_5CH_2OH + C_6H_5COOH$$

Scheme 2.116 Some chemical Cannizzaro reaction

2.14.3 Cannizzaro Reactions Under Sonication

The Cannizzaro reaction under heterogenous conditions catalysed by barium hydroxide is considerably accelerated (Scheme 8) by sonication. The yields are 100% after 10 min, whereas no reaction is observed during this period without ultrasound [107] (Scheme 2.116).

$$RCHO + (CH_2O)_n \xrightarrow[\text{Ba(OH)}_2 \cdot 8 H_2O]{\text{MW or oil bath}} RCH_2OH + RCOOH$$

$$80\text{--}90\% \qquad 1\text{--}20\%$$

Scheme 2.117 Cannizzaro reaction in solid state

2.14.4 Cannizzaro Reactions in Solid State

It has been found that the Cannizzaro reaction proceeds rapidly on a barium hydroxide, Ba(OH)$_2$ 8H$_2$O, surface which demonstrates the first application of this reagent in a solvent-free crossed Cannizzaro reaction [108]. Thus, a mixture of benzaldehyde and paraformaldehyde on mixing with barium hydroxide octahydrate and then irradiation in a microwave oven (100–110 °C) or heated in an oil bath (Scheme 2.117).

2.14.5 Applications (Scheme 2.118)

(*i*) The *o*-methoxybenzyl alcohol is prepared in 79% yield by Cannizzaro reaction [109] of *o*-methoxy benzaldehyde. *o*-Methoxybenzyl alcohol is exclusively obtained by crossed Cannizzaro reaction of *o*-methoxy benzaldehyde with formaldehyde [110].

(*ii*) Benzyl alcohol is obtained [111] by the crossed Cannizzaro reaction of benzaldehyde and formaldehyde.

$$C_6H_5CHO + CH_2O \xrightarrow{\text{30\% NaOH}} C_6H_5CH_2OH + HCO_2H$$

Benzaldehyde   Benzyl alcohol   Formic acid

(*iii*) Internal Cannizzaro reaction of *o*-formyl benzaldehyde gives *o*-carboxybenzyl alcohol [112].

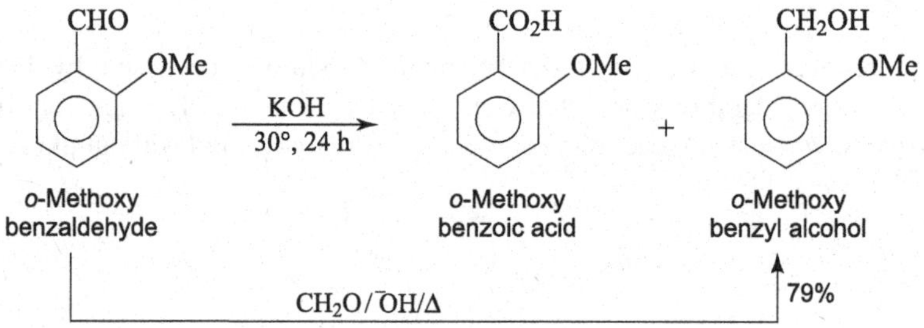

Scheme 2.118 Some applications of Cannizzaro reaction

o-Formyl
benzaldehyde

o-Carboxy
benzyl alcohol

(*iv*) Pentaerythritol, an important industrial product (used for making polymers; the tetranitro derivative is used as explosive under the name PETN) is obtained by crossed Cannizzaro reaction of trihydroxymethylacetaldehyde (which is obtained by aldol condensation of acetaldehyde and formaldehyde in the presence of calcium hydroxide) with formaldehyde.

$$CH_3CHO + 3CH_2O \xrightarrow{\ ^-OH\ } HOCH_2-\underset{\underset{CH_2OH}{|}}{\overset{\overset{CH_2OH}{|}}{C}}-CHO \longrightarrow$$

Acetaldehyde

Trihydroxymethyl
acetaldehyde

$$\xrightarrow[\text{CH}_2\text{O}/^-\text{OH}]{\text{Crossed Cannizzaro reaction}} HOCH_2-\underset{\underset{CH_2OH}{|}}{\overset{\overset{CH_2OH}{|}}{C}}-CH_2OH$$

Pentaerythritol

(*v*) Using Cannizzaro reaction, a number of carboxylic acids and alcohols have been synthesized. Some of these are given below:

(*a*)
$$2\ \underset{\underset{COOH}{|}}{\overset{\overset{CHO}{}}{}} \xrightarrow[\Delta]{NaOH} \underset{\underset{COONa}{|}}{\overset{\overset{CH_2OH}{|}}{}} + \underset{\underset{COONa}{|}}{\overset{\overset{COONa}{|}}{}}$$

Glyoxalic acid   Sod. glycolate   Sod. oxalate

(*b*)

Furfural   Sod. salt of
2-furoic acid   Furyl alcohol

(*c*)

Thiophen-2-
aldehyde   Sod. salt of
thiophen-2-carboxylic
acid   Thiophene-2-carbinol

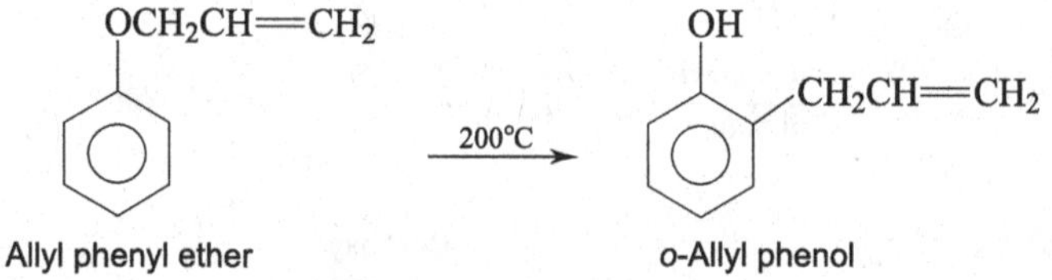

Scheme 2.119 Claisen rearrangement

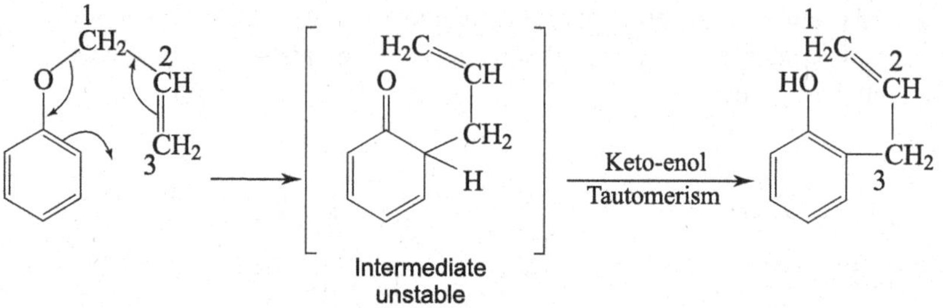

Scheme 2.120 Mechanism of Claisen condensation

2.15 Claisen Rearrangement [113]

Allyl phenyl ether on heating to 200 °C undergoes an intramolecular reaction called Claisen rearrangement. Claisen rearrangement is the earliest record of an organic reaction in solid state. The product is *o*-allylphenol (the allyl group migrates to the ortho position) (Scheme 2.119).

The reaction does not need any catalyst and is an example of **pericyclic reaction.** Other examples of pericyclic reactions are **cope rearrangement** and **Diels–Alder reaction.**

Mechanism [113, 114]

Claisen rearrangement is a [3,3]-sigmatropic rearrangement and proceeds in a concerted manner, in which the bond between C_3 of the allyl group and the *ortho* position of the benzene ring form and at the same time the carbon–oxygen bond of the allyl phenyl ether breaks (Scheme 2.120).

That only C-3 of the allyl group becomes bonded to the benzene ring has been demonstrated by carrying out the rearrangement with allyl phenyl ether containing ^{14}C at C-3. Whole of the product obtained in this reaction has the labelled carbon atom bonded to the ring (Scheme 2.121).

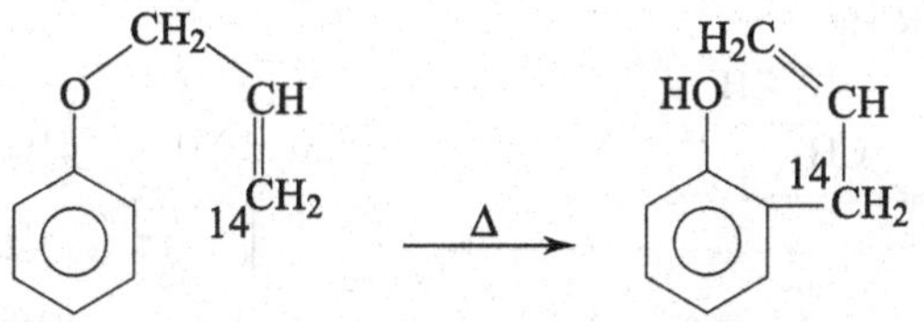

Only product

Scheme 2.121 Evidence for Claisen rearrangement

2-Butenyl phenyl ether   2-(1-Methyl-2-propenyl) phenol

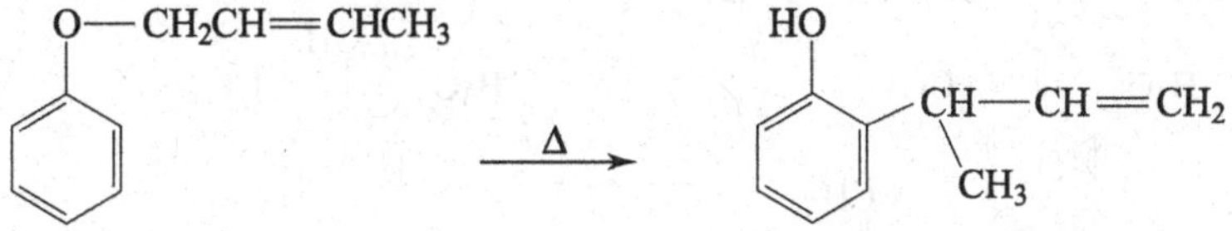

Scheme 2.122 Further evidence for the mechanism of Claisen condensation

Further evidence for the above mechanism (Scheme 2.121) is that if 2-butenyl group is present in an ether, the phenol obtained has a methyl branch on the side chain (Scheme 2.122).

In case both the *ortho* positions are blocked, *p*-substituted phenol is obtained via two successive shifts of allyl group. In this case also, the migration still occurs at the *ortho* position to form *o*-substituted dienone (A). However, the absence of hydrogen at the *ortho* position prevents enolization (or aromatization). So, the allyl group undergoes a second migration through a similar cyclic transition state to form a dienone (B), which aromatizes to a phenol with allyl group at position 4 (Scheme 2.123).

The driving force for the *para* migration (when both the *ortho* positions are occupied) is regaining the aromatic character after allyl migration. Migration to *meta* position has not been observed.

The classic Claisen rearrangement involves aromatic allyl ether. The thermal Claisen rearrangement of allyl vinyl ethers to γ, δ-unsaturated carbonyl compounds was first described[1a] by Claisen in 1912 (Scheme 2.124). This is referred to as **aliphatic Claisen rearrangement**.

The transition state for the above Claisen rearrangement (Scheme 2.124) involves a cycle of six orbitals and six electrons, suggesting that the transition state has aromatic character.

In the case of aliphatic substrate (substituted by allyl vinyl ethers), asymmetry at the terminal methylene carbons is transformed into asymmetry at the two new saturated centres in a sense that suggests [115] a chair-like transition state (Scheme 6). In the case of aromatic Claisen rearrangement, the initially formed dienone undergoes tautomerization to the phenol, so the stereochemistry observable is the transfer of chirality from an optically active starting material with a specific double-bond geometry. This indicates that the aromatic Claisen rearrangement also proceeds

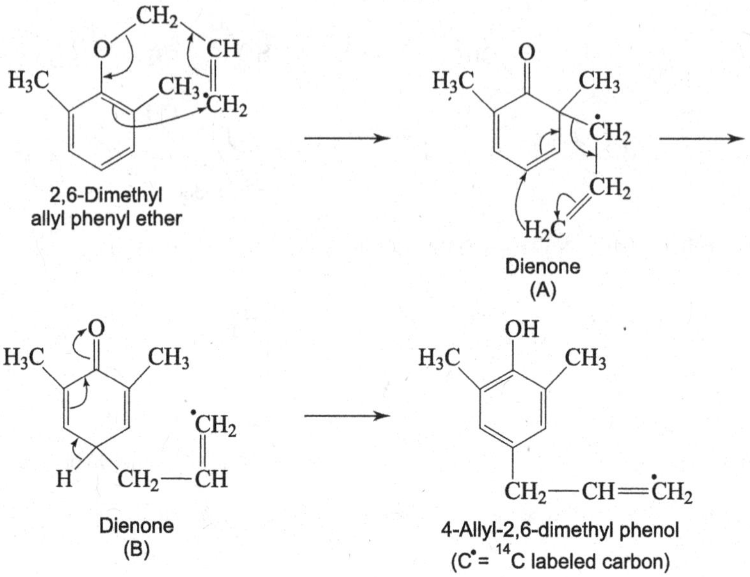

Scheme 2.123 Mechanism of Claisen condensation continued

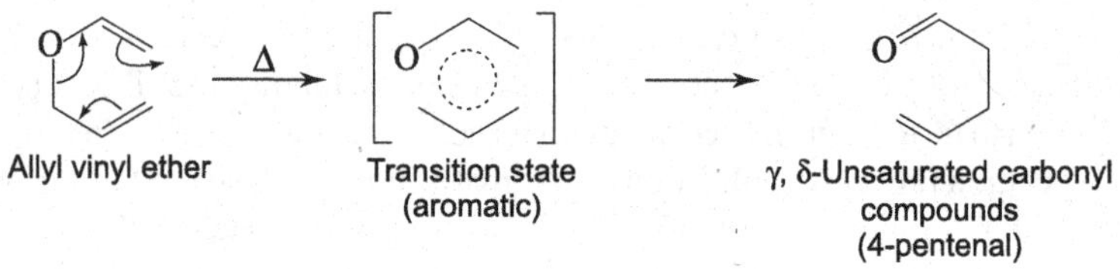

Scheme 2.124 Aliphatic Claisen condensation

with stereochemistry which is in accordance with a chair-like transition state [117] (Scheme 2.125).

2.15.1 Claisen Rearrangement in Water

The use of water in promoting Claisen rearrangement was first recorded [117] in 1970. The first example using pure water for Claisen rearrangement of chorismic acid [121] is given in Scheme 2.126.

Claisen rearrangement of allyl vinyl ether in water gave [118] the aldehyde (4-pentenal, see Scheme 2.126).

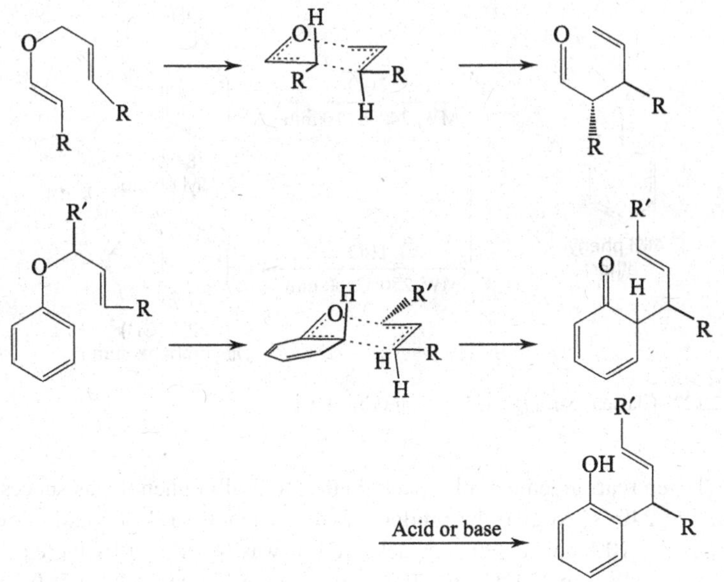

Scheme 2.125 Mechanism of aromatic Claisen condensation

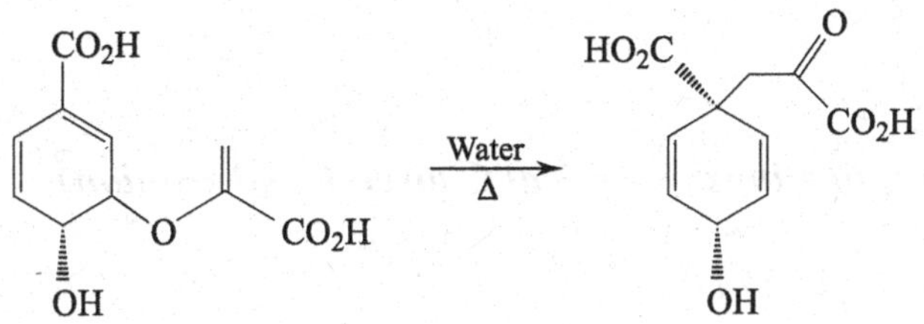

Scheme 2.126 Claisen reaction in water

2.15.2 Claisen Rearrangement in Near Critical Water

Reactions conducted in microwave oven in water above 200 °C exist due to the pressure limit of about 20 bar for most of the commercially available MW instruments. Pure water reaches an autogenic pressure of 50 bar at 250 °C. This condition is called near critical water (NCW) region between 20 and 300 °C [119].

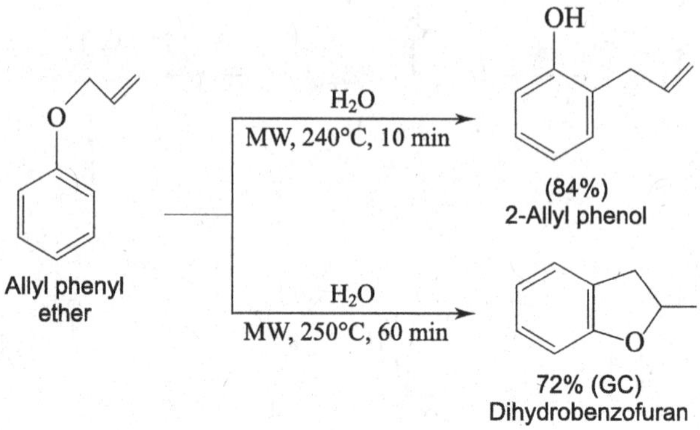

Scheme 2.127   Claisen rearrangement in near critical water

The Claisen rearrangement allyl phenyl ether to 2-allyl phenol was successfully performed at 240 °C to give the product (2-allyl phenol) in 84% yield. However, at 200 °C, only 10% conversion was achieved. It was found that by increasing the temperature and time to 250 °C, only 10% conversion was achieved. It was found that by increasing the temperature and time to 250 °C and 1 h, dihydrobenzofuran was produced in 72% conversion due to involvement of water into the reaction pathology (Scheme 2.127).

2.15.3   Applications (Classical Claisen Condensation)

See Scheme 2.128.

2.15.4   Applications (Aqueous Phase Claisen Rearrangement)

See Scheme 2.129.

(*i*) A simple aliphatic Claisen rearrangement of an allyl vinyl ether gave the aldehyde [123].

The corresponding ester also undergoes similar rearrangement. Carrying out the above reaction in methanol–water (2.1) increased the yield by about 40 times than those in acetone solvent [124].

(*ii*) Claisen rearrangement of a vinyl ether substrate results in the formation of an aldehyde; the reaction was performed by heating 2.5:1 water–methanol solution at 80 °C for 24 h [125].

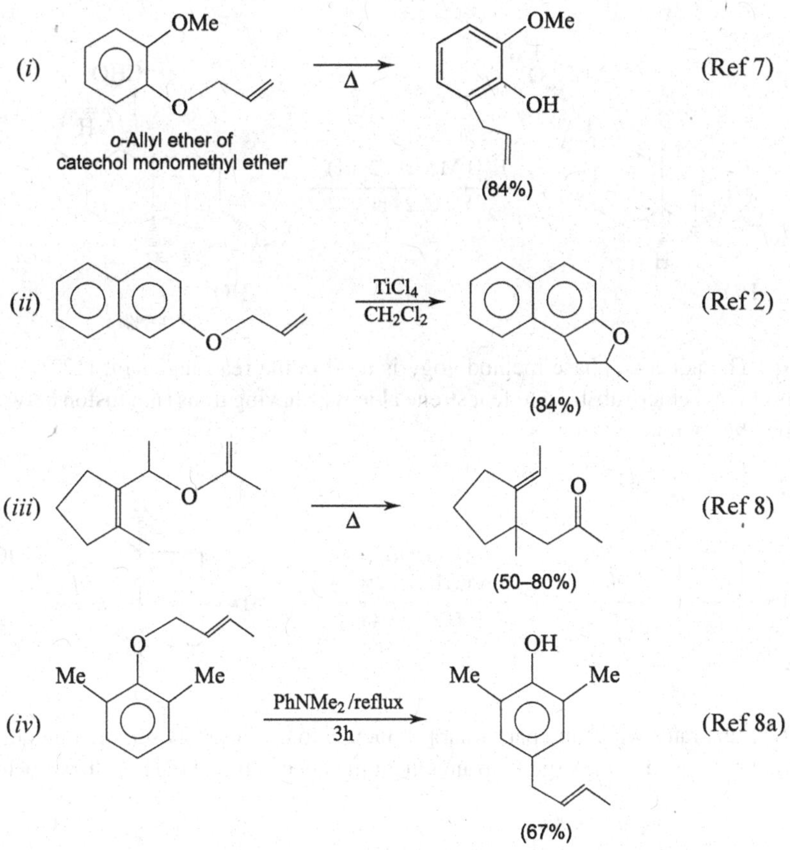

Scheme 2.128 Some applications of classical Claisen rearrangement

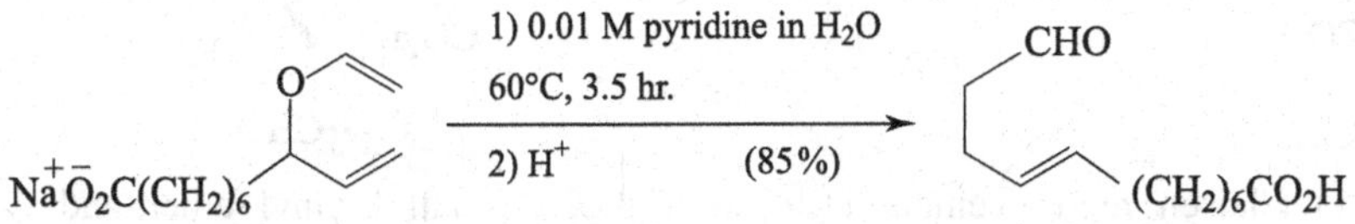

Scheme 2.129 Some applications of aqueous phase Claisen rearrangement

The yield in the above rearrangement is 85% compared to 60% when the diol was protected as the acetonide at 220 °C in the presence of sodium pentoxide [126]. However, in the reaction cited above it is not necessary to protect the diol.
The aldehyde obtained in the above Claisen rearrangement is used in the synthesis of aphidicolin.

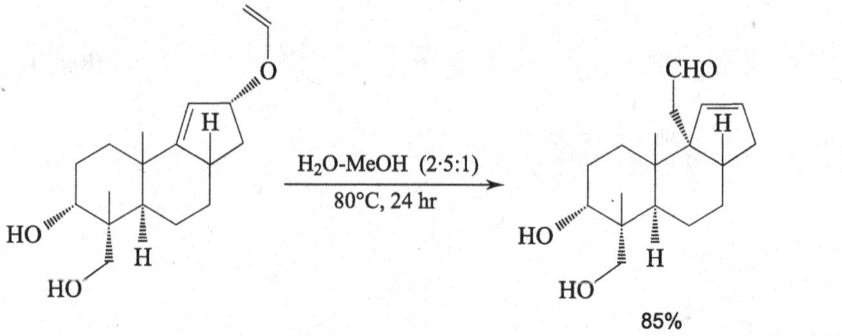

(*iii*) The aqueous phase methodology is used in the rearrangement [127] of the allyl vinyl ether substrate to fenestrene aldehyde having trans ring fusion between the five-membered ring.

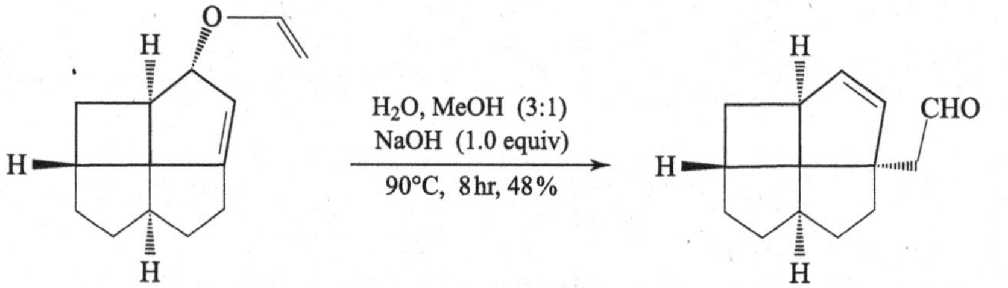

(*iv*) Substrates which are particularly reluctant to undergo Claisen rearrangement can be induced to undergo rearrangement in aqueous base [127] as shown below.

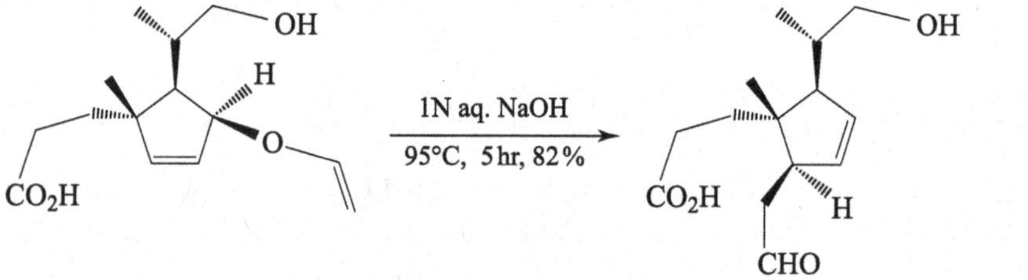

(*v*) Claisen rearrangement [128] of 6-β-glycosylallyl vinyl ether and 6-α-glucosylallyl vinyl ether in water (80 °C, 1 h) was successful. In both the reactions, NaBH$_4$ was added so that the formed aldehyde is converted into the corresponding diol.

Attempts to carry out the above Claisen rearrangements by heating in toluene to higher temperature resulted in destruction of the material.

Claisen rearrangement is an important tool available to synthetic organic chemist and extensive reviews are available [129].

2.16 Claisen–Schmidt Reaction [130]

The base-catalysed condensation of an aromatic aldehyde (without α-hydrogen) with an aliphatic aldehyde or a ketone (having α-hydrogen) to form α,β-unsaturated aldehydes or ketones is known as Claisen–Schmidt reaction or Claisen–Schmidt condensation. It is a type of **crossed aldol condensation** (see aldol condensation). For example, benzaldehyde reacts with acetaldehyde to give cinnamaldehyde (Scheme 2.130).

$$C_6H_5CHO \ + \ CH_3CHO \ \xrightarrow{NaOH} \ C_6H_5CH\!\!=\!\!CHCHO$$

Benzaldehyde   Cinnamaldehyde

Scheme 2.130 Claisen–Schmidt reaction

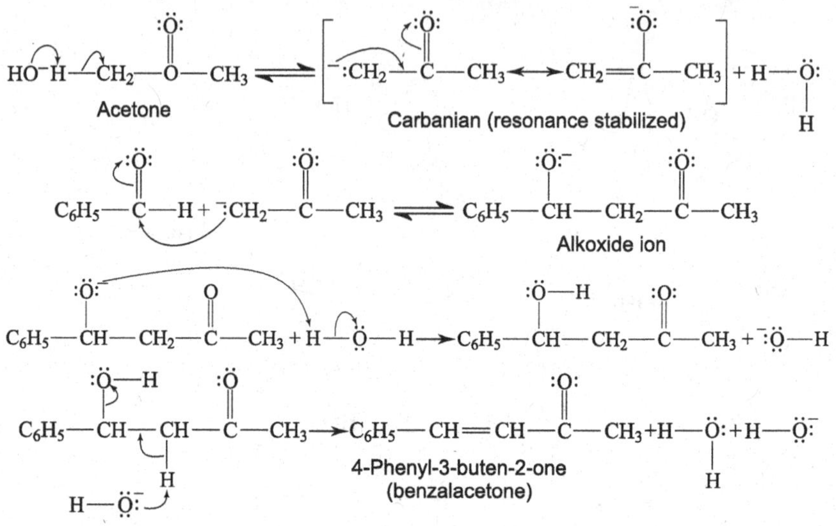

Scheme 2.131 Mechanism of Claisen–Schmidt reaction

Mechanism

The mechanism is similar to **aldol condensation**. Thus, as a first step the base ($^-$OH) removes a proton from the α-carbon of one molecule of ketone to give a resonance stabilized enolate ion, which acts as a nucleophile (as a carbanion) and attacks the carbonyl carbon of the molecule of the aldehyde producing an alkoxide anion, which removes a proton from a molecule of water. Finally, dehydration occurs readily because the double bond that forms is conjugated both with the carbonyl group and with the benzene ring; the conjugated system is thereby extended. Various steps of the mechanism are shown in Scheme 2.131.

2.16.1 *Claisen–Schmidt Reaction in Aqueous Phase*

The Claisen–Schmidt reaction between a ketone and an aldehyde can be conveniently carried out [131] by using silyl enol ether of the ketone in an organic solvent in the presence of TiCl$_4$ (see **Mukaiyama reaction**, Sect. 2.39). These acidic conditions are not suitable for acid-sensitive substrates. In such cases, the reaction is carried out

Table 2.2 Result of the claisen condensation of silyl enol ether of cyclohexanone with benzaldehyde under different conditions

R	R′	Solvent	Temp. (°C)	Time (h)	Reaction conditions	Yield %	Syn/Anti
Me	H	CH_2Cl_2	rt	360	Stirring	0	—
Me	H	CH_2Cl_2	20	2	$TiCl_4$	82	1:3
Me	H	CH_2Cl	60	216	10 Kbar	90	3:1
Me	H	H_2O	20	120	Stirring	23	7:1
Me	H	H_2O–THF	55	24	Stirring	76	2.8:1
t-Bu	H	H_2O–THF	100	16	Stirring	84	1.3:1
Me	NO_2	H_2O–THF	55	36	Sonication	82	2.3:1
Me	OMe	H_2O–THF	55	36	Sonication	29	2.3:1

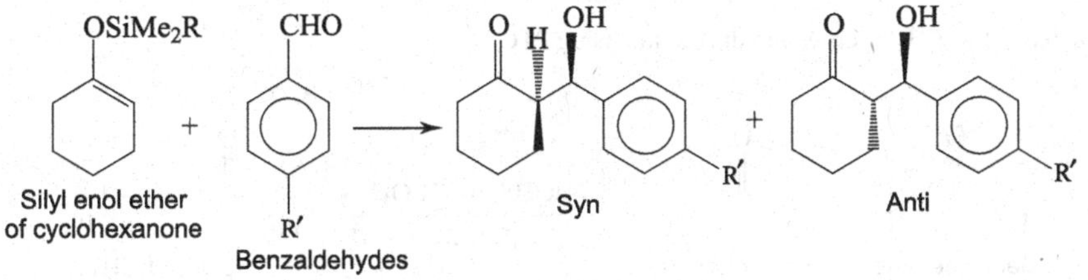

Scheme 2.132 Claisen–Schmidt reaction in aqueous phase

under high pressure (in place of catalyst, $TiCl_4$), but the reaction takes long time [132]. It has now been shown [133] that the Claisen condensation of trimethylsilyl enol ether of cyclohexanone with benzaldehyde can be carried out in water in heterogenous phase at room temperature and atmospheric pressure; no catalyst is required. No reaction takes place in organic solvents (toluene, tetrahydrofuran, dichloromethane, acetonitrile etc.) at room temperature and atmospheric pressure in the absence of catalyst. Table 2.2 gives the result of the Claisen condensation of silyl enol ether of cyclohexanone with benzaldehyde under different conditions (Scheme 2.132).

Phase transfer catalysts such as cetyltrimethylammonium compounds (like CTACl, CTABr, (CTA)$_2$ SO$_4$ and CTAOH) have been successfully used for Claisen–Schmidt reaction of acetophenones with benzaldehydes (Scheme 2.133) under weakly alkaline conditions. This permits the synthesis of biologically interesting compounds, such as chalcones and flavanols in water only [134, 135].

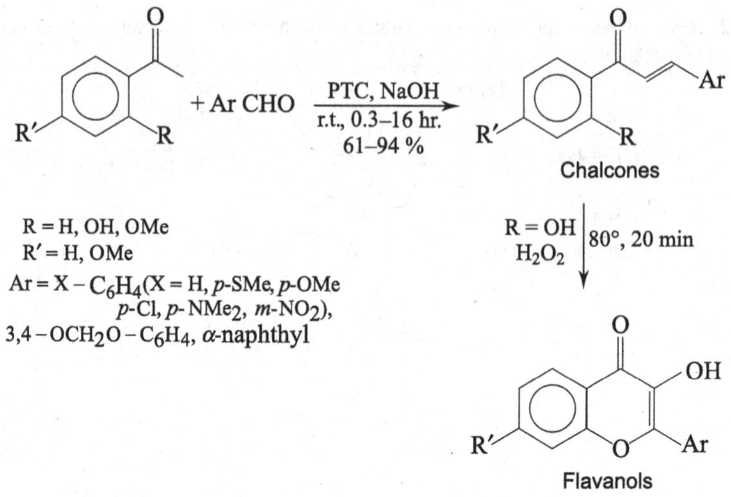

R = H, OH, OMe
R' = H, OMe
Ar = X – C_6H_4(X = H, p-SMe, p-OMe
      p-Cl, p-NMe$_2$, m-NO$_2$),
3,4 – OCH$_2$O – C$_6$H$_4$, α-naphthyl

R = OH | 80°, 20 min
H$_2$O$_2$ |

Scheme 2.133 Claisen–Schmidt reaction using PTC

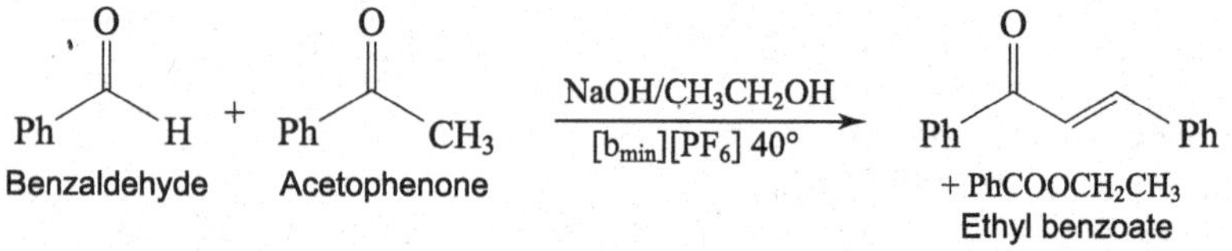

Scheme 2.134 Claisen–Schmidt reaction in ionic liquids

2.16.2 Claisen–Schmidt Reaction in Ionic Liquids

The Claisen–Schmidt reaction between acetophenone and benzaldehyde
(Scheme 2.134) in the presence of ionic liquid, [b$_{min}$][PF$_6$] in the presence of
sodium hydroxide and ethyl alcohol gave low yields [136] of the condensation
product. However, in this case ethylbenzoate was obtained as a by-product and the
base was depleted after two cycles of reuse of the ionic liquid.

It has already been explained that imidazolinium-based ionic liquids are not suit-
able under basic condition (for details, see Baylis–Hillman reaction in ionic liquids,
Sect. 2.8.3). On the basis of studies, it has been shown that in the above reaction
(Scheme 2.135) the source of the ethyl group was sodium ethoxide and not ethanol.

It is best to carry out the above Claisen–Schmidt reaction in the presence of
imidazolium-based ionic liquids containing a phenyl group at the C(2) position [137].
The ionic liquid is represented below.

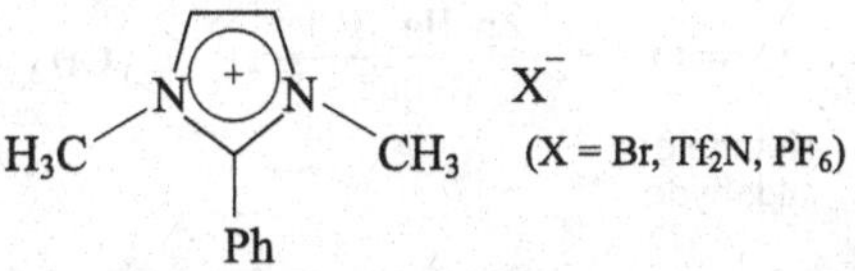

$$\underset{\substack{\text{H}_3\text{C} \qquad\qquad \text{CH}_3 \\ \text{Ph}}}{\text{N}\overset{+}{\diagup}\text{N}} \qquad \text{X}^- \quad (\text{X} = \text{Br, Tf}_2\text{N, PF}_6)$$

Scheme 2.135 An imidazolinium-based ionic liquid

$$\text{C}_6\text{H}_5\text{CHO} + \text{CH}_3\text{CHO} \xrightarrow[\text{8–10 days}]{\text{10\% NaOH, rt}} \underset{\text{Cinnamaldehyde}}{\text{C}_6\text{H}_5\text{CH} = \text{CHCHO}}$$

$$\text{C}_6\text{H}_5\text{CHO} + \text{CH}_3\text{COCH}_3 \xrightarrow[\text{2–3 days (70\%)}]{\text{10\% NaOH, 30°}} \text{C}_6\text{H}_5\text{CH} = \text{CHCOCH}_3$$

Benzylidene acetone
(benzalacetone)
(4-phenyl-3-buten-2-one)

$$\text{C}_6\text{H}_5\text{CHO} + \text{CH}_3\text{COC}_6\text{H}_5 \xrightarrow[< 30°, 85\%]{\text{10\% NaOH}} \text{C}_6\text{H}_5\text{CH} = \text{CHCOC}_6\text{H}_5$$

Benzylidene acetophenone
(benzalacetophenone)

$$\text{C}_6\text{H}_5\text{CHO} + \text{CH}_3\text{COOC}_2\text{H}_5 \xrightarrow[30°]{\text{10\% NaOH}} \text{C}_6\text{H}_5\text{CH} = \text{CHCOOC}_2\text{H}_5$$

Ethyl cinnamate

Scheme 2.136 Some applications of Claisen–Schmidt reaction

2.16.3 Applications [137] (Scheme 2.136)

(*i*) The α,β-unsaturated carbonyl compounds, (e.g., cinnamaldehyde and benzylidene acetone are useful in perfumary) are commercial products.
In the above synthesis, the concentration of alkali is crucial, otherwise **Cannizzaro reaction** may give the major product(s).

(*ii*) The β-ionone, a key intermediate in the synthesis of Vitamin A, is obtained by Claisen–Schmidt reaction of citral or geranial (a naturally occurring aldehyde obtained from lemon grass oil) and acetone gives pseudoionone which is used for the preparation of α- and β-ionones by ring closure using $\text{BF}_3/\text{CH}_3\text{COOH}$.

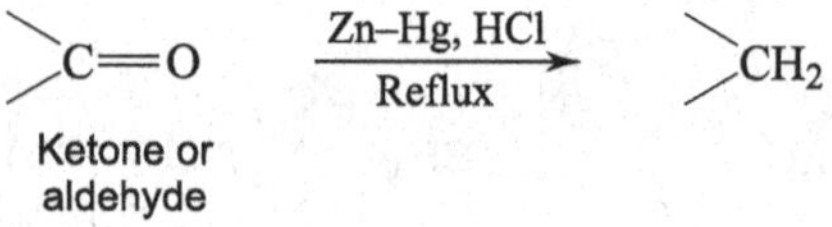

$$\text{C}=\text{O} \xrightarrow[\text{Reflux}]{\text{Zn–Hg, HCl}} \text{CH}_2$$

Ketone or
aldehyde

Scheme 2.137 Clemmensen reduction

Citral or
geranial

$+ \ CH_3COCH_3 \xrightarrow[\text{or Ba(OH)}_2]{\text{NaOEt}}$

Pseudoionone
(49%)

β-Ionone

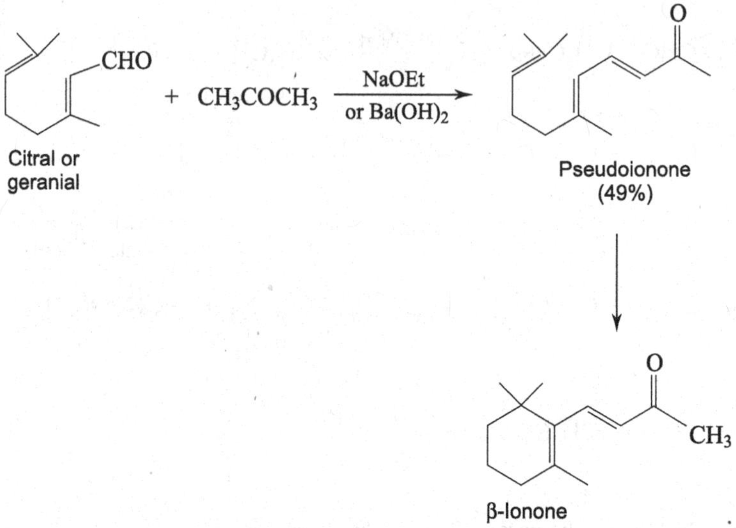

2.17 Clemmensen Reduction

The carbonyl group of aldehydes and ketones on reduction with zinc amalgam and hydrochloric acid give the corresponding hydrocarbons (i.e., the carbonyl group is converted into CH_2 group) (Scheme 2.137) and is known as Clemmensen reduction [138].

Mechanism

The mechanism of Clemmensen reduction is not well-understood. It is, however, clear that in most cases the alcohol is not an intermediate, since Clemmensen reduction do not reduce most alcohols to hydrocarbons. Following mechanism is suggested. It shows that reduction under acidic conditions normally involves protonated species to which the metal offers electron (Scheme 2.138).

Limitations [138] (Scheme 2.139)

(*i*) When the substrate is unstable under acidic condition (but stable under alkaline condition), Clemmensen reduction cannot be used. In such cases, the **Huang-Minlon modification** of the **Wolff–Kishner reduction** can be very conveniently

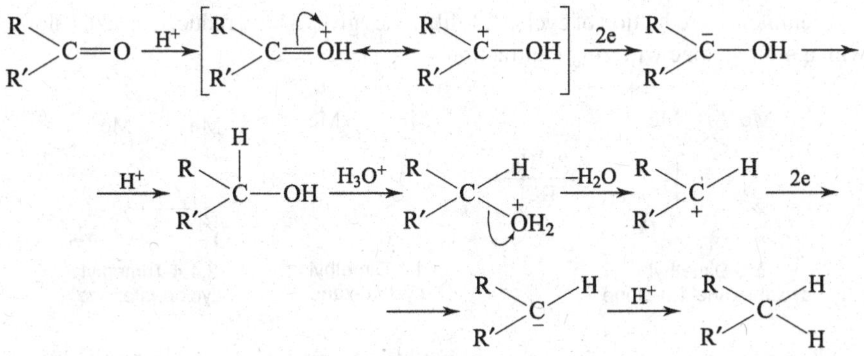

Scheme 2.138 Mechanism of Clemmensen reaction

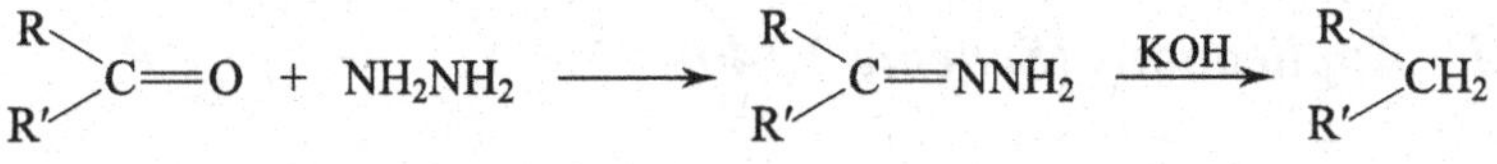

Scheme 2.139 Limitations of Clemmensen reaction

used. Certain compounds like thioacetals can be reduced only in neutral solution; the later procedure is used for compounds that are sensitive to both acids and bases.

It is known that Wolff–Kishner reduction is represented as follows:

In the Huang-Minlon modification diethylene glycol is used as a solvent and isolation of the hydrazone is not necessary.

(ii) Clemmensen reduction is subject to steric effects. For example, hindered ketone (one example shown below) does not undergo reduction.

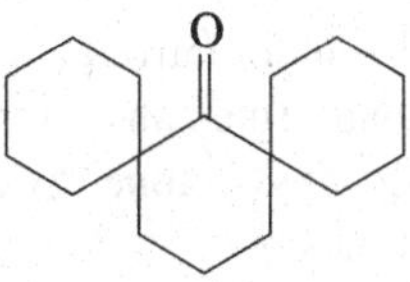

(iii) Certain aldehydes and ketones do not give the normal reduction products only. Thus, α-hydroxy ketones give either ketones through hydrogenolysis of OH group or olefins, and 1,3-diketones give exclusively monoketones accompanied by rearrangement.

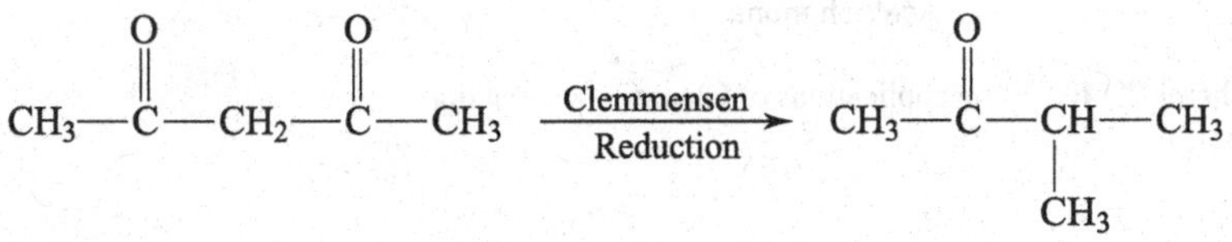

Clemmensen reduction of cyclic 1,3-diketones gives a fully reduced product along with a monoketone with ring contraction.

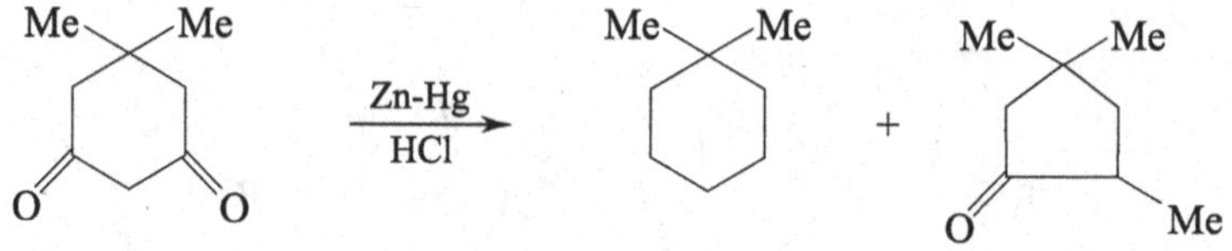

| 5,5-Dimethyl cyclohexane-1,3-dione | 1,1-Dimethyl cyclohexane | 2,4,4-Trimethyl cyclopentanone |

Clemmensen reduction has been reported to be improved by sonication [139].

2.17.1   Applications (Scheme 2.140)

(*i*) Clemmensen reduction is useful for the reduction of aromatic aliphatic ketones [140, 141]. Thus, acetophenone [142] is reduced by Clemmensen method to ethyl benzene.
Similarly,

$$CH_3(CH_2)_5CHO \xrightarrow[HCl]{Zn-Hg} CH_3(CH_2)_5CH_3$$

$$\text{\textit{n}-Heptaldehyde} \qquad\qquad \text{\textit{n}-Heptane}$$

$$C_6H_5COC_3H_7 \xrightarrow[HCl]{Zn-Hg} C_6H_5CH_2C_3H_7$$

$$\text{\textit{n}-Propylphenyl ketone} \qquad \text{\textit{n}-Butylbenzene}$$

Clemmensen reduction is useful for introducing straight chain (without rearrangement) alkyl groups in aromatic rings and subsequent reduction as in the case of acetophenone and *n*-propylphenyl ketone (examples give above).
(*ii*) Reduction of cyclic ketones

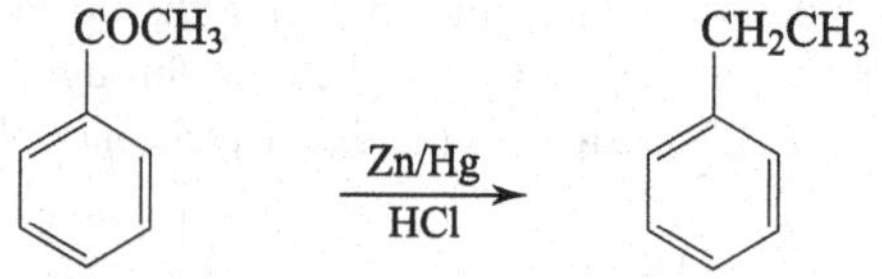

Acetophenone   Ethyl benzene

Scheme 2.140   Some applications of Clemmensen reaction

(a) α-Tetralone

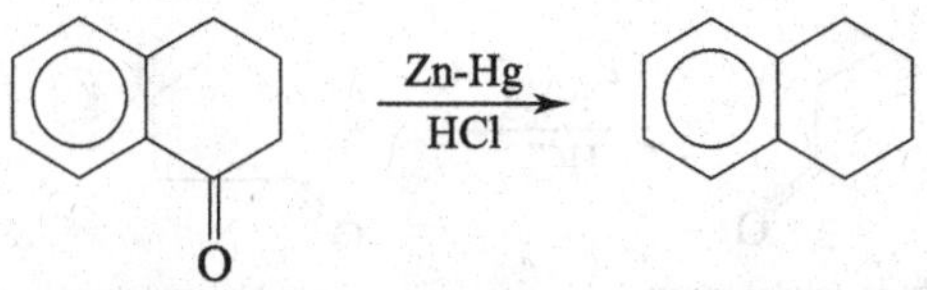

$\xrightarrow[\text{HCl}]{\text{Zn-Hg}}$

(b)

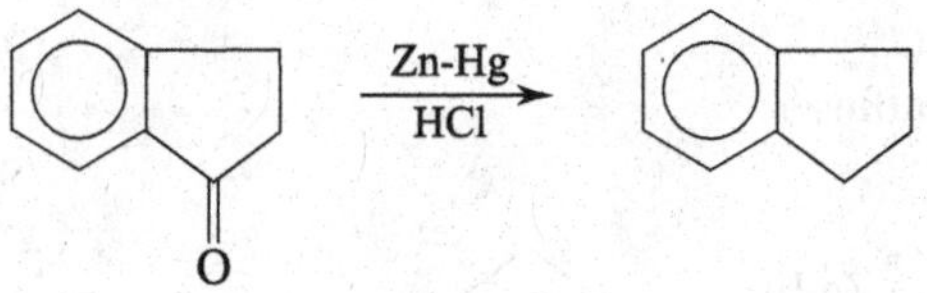

$\xrightarrow[\text{HCl}]{\text{Zn-Hg}}$

(c)

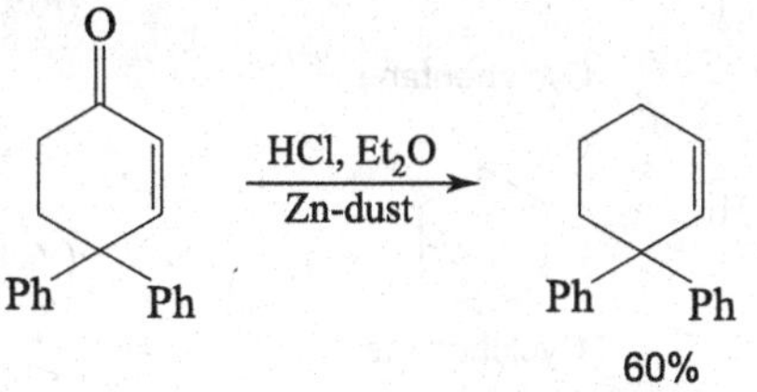

$\xrightarrow[\text{Zn-dust}]{\text{HCl, Et}_2\text{O}}$

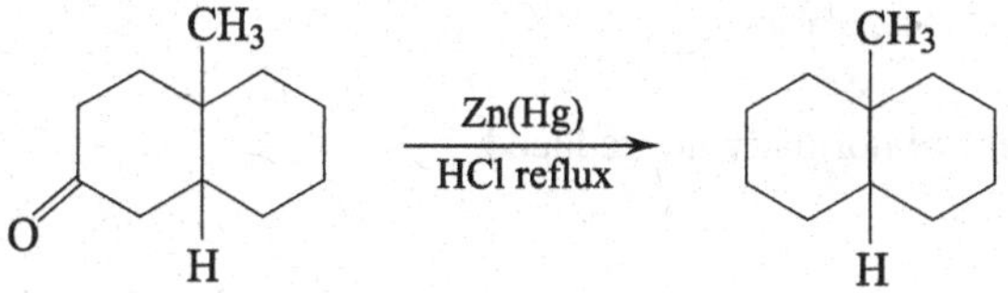

Ph   Ph

Ph   Ph
60%

(Ref 6)

(d)

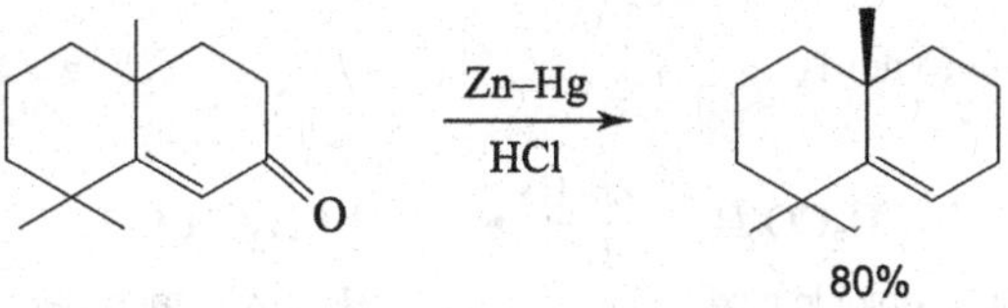

$\xrightarrow[\text{HCl}]{\text{Zn-Hg}}$

80%

(Ref 7)

(e) CH_3

O

H

cis-10-Methyl-2-decalone

$\xrightarrow[\text{HCl reflux}]{\text{Zn(Hg)}}$

CH_3

H

cis-10-Methyldecalin

(Ref 7a)

(*iii*) Reduction with ring expansion

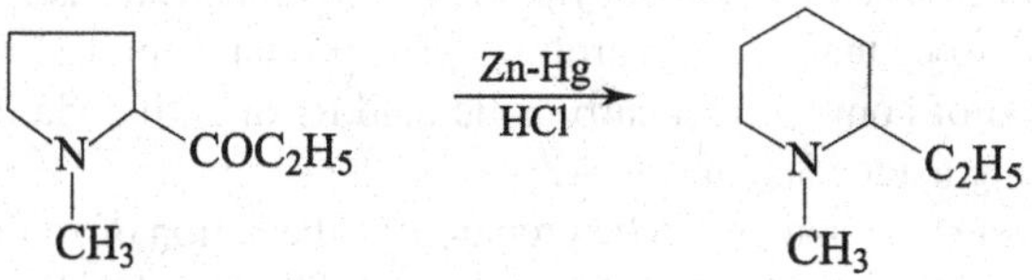

$\xrightarrow[\text{HCl}]{\text{Zn-Hg}}$

N COC_2H_5
|
CH_3

1-Methyl-2-propionyl
pyrrolidine

N C_2H_5
|
CH_3

2-Ethyl-1-methyl-
-piperidine

(*iv*) Reduction with ring contraction

5,5-Dimethylcyclohexane
-1,3-dione

$\xrightarrow[\text{HCl}]{\text{Zn-Hg}}$

2,4,4-Trimethyl-
cyclopentanone

(*v*) Synthesis of cycloparaffins

Cyclopentanone

$\xrightarrow[\text{HCl}]{\text{Zn-Hg}}$

Cyclopentane

(Ref 8, 9)

Cyclohexanone

$\xrightarrow[\text{HCl}]{\text{Zn-Hg}}$

Cyclohexane

(Ref 8, 9)

(*vi*) Reduction of γ-keto acids

$$C_6H_5COCH_2CH_2CO_2H \xrightarrow[\text{HCl}]{\text{Zn–Hg}} C_6H_5CH_2CH_2CH_2CO_2H$$

β-Benzoyl propionic acid
(γ-keto acid)

γ-Phenylbutanoic acid

α- and β-ketones are normally not reduced.

2.18   Curtius Rearrangement

Thermal rearrangement of acid azides in non-aqueous solvents, such as chloroform, benzene or ether to an isocyanate, is known as Curtius rearrangement [146]. The formed isocyanates on hydrolysis give primary amines. So, this is a convenient method of converting a carboxylic acid to an amine via the formation of the corresponding azide (Scheme 2.141).

The starting acid azides can be prepared by the action of sodium azide on an acid chloride or by the action of nitrous acid on acid hydrazides (Scheme 2.142).

The mechanism of Curtius rearrangement is given in Scheme 2.143.

In case of tertiary alkyl azides, Curtius rearrangement takes place via the formation of nitrene intermediate [147] (Scheme 2.144) leading to the formation of imines.

Curtius rearrangement can also be brought about photochemically or under sonication [148] (Scheme 2.145).

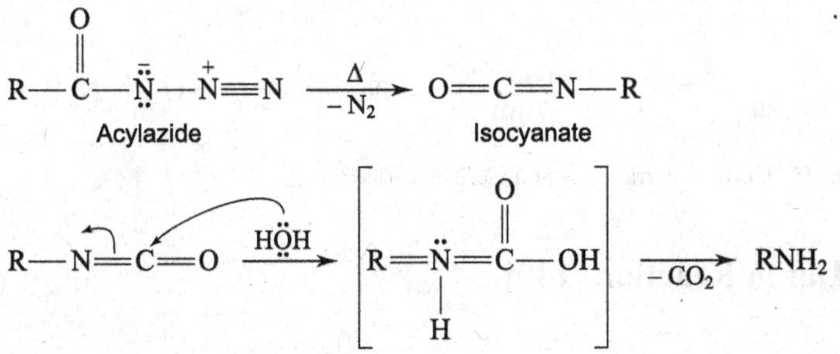

Scheme 2.141 Curtius rearrangement

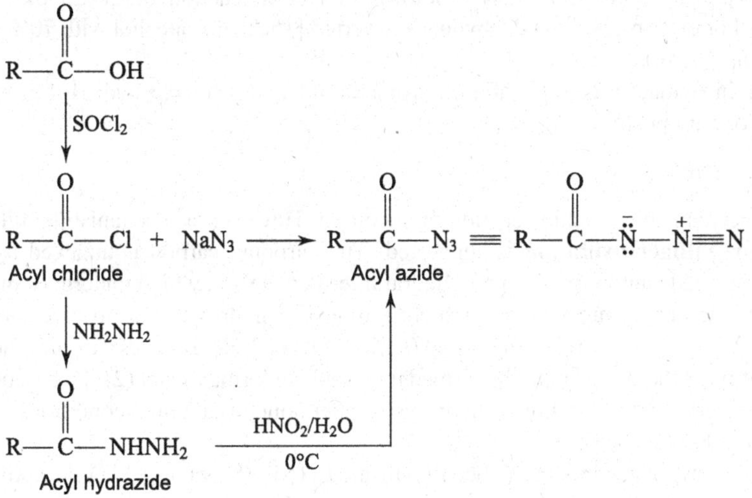

Scheme 2.142 Preparation of acid azides

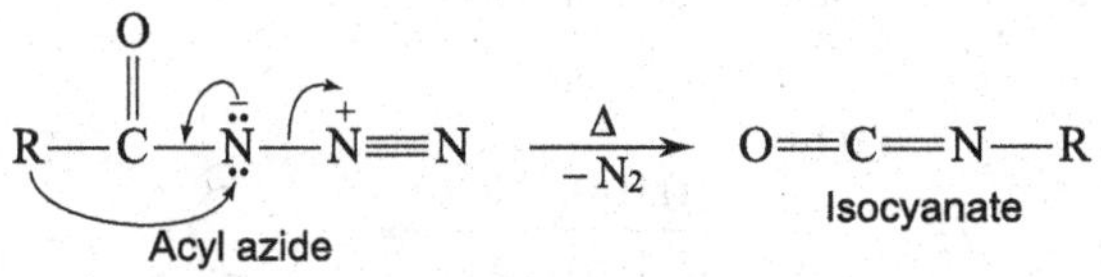

Scheme 2.143 Mechanism of Curtius rearrangement

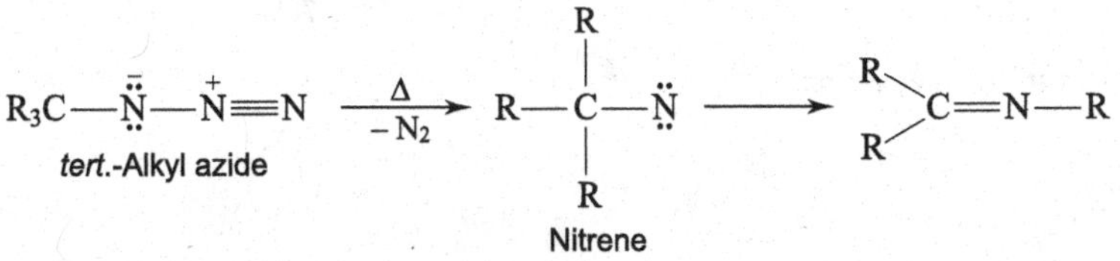

Scheme 2.144 Mechanism of Curtius rearrangement continued

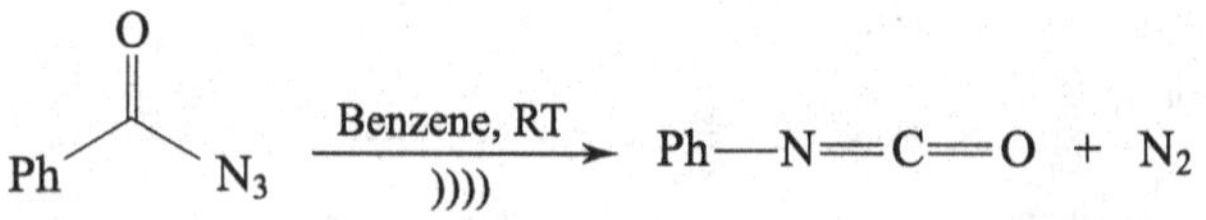

$$\underset{\text{Ph}}{\overset{\overset{\displaystyle O}{\|}}{C}}\diagdown_{N_3} \xrightarrow[\text{))))}]{\text{Benzene, RT}} \text{Ph—N}{=}\text{C}{=}\text{O} \ + \ N_2$$

Scheme 2.145   Curtius rearrangement under sonication

2.19   Dakin Reaction [149]

The oxidation of aldehyde or acetyl group in phenolic aldehydes or ketones by reaction with alkaline hydrogen peroxide results in replacement of –CHO or – COCH$_3$ group with OH. This is known as Dakin reaction or Dakin oxidation [149]. For example, salicylaldehyde is converted [150] into catechol with 70% yield (Scheme 2.146).

Dakin oxidation is applicable in aromatic aldehydes having hydroxyl group in ortho or para positions.

Mechanism

The mechanism of Dakin reaction is uncertain. However, a mechanism similar to **Baeyer–Villiger oxidation** is suggested. The carbonyl carbon is attacked by the hydroperoxide anion to give a tetrahedral intermediate (1), which undergoes migration of the aryl group with subsequent removal of hydroxide ion to give formate ester (2). An electron-releasing group (such as OH or NH$_2$) is necessary for efficient migration of the aryl group. The formed intermediate formate ester (2) can be isolated and converted into catechol on hydrolysis under aqueous alkaline conditions of the reaction (Scheme 2.147).

The above mechanism has been confirmed [151] (Scheme 2.147) by taking o-hydroxybenzaldehyde in which the oxygen of CHO group is labelled by ^{18}O. The product, formic acid had all the labelled oxygen (Scheme 2.148).

Generally, the yields in Dakin reaction are low.

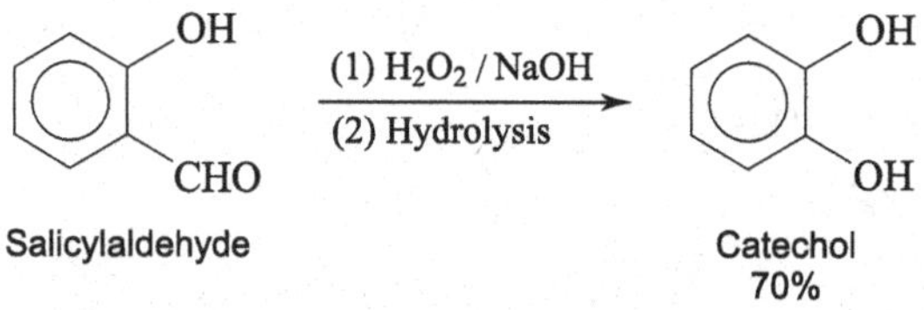

Scheme 2.146   Dakin reaction

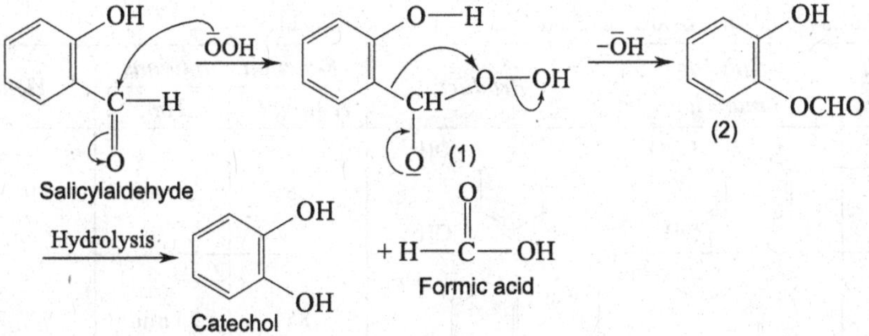

Scheme 2.147 Mechanism of Dakin reaction

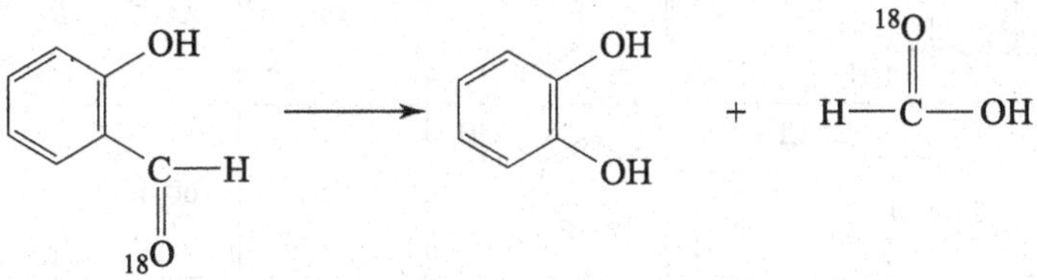

Scheme 2.148 Confirmation of the mechanism of Dakin reaction

2.19.1 Dakin Reaction Under Ultrasonic Irradiation

Dakin reaction has been carried out in high yields using sodium percarbonate (SPC, Na_2CO_3, $1.5H_2O$) in H_2O_2–THF under **ultrasonic irradiation** [152]. Using this procedure the following aldehydes have been oxidized in 85–90% yields: o-hydroxybenzaldehyde, p-hydroxybenzaldehyde, 2-hydroxy-4-methoxybenz-aldehyde, 2-hydroxy-3-methoxybenzaldehyde and 3-methoxy-4-hydroxybenz-aldehyde.

2.19.2 Dakin Reaction in Solid State

Solid-state oxidation of hydroxylated benzaldehyde and acetophenones with urea-hydrogenperoxide adduct is a superior alternative [153] in terms of shorter reaction time, cleaner product formation, ease of manipulation and excellent yield. The reagent hydrogen peroxide–urea complex $\begin{bmatrix} H_2NCONH_2 \\ | \\ HOOH \end{bmatrix}$ (UHP) is commercially available and can also be easily prepared.

The reaction is carried out by stirring a mixture of the hydroxy aldehyde or ketone with urea–hydrogen peroxide adduct (in molar ratio 1:2) and heating in an oil bath at

Table 2.3 Solid-state oxidation of aldehydes and ketone using uHP

S. No.	Starting material	Product	Reaction conditions		% Yield
			Temp (°C)	Time	
1.	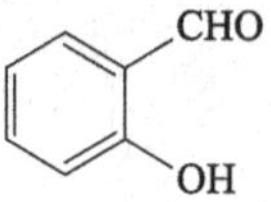 CHO / OH (2-hydroxybenzaldehyde)	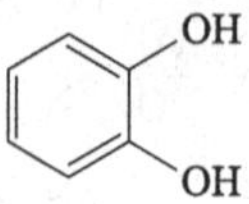 OH / OH (catechol)	85	20 min.	80
	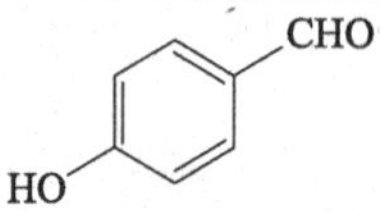 CHO / HO (4-hydroxybenzaldehyde)	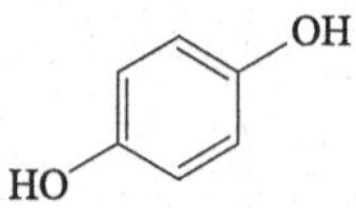 OH / HO (hydroquinone)	85	75 min.	82
	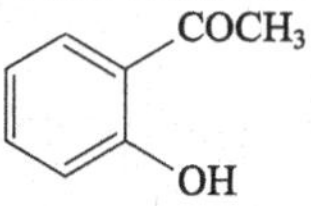 COCH₃ / OH (2-hydroxyacetophenone)	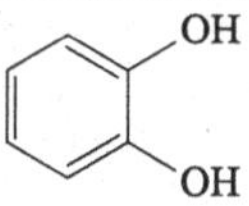 OH / OH	85	60 min.	86
	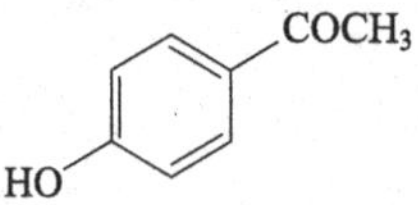 COCH₃ / HO (4-hydroxyacetophenone)	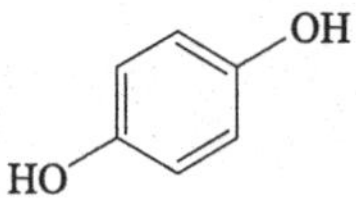 OH / HO	85	60 min.	80
	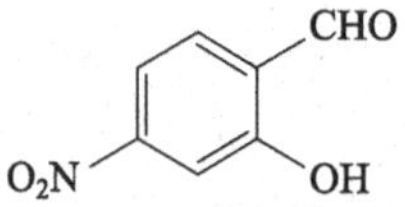 CHO / O₂N...OH	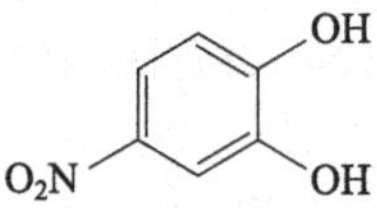 OH / O₂N...OH	85	25 min.	83
	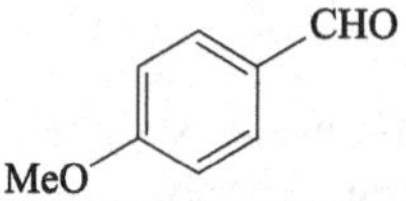 CHO / MeO	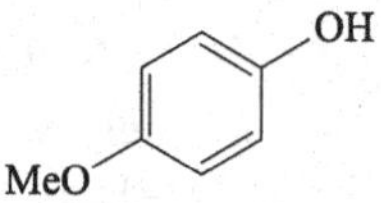 OH / MeO	85	45 min.	80

85 °C for 20 min to 1.5 h and isolating the product by extraction with ethyl acetate. Table 2.3 gives the starting material, product formed, reaction conditions and yield.

2.19.3 Applications

Dakin oxidation has a number of synthetic applications: (Scheme 2.149).

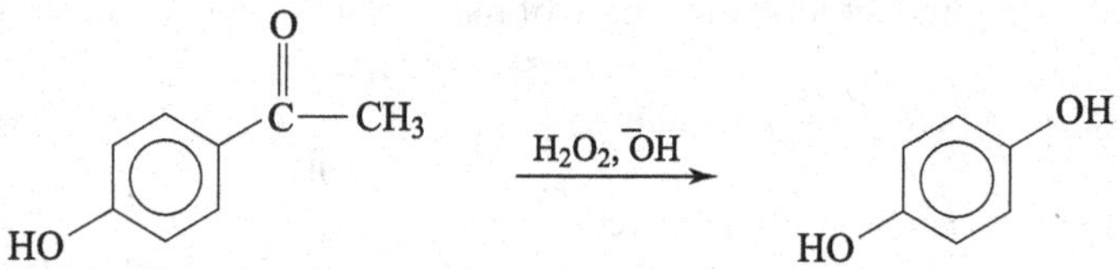

$$\text{p-Hydroxyacetophenone} \xrightarrow{\text{H}_2\text{O}_2, \ \overline{\text{O}}\text{H}} \text{Quinol}$$

Scheme 2.149

(*i*) Synthesis of quinol [154].

(*ii*) Dakin reaction can also be used in case of flavones having acetyl group [155].

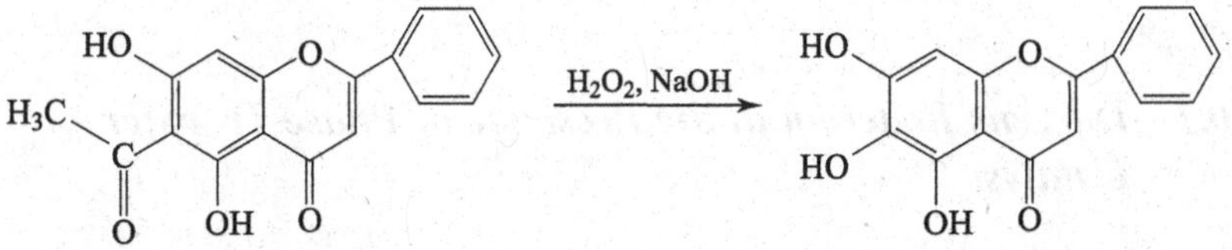

**5,7-Dihydroxy-6-
-acetyl flavone**

5,6,7-Trihydroxy flavone

(*iii*) Synthesis of 3,5-dimethyl catechol [156].

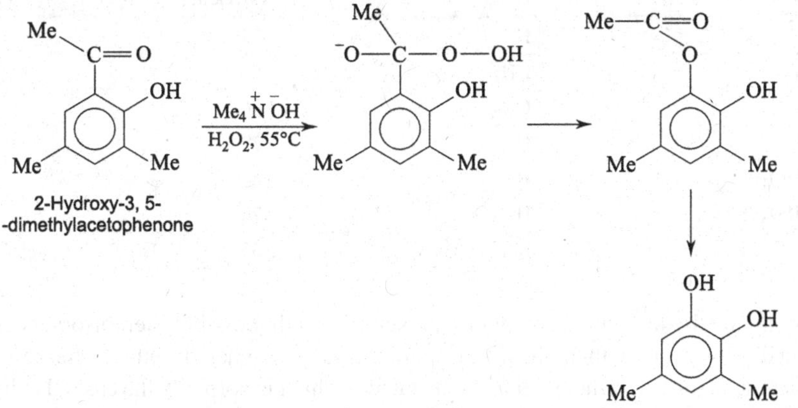

**2-Hydroxy-3, 5-
-dimethylacetophenone**

**25%
3, 5-Dimethyl catechol**

2.20 Darzens Reaction

The condensation of aldehydes or ketones with α-haloester in the presence of a base (like potassium tert. butoxide, a hindered base to avoid the S_N2 displacement of the chloride) gives α,β-epoxy esters called glycidic esters and is known as Darzens reaction [157] or Darzens glycidic ester condensation. As an example, the condensation of acetophenone with ethyl chloroacetate in the presence of base gives the α,β-epoxy ester (Scheme 2.150).

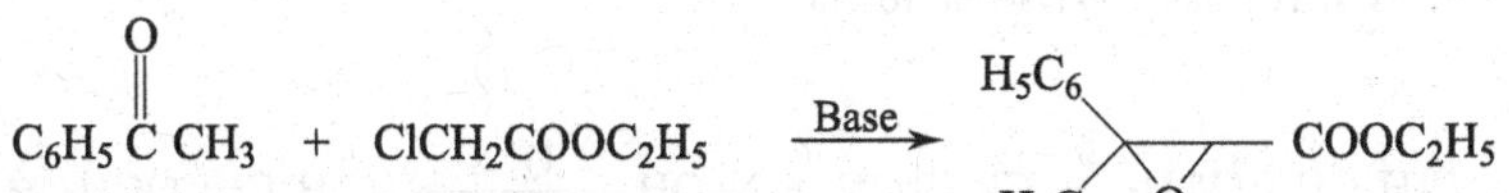

$$C_6H_5 \overset{O}{\underset{}{\overset{||}{C}}} CH_3 \;+\; ClCH_2COOC_2H_5 \;\xrightarrow{\text{Base}}\;$$

Acetophenone   Ethyl chloroacetate

Scheme 2.150 Darzens reaction

In place of α-haloester (ethyl chloroacetate), α-halonitriles (e.g., chloroacetonitrile, ClCH$_2$CN) can also be used.

2.20.1 Darzens Reaction in the Presence of Phase Transfer Catalyst

Darzens reaction has been found to occur in alkaline solution in the presence of a phase transfer catalyst (benzyl triethylammonium chloride) [158] (Scheme 2.151).

R	R'	% Yield
Ph	H	75
CH$_3$	CH$_3$	60
Ph	CH$_3$	80
(CH$_2$)$_3$	H	79
(CH$_2$)$_4$	H	65
(CH$_2$)$_5$	H	78
Ph	Ph	55

With aldehydes and unsymmetrical ketones both possible stereoisomers are formed. However, with more acidic ketones, such as phenylacetone, the ketone carbanion is formed rather than from the nitrile (which is normally the case), leading to alkylation of the ketone (Scheme 2.152).

In the above Darzens reaction using quaternary ammonium salts, it has been shown [159] that the structure of quaternary ammonium salt has virtually no effect on the yield of the glycidonitrile. It is best to perform the reaction at ~20 °C and use sufficient sodium hydroxide (about 3 mol per mole of nitrile).

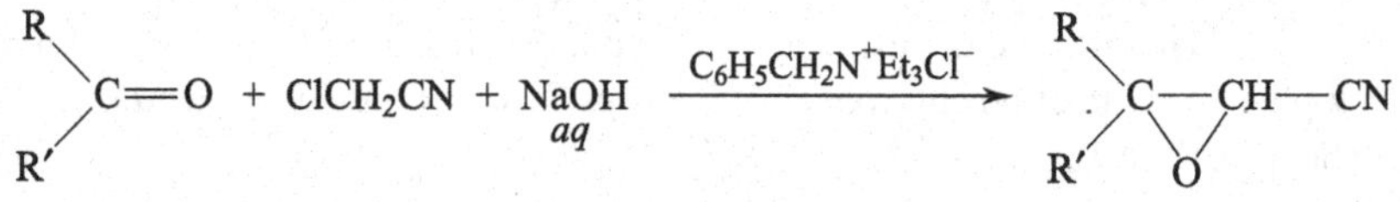

$$\underset{R'}{\overset{R}{\diagdown}}C{=}O \ + \ ClCH_2CN \ + \ \underset{aq}{NaOH} \ \xrightarrow{C_6H_5CH_2N^+Et_3Cl^-} \ \underset{R'}{\overset{R}{\diagdown}}\underset{O}{\overset{}{C}}{-}CH{-}CN$$

Scheme 2.151 PTC catalysed Darzens reaction

$$C_6H_5CH_2COCH_3 \ + \ ClCH_2CN \ + \ \underset{aq}{NaOH} \ \xrightarrow{Q^+X^-} \ \underset{\underset{CH_2CN}{|}}{C_6H_5CHCOCH_3}$$

Phenyl acetone   Chloroaceto
                          nitrile

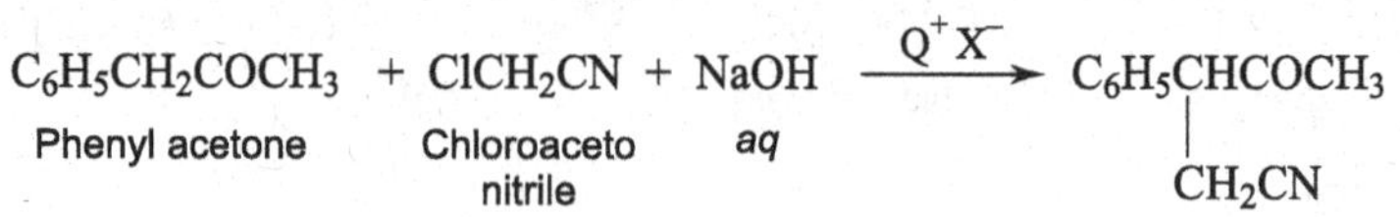

Scheme 2.152 PTC-catalysed Darzens reaction

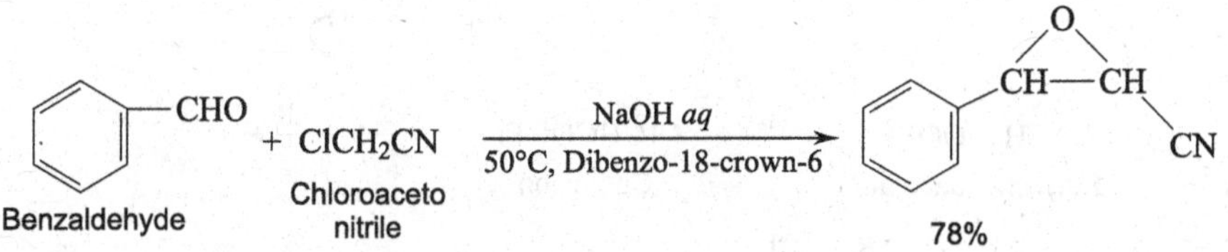

Scheme 2.153 Darzens reaction in the presence of crown ethers

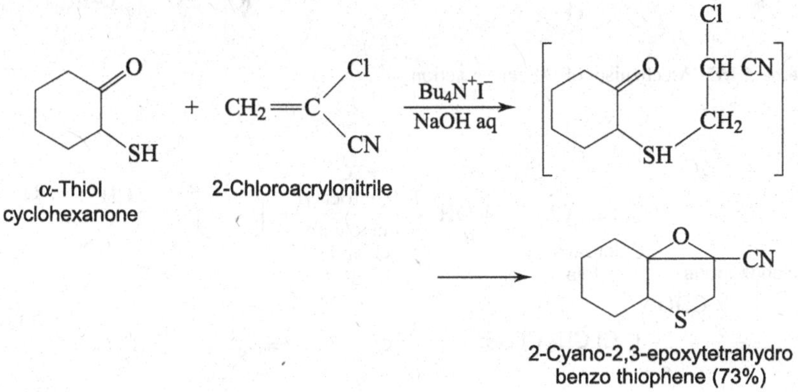

Scheme 2.154 Intramolecular Darzens reaction

In place of quaternary ammonium salts (PTC), crown ethers (e.g., dibenzo-18-crown-6) (1 mol %) can also be used. Thus, the reaction of benzaldehyde with chloroacetonitrile gives 78% of the corresponding glycidonitrile (Scheme 2.153).

The PTC-catalysed condensation of α-thiocarbonyl compounds with 2-chloroacrylonitrile yielded [161] 2-cyano-2,3-epoxytetrahydrothiophenes. The reaction probably involved first the thiol addition followed by an **intramolecular Darzens reaction** (Scheme 2.154).

Mechanism

The reaction of ethyl chloroacetate with a ketone in the presence of a base (potassium *tert.* butoxide) is believed to proceed as follows (Scheme 2.155).

2.20.2 Applications

Some important applications are given in Scheme 2.156.

The glycidic esters (or the glycidonitriles) obtained by the Darzens reaction are important intermediates for organic synthesis. For example, the glycidic ester or glycidonitrile on alkaline hydrolysis gives glycidic acid, which undergoes decarboxylative rearrangement in the presence of acid to give an aldehyde or a ketone.

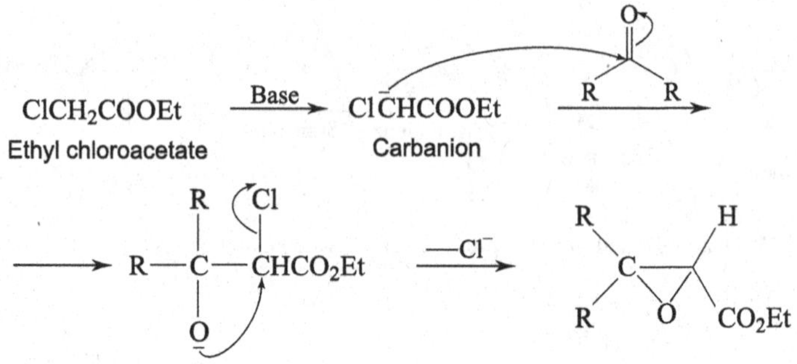

Scheme 2.155 Mechanism of Darzens reaction

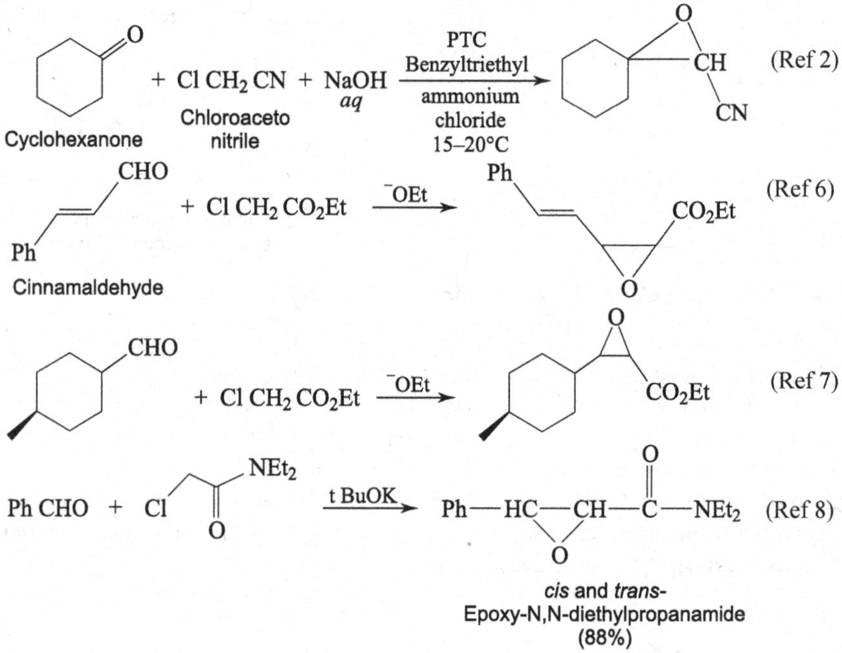

Scheme 2.156 Some applications of Darzens reaction

The overall process involves the addition of one or more carbon atoms, as aldehyde, to a carbonyl group. Thus, an aldehyde, RCHO is converted to a longer chain aldehyde homologue (RCH₂CHO) and a ketone (R₂CO) is converted to a longer chain aldehyde (R₂CHCHO). This is known as chain extension procedure. For example [164], cyclohexanone on Darzens reaction with ethyl chloroacetate followed by hydrolysis of the glycidic ester and subsequent hydrolysis give the epoxy acid; this on decarboxylation produces carboxaldehyde.

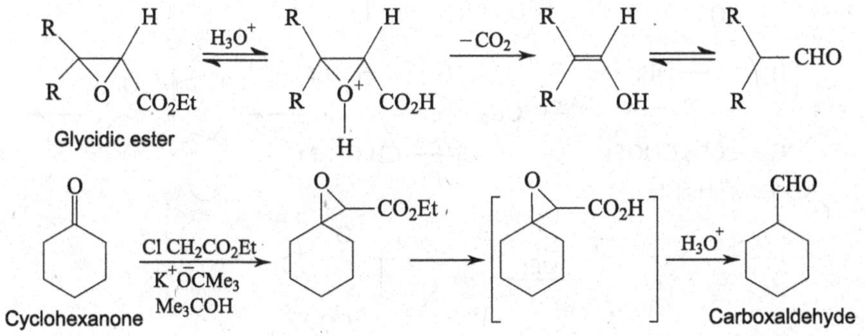

Glycidic ester

Cyclohexanone   Cl CH₂CO₂Et / K⁺ ⁻OCMe₃ / Me₃COH   Carboxaldehyde

2.21 Dieckmann Condensation

Diesters of C_6 and C_7 dibasic acids undergo an **intramolecular Claisen condensation** in the presence of base to give good yields of cyclic β-ketoesters. This is known as Dieckmann condensation [165]. It is of considerable value in the synthesis of cyclic compounds. For example, ethyl esters of adipic acid and pimelic acids give 2-carbethoxycyclopentanone and 2 carbethoxycyclohexanone, respectively (Scheme 2.157).

Dieckmann condensation best proceeds with esters having 6, 7 or 8 carbon atoms and gives stable rings with 5, 6 or 7 carbons.

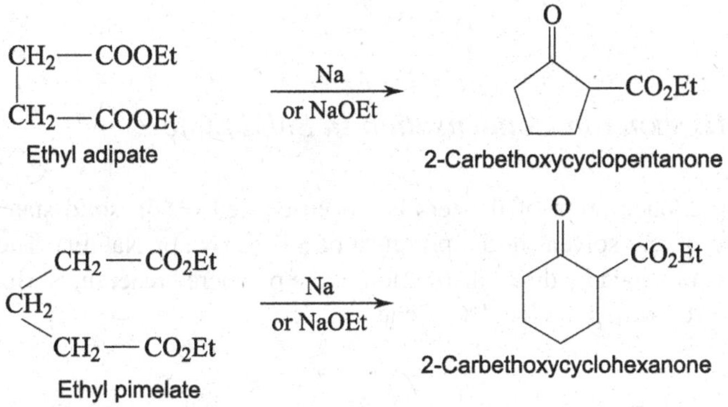

Ethyl adipate   Na or NaOEt   2-Carbethoxycyclopentanone

Ethyl pimelate   Na or NaOEt   2-Carbethoxycyclohexanone

Scheme 2.157 Dieckmann condensation

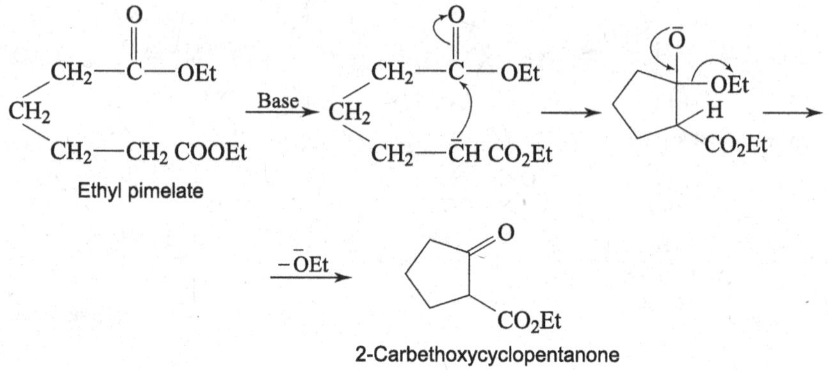

Scheme 2.158 Mechanism of Dieckmann condensation

Diethyl adipate $n = 2$
Diethyl pimelate $n = 3$

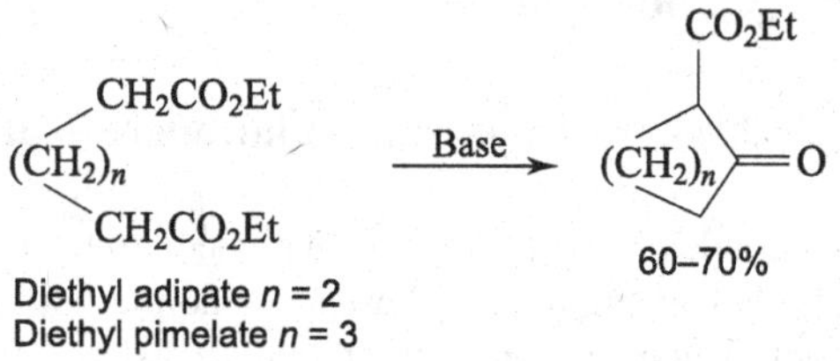

Scheme 2.159 Solid-state Dieckmann condensation

Mechanism

Abstraction of a proton from one of the α-carbons gives the carbanion, which attacks the carbonyl carbon of the other ester group. Finally, the β-ketoester is formed by expulsion of ethoxide anion (OEt) (Scheme 2.158).

2.21.1 Dieckmann Condensation in Solid State

Dieckmann condensation of diesters has been carried out in solid state [166] in the absence of any solvent in the presence of a base (Na or NaOEt). The reaction products are obtained by direct distillation of the powdered reaction mixture, which was neutralized with p-TsOH·H$_2$O (Scheme 2.159).

2.21.2 Dieckmann Condensation Under Sonication

It is found [167] that Dieckmann condensation proceeded very well on sonication in a short time (Scheme 4). On sonication [168], potassium is easily transformed

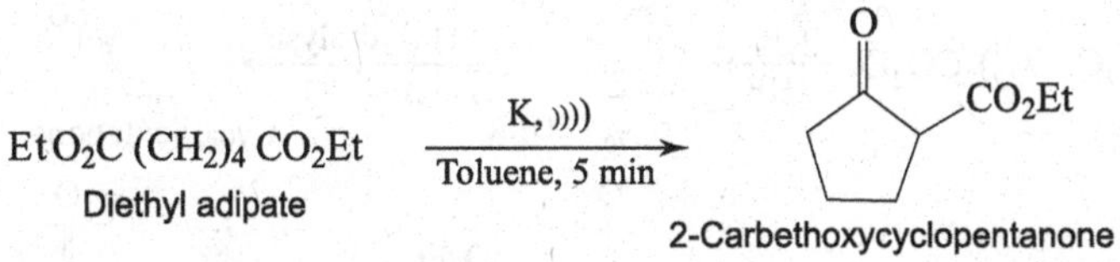

EtO$_2$C (CH$_2$)$_4$ CO$_2$Et $\xrightarrow[\text{Toluene, 5 min}]{\text{K,)))}}$

Diethyl adipate

2-Carbethoxycyclopentanone

Scheme 2.160 Dieckmann condensation under sonication

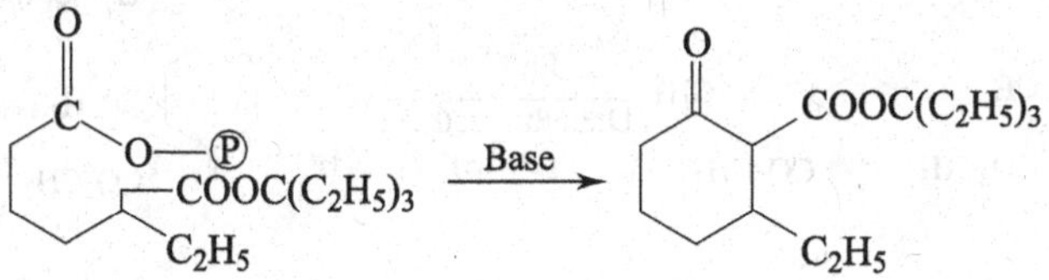

$\xrightarrow{\text{Base}}$

Scheme 2.161 Dieckmann condensation using polymer-supported technique

to a silver blue suspension in toluene. The ultrasonically dispersed potassium is extremely helpful in Dieckmann condensation (cyclization) (Scheme 2.160).

In the above condensation, bases like Bu^t OK, Bu^t ONa, EtOK or EtONa could also be used.

2.21.3   Dieckmann Condensation Using Polymer Support Technique

In Dieckmann cyclization, it is necessary to control the relative rates of two competing reactions, viz., the intramolecular cyclization and the intermolecular reaction; the former must be faster. Normally, this is achieved by having the compound in large volume of the solvent (dilution technique). Similar dilution technique is achieved by using a polymer-supported molecule. The molecule to be cyclized can be diluted by anchoring them at distance far enough apart within the polymeric matrix so that the intermolecular reaction is prevented. This technique has been used successfully for the Dieckmann cyclization of mixed esters of dicarboxylic acids [169–173] (Scheme 2.161).

2.21.4   Applications

Dieckmann condensation has been used for the synthesis of various cyclopentane and cyclohexane derivatives. Some of the important applications of Dieckmann condensation are given in Scheme 2.162.

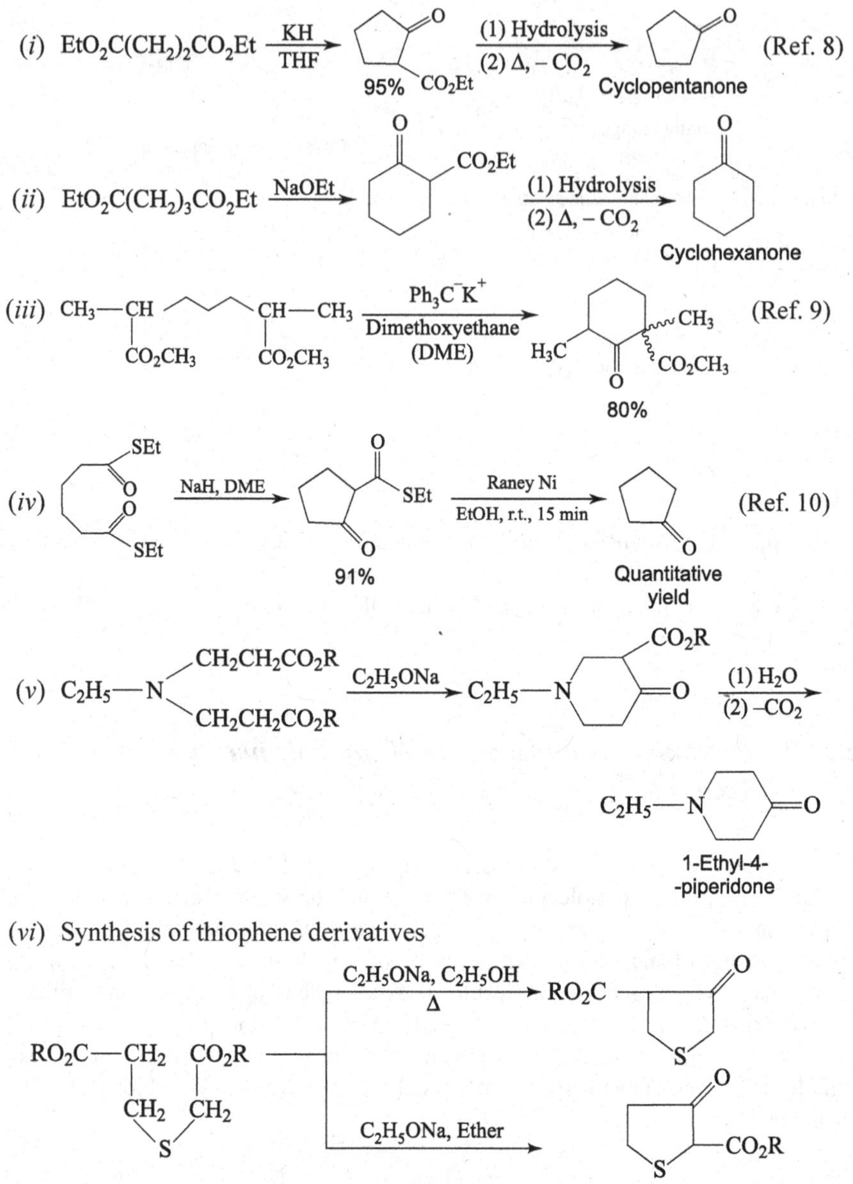

(*i*) EtO$_2$C(CH$_2$)$_2$CO$_2$Et $\xrightarrow{\text{KH} \atop \text{THF}}$ 95% $\xrightarrow{\text{(1) Hydrolysis} \atop \text{(2) }\Delta,\,-\,CO_2}$ Cyclopentanone   (Ref. 8)

(*ii*) EtO$_2$C(CH$_2$)$_3$CO$_2$Et $\xrightarrow{\text{NaOEt}}$ $\xrightarrow{\text{(1) Hydrolysis} \atop \text{(2) }\Delta,\,-\,CO_2}$ Cyclohexanone

(*iii*) $\xrightarrow{\text{Ph}_3\text{C}^-\text{K}^+ \atop \text{Dimethoxyethane} \atop \text{(DME)}}$ 80%   (Ref. 9)

(*iv*) $\xrightarrow{\text{NaH, DME}}$ 91% $\xrightarrow{\text{Raney Ni} \atop \text{EtOH, r.t., 15 min}}$ Quantitative yield   (Ref. 10)

(*v*) $\xrightarrow{\text{C}_2\text{H}_5\text{ONa}}$ $\xrightarrow{\text{(1) H}_2\text{O} \atop \text{(2) }-\text{CO}_2}$

1-Ethyl-4-piperidone

(*vi*) Synthesis of thiophene derivatives

Scheme 2.162 Some applications of Dieckmann condensation

(*vii*) In the synthesis of steroids, five or six-membered ring is built up.

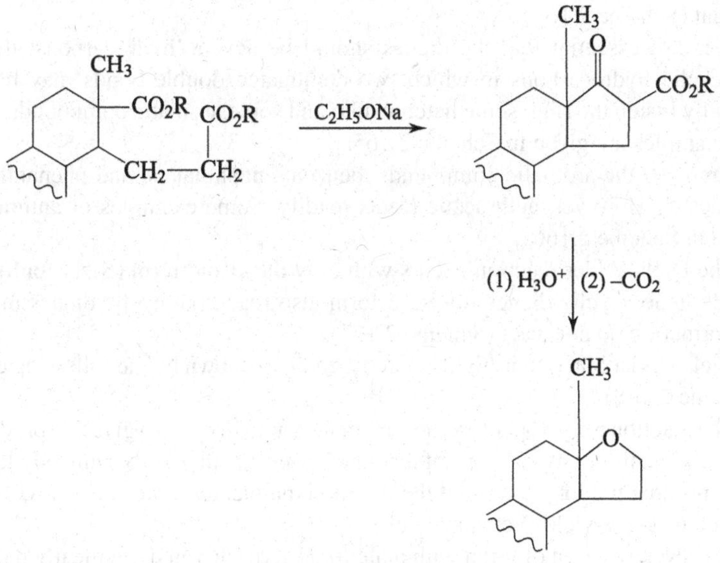

Scheme 2.162 (continued)

2.22 Diels–Alder Reaction [176]

A typical reaction for carbon–carbon bond formation is the well-known Diels–Alder reaction. It is a [4 + 2] cycloaddition reaction between a conjugated diene (4π-electron system) and a compound having a double bond or triple bond called the dienophile (2π-electron system) to form an adduct. A typical example is the reaction between butadiene (diene) and ethylene (dienophile) by heating both the components alone or in an inert solvent. In place of ethylene, acetylene can also be used (Scheme 2.163).

In Diels–Alder reaction, two new σ-bonds are formed at the expense of two π-bonds in the starting materials. The reaction is enhanced if electron-withdrawing

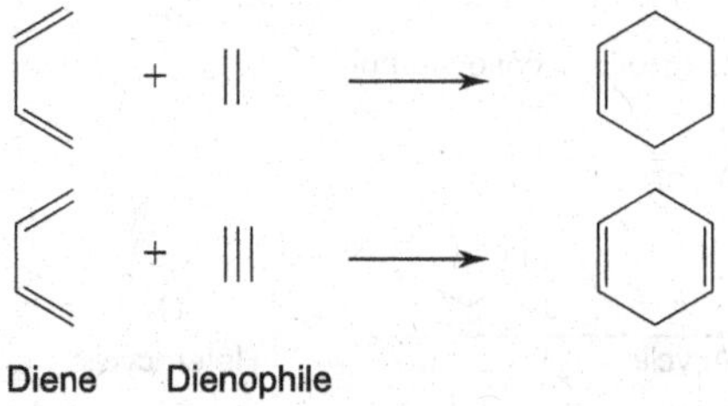

Diene   Dienophile

Scheme 2.163 Diels–Alder reaction

substituents (e.g., $>C{=}O$, $-CHO$, $-COOR$, $-CN$, $-NO_2$ etc.) in the dienophile are present (Scheme 2.164).

It is not essential that the dienes should be acyclic hydrocarbons; these may be acyclic hydrocarbons in which two conjugated double bonds may be present partially inside the ring, some heterocyclic and some aromatic compounds. Some of the examples are given in Scheme 2.165.

Amongst the aromatic compounds, benzene, naphthalene and phenanthrene are unreactive. However, anthracene reacts readily. Some examples of anthracene are given in Scheme 2.166.

The Diels–Alder reaction occurs with only the cisoid form (S-cis conformation) of the diene. Cyclic dienes in cisoid form also react, but cyclic dienes in transoid conformation do not react (Scheme 2.167).

Diels–Alder reaction is highly stereospecific as shown by the following examples (Scheme 2.168).

The reaction of cyclopentadiene with maleic anhydride may give two products, exo and endo. In this case and most of the other cases, the thermodynamically less stable endo-product predominates and the thermodynamically more stable exo-product is formed in 1–2% yield (Scheme 2.169).

However, reaction of furan with maleimide gives thermodynamically more stable exo-product. In this case, the initially formed endo-product gives the more stable exo-product.

The Diels–Alder reaction is regiospecific as is indicated by the following examples (Scheme 2.170).

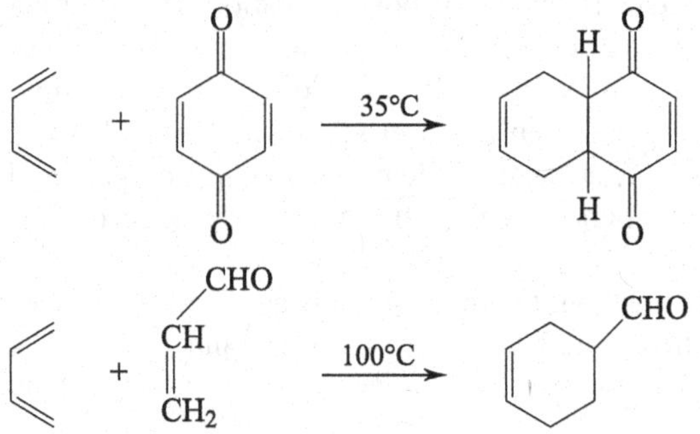

Scheme 2.164 Diels–Alder reaction condensation

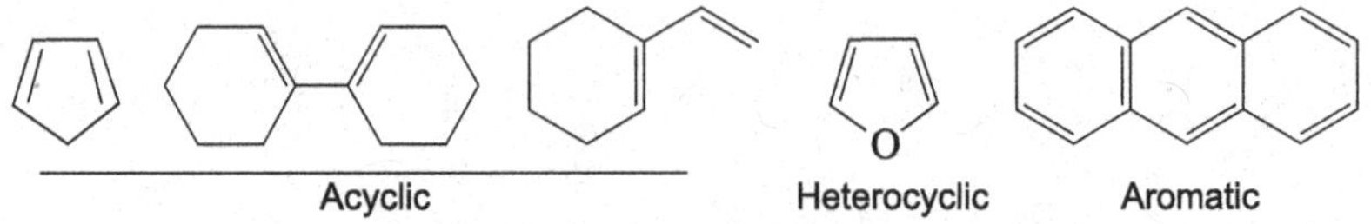

Acyclic   Heterocyclic   Aromatic

Scheme 2.165 In Diels–Alder reaction different types of dienes can be used

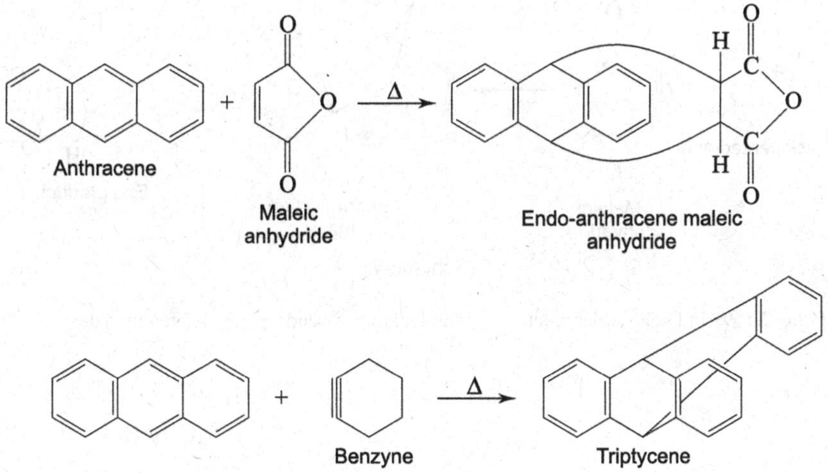

Scheme 2.166 Diels–Alder reactions of anthracene

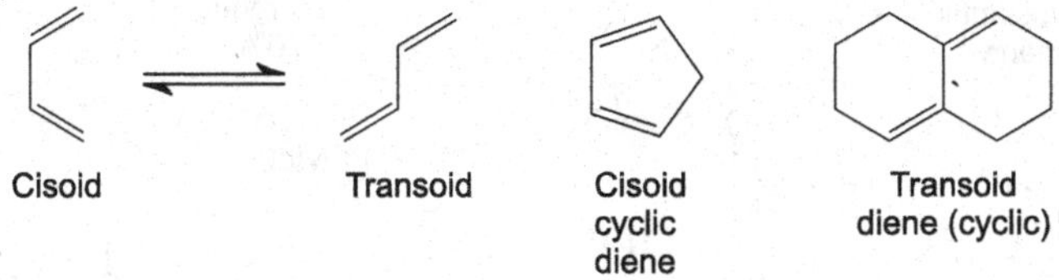

Scheme 2.167 In Diels–Alder reaction the cyclic dienes must be in cisoid form

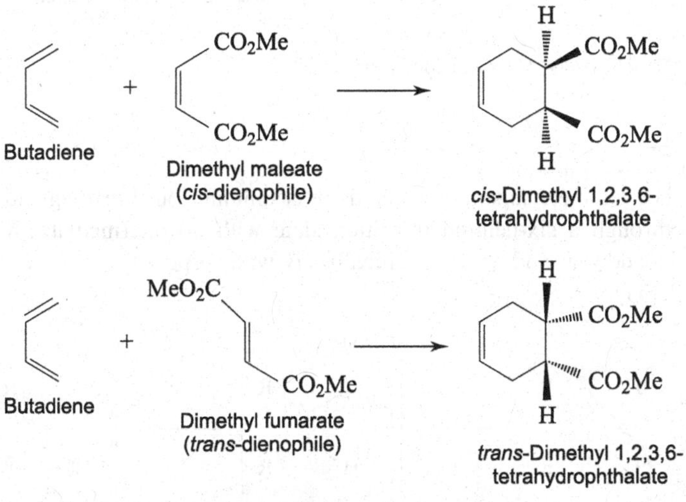

Scheme 2.168 Some examples of Diels–Alder reaction

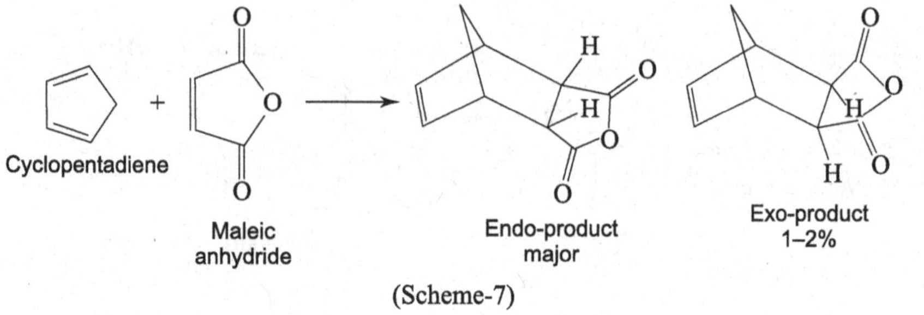

(Scheme-7)

Scheme 2.169 In Diels–Alder reaction thermodynamically endo-products predominate

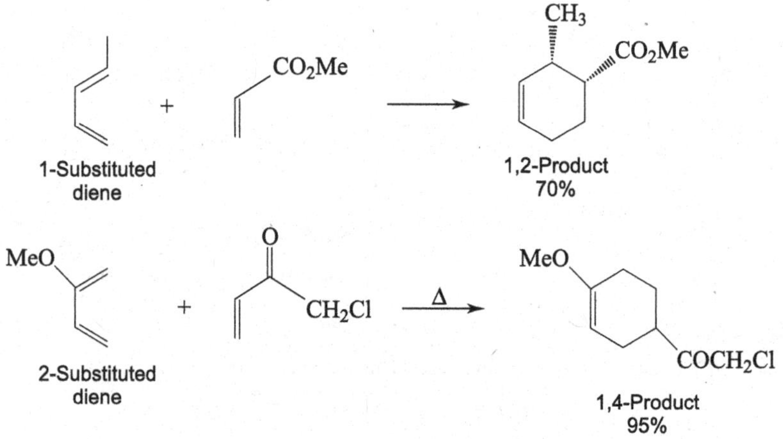

Scheme 2.170 The Diels–Alder reaction is regiospecific

Mechanism

It has a concerted mechanism, in which there is simultaneous breaking and making of bonds through a six-centred transition state with no intermediate. It is a (4 + 2)π cycloaddition and stereospecifically *cis* with respect to both the reactants (Scheme 2.171).

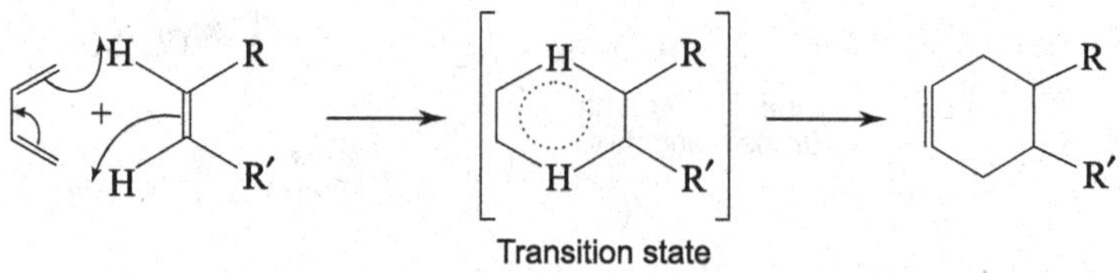

Transition state

Scheme 2.171 Mechanism of Diels–Alder Reaction

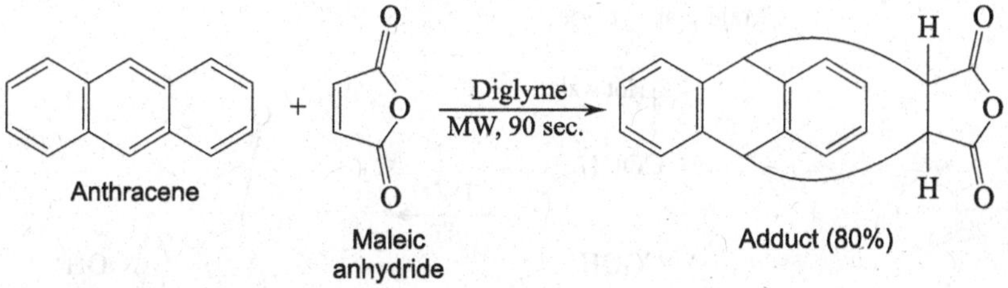

Scheme 2.172 Diels–Alder reaction under microwave irradiation

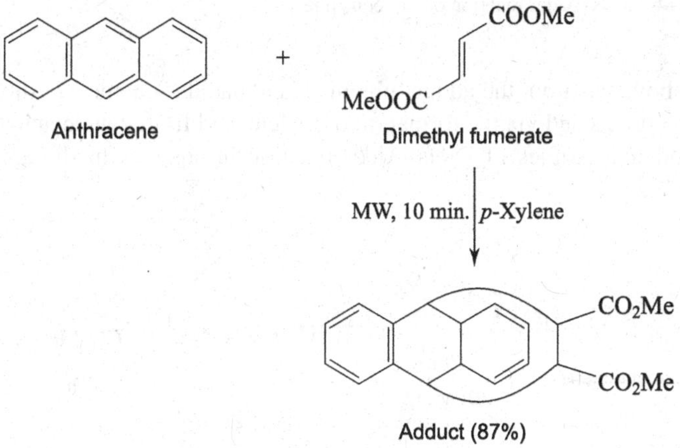

Scheme 2.173 Some other examples of Diels–Alder reaction under microwave irradiation

2.22.1 Diels–Alder Reactions Under Microwave Irradiation

Diels–Alder reactions can be conveniently carried out under microwave irradiation. Thus, anthracene and maleic anhydride in diglyme on irradiation under microwave [177, 178] yielded the adduct in 80% yield. The reaction is complete in 90 s compared to 90 min refluxing under usual conditions [179] (Scheme 2.172).

In a similar way, anthracene reacts with dimethyl fumarate within 10 min in p-xylene to afford [180] the adduct in 87% yield (Scheme 2.173), whereas conventional heating conditions gave only 67% yield in 5 h.

2.22.2 Diels–Alder Reactions in Aqueous Phase

The Diels–Alder reaction in aqueous media was first carried out in the beginning of nineteenth century [181]. Thus, furan reacted with maleic anhydride in hot water to give the adduct (Scheme 2.174).

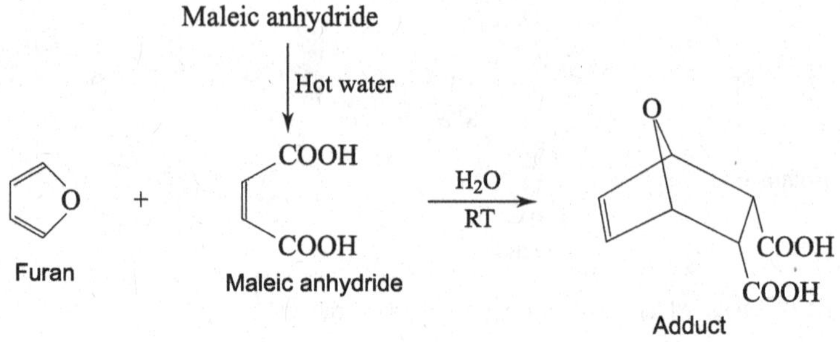

Maleic anhydride

Scheme 2.174 Diels–Alder reaction on aqueous phase

In the above reaction, the adduct obtained is a diacid (Scheme 12) showing that the reaction proceeded via the formation of maleic acid from maleic anhydride.

Some other examples of Diels–Alder reaction in aqueous media are given in Scheme 2.175.

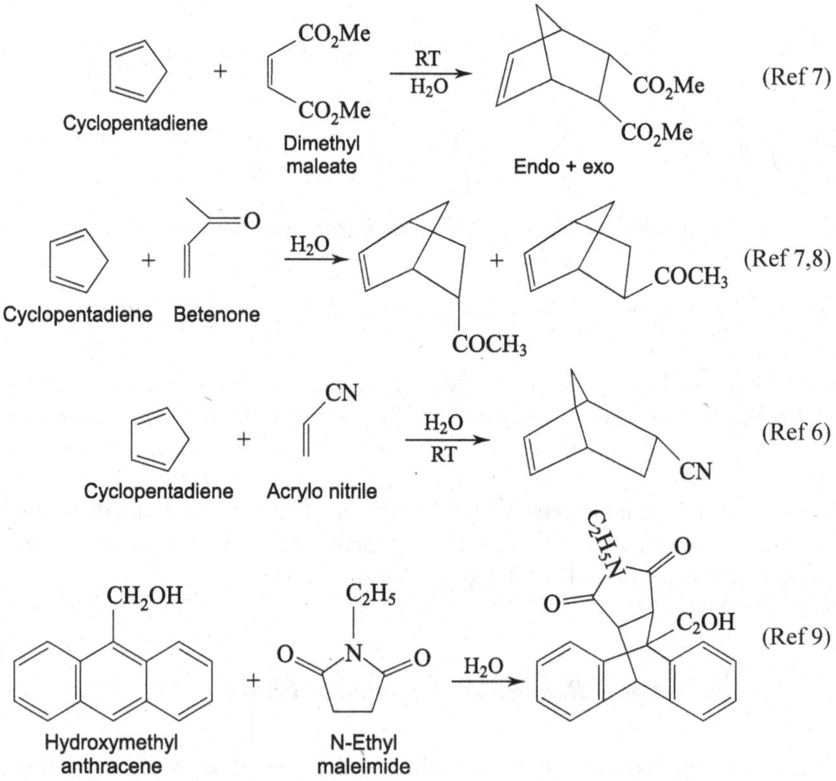

Scheme 2.175 Some examples of Diels–Alder reaction in aqueous phase

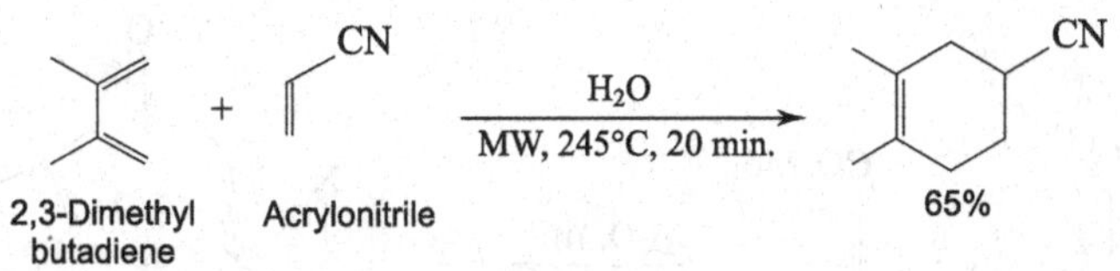

Scheme 2.176 Diels–Alder reaction in high temperature and supercritical N_2O

In the last example of the reaction of hydroxymethyl anthracene and N-ethylmaleimide, the β-cyclodextrin became an inhibitor. A slight deactivation was also observed with a salting in salt solution, such as guanidine chloride in aqueous solution.

2.22.3 Diels–Alder Reactions in High Temperature Water and Supercritical Water

A number of different dienes/dienophile combinations have been used for Diels–Alder reactions in high temperature [185] water and in supercritical water [185].

It should be noted [185] that water near or above its critical point (374 °C, 218 atms.) offers environmental advantages. Water near its critical point possesses very different properties from those of ambient liquid water, and the number and persistence of hydrogen bonds are both diminished. So, high temperature water behaves like many organic solvents in that organic compounds have high solubility in near critical water and completely miscible with supercritical water (SCW).

Thus, the reaction of 2,3-dimethyl butadiene and acrylonitrile by heating with water in a MW oven for 20 min (using MW generated NCW) gave the adduct [186] (Scheme 2.176).

2.22.4 Diels–Alder Reaction Under Sonication

Sonication facilitates Diels–Alder reactions. Thus, the addition of dimethyl acetylene dicarboxylate to furan in water at 22–45 °C gives quantitative yield of the product on sonication [187] (Scheme 2.177).

The Diels–Alder cycloaddition reaction of various dienes, mostly belonging to 1-vinyl cyclohexenes, with o-quinone proceeds very well giving the expected products with 59% yield [188] (Scheme 2.178). The yields are significantly higher and the regioselectivity increases. Better results are obtained by irradiation of the neat mixture of reagents.

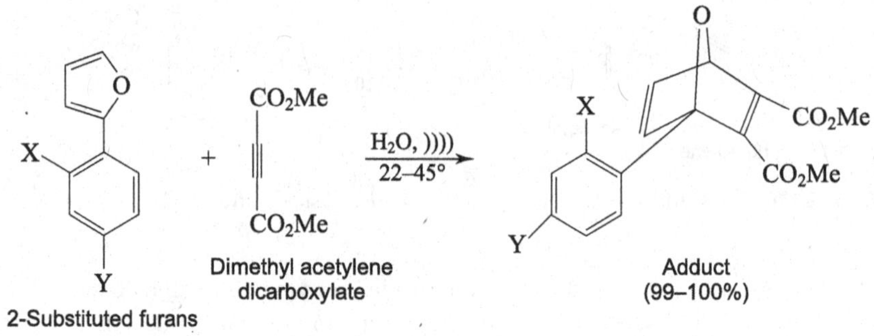

Scheme 2.177 Diels–Alder reaction under sonication

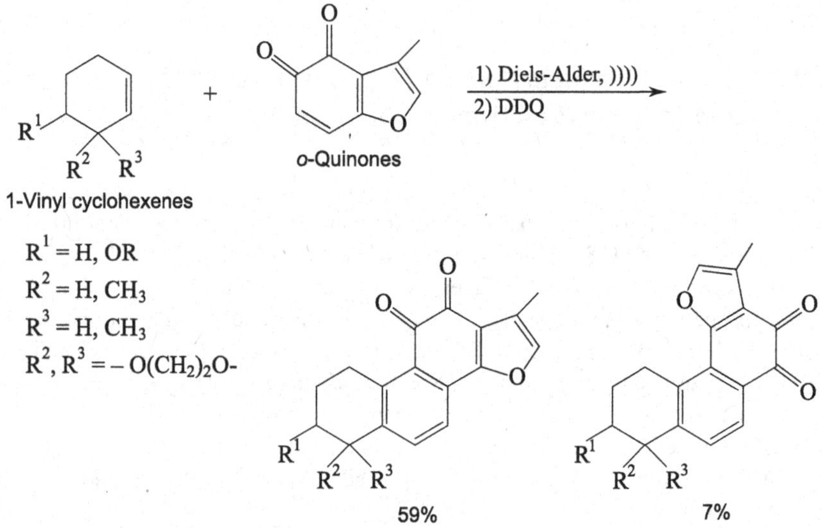

$R^1 = H, OR$
$R^2 = H, CH_3$
$R^3 = H, CH_3$
$R^2, R^3 = - O(CH_2)_2O-$

Scheme 2.178 Diels–Alder cycloaddition reaction

2.22.5   Diels–Alder Reaction Using Ionic Liquids

Use of ionic liquids, such as $[b_{min}][BF_4]$, $[b_{min}][ClO_4]$, $[e_{min}][CF_3SO_3]$ and $[e_{min}][PF_6]$ for Diels–Alder reaction between cyclopentadiene and methyl acrylate results in rate enhancement, high yields and strong endoselectivities (Scheme 2.179) [189, 190].

2.22.6   Diels–Alder Reaction in Supercritical Carbon Dioxide

Following are some of the Diels–Alder reactions which have been conducted [191]:

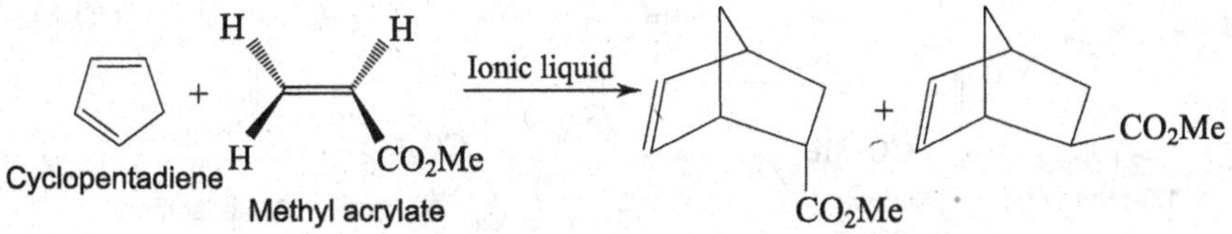

Scheme 2.179 Diels–Alder reaction using ionic liquids

(a) Diels–Alder reaction of isoprene and methylacrylate to produce para and meta isomers (Scheme 2.180).

Following table gives the ratio of para and meta isomers along with the total yield obtained in conventional solvent (toluene) and supercritical CO_2.

Condition	Yield (%)[a]	Ratio[b]	1:2
$PhCH_3$, 145 °C, 15 h	78	71:29	(71:29)
$PhCH_3$, 50 °C, 3d	(7)	69:31	(72:28)
CO_2, 49.5 bar, 50 °C, 3d	(11)	69:31	(73:27)
CO_2, 74.5 bar, 50 °C, 3d	(5)	67:33	(73:27)
CO_2, 95.2 bar, 50 °C, 7d	(4)	71:29	(73:27)
CO_2, 117 bar, 50 °C, 3d	(3)	70:30	(72:28)

[a]Isolated yield (estimated by ^{1}HNMR)
[b]Ratio of isomers determined by ^{1}HNMR and GC analyse data taken from reference 15

(b) Diels–Alder reaction of 2-t-butyl-1,3-butadine and methyl acrylate to produce [191] para and meta products (Scheme 2.181).

Following table gives the ratio of para (1) and meta (2) isomers along with the total yield obtained in conventional procedure and supercritical –CO_2.

Condition	Yield (%)[a]	Ratio[b]	1:2
neat, 185 °C, 16 h	78	63:37	(63:37)
$PhCH_3$, 50 °C, 3d	(19)	69:31	(68:32)
CO_2, 87 bar, 50 °C, 3d	(5)	71:29	(68:32)
CO_2, 117 bar, 50 °C, 3d	(4)	69:31	(69:31)
CO_2, 117 bar, 150 °C, 24 h	54	65:35	(64:36)

[a]Isolated yield (estimated by ^{1}HNMR)
[b]Ratio of isomers determined by ^{1}HNMR and GC analyse data taken from ref. 15

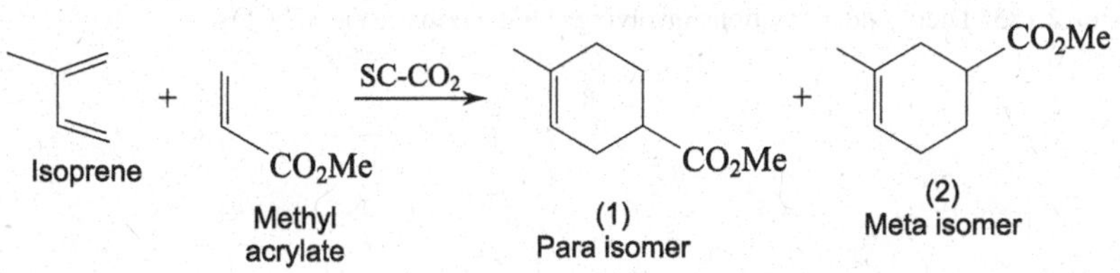

Scheme 2.180 Diels–Alder reaction in SC CO_2

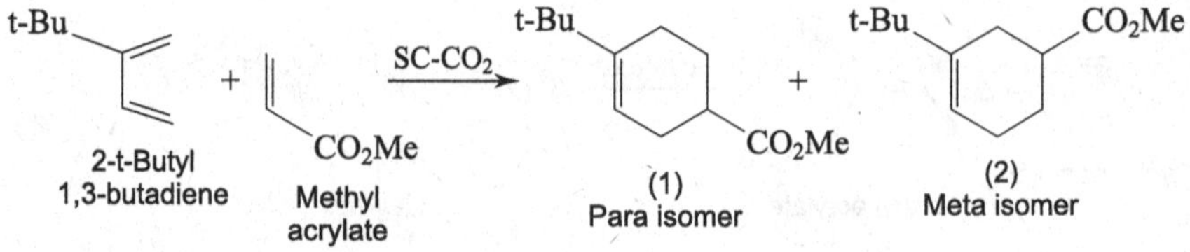

Scheme 2.181 Diels–Alder reaction in SC-CO$_2$

(c) Some of the most important results have been observed in Diels–Alder reaction conducted in SC-CO$_2$ in the presence of Lewis acid catalyst, such as scandiumtrifluoromethane sulfonate, and optimum selectivity (in this case, endo:exo ratio) was usually observed near the critical point of the reaction mixture [192].

The SC(OT$_F$)$_3$-catalysed Diels–Alder reaction of n-butylacrylate with cyclopentadiene in SC-CO$_2$ is represented in Scheme 2.182.

The catalyst is commercially available and is relatively inexpensive.

(d) Diels–Alder reactions involving chiral auxiliaries have also been reported in SC-CO$_2$. Thus, a selection of rare earth triflates were shown to catalyse the reaction of cyclopentadiene with a chiral dinophile (Scheme 2.183). The reaction proceeded rapidly in SC-CO$_2$ to give endo adduct with higher diastereoselectivity than obtained in dichloromethane [193].

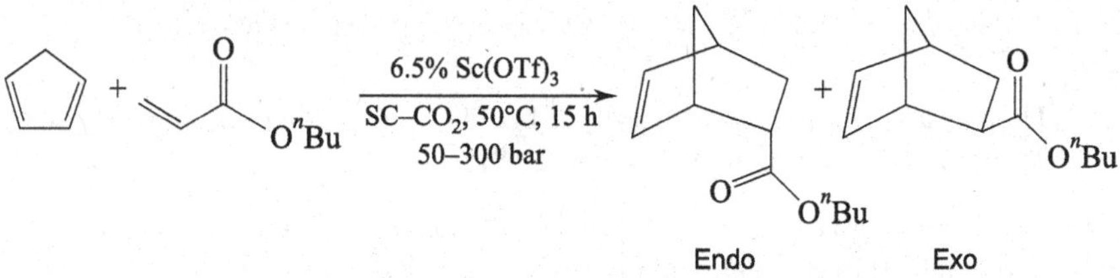

Scheme 2.182 SC(OT$_F$)$_3$-catalysed Diels–Alder reaction

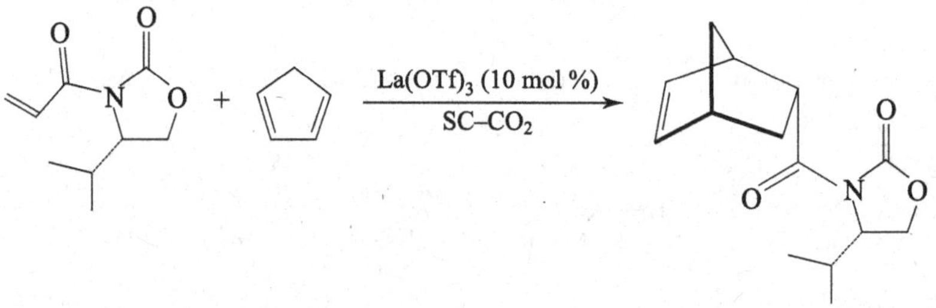

Scheme 2.183 Diels–Alder reactions involving chiral auxiliaries in SC-CO$_2$

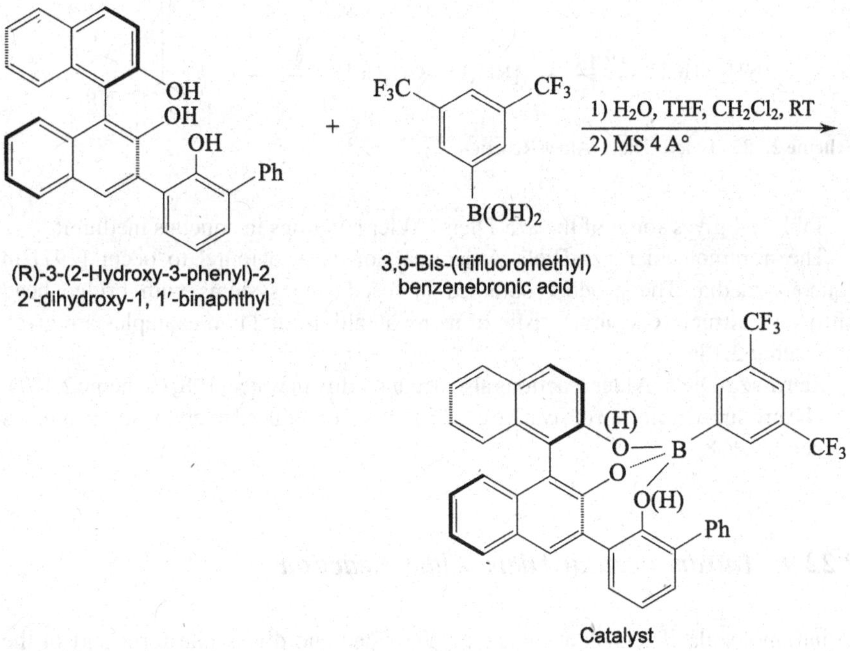

Scheme 2.184 Asymmetric Diels–Alder reaction in H_2O

2.22.7 Asymmetric Diels–Alder Reactions in Water

Catalytic asymmetric Diels–Alder reactions have been conducted in water. The catalyst used is a Brönsted acid-assisted chiral (BLA) Lewis acid prepared (Scheme 2.184) from (R)-3-(2-hydroxy-3-phenyl)-2,2′-dihydroxy-1,1′-binaphthyl and 3,5-bis (trifluoromethyl)-benzeneboronic acid, which is effective in enantioselective Diels–Alder reaction between both α,β-enals and various dienes [194].

Use of the catalyst (prepared as given in Scheme 2.184) in the Diels–Alder reaction of methacrolein and cyclopentadiene give the adduct with 99% ee.

2.22.8 Hetero-Diels–Alder Reactions

Hetero-Diels–Alder reactions with nitrogen or oxygen containing dienophiles are of special interest for the synthesis [195] of heterocyclic compounds. The earliest example of hetero-Diels–Alder reaction with nitrogen containing dienophiles in aqueous medium was reported [196] in 1985. In this procedure, simple iminium salts, generated in situ under Mannich-like conditions, reacted with dienes in water to give aza-Diels–Alder reaction products (Scheme 2.185). This procedure has the potential for the synthesis of alkaloids.

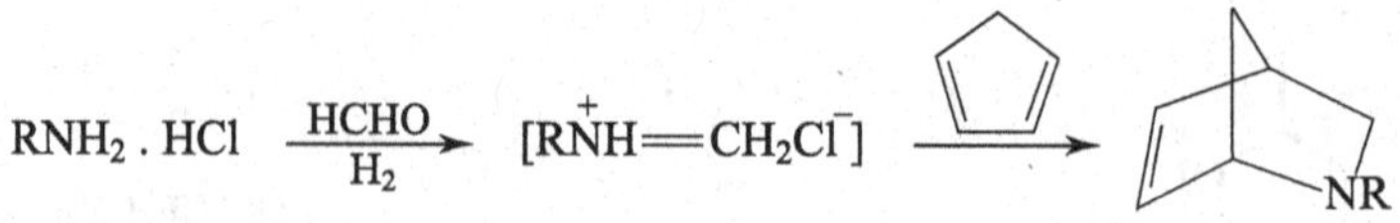

$$RNH_2 \cdot HCl \xrightarrow[H_2]{HCHO} [R\overset{+}{N}H={=}CH_2Cl^-] \longrightarrow$$

Scheme 2.185 Hetero-Diels–Alder Reaction

Table 2.4 gives some of the aza-Diels–Alder reactions in aqueous medium:

The intramolecular aza-Diels–Alder reaction is also found to occur [197] in aqueous media. The product obtained is fused ring systems with bridge head nitrogen, a structure, characteristic of many alkaloids, and two examples are given in Scheme 2.186.

Retro-aza-Diels–Alder reactions also occur readily in water [198] (Scheme 2.187).

The produced iminium derivative can be trapped or reduced to give primary amines (Scheme 2.188).

2.22.9   Intramolecular Diels–Alder Reaction

In intramolecular Diels–Alder reaction, the diene and dienophile form part of the same molecule. It has been widely used in the synthesis of natural products in the alkaloid, steroid and terpenoid series [199–202]. A typical example in the intramolecular Diels–Alder reaction is of *trans*-dienyl-acrylic ester to give the *cis*-hydrindane (Scheme 2.189).

A valuable feature of the intramolecular Diels–Alder reactions is their stereoselectivity. Thus, the triene (A) forms the indane derivative (C) in a reaction in which four new chiral centres are set up selectively in one step by way of sterically preferred transition state [203–205] (Scheme 2.190).

There is no record of intramolecular Diels–Alder reaction conducted in aqueous medium. However, for hetero-Diels–Alder reactions with an oxygen containing dienophile, cyclopentadiene or cyclohexadiene reacted with an aqueous solution of glyoxylic acid to give α-hydroxyl-γ-lactones arising from the rearrangement of the cycloadducts (Scheme 2.191).

The 5,5-fused systems generated have been used in the total synthesis of several bioactive compounds.

As in the case of aza-Diels–Alder reactions, the retro Diels–Alder reaction are also known. The Diels–Alder reactions are reversible and on heating many adducts dissociate into the starting components under mild conditions. This has been made use of for the separation of anthracene derivatives from mixtures with other hydrocarbons through their adducts with maleic anhydride and also in the preparation of pure vitamin D from the mixtures obtained by irradiation of the provitamins.

Table 2.4 Aza-Diels–Alder reactions in aqueous medium

Diene	Amine + Carbonyl comound	Product	Yield %
	BnNH$_2$. HCl + HCHO	NBn	41
	BnNH$_2$. HCl + HCHO	NBn	69
	BnNH$_2$. HCl + HCHO	NBu	59
	BnNH$_2$. HCl + HCHO	NBu	62
	BnNH$_2$. HCl + HCHO	NBu	49
	MeNH$_2$. HCl + HCHO	NMe	82
	NH$_4$Cl + HCHO	NH . HCl	44
	NH$_4$Cl + HCHO	NH . HCl	40
	BnNH$_2$. HCl + MeCHO	NBn	47

2.23 Fischer-Indole Synthesis

It is one of the most important methods for synthesizing indole derivative. It consists
[206] of heating arylhydrazones of aldehydes or ketones with catalysts, such as
polyphosphoric acid (PPA), zinc chloride or boron trifluoride (Scheme 2.192).

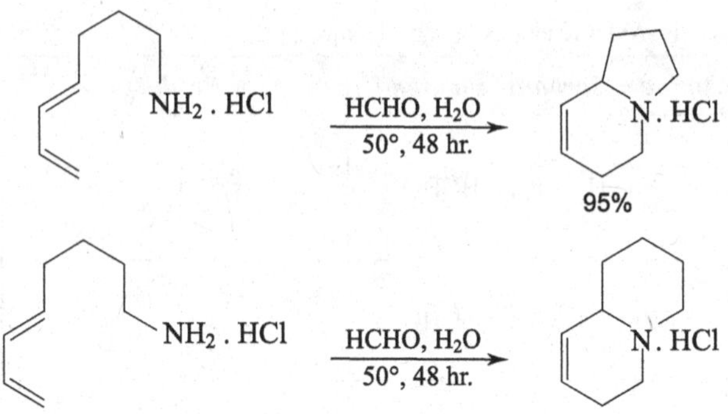

Scheme 2.186 Intramolecular aza-Diels–Alder reaction in H₂O

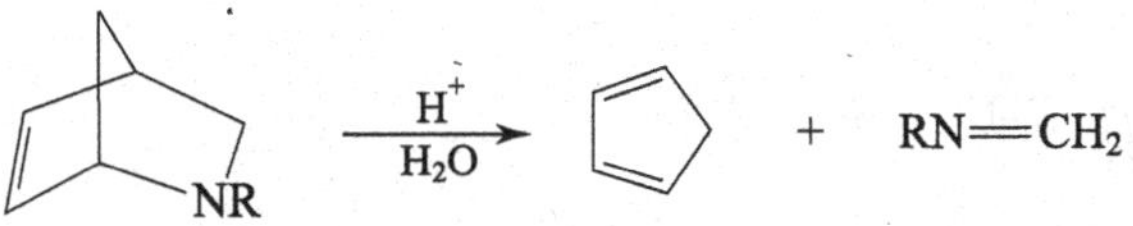

Scheme 2.187 Retro-aza-Diels–Alder reaction in water

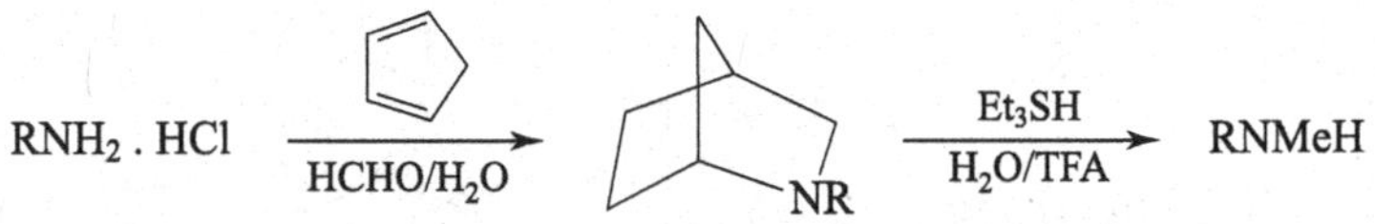

Scheme 2.188 Synthesis of primary amines

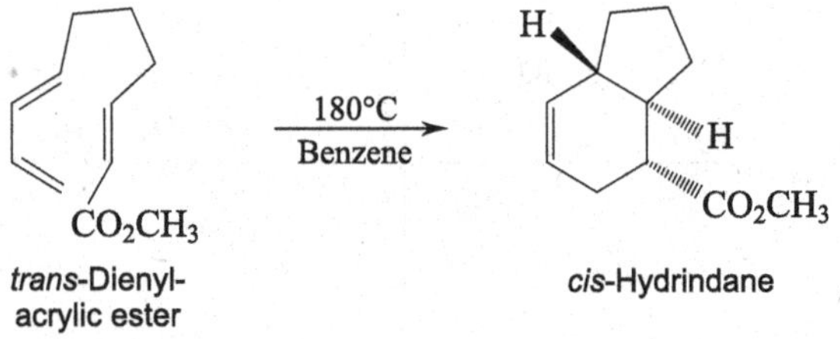

Scheme 2.189 Intramolecular Diels–Alder reaction

The mechanism of the reaction is given in Scheme 2.193.

The above mechanism is confirmed by using ^{15}N-labelled precursors (Scheme 2.194).

Indole itself cannot be prepared by the cyclization of acetaldehyde phenyl hydrazene. It is conveniently prepared by the decarboxylation of indole-2-carboxylic acid (by heating with copper chromite in quinoline solution), which is prepared by cyclizing pyruvic acid phenylhydrazone (Scheme 2.195).

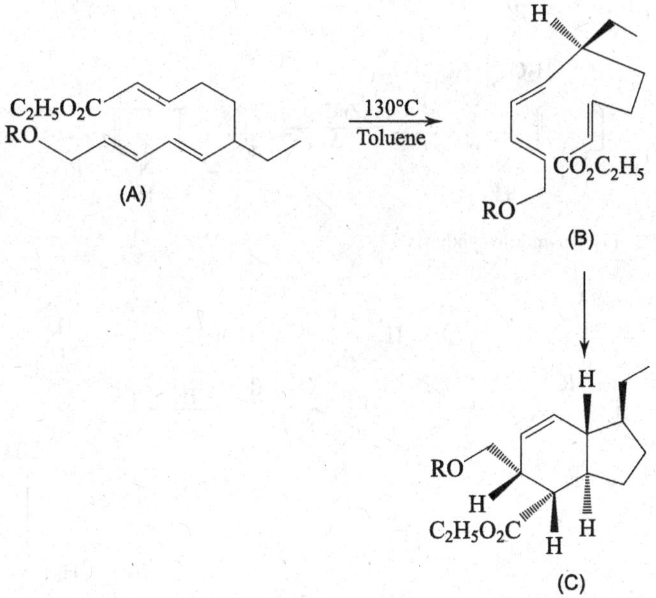

Scheme 2.190 Intramolecular Diels–Alder-Reaction

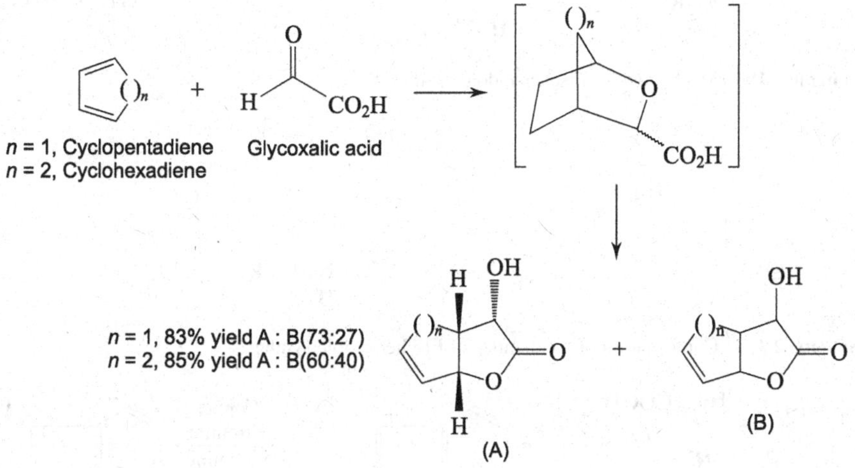

n = 1, 83% yield A : B(73:27)
n = 2, 85% yield A : B(60:40)

Scheme 2.191 Hetero Diels–Alder reaction in water

The last step of decarboxylation can be conveniently affected by microwave irradiation by using quinoline as the solvent [207].

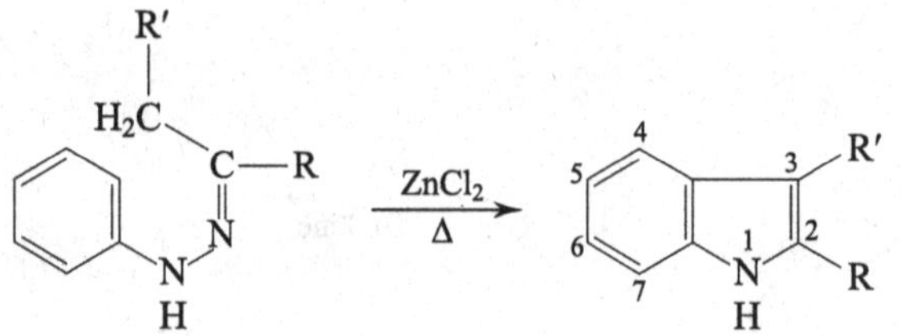

Scheme 2.192 Fischer-indole synthesis

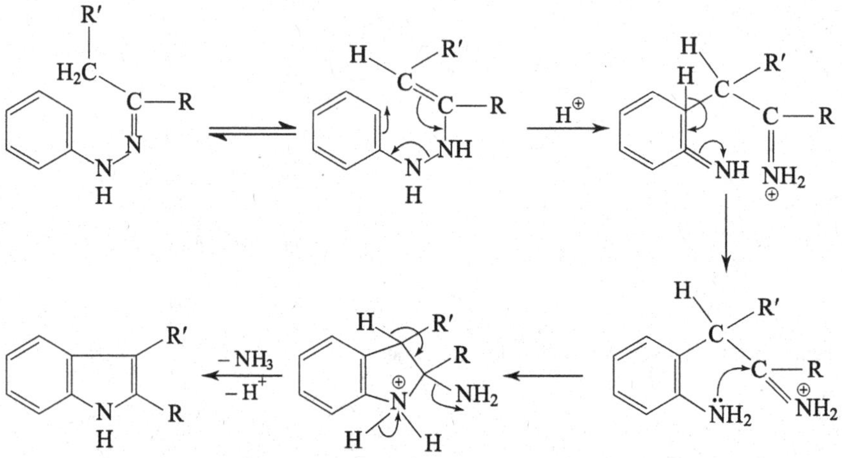

Scheme 2.193 Mechanism of Fischer-indole synthesis

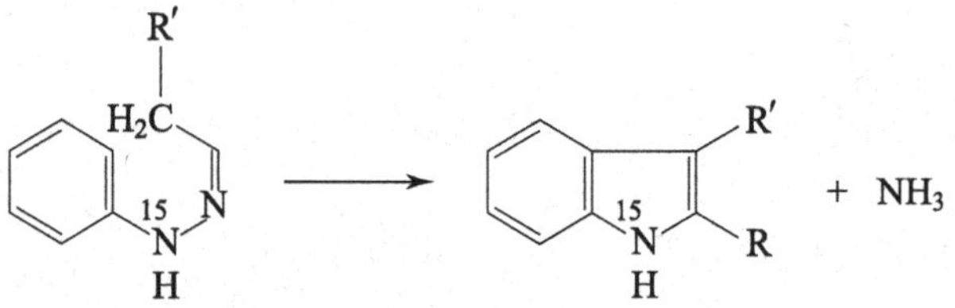

Scheme 2.194 Confirmation of mechanism of Fischer-indole synthesis

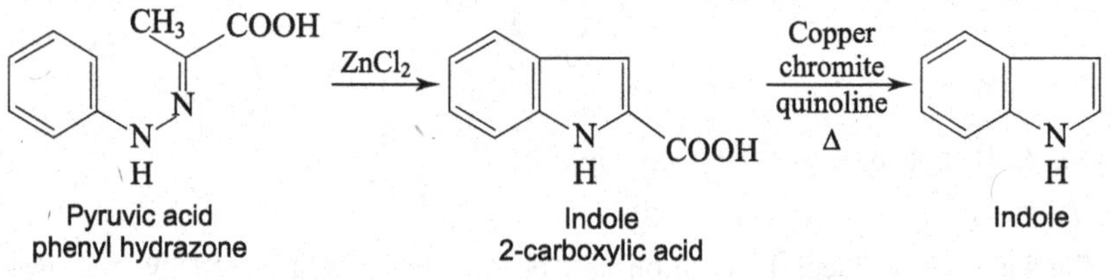

Scheme 2.195 Synthesis of indole

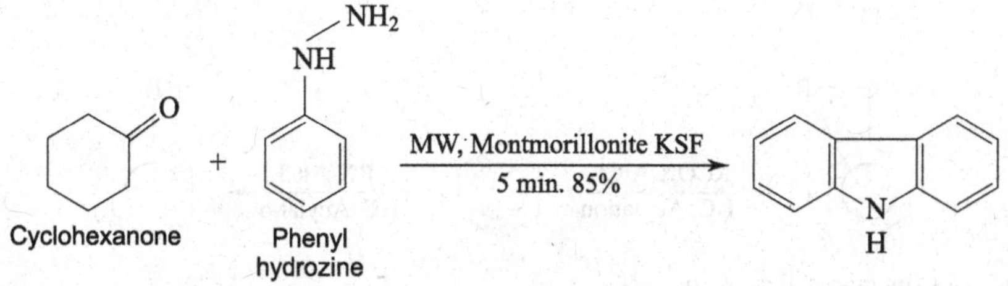

Scheme 2.196 Fischer-indole synthesis under dry conditions

2.23.1 Fischer-Indole Synthesis in Dry Conditions

Fischer-indole synthesis has been carried out [208] using cyclohexanone and phenyl hydrazine in the presence of montmorillonite KSF clay using microwave irradiation (Scheme 2.196).

2.23.2 Fischer-Indole Synthesis in Water

The Fischer-indole synthesis of 2,3-dimethylindole has been accomplished with 67% yield in water from phenylhydrazine and ethylmethyl ketone at 220 °C within 30 min, [209] thus circumventing the use of preformed hydrazone or acid catalyst (Scheme 2.197).

2.24 Friedel–Crafts Reaction

Friedel–Crafts alkylation and acylation are two examples of the well-known Friedel–Crafts reaction [210] (Scheme 2.198).

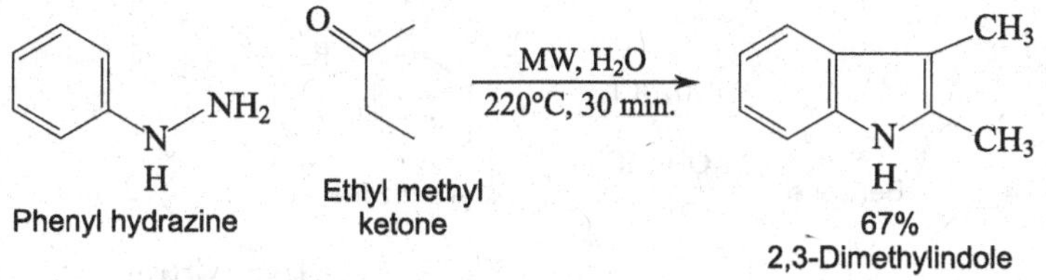

Scheme 2.197 Fischer-indole synthesis in water

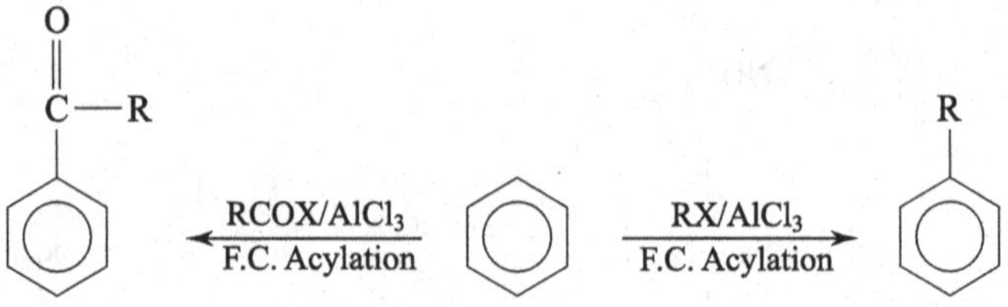

Scheme 2.198 Friedel–Crafts Reaction

2.24.1 Friedel–Crafts Alkylation

This reaction is of great synthetic utility for carbon–carbon bond formation. It is used for introducing alkyl group in the benzene ring. Alkylation can be affected by reacting benzene with alkyl halide, alcohol or olefins in the presence of anhydrous aluminium chloride. The reactions are usually performed in carbon disulphide or nitrobenzene (Scheme 2.199).

In Friedel–Crafts alkylation, the function of $AlCl_3$ (or any other Lewis acid like BF_3, $AlBr_3$, $GaCl_3$ etc.) is to provide an electrophilic agent which attacks the aromatic ring. Further steps of the mechanism are shown ahead (Scheme 2.200).

Similar carbocation in obtained in case of Friedel–Crafts reaction using alcohol or alkene (Scheme 2.201).

It is of interest to note that in Friedel–Crafts reaction, aryl halides cannot be used in place of alkyl halides. Also, nitrobenzene cannot be alkylated due to the deactivating influence of nitrogroup in the ring. In fact, nitrobenzene, as already stated is used as solvent in many Friedel–Crafts reactions. Also, aniline and other aromatic amines do not undergo Friedel–Crafts reaction due to the formation of a complex between NH_2 group and Lewis acid. Since this complex has a positive charge on nitrogen, it deactivates the ring (Scheme 2.202).

Another limitation of Friedel–Crafts alkylation is isomerization. Thus, use of propyl alcohol in the alkylation of benzene gives isopropylbenzene and not n-propyl benzene. This is because the primary carbocation ($CH_3CH_2\overset{\oplus}{C}H_2$) gets converted into more stable secondary carbocation [$(CH_3)_2\overset{\oplus}{C}H$].

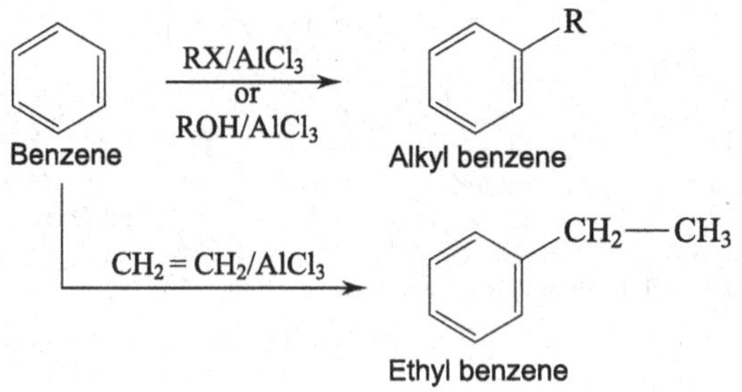

Scheme 2.199 Friedel–Crafts alkylation

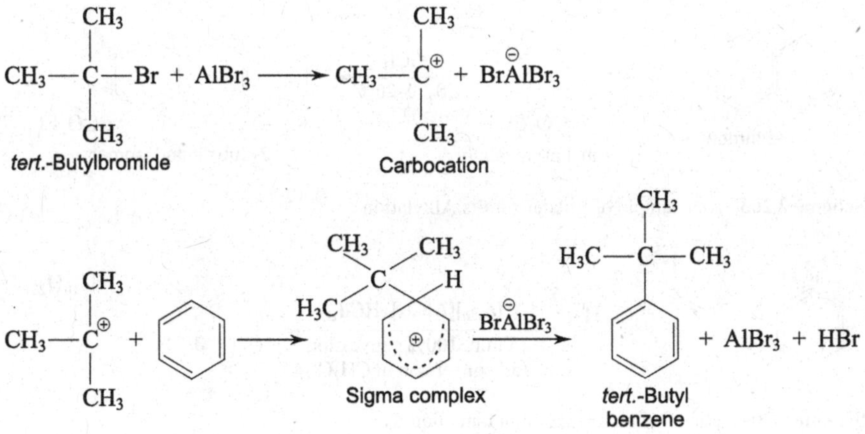

Scheme 2.200 Mechanism of Friedel–Crafts alkylation

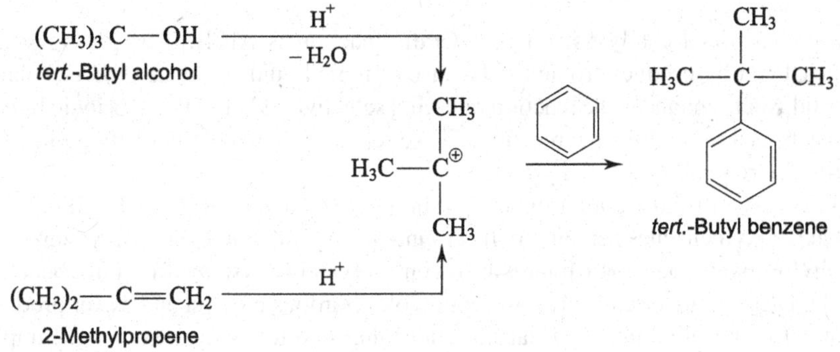

Scheme 2.201 Mechanism of Friedel–Crafts alkylation

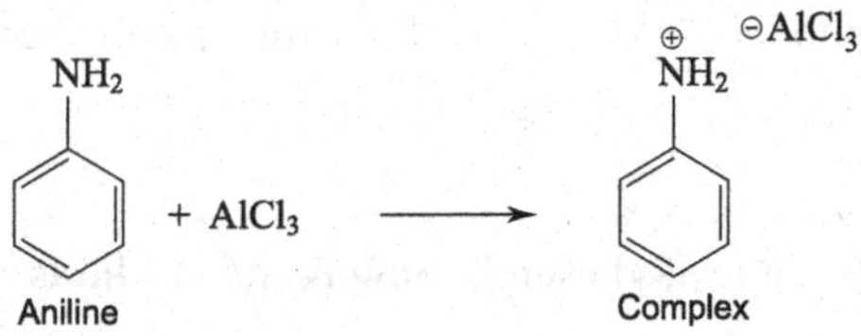

Scheme 2.202 Some observations in Friedel–Crafts alkylation

Some applications of Friedel–Crafts alkylation are given in Scheme 2.203.

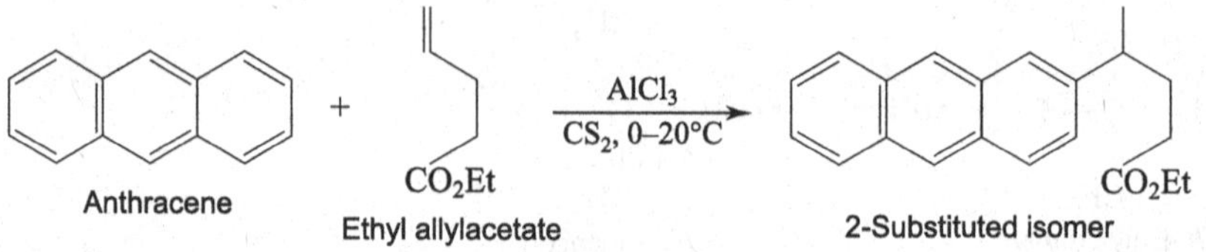

Scheme 2.203 An example of Friedel–Crafts Alkylation

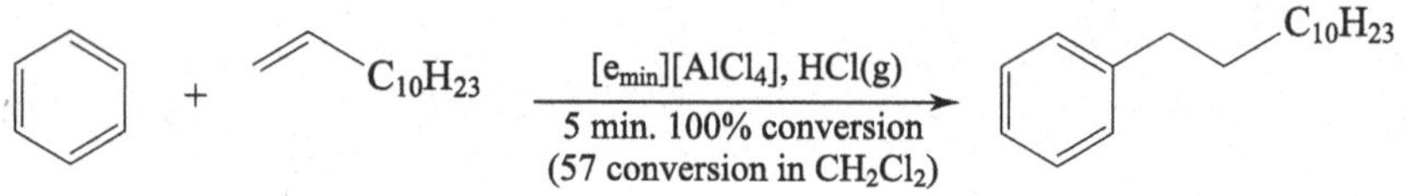

Scheme 2.204 Friedel–Crafts reaction in ionic liquids

2.24.1.1 Friedel–Crafts Alkylation in Ionic Liquids

The conventional catalyst in Friedel–Crafts reaction is $AlCl_3$. This gives rise to disposal and by-product problems. The use of ionic liquid [e_{min}] Cl–$AlCl_3$ in place of solid $AlCl_3$ enhances the reaction rates and selectivity [211–217], this ionic liquid also acts as a solvent for the reaction. These reactions worked efficiently giving the stereoelectronically favoured products.

Friedel–Crafts alkylation is a key step in the manufacture of about 3 million tons of linear alkylbenzenes per annum. It is found that production of these alkylbenzenes using chloroaluminate ionic liquids as solvent and catalyst system offers huge benefits [218]. These include reduced consumption of hazardous catalyst and easier product separation and elimination of caustic quenching due to catalyst leaching. Simple examples of Friedel–Crafts alkylation are shown in Scheme 2.204.

Friedel–Crafts alkylation of aromatic compounds with alkenes using Sc(OTf)$_3$-ionic liquid system give the benefits of simple procedure, easy recovery and reuse of catalyst, contributing to the development of environmentally benign and waste-free process [219].

2.24.1.2 Friedel–Crafts Alkylation in Supercritical Fluids

Friedel–Crafts alkylation of aromatics has been performed in supercritical fluids (SCF) media [220]. Thus, alkylation of mesitylene in SC-propane ($T_C = 91.9$ °C, $P_C = 46.0$ bar) acted both as solvent and alkylating agent (Scheme 2.205). A mixture of three products, viz., mono-, di- and trialkylated products was obtained in 25%, 6%, minor, respectively. The selectivity is much improved if SC–CO_2 is used as the reaction medium with propan-2-ol as alkylating agent. At a molar ratio of mesitylene to propane-2-ol 2:1, a pressure of 200 bar, catalyst temperature of 250 °C and a flow

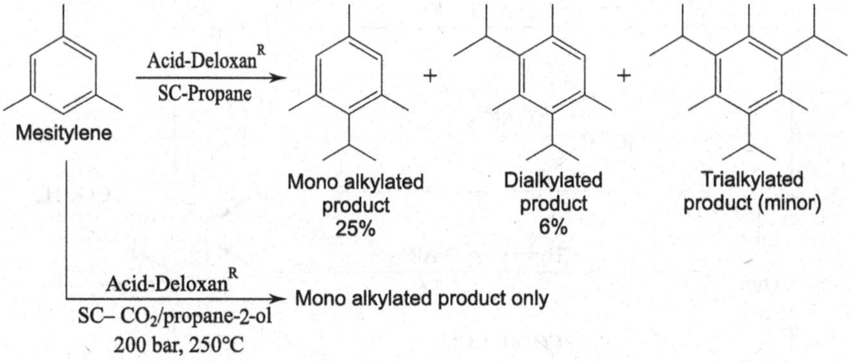

Scheme 2.205 Friedel–Crafts alkylation in SC fluids

rate of 0.60 g/min, the mono-alkylated product was obtained as the only product with a conversion of 42%.

2.24.1.3 Friedel–Crafts Alkylation Using Microencapsulated Lewis Acid

Microencapsulated Lewis acid has replaced traditional corrosive monomeric Lewis acid in Friedel–Crafts alkylation [221]. One such example is given in Scheme 2.206.

Friedel–Crafts alkylation reactions have also been accomplished [222, 223] in high temperature water.

2.24.2 Friedel–Crafts Acylation

Like Friedel–Crafts alkylation, this reaction also has wide scope and is used for introducing a keto group in the benzene ring. The method consists in treating benzene with an acid chloride in the presence of anhydrous aluminium chloride. In place of acid chlorides, acid anhydrides, ketones and carboxylic acids can also be used (Scheme 2.207).

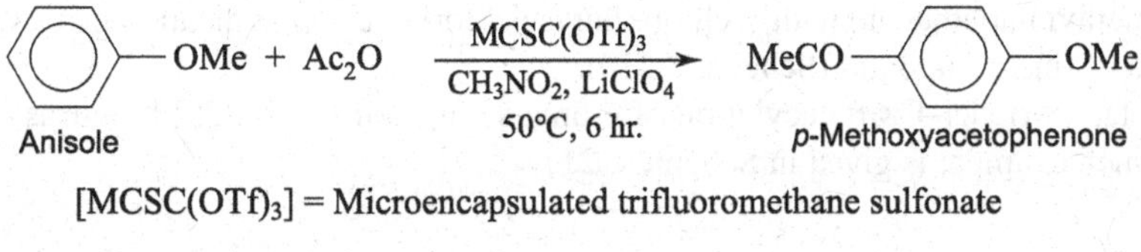

$[MCSC(OTf)_3]$ = Microencapsulated trifluoromethane sulfonate

Scheme 2.206 Friedel–Crafts reaction using microencapsulated Lewis acids

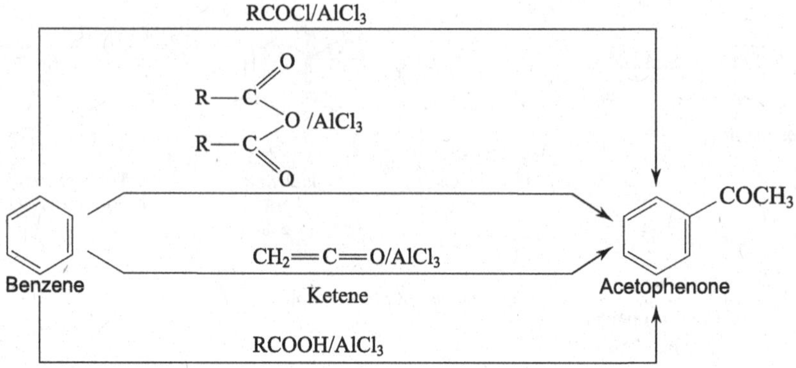

Scheme 2.207 Some applications of Friedel–Crafts acylation

$$CH_3COCl + AlCl_3 \longrightarrow CH_3\overset{\oplus}{C}O \cdots \overset{\ominus}{A}lCl_4$$

$$(CH_3CO)_2O + AlCl_3 \longrightarrow CH_3\overset{\oplus}{C}O \cdots \overset{\ominus}{A}lCl_4 + CH_3COOAlCl_2$$

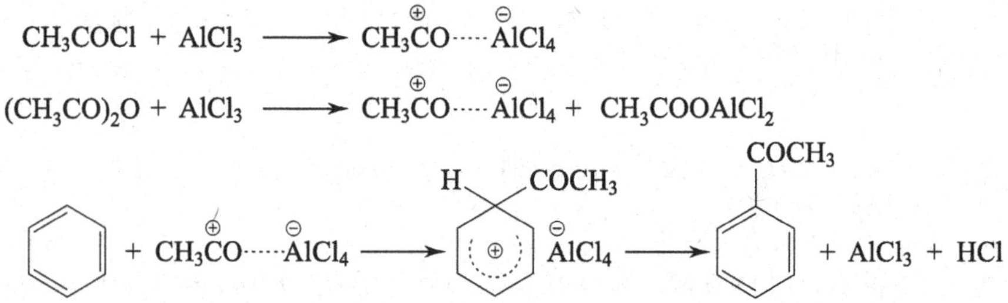

Scheme 2.208 Mechanism of Friedel–crafts acylations

In the Friedel–Crafts acetylation, the acylating entity is an acylium ion produced by the reaction of acid halide or anhydride with a Lewis acid. Subsequent steps of the mechanism are shown in Scheme 2.208.

Unlike alkylations, acylation reactions are not accompanied by rearrangement within the acyl group.

Some of the applications of Friedel–Crafts acylation are given in Scheme 2.209.

2.24.2.1 Friedel–Crafts Acylation Under Sonication

An interesting application of Friedel–Crafts acylation in the reaction of 2,4,5,7 tetramethyl naphthalene with β-chloro butyrylchloride under sonication gives [224] an interesting compound (Scheme 2.210).

In fact, Friedel–Crafts acylation of aromatics is facilitated [225] by ultrasound. One such example is given in Scheme 2.211.

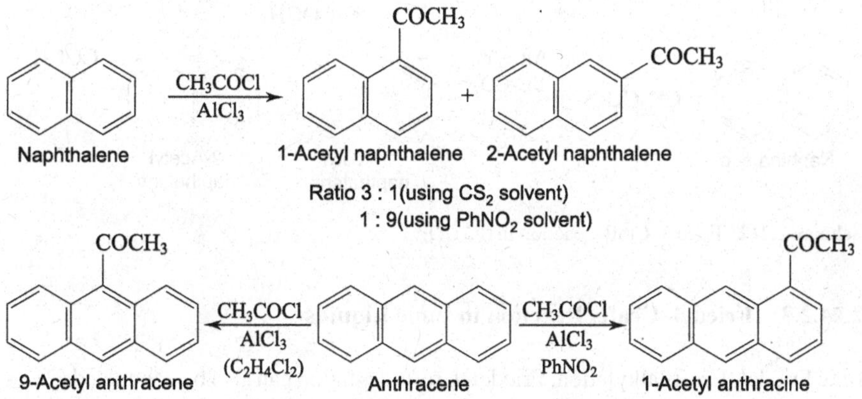

Scheme 2.209 Some examples of Friedel–Crafts application

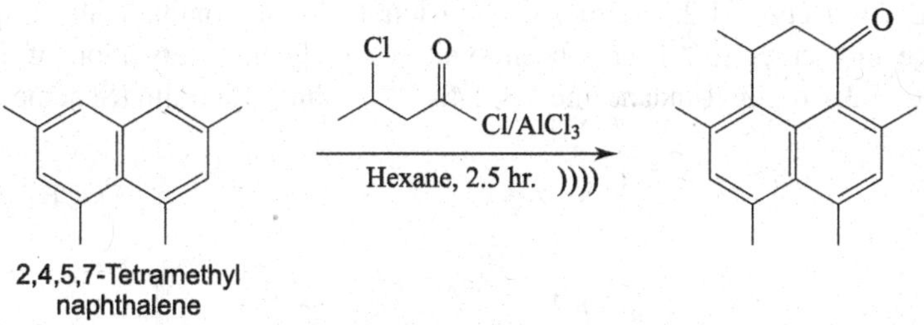

Scheme 2.210 Friedel–Crafts acylation under sonication

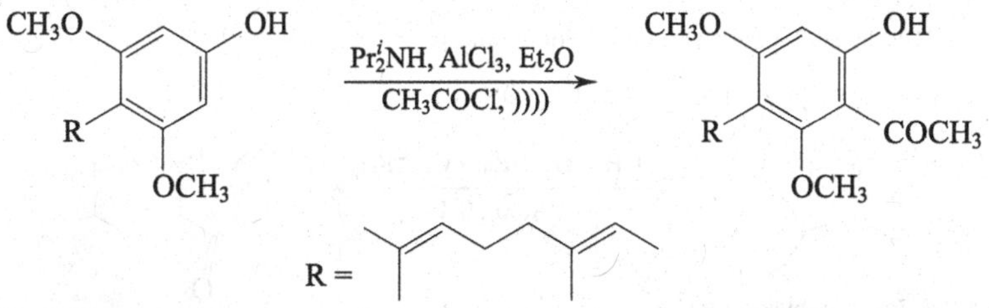

Scheme 2.211 Friedel–Crafts acylation under sonication continued

2.24.2.2 Friedel–Crafts Acylation in Supercritical Carbon Dioxide

Using Friedel–Crafts acylation, the following transformation has been achieved [226] in SC-CO$_2$ (Scheme 2.212).

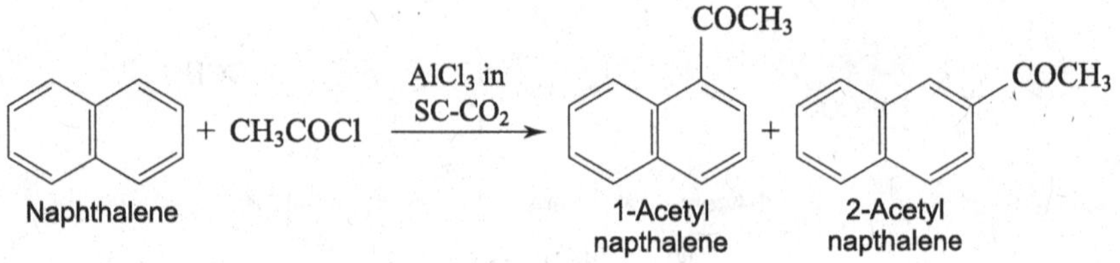

Scheme 2.212 Friedel–Crafts reaction in SC CO$_2$

2.24.2.3   Friedel–Crafts Reaction in Ionic Liquids

Like Friedel–Crafts alkylation, Friedel–Crafts acylation can also be conducted [211–217] in ionic liquid. Thus, in acetylation of naphthalene, the major product is the thermodynamically, unfavoured 1-isomer (Scheme 2.213).

Using ionic liquids, Todalid (5-acetyl-1,1,2,6-tetramethyl-3-isopropylindane) and traseolide (6-acetyl-1,1,2,4,4,7-hexamethyl tetralin, both commercially important fragrance molecules [228] have been synthesized by the acetylation of 1,1,2,6-tetramethyl-3-isopropyl-indane and 1,1,2,4,4,7-hexamethyltetralin (Scheme 2.214).

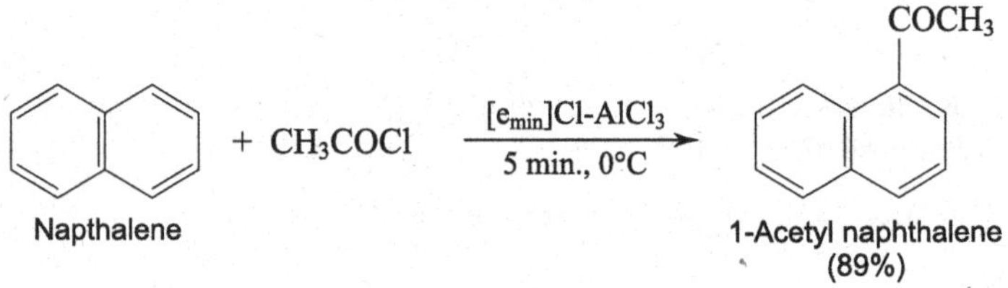

Scheme 2.213 Friedel–Crafts reaction in ionic liquids

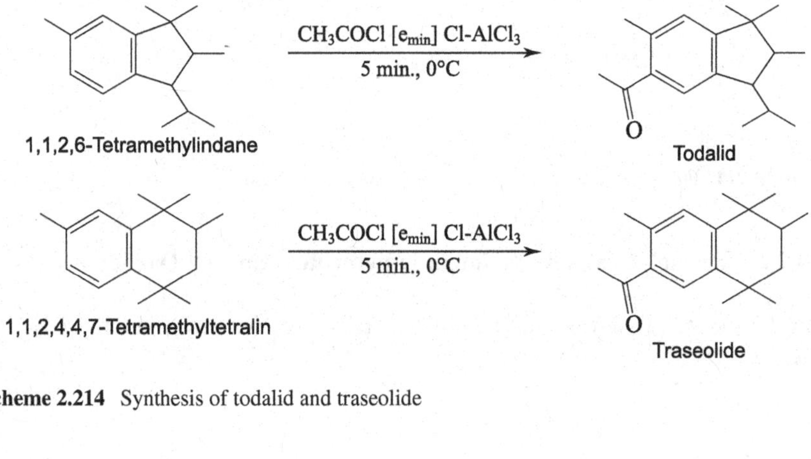

Scheme 2.214 Synthesis of todalid and traseolide

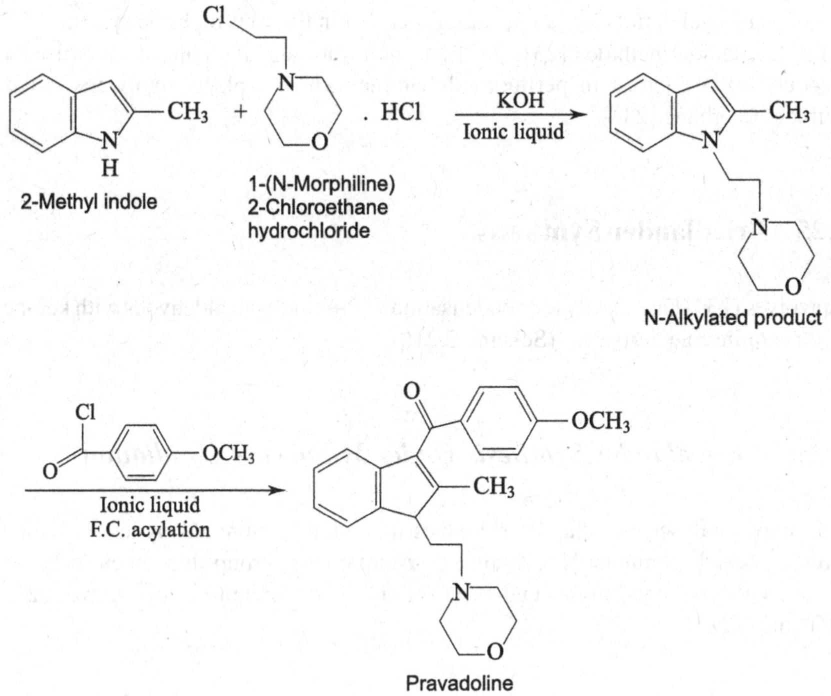

Scheme 2.215 Synthesis of pravadoline

Pravadoline, an important pharmaceutical, was synthesized using ionic liquid and a combination of Friedel–Crafts acylation reaction and a nucleophilic displacement reaction. Thus, the alkylation of 2-methyl indole with 1-(N-morpholine)-2-chloroethane in ionic liquid, 1-butyl-2,3-dimethylimidazolium hexafluorophosphate [b_{min}][PF_6] and KOH as base. The alkylation product obtained in 99% yield on Friedel–Crafts acylation in chloroaluminate (III) ionic liquid at 0 °C gave the required pravadoline (Scheme 2.215) [219].

2.24.2.4   Friedel–Crafts Acylation Using Fluorous Phase Technique

In fluorous phase technique, the organic substrate is dissolved in an organic solvent (e.g., toluene or dichloromethane) and the fluoro-tagged catalyst, or reagents dissolved in perfluoroalkenes (as FC-72, a mixture of perfluorohexanes) are added. At elevated temperatures, the biphasic heterogeneous mixture becomes monobasic, permitting the reaction to take place. After the reaction has taken place (as indicated by TLC), the resultant liquid is cooled when the two phases are again formed. The formed product of the reaction is in the organic phase and is recovered by decantation of the upper organic phase. The catalyst remains in the lower fluorous phase and can be reused. For more details about fluorous phase technique, See Reference [231].

Thus, Friedel–Crafts acylation was carried out in fluorous biphasic system (FBS) using lanthanide methides [231, 232] and non-fluorous zinc chloride was used as Friedel–Crafts catalyst in perfluoroethylamine, which replace highly toxic sym. Tetrachloroethane [233].

2.25   Friedlander Synthesis

It involves [234] base-catalysed condensation of 2-aminobenzaldehydes with ketones to form quinoline derivative (Scheme 2.216).

2.25.1   Friedlander Synthesis Under Microwave Irradiation

KSF clay-catalysed Friedlander synthesis involving 2-aminoaldehydes or ketones with carbonyl compounds containing α-methylene group has been achieved in solvent-free conditions under MWI to give quinoline derivative [235] (Scheme 2.217).

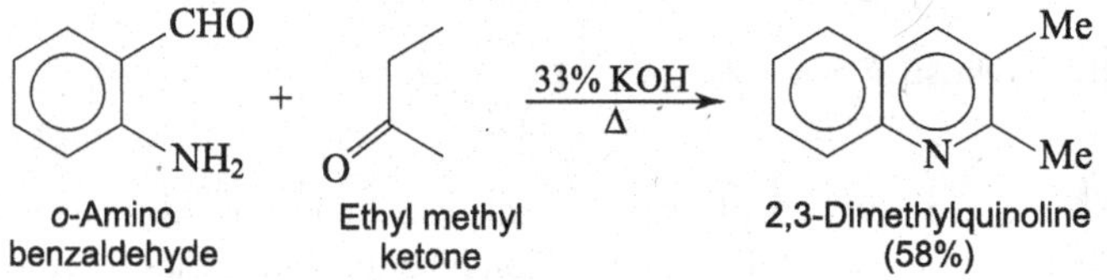

Scheme 2.216   Friedlander synthesis

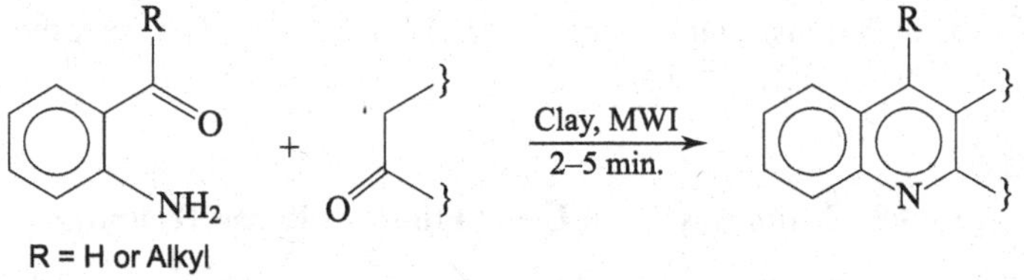

Scheme 2.217   Friedlander synthesis under microwave irradiation

2.26 Fries Rearrangement

Esters of phenols on heating with anhydrous aluminium chloride (Lewis acid) undergo rearrangement to give phenolic ketones. This is known as Fries rearrangement [236] (Scheme 2.218).

In general, low temperature (<60 °C) favours the formation of the *p*-isomer and higher temperature (>160 °C) favours the formation of the *o*-isomer. The mixture of *o*- and *p*-isomers can be separated by steam distillation. The *o*-isomers are intramolecularly hydrogen bonded (chelated), and have greater volatility.

The mechanism of the reaction is not certain. Two mechanisms (intermolecular and intramolecular) have been proposed [237].

Intermolecular

The mechanism involves the formation of acylium ion followed by Friedel–Crafts acylation at *ortho*- or *para*-position (Scheme 2.219).

Intramolecular

2.26.1 Photo-Fries Rearrangement [237, 238]

Photolysis of phenolic esters in solution gives a mixture of o- and p-acetylphenols. Unlike the normal Fries rearrangement, photo Fries rearrangement does not need a catalyst and is predominantly an intramolecular free radical process.

2.26.2 Fries-Rearrangement Under Microwave Irradiation [240]

There is considerable rate enhancement of Fries rearrangement by commercial microwave ovens over conventional methods. Thus, a mixture of p-cresylacetate

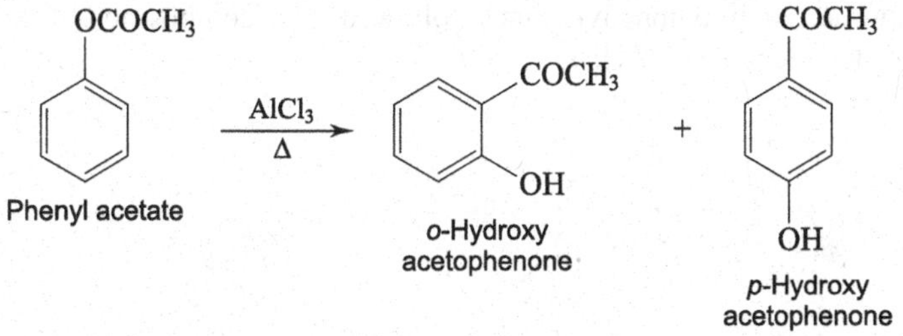

Scheme 2.218 Fries rearrangement

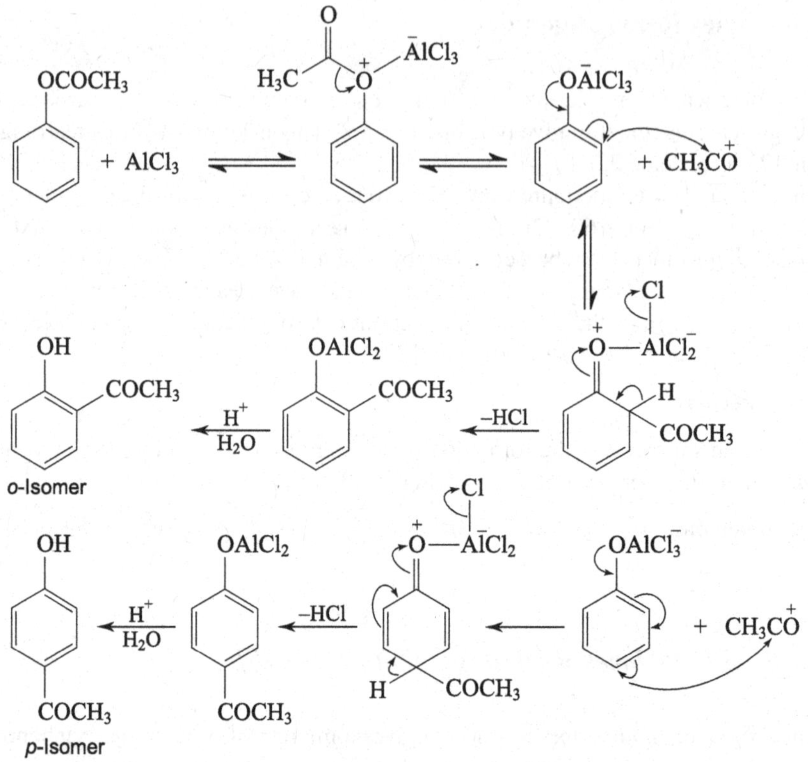

Scheme 2.219 Intermolecular and intramolecular mechanism of Fries rearrangement

and anhydrous aluminium chloride is heated in dry chlorobenzene in a sealed tube in a microwave oven for 2 min to give 85% yield of the product (Scheme 2.220).

2.27   Graebe-Ullman Synthesis

This synthesis deals with the formation of carbazoles [241] by the action of nitrous acid on 2-aminodiphenylamines, followed by decomposition of the formed benzotriazoles (Scheme 2.221).

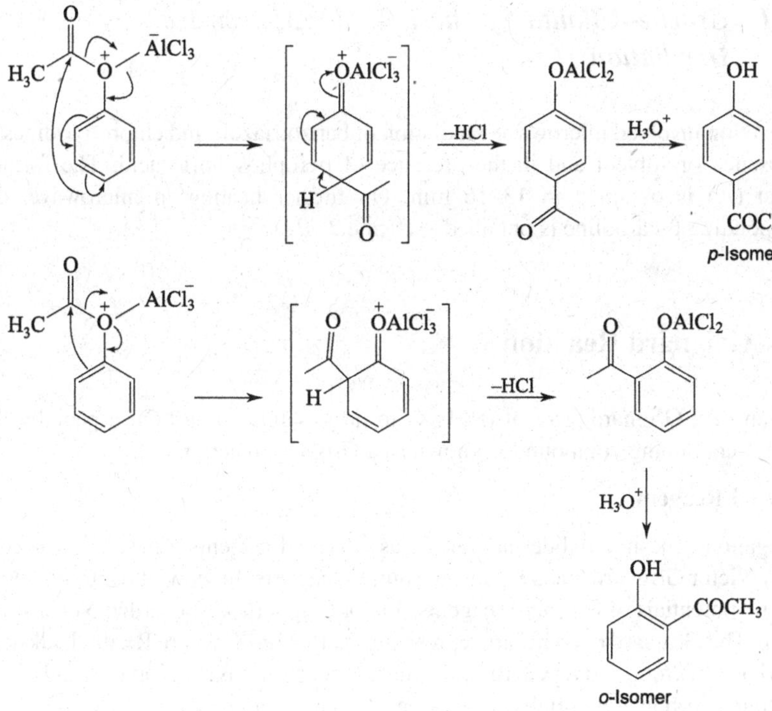

Scheme 2.219 (continued)

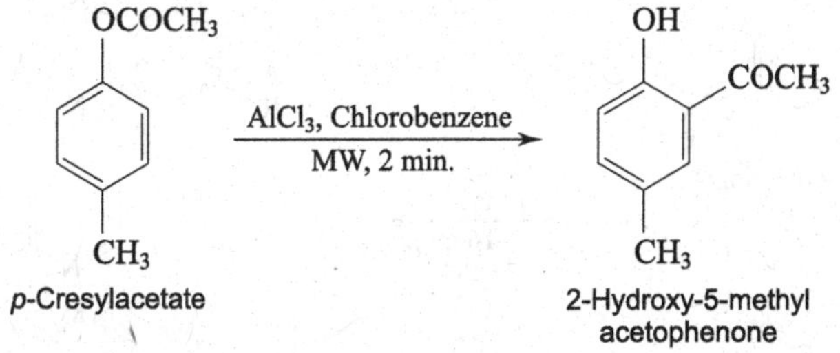

p-Cresylacetate

2-Hydroxy-5-methyl
acetophenone

Scheme 2.220 Fries rearrangement under microwave irradiation

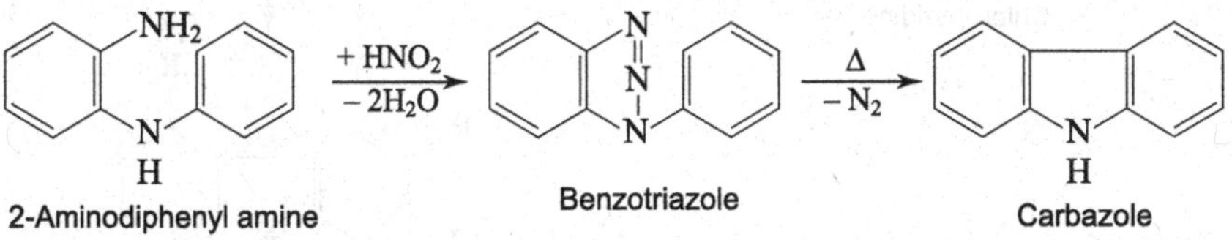

2-Aminodiphenyl amine

Benzotriazole

Carbazole

Scheme 2.221 Graebe–Ullman synthesis

2.27.1 Graebe–Ullman Synthesis Under Microwave Irradiation

The reaction involved microwave irradiation of benzotriazole and chloropyridines in the absence of solvent and in the presence of pyrophosphoric acid. The formed product (A) is obtained in 13–16 min. On further heating in microwave, the corresponding β-carboline is obtained (Scheme 2.222).

2.28 Grignard Reaction

Addition of a Grignard reagent (RMgX) to an unsaturated carbon, especially the carbonyl-containing compound is known as a Grignard reaction [243].

Grignard Reagent

The organomagnesium halides are known as Grignard reagents. These were discovered by Victor Grignard and so named Grignard reagents. In view of the tremendous synthetic potentials of Grignard reagents, Victor Grignard was awarded Nobel Prize in 1912. The Grignard reagents are represented as R–Mg–X, where R=alkyl, alkenyl, alkynyl or aryl group and X=Cl, Br, I. Grignard reagents act as nucleophiles and attack unsaturated carbon—especially the carbon of a carbonyl group.

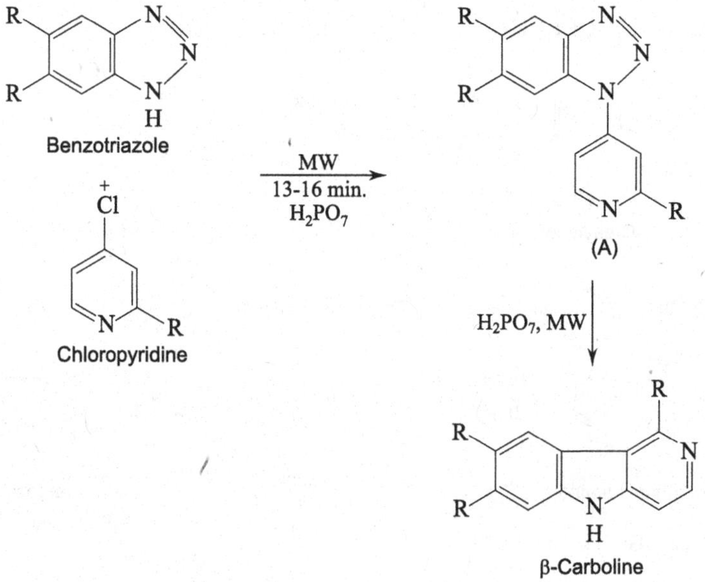

Scheme 2.222 Graebe–Ullman synthesis under microwave irradiation

$$R - X + Mg \xrightarrow[\text{Reflux}]{\text{Ether}} RMgX$$

$R = CH_3$ or C_2H_5 Grignard reagent

$X = I$

Scheme 2.223 Preparation of Grignard reagents

The Grignard reagents are prepared by the action of magnesium on alkyl halide in anhydrous ether (Scheme 2.223).

For a given halogen, the reactivity of an alkyl group is $CH_3 > C_2H_5 > C_3H_7$ or in other words with the increase in carbon atoms, the formation of Grignard reagent becomes difficult. Also, for an alkyl halide, the ease of formation of Grignard reagent is of the order $RI > RBr > RCl$.

2.28.1 Grignard Reaction Under Sonication

The reaction is performed under anhydrous conditions. In case the reaction is slow or sluggish, small amount of iodine is added to start the reaction.

It is found that it is best to activate magnesium by sonication. This activated magnesium finds applications [244] in the synthesis of Grignard reagents without the use of activators (Scheme 2.224).

Use of sonication makes the preparation of Grignard reagents very convenient. Using this procedure even halides containing olefinic groups can be made to react with activated magnesium to give [245] the corresponding Grignard reagents.

Ary halides react with magnesium to form arylmagnesium halides (e.g., phenyl magnesium bromide, C_6H_5MgBr). In a similar way, vinyl halides ($CH_3CH{=}{=}CHX$) react with magnesium to form vinyl Grignard reagents (e.g., $CH_3CH{=}{=}CHMgBr$).

In addition to the method described above, the Grignard reagents are prepared from substrates that have acidic hydrogens. Some examples are shown in Scheme 2.225.

Stucture of Grignard Reagent

A Grignard reagent is represented as RMgX. Its exact structure has been a matter of discussion for a long time. It is believed that a Grignard reagent exists as a coordination complex with ether as shown below:

$$R - X + Mg \xrightarrow[))))]{\text{Ether}} R\ Mg\ X$$

90%

Scheme 2.224 Grignard reaction under sonication

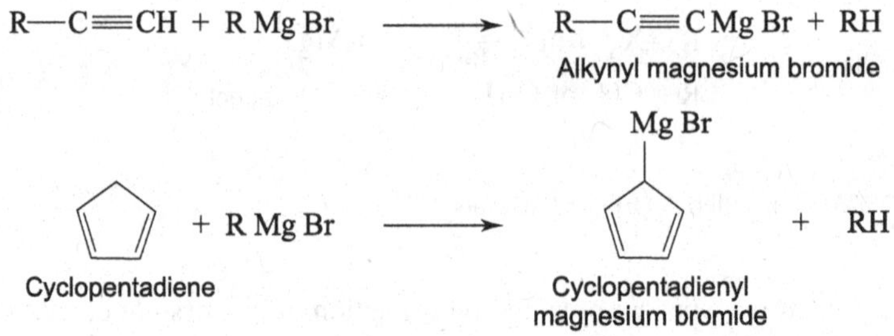

$$R—C≡CH + R\,Mg\,Br \longrightarrow R—C≡C\,Mg\,Br + RH$$

Alkynyl magnesium bromide

Cyclopentadiene $\quad + R\,Mg\,Br \longrightarrow \quad$ Cyclopentadienyl magnesium bromide $\quad + \quad RH$

Scheme 2.225 Preparation of Grignard reagents

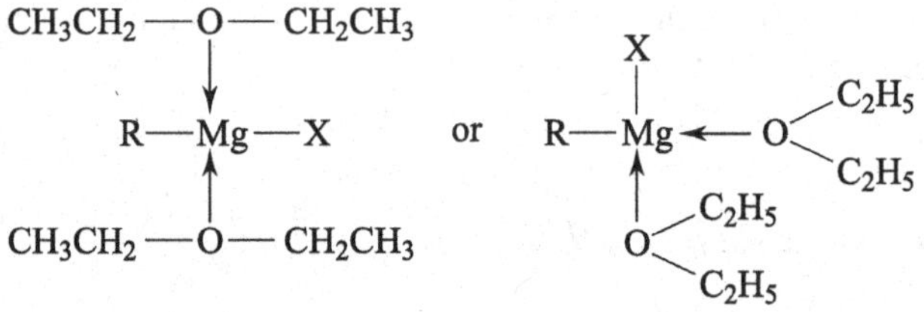

$$CH_3CH_2—O—CH_2CH_3$$
$$R—Mg—X \quad \text{or} \quad R—Mg \leftarrow O$$
$$CH_3CH_2—O—CH_2CH_3$$

The structure of phenyl magnesium bromide as given below has been established by X-ray diffraction studies (Scheme 2.226).

Grignard Reaction

The Grignard reagents react with carbonyl compounds (aldehydes and ketones) to give alcohols (Scheme 2.227). This reaction is known as Grignard reaction.

Reaction Mechanism

It is believed that the Grignard reaction occurs in the following ways:

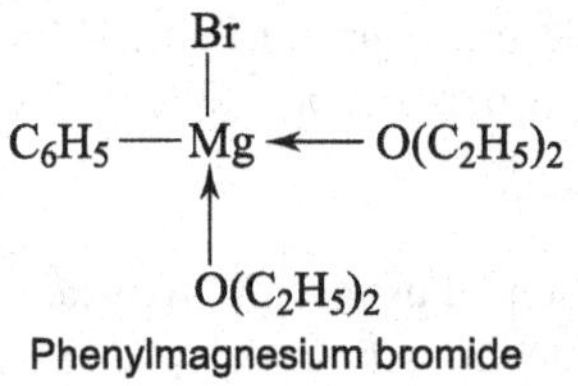

$$C_6H_5—Mg \leftarrow O(C_2H_5)_2$$

Phenylmagnesium bromide

Scheme 2.226 Structure of Grignard reagent

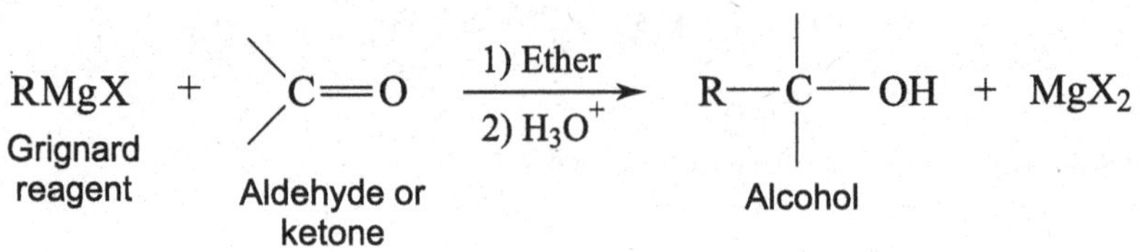

$$\text{RMgX} + \underset{\text{Aldehyde or ketone}}{\overset{\displaystyle \searrow}{\underset{\displaystyle \nearrow}{C}}=O} \xrightarrow[\text{2) } H_3O^+]{\text{1) Ether}} R—\overset{|}{\underset{|}{C}}—OH + MgX_2$$

Grignard reagent   Alcohol

Scheme 2.227 Grignard reaction

(*i*) The Grignard reagent being strongly nucleophilic uses by its electron pair to form a bond to the carbon atom of the carbonyl group. One electron pair of the carbonyl group shifts out to the oxygen. This reaction is a nucleophilic addition to the carbonyl group, and it results in the formation of an alkoxide ion associated with Mg^{2+} and X^- the adduct is called halomagnesium alkoxide.

(*ii*) The next step is the protonation of the alkoxide ion by the addition of aqueous hydrogen halide (HX). This results in the formation of the alcohol and MgX_2. The various steps are represented as shown in Scheme 2.228.

Limitations

Though Grignard synthesis is one of the most valuable of all general synthetic procedures, it has some limitations (Scheme 2.229):

(*i*) Since Grignard reagent contains a carbanion (and so is a very powerful base), it is not possible to prepare a Grignard reagent from an organic group that contains an acidic hydrogen, i.e., hydrogen atom which is more acidic than the hydrogen

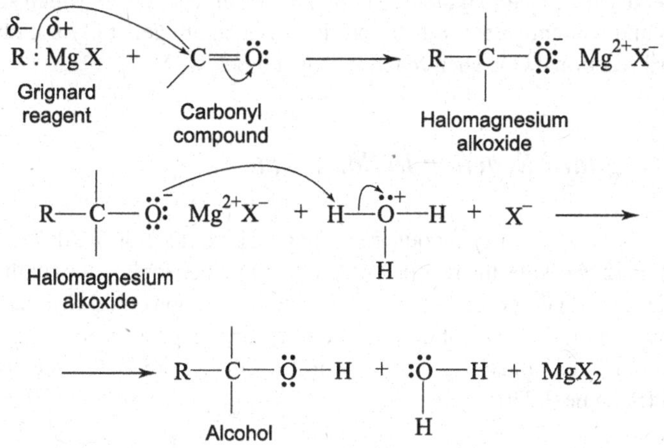

Scheme 2.228 Mechanism of Grignard reaction

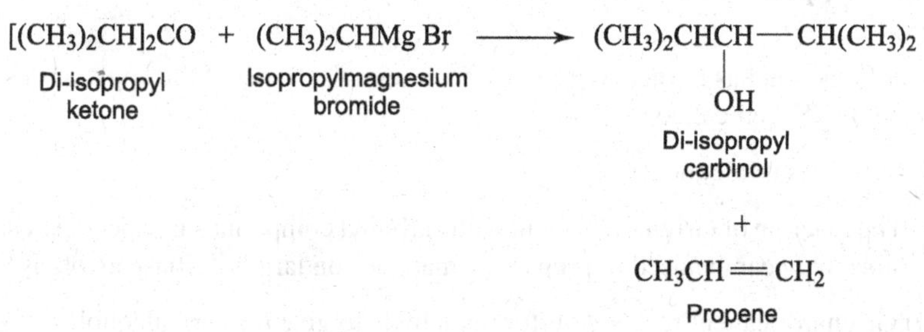

Scheme 2.229 Some limitations of Grignard reagents

atoms of an alkane or alkene. It is thus not possible to prepare a Grignard reagent from a compound containing an –OH group, —NH_2 group, —SH group, —COOH group or an —SO_3H group.

(*ii*) Since Grignard reagents are powerful nucleophiles, it is not possible to prepare a Grignard reagent from any halide that contains a carboxyl, epoxy, nitro or cyano group.

This implies that the preparation of Grignard reagents is limited to alkyl halides or to analogous organic halides containing carbon–carbon double bonds, internal triple bonds, ether linkages and —NR_2 group.

(*iii*) A number of cases are known where a Grignard reagent is made to react with a ketone, which does not give the usual product (i.e., *tert.* alcohol). Instead, 'abnormal' products are obtained. It is found that branching of the carbon chain near the carbonyl group hinders the nucleophilic addition of the reagent due to steric hinderance. Also, in case the Grignard reagent has bulky alkyl or aryl group, it does not attack the electrophilic centre of the carbonyl compound. Thus, even methylmagnesium bromide does not react with di-*tert.*-butyl ketone. In a similar way, methyl isopropyl ketone reacts with methylmagnesium halides, but not with *tertiary*-butylmagnesium halides. There are certain reactions in which substituted ketones and reagents react to form products which are generally not obtained in Grignard reactions. One such reaction is given below:

2.28.2 *Grignard Reaction in Solid State*

The results obtained by carrying out the usual Grignard reaction are different than in the solid state [246]. Thus the reaction of ketone (e.g., benzophenone) with Grignard reagent [the reaction is carried out by mixing ketone and powdered Grignard reagent, obtained by evaporating the solution of the Grignard reagent (prepared as usual) in vacuo (caution)] in solid state gives more of the reduced product of the ketone than the adduct (Scheme 2.230).

2.28.3 *Applications*

Grignard reaction has tremendous synthetic potentials. Some of the applications are given below (Scheme 2.231):

(*i*) Synthesis of alcohols

The reaction of Grignard reagent with carbonyl compounds is especially useful since they can be used to prepare primary, secondary or tertiary alcohols.

(*a*) Grignard reagent reacts with formaldehyde to give primary alcohol.
If R = C_6H_5, benzyl alcohol is obtained with 90% yield.

$$Ph_2CO \; + \; RMgX \xrightarrow[\text{Solid}]{0.5\,\text{hr.}} Ph_2RCOH \; + \; Ph_2CHOH$$

Benzophenone   Grignard   Adduct   Reduced
                reagent   (A)   product of the
                                                     ketone (B)

Grignard reagent RMgX		% Products obtained in solid state	
R	X	(A)	(B)
Me	I	No reaction	
Et	Br	30	31
iPr	Br	2	20
Ph	Br	59	—

Scheme 2.230 Grignard reaction in solid state

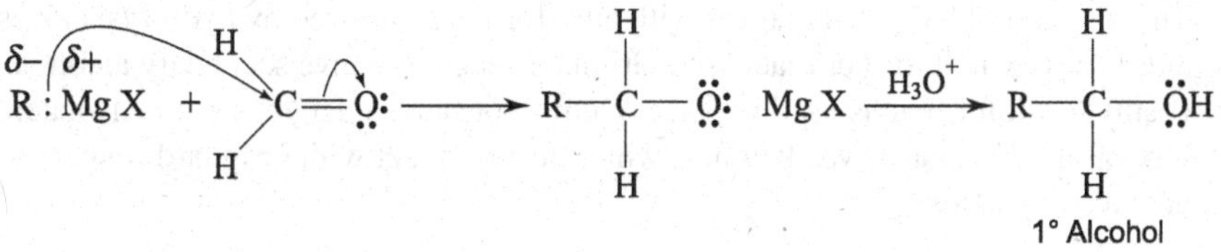

1° Alcohol

Scheme 2.231 Some applications of Grignard reagents

Grignard reagent also reacts with an epoxide (ethylene oxide) to give primary alcohol with lengthening of the carbon chain by two carbon atoms.

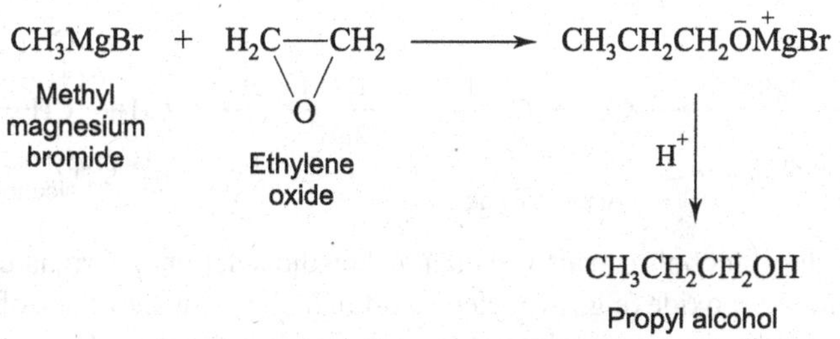

CH_3MgBr + H_2C—CH_2 $\longrightarrow$ $CH_3CH_2CH_2\overset{+}{\overset{-}{O}}MgBr$

Methyl magnesium bromide

Ethylene oxide

H^+

$CH_3CH_2CH_2OH$

Propyl alcohol

(b) All aldehydes except formaldehyde react with Grignard reagent to give secondary alcohols.

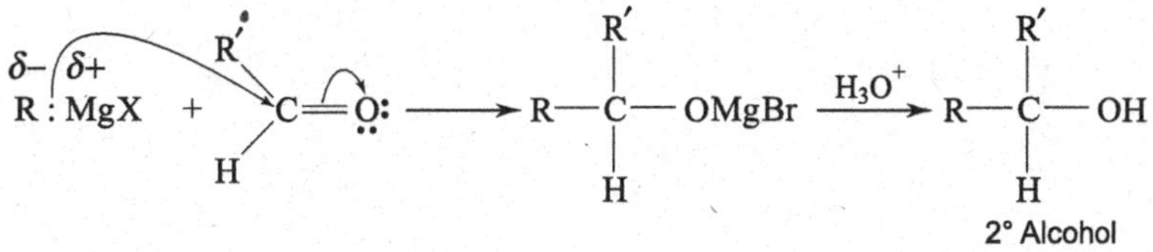

2° Alcohol

Thus, 2-butanol is obtained with 80% yield by the reaction of ethylmagnesium bromide with acetaldehyde.

$$CH_3CH_2MgBr \quad + \quad \underset{H}{\overset{CH_3}{\diagdown}}C{=}O \quad \xrightarrow{Et_2O} \quad CH_3CH_2{-}\underset{H}{\overset{CH_3}{\underset{|}{\overset{|}{C}}}}{-}O\ Mg\ Br$$

Ethyl magnesium bromide   Acetaldehyde

$$\Big\downarrow H_3O^+$$

$$CH_3CH_2\underset{|}{\overset{}{C}}HCH_3$$
$$OH$$

2-Butanol
(80%)

The reaction of Grignard reagent with ethylformate followed by hydrolysis gives aldehydes, which further react with Grignard reagent to give secondary alcohols. If ethyl orthoformate is used in place of ethyl formate, aldehydes can be isolated. Use of ethyl acetate gives ketones, which further react with Grignard reagent to give tertiary alcohol.

$$\overset{\delta-}{\underset{CH_3MgI}{\diagup}}\overset{\delta+}{\diagdown} \ + \ HC\overset{\diagup O}{\underset{OEt}{\diagdown}} \quad \longrightarrow \quad \left[\underset{H_3C}{\overset{H}{\diagdown}}\underset{OEt}{\overset{\overset{+}{O}\ MgI}{C}} \right] \quad \longrightarrow$$

Methyl magnesium bromide   Ethyl formate

$$\xrightarrow{-Mg(OEt)I} \quad CH_3{-}C\overset{\diagup O}{\underset{H}{\diagdown}} \quad \xrightarrow[2)\ H^+]{1)\ CH_3MgI} \quad CH_3{-}\underset{}{\overset{CH_3}{\underset{|}{\overset{|}{C}}}}H{-}OH$$

Acetaldehyde   Isopropyl alcohol
2° alcohol

Secondary alcohols can also be obtained by the reaction of Grignard reagent with substituted epoxide (e.g., propylene oxide). In this case, the attack of the Grignard reagent is from the less substituted ring carbon of the epoxide.

$$C_6H_5MgBr \ + \ H_2C\!\!-\!\!CHCH_3 \longrightarrow C_6H_5CH_2\!\!-\!\!CH\!\!-\!\!\overset{-}{O}\overset{+}{M}gBr$$

Phenyl magnesium   O |
bromide   CH$_3$
                      Propylene
                        oxide

$$\Big\downarrow H^+$$

$$C_6H_5CH_2\overset{|}{C}HOH$$
                    |
                   CH$_3$
                2° Alcohol

(c) Grignard reagents react with ketones to give tertiary alcohols.

$$\underset{\text{Grignard reagent}}{\overset{\delta-}{R}\,\overset{\delta+}{:}\,MgX} \ + \ \underset{\text{Ketone}}{\overset{R'}{\underset{R''}{>}}C\!\!=\!\!\ddot{O}} \longrightarrow R\!\!-\!\!\overset{R'}{\underset{R''}{\overset{|}{C}}}\!\!-\!\!\overset{-}{O}\overset{+}{M}gBr \ \xrightarrow[H_2O]{NH_4Cl} \ R\!\!-\!\!\overset{R'}{\underset{R''}{\overset{|}{C}}}\!\!-\!\!OH$$

3° Alcohol

Thus, butylmagnesium bromide on reaction with acetone gives 2-methyl-2-hexanol with 92% yield.

$$\underset{\substack{\text{Butyl magnesium}\\\text{bromide}}}{CH_3CH_2CH_2CH_2MgBr} \ + \ \underset{\text{Acetone}}{\overset{CH_3}{\underset{CH_3}{>}}C\!\!=\!\!O} \longrightarrow CH_3CH_2CH_2CH_2\!\!-\!\!\overset{CH_3}{\underset{CH_3}{\overset{|}{C}}}\!\!-\!\!\overset{-}{O}\overset{+}{M}gBr$$

$$\Big\downarrow NH_4Cl\ |\ H_2O$$

$$CH_3CH_2CH_2CH_2\!\!-\!\!\overset{CH_3}{\underset{OH}{\overset{|}{C}}}\!\!-\!\!CH_3$$

2-Methyl-2-hexanol
92%

Tertiary alcohols are also obtained by the reaction of esters with Grignard reagent (2 equivalent); in this case, the initial product is a ketone, which further reacts (ketones are more reactive than esters) with Grignard reagent to give 3° alcohol.

$$\delta- \overset{\delta+}{R\!:\!Mg\,X} + \overset{R'}{\underset{R''\overset{..}{\underset{..}{O}}}{C}}\!\!=\!\!\overset{..}{\underset{..}{O}}: \longrightarrow \left[R\!-\!\overset{R'}{\underset{\underset{:O\!-\!R''}{|}}{C}}\!-\!\overset{..}{\underset{..}{O}}\!-\!Mg\,X \right] \xrightarrow[\text{Spontaneous}]{-R''OMgX}$$

$$\left[\overset{R'}{\underset{R}{C}}\!\!=\!\!\overset{..}{\underset{..}{O}} \right] \xrightarrow{RMgX} R\!-\!\overset{R'}{\underset{R}{\overset{|}{\underset{|}{C}}}}\!-\!\overset{..}{\underset{..}{O}}\,MgX \xrightarrow[\text{H}_2\text{O}]{\text{NH}_4\text{Cl}} R\!-\!\overset{R'}{\underset{R}{\overset{|}{\underset{|}{C}}}}\!-\!OH$$

Ketone   3° Alcohol

Thus, by the reaction of ethylmagnesium bromide with ethyl acetate (2 mol), 3-methyl-3-pentanol is obtained with 67% yield.

$$CH_3CH_2MgBr \;+\; \overset{H_3C}{\underset{C_2H_5O}{C}}\!\!=\!\!O \longrightarrow CH_3\!-\!CH_2\!-\!\overset{CH_3}{\underset{OH}{\overset{|}{\underset{|}{C}}}}\!-\!CH_2\!-\!CH_3$$

Ethyl magnesium bromide   Ethyl acetate   3-Methyl-3-pentanol 67%

Tertiary alcohols are also obtained by the reaction of cyclic ketones with Grignard reaction.

Cyclohexanone   3° Alcohol

In the above reaction, the formed carbinol loses a molecule of water on treatment with conc. H_2SO_4 to give the corresponding cyclohexene, which on treatment with per acid gives the epoxide.

Carbinol   1-Ethyl cyclohexene   Epoxide

The above sequence of reaction is a convenient route for the synthesis of 1,2-epoxy-1-ethycylclohexane.

Tertiary alcohols are also obtained by the reaction of acid chloride, acid anhydride or amides with Grignard reagents.

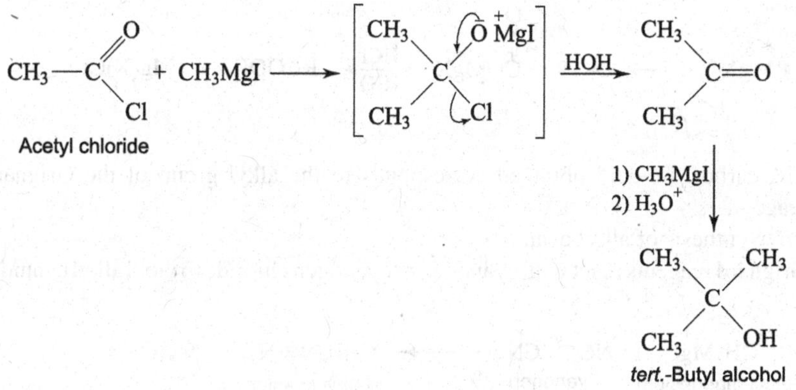

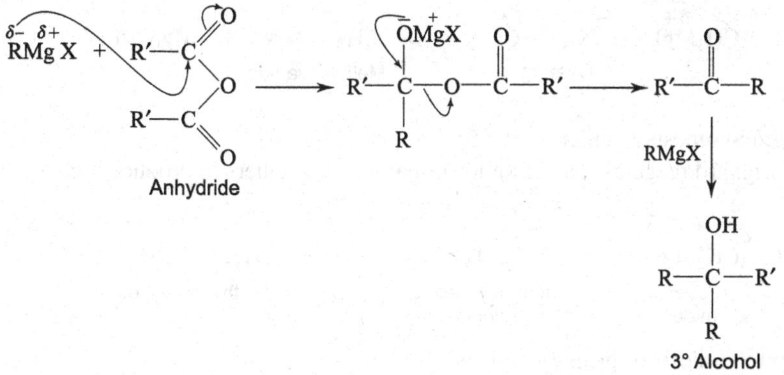

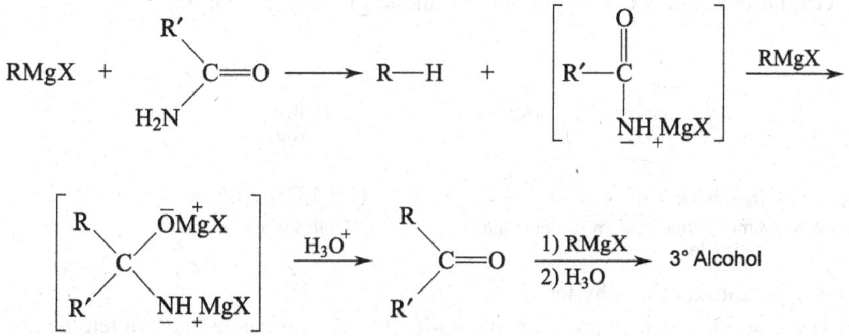

Acetyl chloride

tert.-Butyl alcohol

Anhydride

3° Alcohol

By using Grignard reaction (or synthesis) skilfully, any alcohol can be synthesized using correct Grignard reagent and the correct aldehyde, ketone, ester or epoxide.
(*ii*) Synthesis of carboxylic acids
Grignard reagents on treatment with solid carbon dioxide followed by hydrolysis yields carboxylic acids. Alternatively, a good yield of carboxylic acid can be obtained by passing CO_2 into the Grignard solution at low temperature (0 °C).

$$\delta- \overset{\delta+}{RMgX} + \underset{O}{\overset{O}{\underset{\|}{\overset{\|}{C}}}} \longrightarrow R-\overset{O}{\overset{\|}{C}}\overset{+}{OMgX} \xrightarrow[H_2O]{HCl} RCOOH + Mg(X)Cl$$

The carboxylic acid obtained corresponds to the alkyl group of the Grignard reagent.

(*iii*) Synthesis of alkyl cyanides

Grignard reagents react with cyanogen or cyanogen chloride to yield alkyl cyanide.

$$\delta-\overset{\delta+}{CH_3MgI} + NC-CN \longrightarrow CH_3C\equiv N + Mg(CN)I$$

Methyl magnesium   Cyanogen   Methyl cyanide
iodide

$$\delta-\overset{\delta+}{CH_3MgI} + NC-Cl \longrightarrow CH_3C\equiv N + Mg(Cl)I$$

Cyanogen   Methyl cyanide
chloride

(*iv*) Synthesis of ethers

Grignard reagents react with lower halogenated ethers to produce higher ethers.

$$\delta-\overset{\delta+}{CH_3CH_2MgBr} + CH_3OCH_2-Cl \longrightarrow CH_3CH_2CH_2OCH_3$$

Ethyl magnesium   Methoxy methyl   Methyl propyl ether
bromide   chloride

(*v*) Synthesis of primary amines

Grignard reagents react with chloramine to give primary amines.

$$\delta-\overset{\delta+}{CH_3MgI} + H_2N-Cl \longrightarrow CH_3NH_2 + Mg(Cl)I$$

Methyl magnesium   Chloramine   Methyl
iodide   amine

$$\delta-\overset{\delta+}{(CH_3)_2CHCH_2MgI} + H_2N-Cl \longrightarrow (CH_3)_2CH CH_2NH_2 + Mg(Cl)I$$

Isopropyl methyl magnesium Chloramine   2-Methyl propyl amine
iodide

(*vi*) Synthesis of aldehydes

The reaction of Grignard reagents with ethyl orthoformate gives aldehyde as the main product. In case of ethyl orthoformate, the formation of secondary alcohol is not possible due to the formation of acetal. The acetal can be isolated and treated with acid to give aldehyde.

$$\begin{array}{ccc}
\overset{\delta-}{CH_3}\overset{\delta+}{MgI} & + & H-C-OC_2H_5 \\
& & | \\
& & OC_2H_5
\end{array} \longrightarrow CH_3CH(OC_2H_5)_2 + Mg(OC_2H_5)I$$

Methyl magnesium OC_2H_5 Acetaldehyde acetal
iodide

Ethyl orthoformate

$$\Big\downarrow H^+$$

$$CH_3CHO + 2C_2H_5OH$$

Acetaldehyde

This procedure is useful for the synthesis of formyl compounds which are difficult to prepare by usual methods.

45%

(vii) Synthesis of ketones

Grignard reagents react with nitriles (alkyl cyanides) to give ketones. (However, use of HCN in place of alkyl cyanide gives an aldehyde.) The reaction proceeds via the formation of a ketimine, which on acid hydrolysis gives ketone (*see* also reaction of Grignard reagent with acid chlorides and esters, which yield alcohols).

$$R\,MgX + R'C \equiv N \longrightarrow RR'C = N\,MgX$$

$$\Big\downarrow H_2O$$

$$RCOR' \xleftarrow{\ H_3O^+\ } [RR'C = NH]$$

Ketone   Ketimine

The above procedure is used for the preparation of ketones which are difficult to prepare by conventional methods.

(viii) Synthesis of hydrocarbons

(*a*) Alkanes are obtained by the reaction of Grignard reagent with compounds containing active hydrogen like water, alcohol, ammonia or amines.

$$CH_3MgI + HOH \longrightarrow CH_4 + Mg(OH)I$$

Methylmagnesium   Water   Methane
iodide

$$CH_3MgI + R-OH \longrightarrow CH_4 + Mg(OR)I$$

Methylmagnesium   Alcohol
iodide

$$CH_3MgI + NH_3 \longrightarrow CH_4 + Mg(NH_2)I$$

Methylmagnesium   Ammonia
iodide

$$CH_3MgI + RNH_2 \longrightarrow CH_4 + Mg(NHR)I$$

Methylmagnesium   Amine
iodide

This method forms the basis for the **Zerewitinoff determination of active hydrogens**. Thus, by measuring the amount of methane liberated from a known weight of the compound, containing active hydrogens, the number of active hydrogens present in a molecule can be obtained.

Alkanes can also be obtained by the reaction of Grignard reagents with alkyl halides.

$$CH_3MgBr + CH_3CH_2Br \longrightarrow CH_3CH_2CH_3 + Mg(Br)I$$

Methyl magnesium   Ethyl bromide   Propane
bromide

(*b*) Alkenes can be similarly obtained by the reaction of Grignard reagents with unsaturated alkyl halides.

$$CH_3MgI + CH_2=CHCH_2Br \longrightarrow CH_2=CHCH_2CH_3 + Mg(Br)I$$

Methyl magnesium   Allyl bromide   1-Butene
iodide

(*c*) Higher alkynes are obtained by the treatment of terminal alkynes with Grignard reagents followed by treatment of the formed alkynylmagnesium halide with alkyl halides.

$$CH_3C{\equiv}CH + CH_3MgI \longrightarrow CH_3C{\equiv}C-MgI$$

Propyne   Methyl magnesium   Propyenyl magnesium
iodide   iodide

$$CH_3I \Big| S_N2$$

$$CH_3C{\equiv}CCH_3 + MgI_2$$

2-Butyne

(*d*) Cycloalkanes are obtained by the treatment of cyclobutylmagnesium bromide with water.

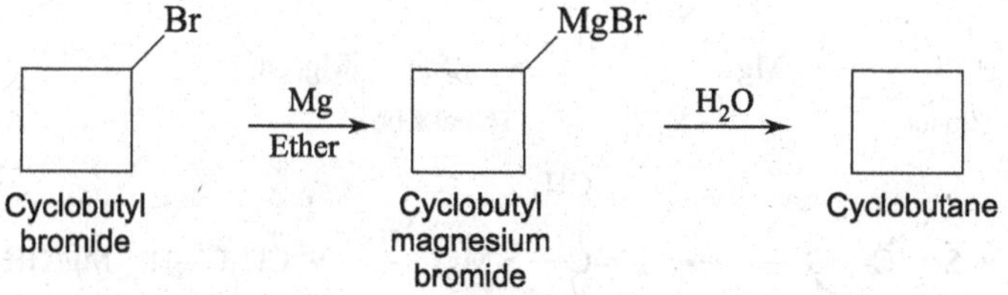

Cyclobutyl bromide — Mg / Ether → Cyclobutyl magnesium bromide — H₂O → Cyclobutane

The above procedure is used for the replacement of a halogen with hydrogen in alkyl halides.

$$R—X \xrightarrow[\text{Ether}]{\text{Mg}} RMgX \xrightarrow{\text{H}_2\text{O}} RH + Mg(X)OH$$

(e) Deuterated hydrocarbons

The reaction of Grignard reagent with D_2O gives deuterated hydrocarbons.

$$\underset{\text{Alkyl halide}}{R—I} \xrightarrow[\text{Ether}]{\text{Mg}} \underset{\substack{\text{Grignard}\\\text{reagent}}}{RMgI} \xrightarrow{\text{D}_2\text{O}} RD + Mg(I)OD$$

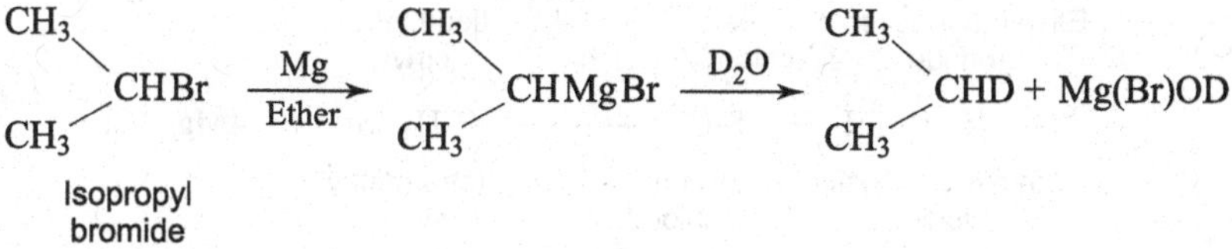

Isopropyl bromide

(ix) Synthesis of alkyl iodides
The reaction of Grignard reagents (RMgCl or RMgBr) with iodine in acetone gives the corresponding alkyl iodides.

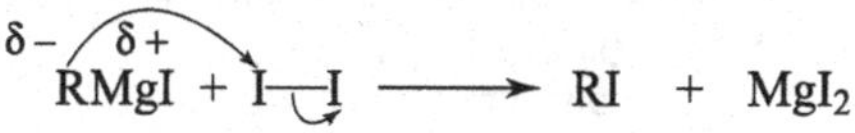

$$RMgI + I—I \longrightarrow RI + MgI_2$$

$$\underset{\substack{\text{Neopentyl}\\\text{chloride}}}{(CH_3)_3CCH_2Cl} \xrightarrow[\text{Ether}]{\text{Mg}} (CH_3)_3CCH_2MgCl$$

$$\downarrow I_2/\text{Acetone}$$

$$\underset{\substack{\text{Neopentyl}\\\text{iodide}}}{(CH_3)_3CCH_2I}$$

(x) Synthesis of thioalcohols, dithioacetic acids and sulfinic acids
Grignard reagents react with sulphur, carbon disulphide and sulphur dioxide to give the corresponding thioalcohols, dithioacetic acids and sulfinic acids, respectively.

$$\text{RMgX} + \text{S} \longrightarrow \text{Mg} \overset{\text{SR}}{\underset{\text{X}}{\diagup}} \xrightarrow{\text{H}_3\text{O}^+} \text{RSH} + \text{Mg(OH)X}$$

Sulfur   Thioalcohols

$$\overset{\delta-}{\text{CH}_3}\overset{\delta+}{\text{MgI}} + \text{S}=\text{C}=\text{S} \longrightarrow \text{S}=\overset{\text{CH}_3}{\underset{}{\text{C}}}-\bar{\text{S}} \overset{+}{\text{MgI}} \xrightarrow{\text{H}_3\text{O}^+} \text{CH}_3\text{CS}_2\text{H} + \text{Mg(OH)I}$$

Carbon   Dithioacetic
disulphide   acid

$$\text{RMgX} + \text{SO}_2 \longrightarrow \text{R}-\overset{\bar{\text{O}}\,\overset{+}{\text{MgX}}}{\underset{\text{O}}{\text{S}}} \xrightarrow{\text{H}_3\text{O}^+} \text{R}-\overset{}{\underset{\text{O}}{\text{S}}}-\text{OH} + \text{Mg(OH)X}$$

Sulfur   Sulfinic
dioxide   acid

(*xi*) Synthesis of organometallic compounds
Grignard reagents react with inorganic halides to give organometallic compounds.
Some examples are.

$$4\text{C}_2\text{H}_5\text{MgBr} + \text{PbCl}_4 \longrightarrow (\text{C}_2\text{H}_5)_4\text{Pb} + 4\text{Mg(Br)Cl}$$

Ethyl magnesium   Lead   Lead tetra
bromide   chloride   ethyl

$$4\text{CH}_3\text{MgI} + \text{SiCl}_4 \longrightarrow (\text{CH}_3)_4\text{Si} + 4\text{Mg(I)Cl}$$

Methyl magnesium   Silicon   Tetra methyl
iodide   tetra chloride   silane

$$3\text{CH}_3\text{MgBr} + \text{PCl}_5 \longrightarrow (\text{CH}_3)_3\text{P} + 3\text{Mg(Cl)Br}$$

Methyl magnesium Phosphorus   Trimethyl
bromide   pentachloride   phosphene

$$3\text{CH}_3\text{MgBr} + \text{CdCl}_2 \longrightarrow (\text{CH}_3)_2\text{Cd} + 2\text{Mg(Cl)Br}$$

Methyl magnesium   Cadmium   Dimethyl
chloride   chloride   cadmium

(*xii*) Miscellaneous applications of Grignard reaction
(a) **Synthesis of hydroperoxides**
Oxygenation of Grignard reagent at low temperature provides an excellent method
for the synthesis of hydroperoxides.

$$\text{RMgX} + \text{O}_2 \xrightarrow{-70°} \text{ROOMgX} \xrightarrow{\text{H}^+} \text{ROOH}$$

To prevent the formation of excessive amounts of alcohol, inverse addition is
best, i.e., a solution of Grignard reagent is added to ether through which oxygen
is bubbled rather than have the oxygen bubble through a solution of the Grignard
reagent.
(*b*) **Synthesis of stilbene**
The reaction of benzylmagnesium bromide with benzaldehyde followed by
dehydration of the formed alcohol gives stilbene.

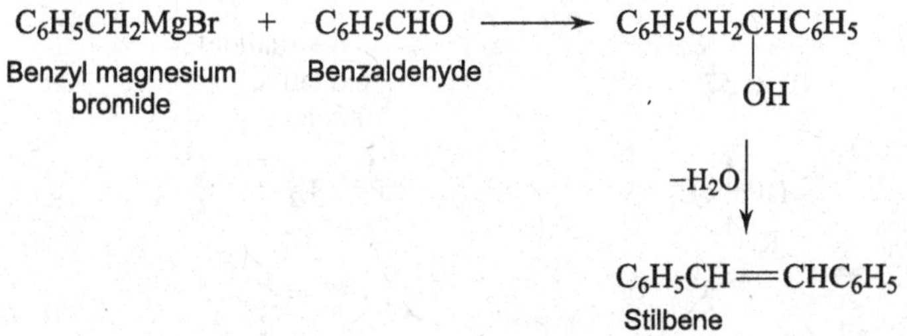

$$C_6H_5CH_2MgBr \quad + \quad C_6H_5CHO \quad \longrightarrow \quad C_6H_5CH_2CHC_6H_5$$

Benzyl magnesium   Benzaldehyde
bromide

$$\overset{|}{\underset{OH}{}}$$

$-H_2O$

$$C_6H_5CH{=\!=}CHC_6H_5$$
Stilbene

(c) **Synthesis of 1,2-divinyl cyclohexanol** [247].

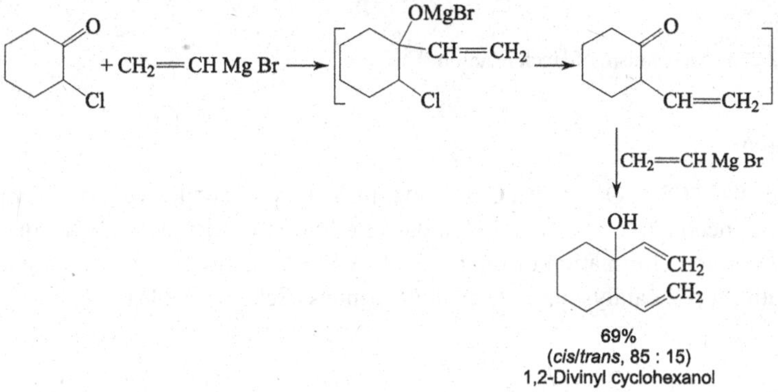

69%
(cis/trans, 85 : 15)
1,2-Divinyl cyclohexanol

2.29 Heck Reaction

It is a very useful carbon–carbon bond formation reaction for organic synthesis. The coupling of an alkene with a halide or triflate in the presence of Pd(O) catalyst to form a new alkene is known as Heck reaction [248] (Scheme 2.232).

R = aryl, vinyl or alkyl group without β-hydrogen on a sp^3 carbon atom

X = halide or triflate (OSO_2CF_3).

The base used in Heck reaction is a mild base like Et_3N or anions like ^-OH, $^-OCOCH_3$, $CO_3{}^{2-}$ etc., and the reaction is carried out in anhydrous polar solvents like DMF, MeCN etc.

Heck reaction is one of the most synthetically useful palladium-catalysed reaction. It is used for the alkylation or arylation of alkenes.

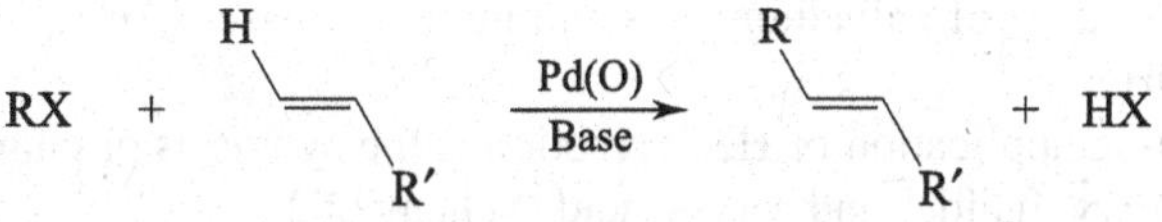

Scheme 2.232 Heck reaction

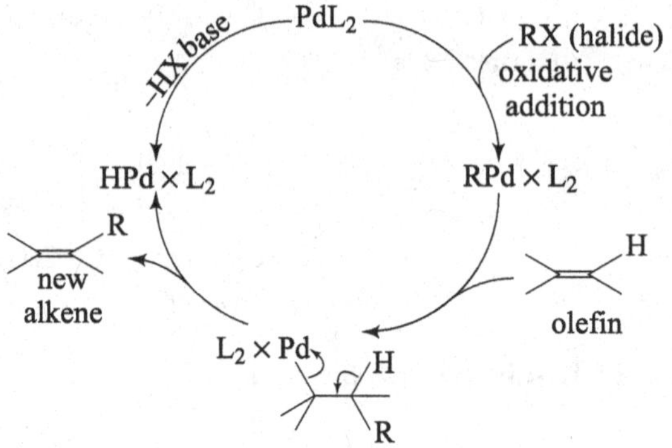

Scheme 2.233 Mechanism of Heck reaction

Mechanism

The bases like ^-OH, $^-OCOCH_3$, $CO_3{}^{2-}$ etc. may serve as ancillary ligands for palladium. The mechanism involves the oxidative addition of the halide (RX), insertion of the olefin and elimination of the product by β-elimination. The base regenerates the palladium (O) catalyst and the cycle continues (Scheme 2.233).

2.29.1   Heck Reaction in Aqueous Phase

Though traditionally Heck reaction uses anhydrous solvents (*e.g.*, DMF and MeCN), it has been found that the reaction can proceed very well *in water*. The role of water in the Heck reaction, as well as other reactions catalysed by Pd(O) in the presence of phosphine ligand is transformation of catalyst precursor into Pd(O) species, and the generation of zero-valent palladium species capable of oxidative addition by oxidation of phosphine ligands by the Pd(II) precursor can be affected by the water content of the reaction mixture.

It has been found that the Heck reaction can be successfully carried out under PTC conditions [249] with inorganic carbonates as base under very mild conditions even at room temperature. Using this procedure, even substrates such as methyl vinyl ketone that is unstable under the conventional condition of Heck arylation react (action of base at high temperature). Subsequently, it was shown that water and aqueous organic solvents can be used successfully for carrying out Heck reaction in the presence [250] of palladium salts and inorganic bases like K_2CO_3, Na_2CO_3, $NaHCO_3$, KOH etc.

An interesting application of Heck reaction is the synthesis of cinnamic acid by the reaction of aryl halides and acrylic acid (Scheme 2.234).

Use of acrylonitrile in place of acrylic acid in the above reaction gives the corresponding cinnamonitriles. However, in case of acrylonitrile a mixture of (E) and (Z)

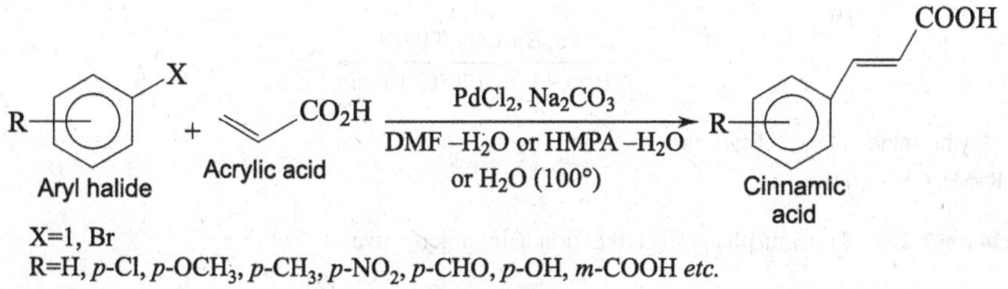

Scheme 2.234 Friedel–Crafts acylation under sonication continued

isomers in the ratio 3:1, close to that observed under conventional conditions [251] is obtained in comparison to almost exclusively (E) isomers in the Heck reactions.

Heck reaction can also be carried out at room temperature [252] if diaryliodonium salts are taken as the arylating agent in water. At room temperature, only one aryl group of the iodonium salt is transferred to the product. However, at 100 °C, both aryl groups of the iodonium salt are utilized (Scheme 2.235).

The Heck reaction has also been performed under mild conditions by the addition of acetate ion as given in Scheme 2.236.

A large number of other applications of the Heck reaction have been described in the literature [253].

Heck coupling reaction in water using microwave heating (Scheme 2.237) give the coupled products [254].

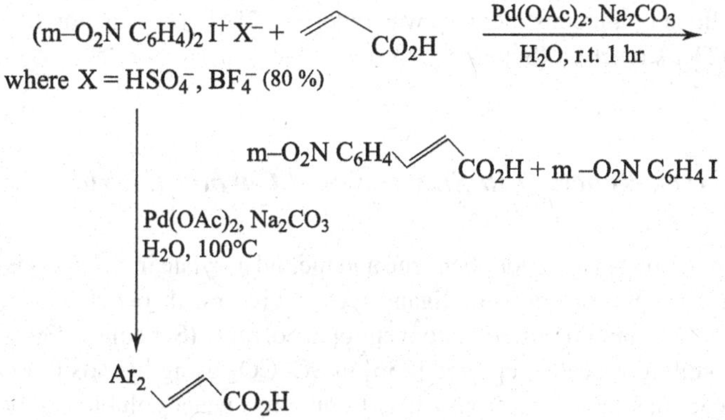

Scheme 2.235 Heck reaction continued

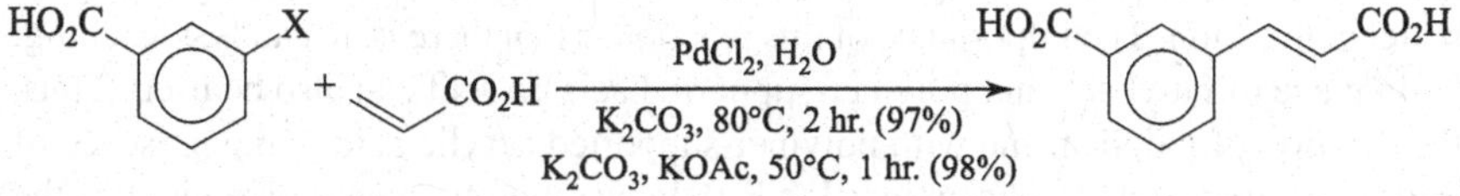

Scheme 2.236 Heck reaction under mild condition

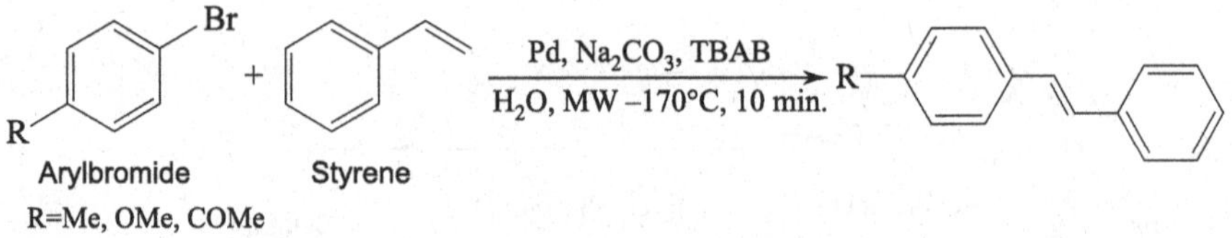

Scheme 2.237 Aqueous phase Heck reaction using microwave

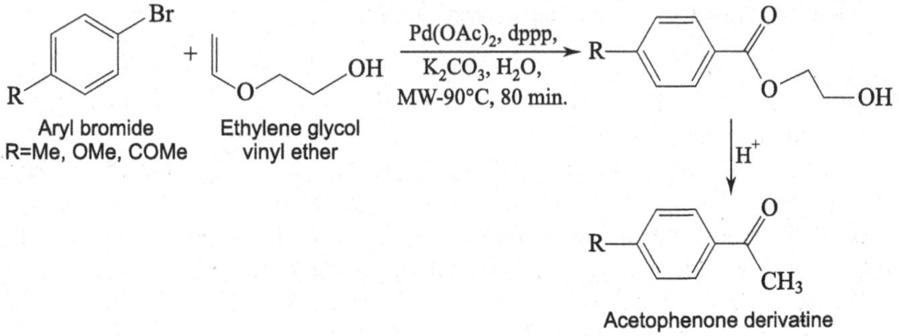

Acetophenone derivatine

Scheme 2.238 Internal Heck reaction

It was found that in Heck reaction Pd-catalyst concentration as low as 500 ppb was sufficient for these reactions with good product yield.

Regioselective Pd(O) catalysed internal R-arylation of ethylene glycol vinyl ether and arylhalides in aqueous medium was reported [255]. The reaction is referred to as internal Heck reaction (Scheme 2.238).

2.29.2 Heck Reaction in Supercritical Carbon Dioxide

The Heck reaction between iodobenzene and methyl acrylate in SC–CO_2 is catalysed by Pb(OAc)$_2$ in the presence of ligand (A) to give much better yield of methyl cinnamate (92%) than reported in conventional solvents (Scheme 2.239).

Heck coupling was also reported [256] in SC–CO_2 using ligands B and C.

Heck reactions have also been carried out using water-soluble catalyst in SC–CO_2 (water-biphasic systems) [257]. Thus, the coupling of iodobenzene with butyl acrylate in SC–CO_2 using Pd(OAc)$_2$ and triphenyl phosphine trisulfonate sodium salt (TPPTS) (Scheme 2.240) gave butyl cinnamate. The use of ethylene glycol gave better conversion. The importance of this reaction is that there is no catalyst leaching.

In place of butyl acetate, polymer-supported acrylic acid can also be used. Thus, the reaction of iodobenzene with polymer-supported acrylic acid in the presence of Pd(OAc)$_2$, tri-*tert.*-butylphosphine [P(t-Bu)$_3$] gave merely quantitative yield of the product [258] (Scheme 2.241).

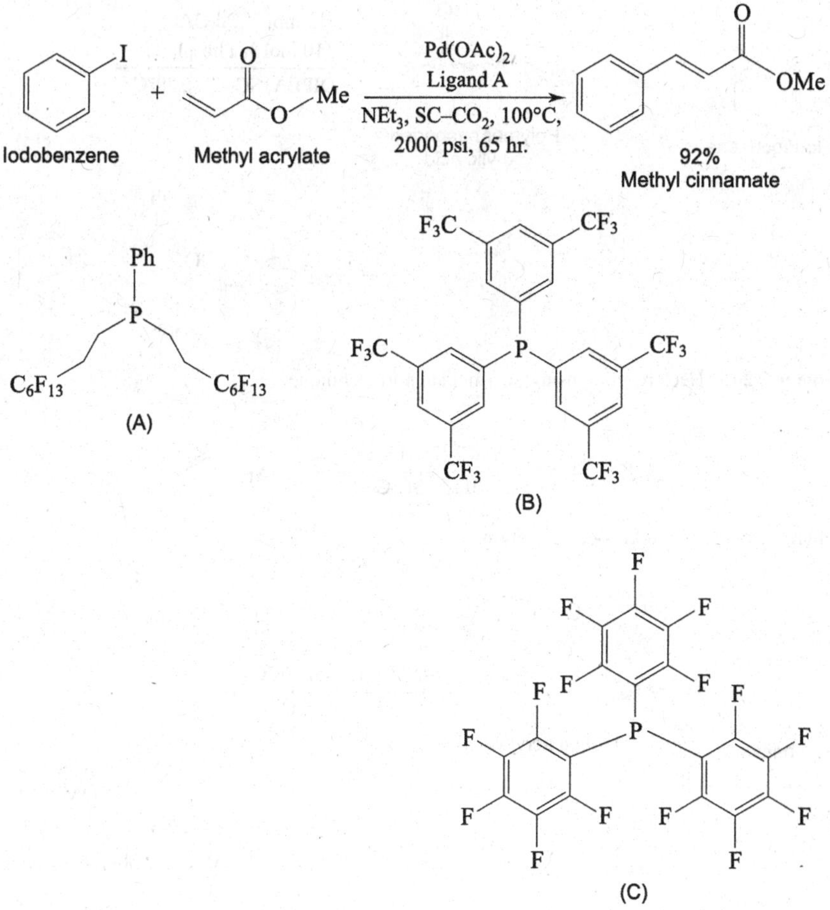

Scheme 2.239 Heck reaction in SC-CO$_2$

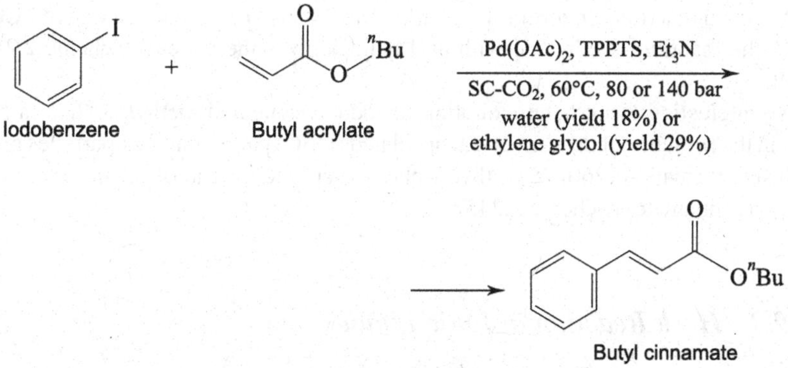

Scheme 2.240 Friedel–Crafts acylation under sonication continued

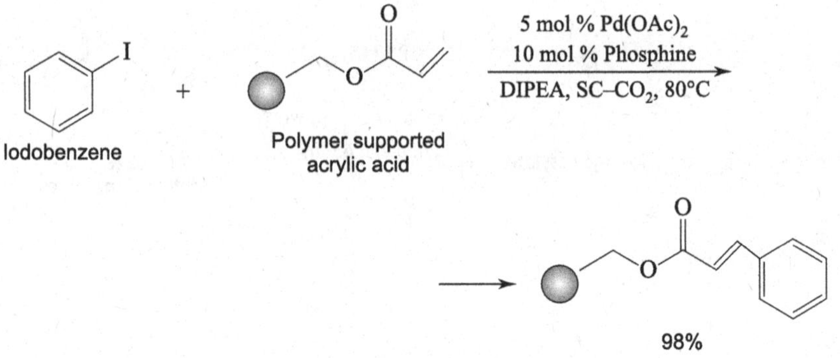

Scheme 2.241 Heck reaction using polymer support technique

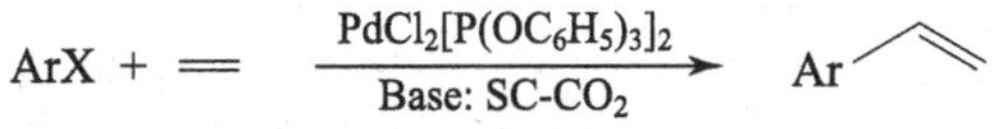

$$ArX + \;=\!=\; \xrightarrow[\text{Base: SC-CO}_2]{\text{PdCl}_2[\text{P(OC}_6\text{H}_5)_3]_2} \; Ar\diagup\!\!\!\diagdown$$

Scheme 2.242 Mizoroki–Heck arylation

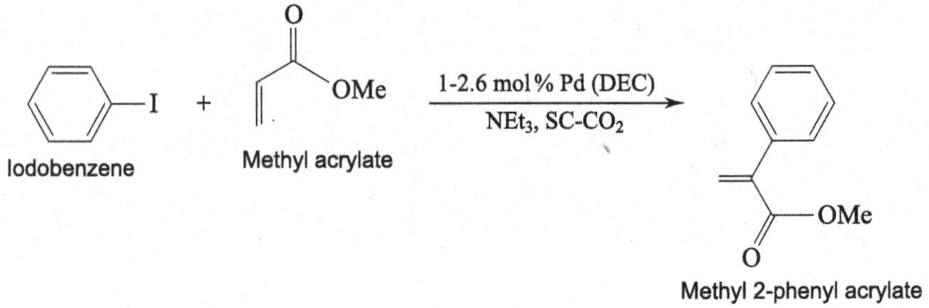

Scheme 2.243 Heck homocouplings

A related reaction, Mizoroki–Heck arylation has also been conducted in $SC\text{-}CO_2$. Thus, the reaction of ethylene with aryl halides gives the styrene (Scheme 2.242) [259].

An interesting Heck heterocoupling of iodobenzene and methyl acylate in SC-CO_2 in the presence of dendrimer-encapsulated (DEC) palladium nanoparticles gives exclusive formation [260] of methyl 2-phenyl-acrylate instead of the usual product (methyl cinnamate) (Scheme 2.243).

2.29.3 Heck Reaction in Ionic Liquids

The Heck reaction has been performed in ionic liquids which are excellent solvents. Use of ionic liquids allows recycling of the catalyst [261], which is not possible under usual conditions. Thus, in the Heck reaction of iodobenzene with ethylacry-late in both N-hexylpyridinium, [C$_6$py] and N,N-dialkyl imidazolium-based ionic liquid (Scheme 2.244), higher yields are obtained in the former ionic liquid than the corresponding reaction in the imidazolium salts.

The low yield in imidazolium ionic liquid is due to the formation of carbene, which reacts with palladium to form a mixture of palladium carbene complex (for more details, *see* Baylis–Hillman reaction, Sect. 2.8.3).

The Heck reaction has also been carried out by the reaction of aryl halides with acrylates as well as with styrene in the presence of Pd(OAc)$_2$ in ionic liquid [b$_{min}$][BF$_4$] or [b$_{min}$][Br] (Scheme 2.245) [262].

Better selectivity and conversion were observed in reactions carried out in [b$_{min}$][Br]. The low yield in case of [b$_{min}$][BF$_4$], as already stated, is due to the formation of carbene intermediate which complexes with palladium to give mixtures of palladium carbene complex.

It has been suggested [262] that the active catalyst in the Heck reaction is a palla-dium nanoparticle generated *in situ* from palladium-carbene species. In fact, solution of ammonium stabilized Pd clusters is a useful catalyst [263] for the Heck reaction.

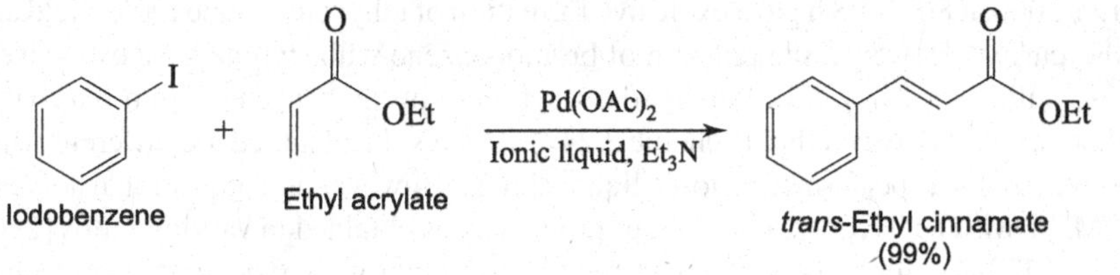

Scheme 2.244 Heck reaction in ionic liquids

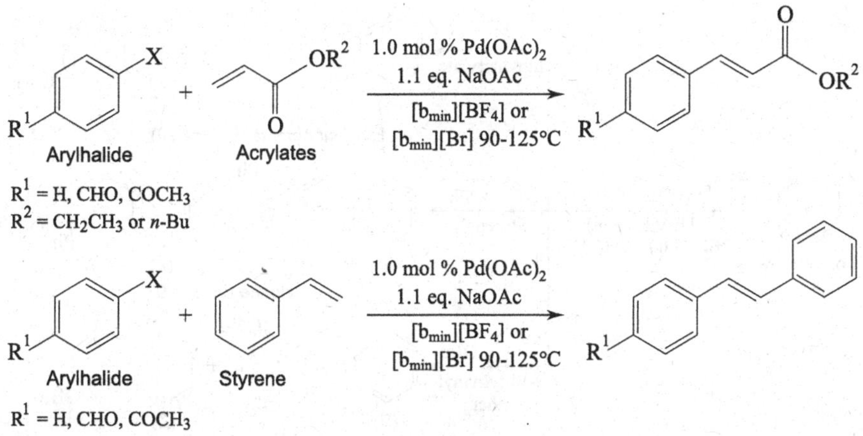

Scheme 2.245 Some more examples of Heck reaction

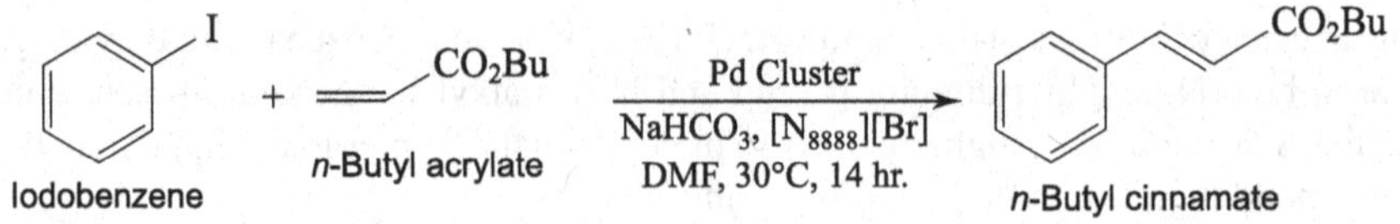

Scheme 2.246 Active catalyst in Heck reaction

Thus, the Heck reaction between iodobenzene and n-butyl acrylate in the presence of [N$_{8888}$][Br] stabilized 3 nm Pd clusters to afford n-butyl cinnamate (Scheme 2.246).

2.29.4   Heck Reaction in Polyethylene Glycol

Polyethylene glycol having molecular weight 2000 (or lower) has been used [264] as an efficient reaction medium for the Heck reaction. The stereo and regioselectivities are also different from those with conventional ionic liquids [265].

Thus, the reaction of bromobenzene with ethyl acrylate, Pd(OAc)$_2$ and TEA in PEG-2000 at 80° for 8 h gave exclusive formation of ethyl cinnamate (90% yield and 90% purity). However, the reaction of bromobenzene with styrene gave exclusively *trans* stilbene (93% yield). Finally, the reaction of bromobenzene with n-butylvinyl ether gave butylstyryl ether (Scheme 2.247). The results obtained are different when the reaction was performed in ionic liquids [265]. However, in conventional solvents (DMF, DMSO, CH$_3$CN), a mixture of products was obtained in varying ratios [266]. Thus, PEG is unique in obtaining a single regio isomer with good diastereoselection (80/20 E/Z).

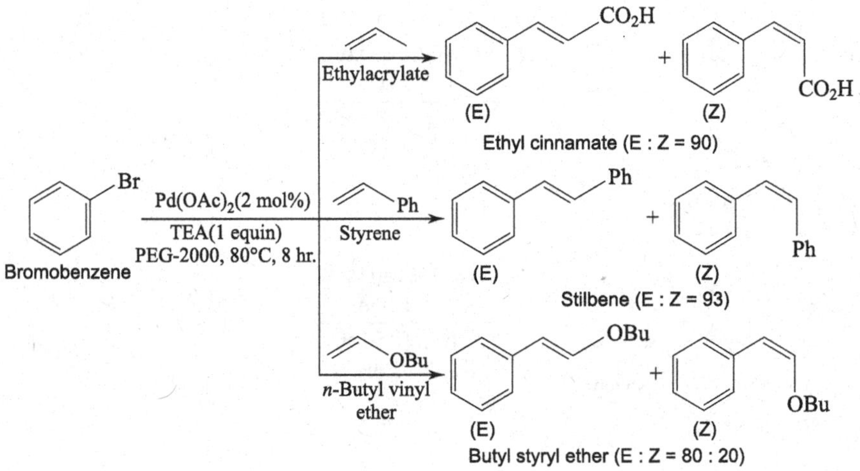

Scheme 2.247 Heck Reaction in PEG

The Heck reaction of 4-bromoanisole with all the three olefins, viz., ethyl acrylate, styrene and *n*-butyl vinyl ether gave excellent regio and stereoselectivities of 4-methoxy ethyl cinnamate (E:Z 91), 4-methoxy stilbene (E:Z 85) and butyl-*p*-bromostyryl ether (E:Z 70:30), respectively. In the Heck reaction of 4-chlorobromobenzene with butyl vinyl ether, there is exclusive formation of E-geometrical olefin. On the basis of the results obtained, it was concluded [264] that electronic factors in the aryl system control the geometry of the olefins to a certain extent; the other olefins, viz., styrene and ethyl acrylate behave normally. In fact, PEG besides being an efficient solvent also acted as a PTC for C–C bond formation.

2.29.5 Heck Reaction Using Fluorous Phase Technique

As already mentioned, fluorous phase technique has been used for Friedel–Crafts acetylation (see Sect. 2.24.2.4). Using the technique, palladium-catalysed Heck reaction was also carried out. In this reaction, there is advantage of product isolation and catalyst recovery. Now reaction kits for Pd-catalysed C–C cross-coupling reactions in FBS are commercially available. These kits have been developed by Schneider and Bannwarth Fluka Company Ltd., Buchs, Switzerland.

2.30 Hantzsch Pyridine Synthesis

This involves [267] the condensation of two molecules of β-ethyl acetoacetate with one mole of an aldehyde in the presence of ammonia. The resulting dihydropyridine is dehydrogenated with an oxidizing agent (Scheme 2.248).

The reaction is believed to involve Michael type addition of β-amino-α,β-unsaturated carbonyl compound (1) (formed from β-keto ester and ammonia) to alkylidene-1,3-dicarbonyl compound (2) (formed from β-ketoester and aldehyde). The condensation of (1) and (2) gives the intermediate (3), which cyclizes to dihydropyridine (4). Oxidation of (4) gives pyridine derivative (5) (Scheme 2.249).

2.30.1 Hantzsch Pyridine Synthesis Under Microwave Irradiation

It involves the reaction of 1,3-dicarbonyl compounds and aryl aldehyde, and ammonium nitrate (as ammonium source) under microwave irradiation [268] (Scheme 2.250).

In the above synthesis, the required pure substituted pyridines were obtained by extraction by organic solvents from the support.

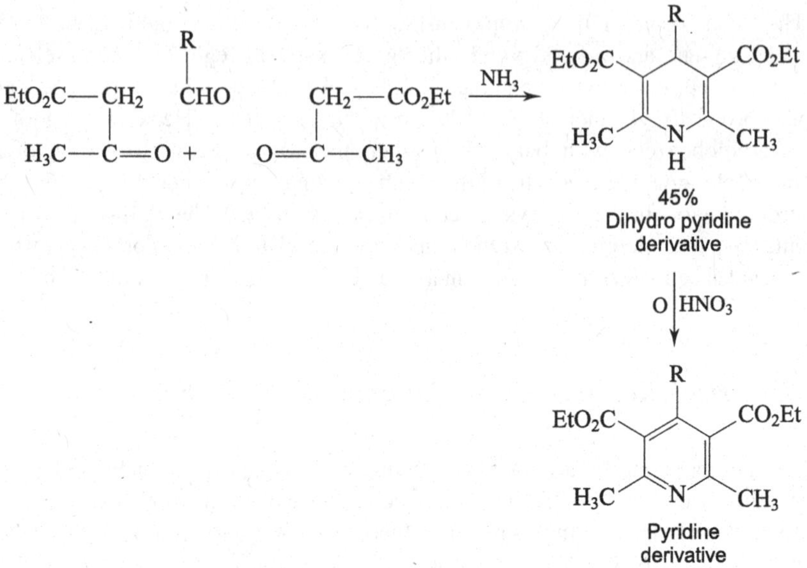

Scheme 2.248 Hantzsch pyridine synthesis

2.31 Henry Reaction

It is the aldol condensation of nitroalkanes with aldehydes [269] (Scheme 2.251). (*see* also Sect. 2.2.3).

2.31.1 Henry Reaction Under Microwave Irradiation

The condensation of nitroalkanes with carbonyl compounds under MW irradiation in the presence of catalytic amount of ammonium acetate [270] (to yield unsaturated alkenes) has been accomplished, thus avoiding the use of large excess of polluting nitrohydrocarbons usually employed in these reactions (Scheme 2.252).

Henry Reaction using Ionic Liquids

Henry reactions can be accelerated in chloroaluminate ionic liquids [271]. The tetra methyl guanidine (trifluoroacetate and lactate)-based ionic liquids has been reported as a recyclable catalyst for Henry reaction to produce 2-nitroalcohols (Scheme 2.253) [272].

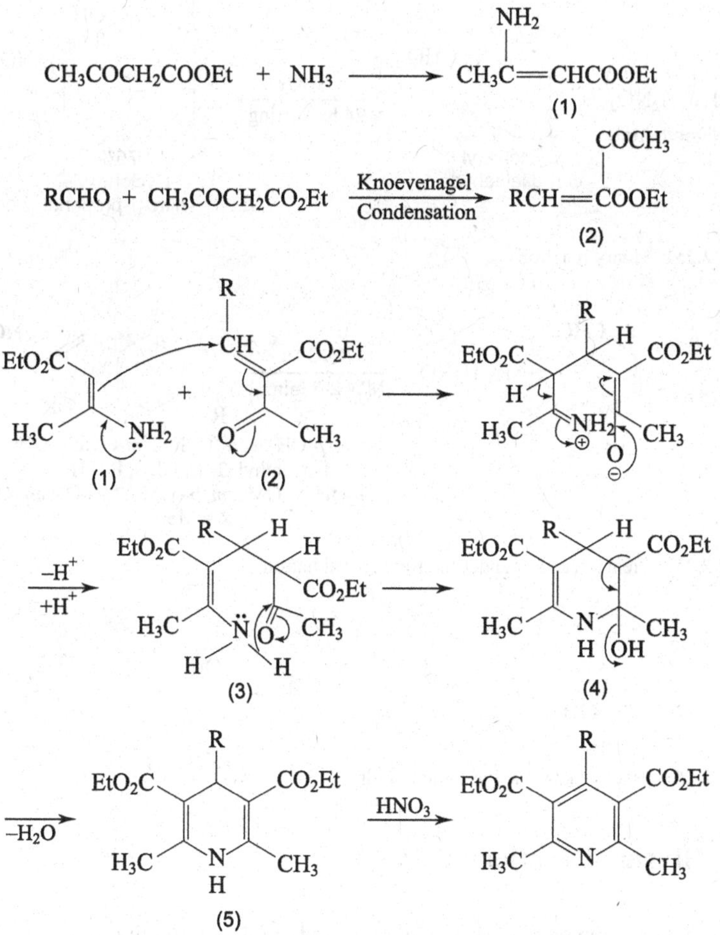

$$CH_3COCH_2COOEt \ + \ NH_3 \longrightarrow CH_3\overset{\overset{\displaystyle NH_2}{|}}{C}=CHCOOEt$$

(1)

$$RCHO \ + \ CH_3COCH_2CO_2Et \xrightarrow[\text{Condensation}]{\text{Knoevenagel}} RCH=\overset{\overset{\displaystyle COCH_3}{|}}{C}COOEt$$

(2)

Scheme 2.249 Mechanism of Hantzsch reaction

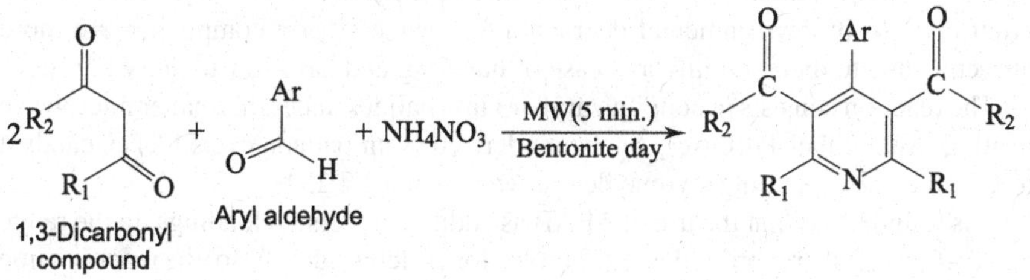

Scheme 2.250 Hantzsch pyridine synthesis under microwave irradiation

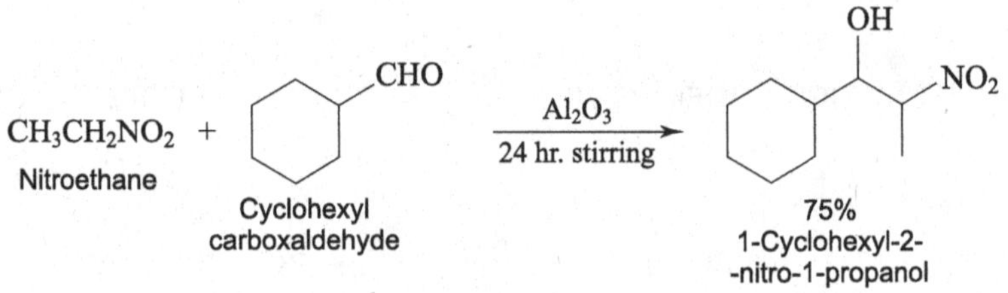

Scheme 2.251 Henry reaction

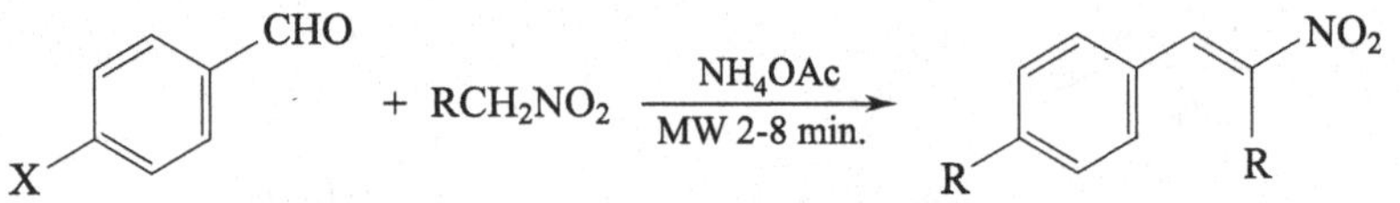

X = H, p-OH, m, p-(OMe)2, m-OMe-p-OH,
1-naphthyl, 2-naphthyl; R = H
R = H, p-OH, m-OMe, m, p-(OMe)2, m-OMe-p-OH;
R = Me

Scheme 2.252 Henry reaction under microwave irradiation

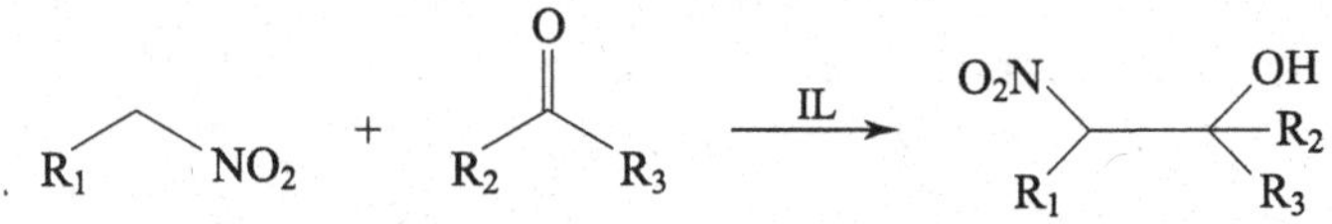

Scheme 2.253 Heck reaction using ionic liquids

2.32 Hiyama Reaction [273]

It is a carbon–carbon bond-forming reaction and involves palladium-catalysed cross-coupling of organosilicon compounds with organic halides. It is a good alternative to other coupling reactions which employ different organometallic reagents from the point of view of environmental degradation. Organo silicon compounds are more attractive due to their stability and ease of handling and have low toxicity [273].

The reaction consists in coupling aryl or vinyl halides under thermal or microwave heating using either Pd(OAc)$_2$ or oxime derived from pallada cycle (A) as catalyst to give the corresponding styrene derivatives (Scheme 2.254).

It is found [274] that the use of TBAB as additive lowers Pd loadings, in the range of 0.5–1 mol% of aryl bromides and 2 mol% for arylchlorides. Also, the reaction time can be reduced from about 24 h under thermal heating in a pressure tube under the same conditions to 10–25 min by using MW irradiation. The reaction of activated aryl chlorides with vinyltrimethoxysilane to the corresponding styrene proceeded only under microwave conditions. It is believed [274] that the active catalytic species could be Pd nanoparticles.

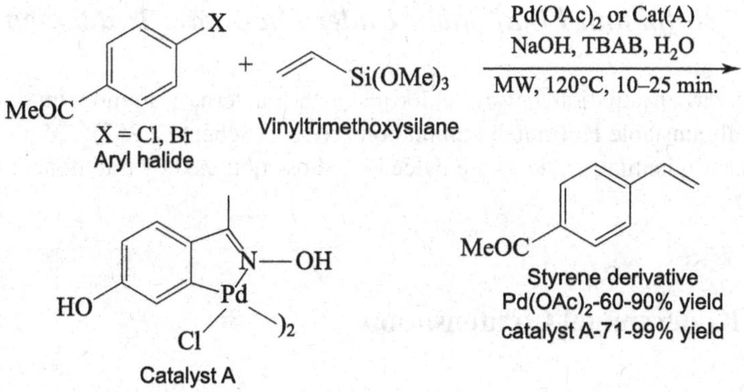

Scheme 2.254 Hiyama reaction

2.33 Hofmann Elimination [275]

It involves the formation of olefins by heating quaternary ammonium hydroxides (Scheme 2.255).

The mechanism of this involves abstraction of a proton from the β-carbon atom by hydroxide ion with the simultaneous expulsion of a tertiary base from the α-carbon atom. Thus, this reaction results in the formation of a double bond between α- and β-carbon atom (Scheme 2.256).

These eliminations are governed by the well-known Hoffmann rule, which states that the charged substrates yield the least substituted olefins.

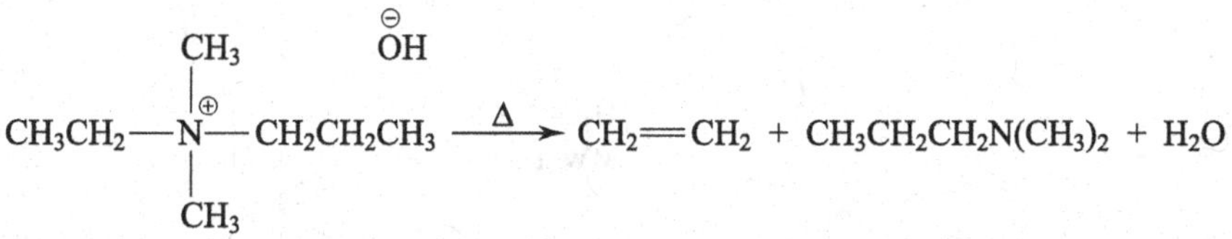

Scheme 2.255 Heck homocouplings

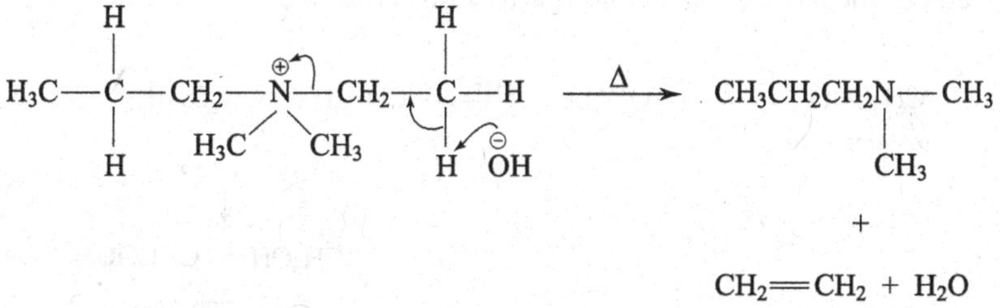

Scheme 2.256 Mechanism of Hofmann elimination

2.33.1   *Hoffmann Elimination Under Microwave Irradiation*

On microwave irradiation in water-chloroform, the quaternary ammonium salts yield a thermally unstable Hoffmann elimination product (Scheme 2.257).

The near quantitative yields are twice than those obtained by traditional methods [276].

2.34   Knoevenagel Condensation

Base-catalysed condensation between an aldehyde or ketone, with any compound having an active methylene group (especially malonic ester) is called Knoevenagel condensation [277] (Scheme 2.258).

The base used in the above condensation is a weak base like ammonia or amine (primary or secondary). However, when condensation is carried out in the presence of pyridine as a base, decarboxylation usually occurs during the condensation. This is known as **Doebner modification** [278] (Scheme 2.259).

The Knoevenagel reaction is more useful with aromatic aldehydes, since with aliphatic aldehydes, the product obtained undergoes **Michael condensation**. As an example, tetraethyl propane-1,1,3,3-tetracarboxylate is obtained by Knoevenagel condensation of formaldehyde with diethyl malonate in the presence of diethylamine, followed by Michael addition reaction to yield the final product (Scheme 2.260).

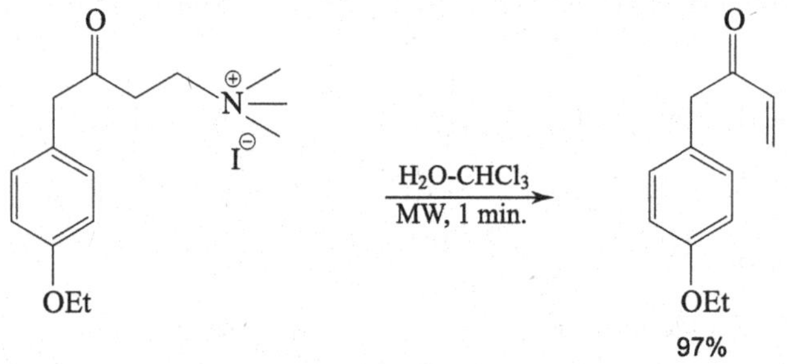

$$\frac{H_2O\text{-}CHCl_3}{MW, 1\ min.}$$

97%

Scheme 2.257   Hoffmann elimination under microwave irradiation

$$CH_3CHO\ +\ CH_2(COOH)_2\ \xrightarrow{\text{Base}}\ CH_3CH{=\!=}C(COOH)_2$$

Acetaldehyde

$$\Delta \Big| {-}CO_2$$

$$CH_3CH{=\!=}CHCOOH$$

Crotonic acid

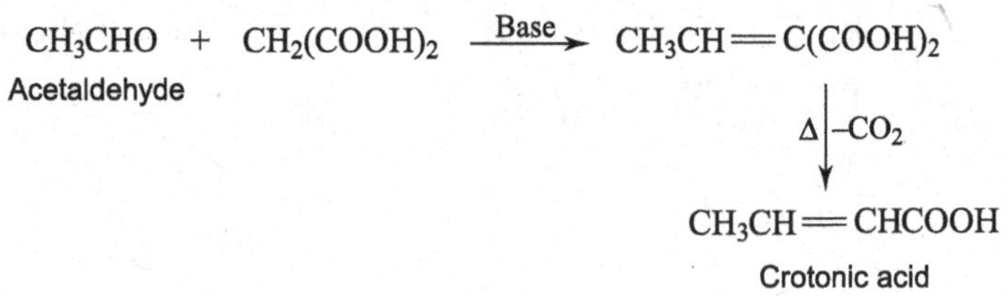

Scheme 2.258   Knoevenagel condensation

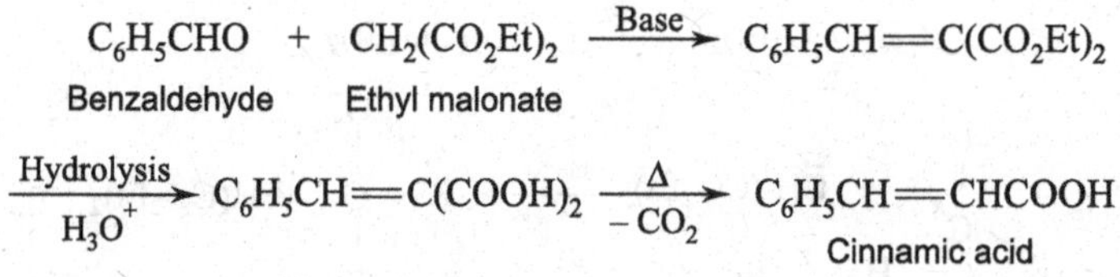

$$C_6H_5CHO + CH_2(CO_2Et)_2 \xrightarrow{\text{Base}} C_6H_5CH\!\!=\!\!C(CO_2Et)_2$$

Benzaldehyde   Ethyl malonate

$$\xrightarrow[\text{H}_3\text{O}^+]{\text{Hydrolysis}} C_6H_5CH\!\!=\!\!C(COOH)_2 \xrightarrow[-CO_2]{\Delta} C_6H_5CH\!\!=\!\!CHCOOH$$

Cinnamic acid

Scheme 2.259 Doebner modification of Knoevenagel condensation

$$O\!\!=\!\!CH_2 + \bar{C}H(CO_2Et)_2 \rightleftharpoons \xrightarrow{Et_2NH} HOCH_2\!\!-\!\!CH(CO_2Et)_2$$

Formaldehyde

$$\Big\downarrow -H_2O$$

$$(EtO_2C)_2CHCH_2CH(CO_2Et)_2 \xleftarrow[\text{Michael addn.}]{\bar{C}H(CO_2Et)_2} CH_2\!\!=\!\!C(COOEt)_2$$

Adduct
Tetraethyl propane-
-1,1,3,3-tetracarboxylate

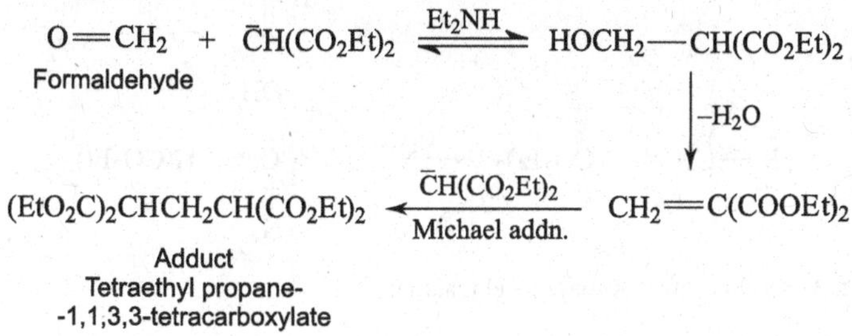

Scheme 2.260 Synthesis of tetraethyl propane 1,1,3,3-tetracarboxylate

The effectiveness of various activating groups in the active methylene compounds is found to be in the order (Scheme 2.261).

Ketones do not undergo Knoevenagel condensation with malonic ester, but can react with more active cyanoacetic acid or its ester. For example, acetone forms [279] isopropylidene cyanoacetic ester when condensed with ethyl cyanoacetate (Scheme 2.262).

The Knoevenagel reaction is reversible and the equilibrium normally lies towards left giving low yields. The yield can be improved by carrying out the reaction in benzene and removal of the formed water by azeotropic distillation using a Dean stark apparatus. This is referred to as **Cope-Knoevenagel reaction**.

$$NO_2 > CN > COCH_3^- > COC_6H_5 > COOC_2H_5$$

Scheme 2.261 Activating groups in Knoevenagel condensation

$$(CH_3)_2C\!\!=\!\!O + CH_2\!\!\underset{CO_2Et}{\overset{CN}{\big\langle}} \longrightarrow (CH_3)_2C\!\!=\!\!C\!\!\underset{CO_2Et}{\overset{CN}{\big\langle}}$$

Acetone   Ethyl cyanoacetate   Isopropylidene
                                                    cyanoacetic ester

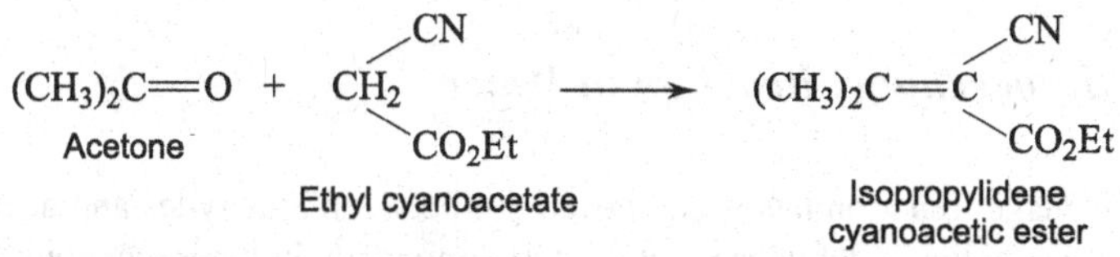

Scheme 2.262 Synthesis of isopropylidene cyanoacetic ester

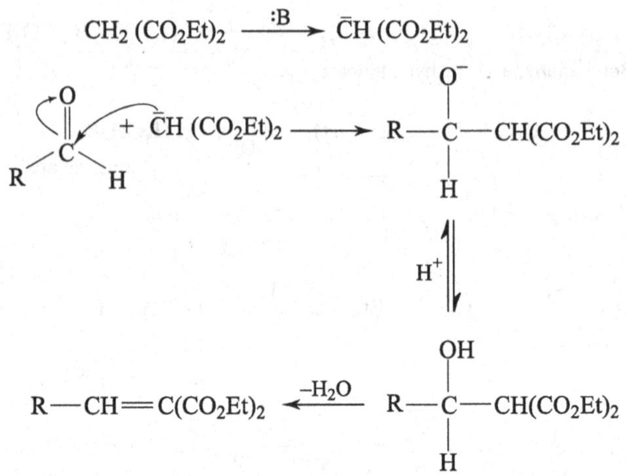

$$CH_2 (CO_2Et)_2 \xrightarrow{\ :B\ } \bar{C}H (CO_2Et)_2$$

$$R-CH=C(CO_2Et)_2 \xleftarrow{-H_2O} R-\underset{\underset{H}{|}}{\overset{\overset{OH}{|}}{C}}-CH(CO_2Et)_2$$

Scheme 2.263 Mechanism of Knoevenagel reaction

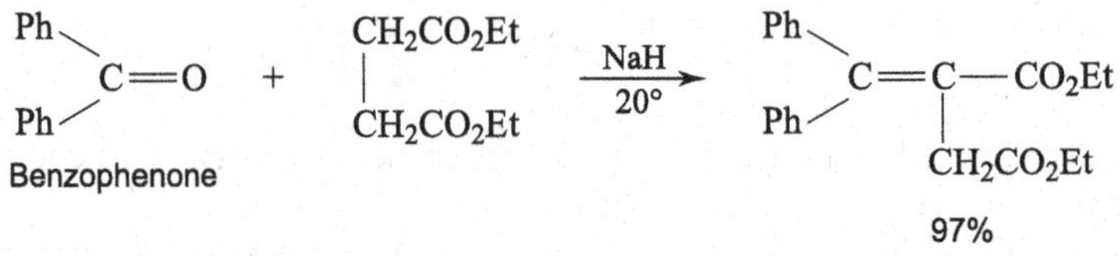

Scheme 2.264 Reaction of Ph_2CO with succinic ester

Mechanism

The base removes a proton from the active methylene compound to give a carbanion (which is resonance stabilized). The carbanion then attacks the carbonyl carbon of the aldehyde or ketone. Subsequent protonation of the anion followed by dehydration yields the product (Scheme 2.263).

Ketones or aldehydes also react with a succinic ester in the presence of sodium hydride to give the corresponding condensation product (Scheme 2.264).

This condensation is known as **Knoevenagel–Stobbe condensation** [280].

2.34.1 Knoevenagel Reaction in Water

The Knoevenagel reaction has been carried out between aldehydes and acetonitrile in water. Thus, salicylaldehydes react with malononitrile in heterogeneous aqueous alkaline medium at room temperature to give *o*-hydroxy benzylidenemalononitrile [281], which are converted by acidification and heating to give 3-cyanocoumarins in good yield (Scheme 2.265).

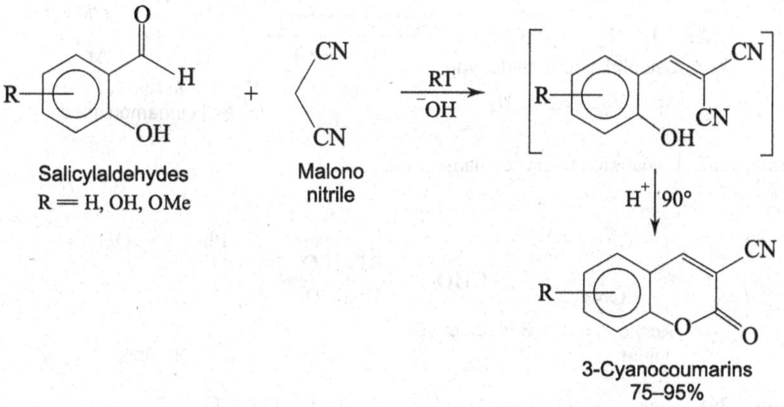

Scheme 2.265 Knoevenagel reaction in water

The condensation of substituted acetonitriles with salicylaldehydes requires the presence of catalytic amount of CTABr. It is found that the aqueous phase reaction [281] gives better yield (Scheme 2.266).

The condensation of benzaldehyde with aryl acetonitrile does not take place in water, but requires the presence of CTACl or TBACl to give high yields of aryl cinnamonitriles [282] (Scheme 2.267).

Knoevenagel-type addition products can be obtained [283] by the reaction of acrylic derivatives in the presence of 1,4-diazabicyclo [2.2.2] octane (DABCO) (Scheme 2.268).

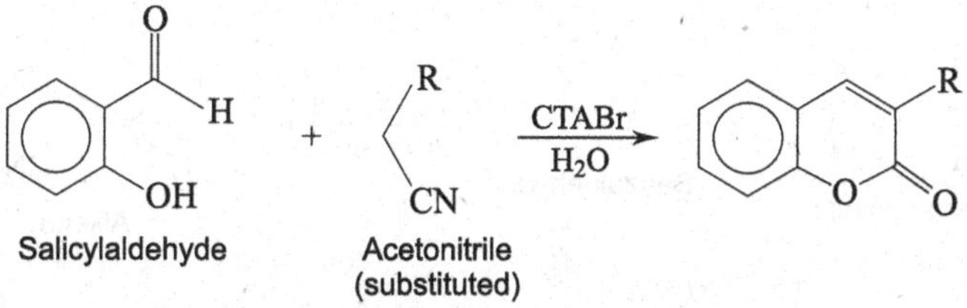

R	Yield of Curmarin	
	in water	in ethanol
CN	90	70
CO$_2$Et	66	35
NO$_2$	87	80
2-Py	98	55

Scheme 2.266 Knoevenagel condensation in the presence of CTABr aqueous phase

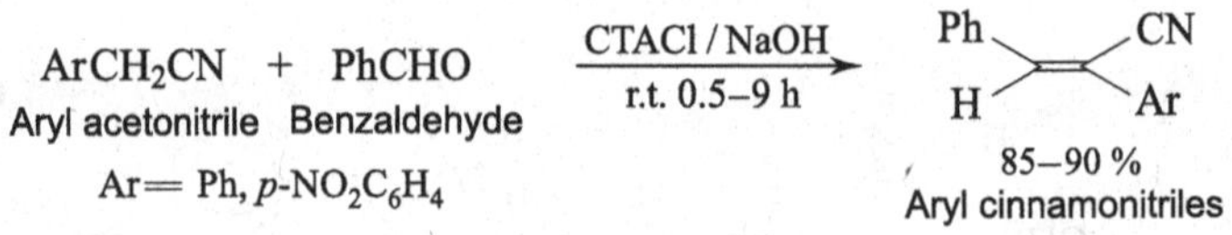

ArCH₂CN   +   PhCHO
Aryl acetonitrile   Benzaldehyde
Ar= Ph, p-NO₂C₆H₄

$$\xrightarrow[\text{r.t. 0.5–9 h}]{\text{CTACl / NaOH}}$$

Ph   CN
   ⟍   ⟋
   ‖
   ⟋   ⟍
H   Ar
85–90 %
Aryl cinnamonitriles

Scheme 2.267 Preparation of aryl cinnamonitriles

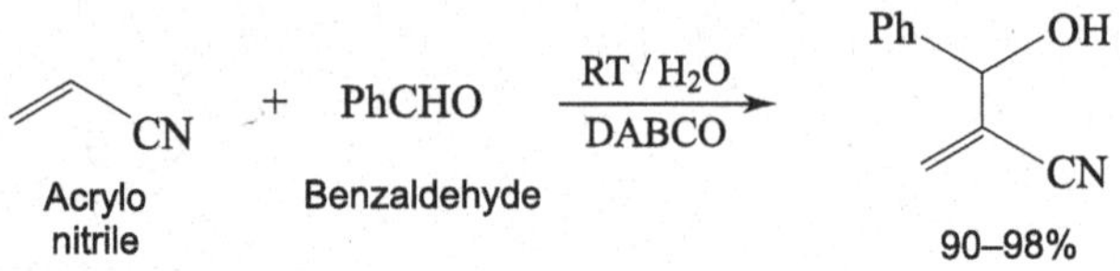

CN   +   PhCHO
Acrylo   Benzaldehyde
nitrile

$$\xrightarrow[\text{DABCO}]{\text{RT / H}_2\text{O}}$$

Ph   OH
   ⟍   ⟋
   ‖
   ⟋   ⟍
        CN
90–98%

Scheme 2.268 Knoevenagel type addition in the presence of DABCO

2.34.2   Knoevenagel Condensation Under Microwave Irradiation

The Knoevenagel condensation reaction involving active methylene compounds and carbonyl compounds has been reported [284] using MW irradiation (Scheme 2.269). The reactions are conducted in open vessels that lead to the efficient removal of water, thus circumventing the use of Dean-Stark apparatus.

Salicylaldehydes undergo Knoevenagel condensation with a variety of ethyl acetate derivatives under basic conditions (piperidine) on MW irradiation to yield coumarins [285] (Scheme 2.270).

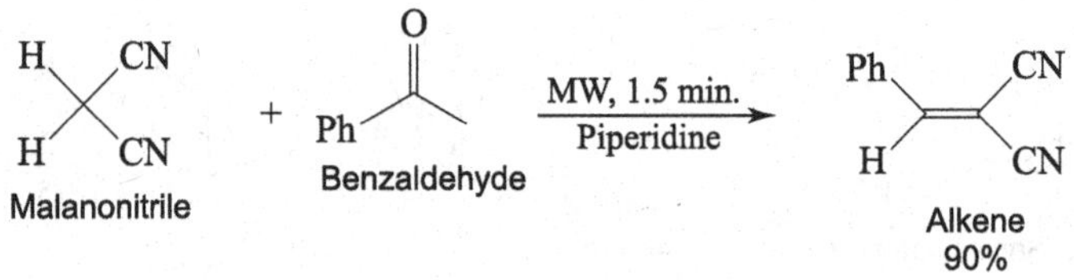

H   CN
 ⟍ ⟋
  ⟋⟍
H   CN
Malanonitrile
   +   Ph—C(=O)—
        Benzaldehyde

$$\xrightarrow[\text{Piperidine}]{\text{MW, 1.5 min.}}$$

Ph   CN
   ⟍   ⟋
   ‖
   ⟋   ⟍
H   CN
Alkene
90%

Scheme 2.269 Knoevenagel condensation under microwave irradiation

Salicylaldehydes   +   Alkyl substituted ethylacetate (R_2, CO₂Et)   → (Pipereni MW) → Coumarins

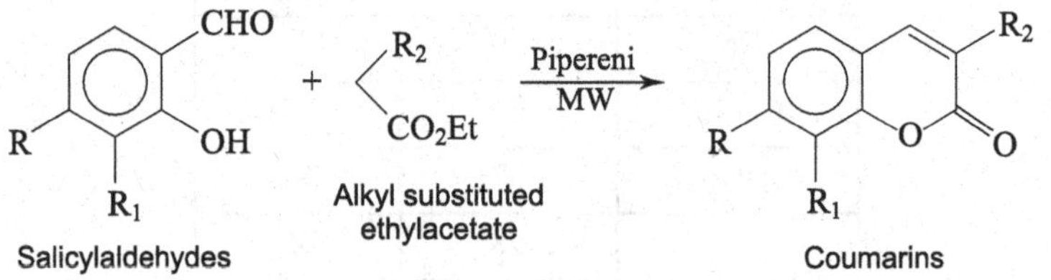

Scheme 2.270 Knoevenagel condensation under basic condition and microwave irradiation

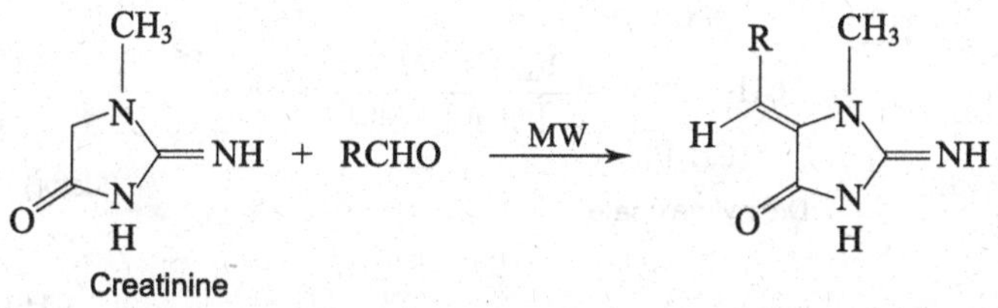

Scheme 2.271 Knoevenagel condensation using focused microwave irradiation under solvent-free conditions

2.34.3 Knoevenagel Condensation in Solid State

Knoevenagel reaction has been carried out in dry media [286]. The method consists of adding solid inorganic support to a solution of aromatic aldehyde and diethylmalonate in acetone. The adsorbed material was mixed properly, dried in air (beaker) and placed in an alumina bath [287] inside the microwave oven for 2–3 min at medium power level (600 W) intermittently at 0.5 min intervals, at 102 °C. The product was isolated by extraction of the reaction mixture with alcohol.

An expedite Knoevenagel condensation of creatinine with aldehydes has been achieved using focused MW irradiation (40–60 W) under solvent-free reaction conditions at 160–170 °C (Scheme 2.271) [288].

2.34.4 Knoevenagel Condensation in Ionic Liquids

The Knoevenagel condensation of benzaldehyde with malononitrile in the presence of KOH dissolved in $[b_{min}][PF_6]$ gave only low yield of the styrene derivative (Scheme 2.272) [289].

The low yield of the product was due to the formation of anion of the ionic liquid, which reacted with benzaldehyde (for details, see Sect. 2.8.3). The yield could be increased as the substrate concentration was increased and the ionic liquid was reused. The ionic liquid could be reused up to five times without the need of additional base.

Knoevenagel condensation can also be carried out by using chloroaluminate ionic liquids [290], which have a variable Lewis activity, such as 1-butyl-3-methylimidazolium chloroaluminate, $[b_{min}]Cl. AlCl_3$, $X(AlCl_3) = 0.67$, where X is the mol fraction and 1-butyl pyridinium chloroaluminate [bpy]Cl. $AlCl_3$, $X(AlCl_3)$

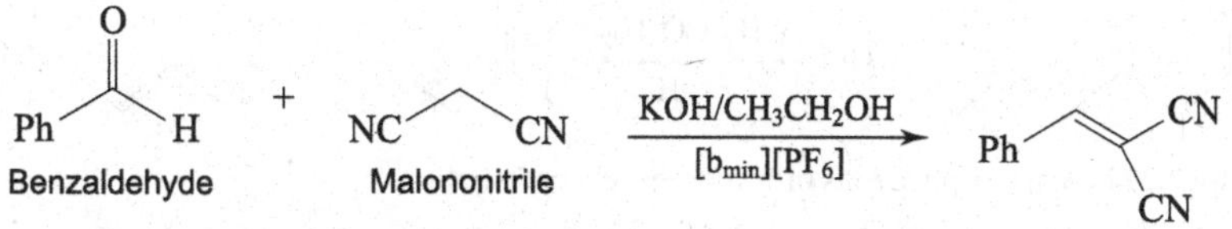

Scheme 2.272 Knoevenagel condensation in ionic liquids

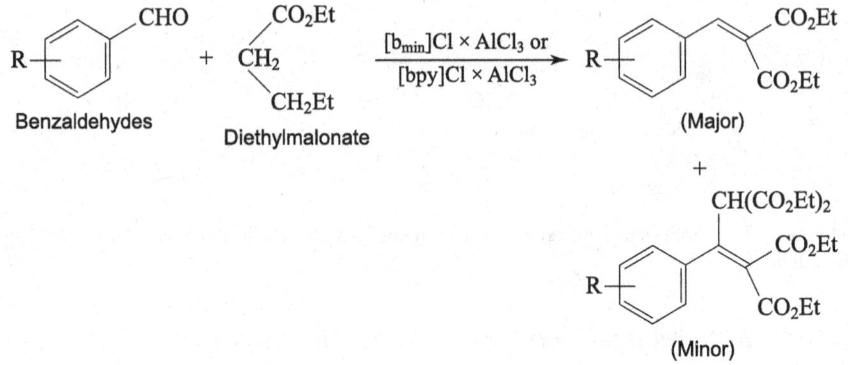

Scheme 2.273 Knoevenagel condensation using chloroaluminate ionic liquids

$= 0.67$. Ionic liquids work as Lewis acid catalyst and solvent in the Knoevenagel condensation. A typical example is given in Scheme 2.273.

2.34.5 Applications

Knoevenagel reactions (or condensation) is of great importance. Some applications are given in Scheme 2.274.

(*i*) Synthesis of α,β-unsaturated carboxylic acids [290].
(*ii*) Synthesis of conjugated carboxylic acids [291].

$$C_6H_5\,CH{=}CH{-}CHO + CNCH_2CO_2Et \xrightarrow{KF} C_6H_5{-}CH{=}CH{-}CH{=}C\begin{smallmatrix} CO_2Et \\ CN \end{smallmatrix}$$

36%

(*iii*) Synthesis of malic acid.

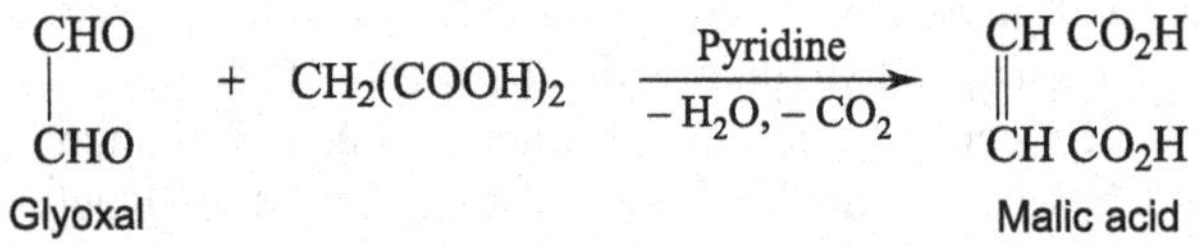

(*iv*) Synthesis of substituted ethyl cinnamates.

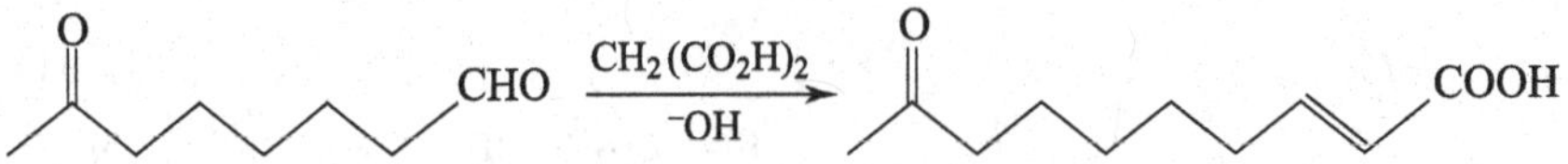

Scheme 2.274 Some applications of Knoevenagel condensation

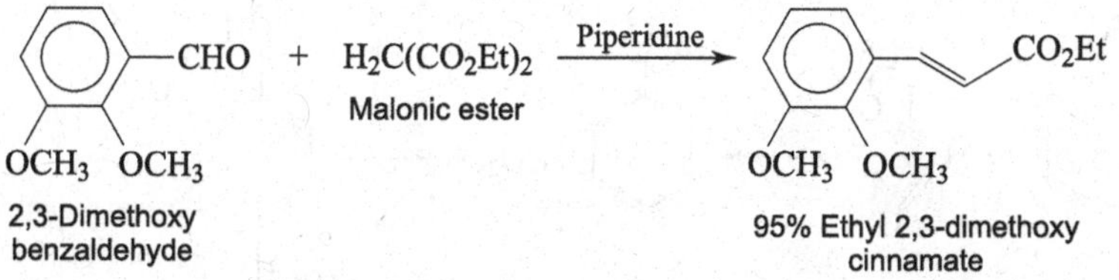

2,3-Dimethoxy benzaldehyde

95% Ethyl 2,3-dimethoxy cinnamate

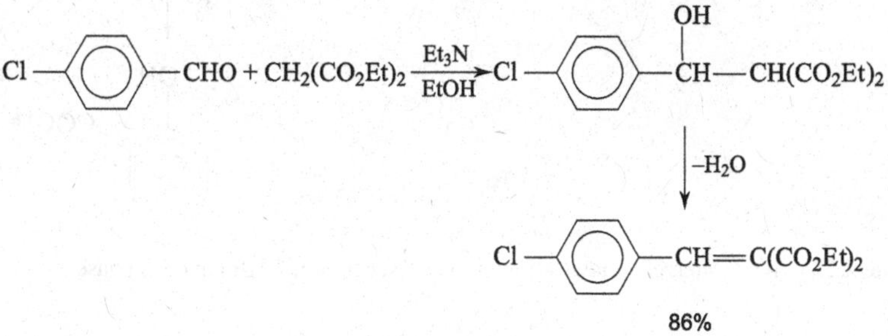

86%

2.35 Kolbe–Schmitt Reaction

The formation of aromatic hydroxy acids by the carbonylation of phenolates, mostly in ortho position by CO_2 is known as Kolbe–Schmitt reaction [292]. A typical example is the preparation of salicylic acid by the reaction of sodium phenoxide with CO_2 at 120 °C under pressure (Scheme 2.275).

The reaction is particularly facile with di or trihydroxy phenols.

The mechanism of this reaction is believed to involve the attack of CO_2 (as electrophile) on benzene nucleus of phenoxide ion. The steps involved in the mechanism are as given in Scheme 2.276.

In case if Kolbe–Schmitt reaction is conducted at higher temperature (230 °C), thermodynamic more stable para product predominates.

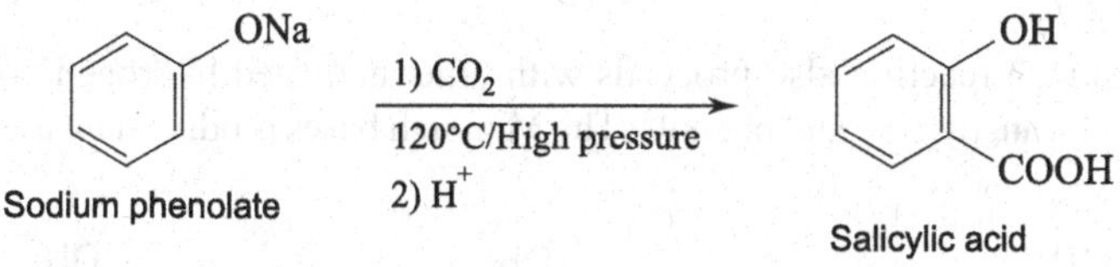

Sodium phenolate

Salicylic acid

Scheme 2.275 Kolbe–Schmitt reaction

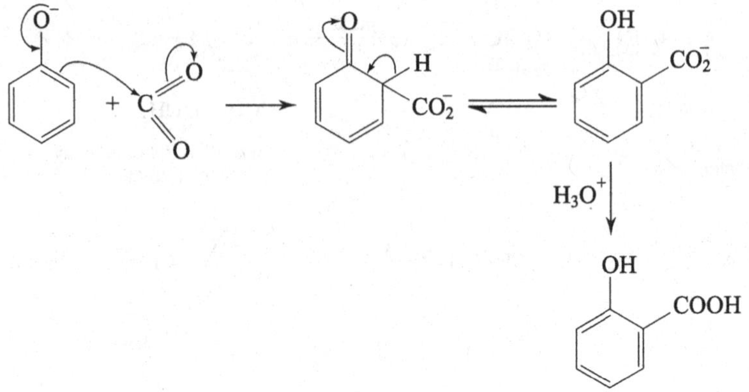

Scheme 2.276 Knoevenagel condensation in the presence of CTABr aqueous phase

2.35.1 Kolbe–Schmitt Reaction in SC CO$_2$

A direct carbonylation reaction occurs in SC CO$_2$ [293, 294]. The *ortho* and *para* selectivity between isomers of hydroxy benzoic acid is of particular interest (Scheme 2.277).

2.36 Mannich Reaction [295]

Compounds containing at least one active hydrogen (ketones, nitroalkanes, β-ketoesters, β-cyanoacids etc.) condense with formaldehyde and primary or secondary amine or ammonia (in the form of hydrochloride) to give product (β-amino carbonyl compounds) known as Mannich base [295]. As an example, dimethylamine on reaction with formaldehyde and acetophenone gives the corresponding Mannich base (Scheme 2.278).

The Mannich reaction also proceeds with other activated hydrogen compounds like indole, furan, pyrrole and phenols. The Mannich bases produced by the Mannich

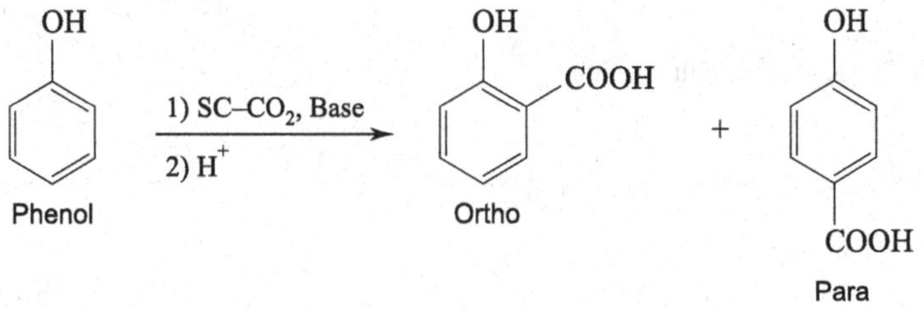

Scheme 2.277 Kolbe–Schmitt Reaction in SC-CO$_2$

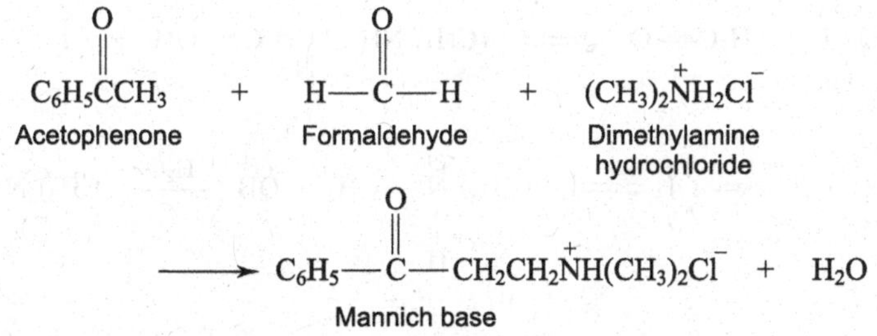

$$\underset{\text{Acetophenone}}{C_6H_5\overset{\displaystyle O}{\overset{\|}{C}}CH_3} \; + \; \underset{\text{Formaldehyde}}{H—\overset{\displaystyle O}{\overset{\|}{C}}—H} \; + \; \underset{\substack{\text{Dimethylamine}\\\text{hydrochloride}}}{(CH_3)_2\overset{+}{N}H_2Cl^-}$$

$$\longrightarrow \; \underset{\text{Mannich base}}{C_6H_5—\overset{\displaystyle O}{\overset{\|}{C}}—CH_2CH_2\overset{+}{N}H(CH_3)_2Cl^-} \; + \; H_2O$$

Scheme 2.278 Mannich reaction

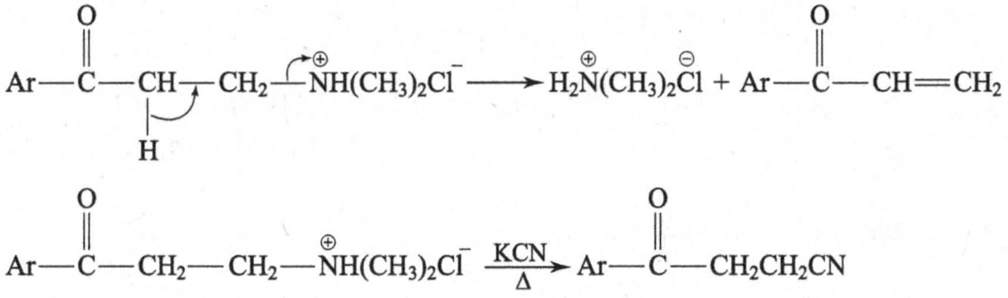

$$\underset{\quad\quad\quad H}{Ar—\overset{\displaystyle O}{\overset{\|}{C}}—CH—CH_2—\overset{\oplus}{N}H(CH_3)_2Cl^-} \longrightarrow H_2\overset{\oplus}{N}(CH_3)_2\overset{\ominus}{Cl} + Ar—\overset{\displaystyle O}{\overset{\|}{C}}—CH{=}CH_2$$

$$Ar—\overset{\displaystyle O}{\overset{\|}{C}}—CH_2—CH_2—\overset{\oplus}{N}H(CH_3)_2Cl^- \xrightarrow[\Delta]{KCN} Ar—\overset{\displaystyle O}{\overset{\|}{C}}—CH_2CH_2CN$$

Scheme 2.279 Mannich reaction of great synthetic importance

reaction are of great synthetic importance. Thus, Mannich base eliminates an amine hydrochloride on heating to yield α,β-unsaturated ketones. Another usefulness of Mannich base is the replacement of dimethylamino group by nitrile on heating with KCN (Scheme 2.279).

The Mannich reaction is believed to involve the formation of the intermediate methylene ammonium salt (1), which condenses either with enol form of the ketone (acid catalyses the conversion of keto into enol form) or with the carbanion derived from the ketone (small amount of amine acting as base abstracts the α-hydrogen) (Scheme 2.280).

2.36.1 Mannich Reaction in Water

The original Mannich reaction needs drastic conditions and gives low yields. An aqueous MW-assisted Mannich reaction has been reported [296] by the reaction of acetophenones, secondary amines in the form of their hydrochlorides and trioxy methylene as a source of formaldehyde (Scheme 2.281).

The above reaction (Scheme 4) can be performed in higher yields in shorter reaction times (20–50 s) by using combined microwave and ultrasound conditions [297].

A novel type of Mannich reaction involving the condensation of an aldehyde, a primary or secondary amine and a terminal alkyne in the presence of Cu(I) iodide,

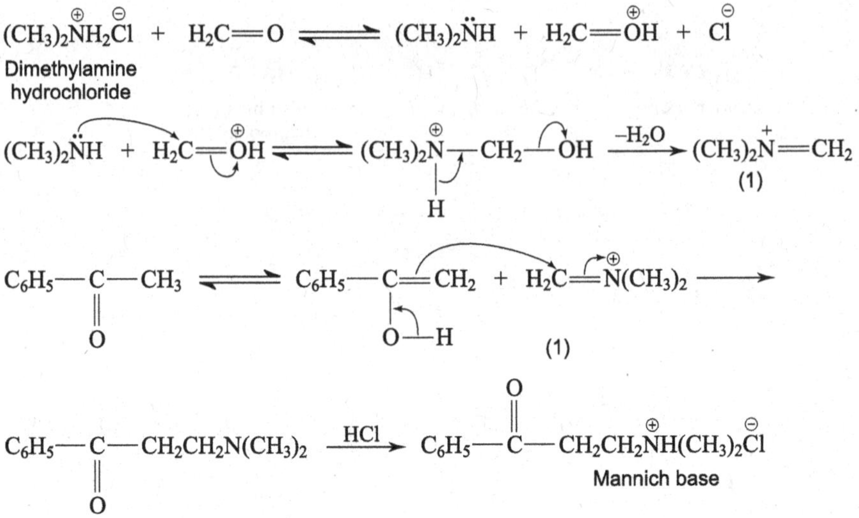

$$(CH_3)_2\overset{\oplus}{N}H_2\overset{\ominus}{Cl} + H_2C{=}O \rightleftharpoons (CH_3)_2\ddot{N}H + H_2C{=}\overset{\oplus}{O}H + \overset{\ominus}{Cl}$$

Dimethylamine
hydrochloride

$$(CH_3)_2\ddot{N}H + H_2\overset{\frown}{C}{=}\overset{\oplus}{O}H \rightleftharpoons (CH_3)_2\underset{\underset{H}{|}}{N}{-}CH_2{-}\overset{\frown}{O}H \xrightarrow{-H_2O} (CH_3)_2\overset{+}{N}{=}CH_2 \quad (1)$$

$$C_6H_5{-}\underset{\underset{O}{\|}}{C}{-}CH_3 \rightleftharpoons C_6H_5{-}\underset{\underset{O{-}H}{|}}{C}{=}CH_2 + H_2\overset{\frown}{C}{=}\overset{\oplus}{N}(CH_3)_2 \longrightarrow$$

$$(1)$$

$$C_6H_5{-}\underset{\underset{O}{\|}}{C}{-}CH_2CH_2N(CH_3)_2 \xrightarrow{HCl} C_6H_5{-}\underset{\underset{O}{\|}}{C}{-}CH_2CH_2\overset{\oplus}{N}H(CH_3)_2\overset{\ominus}{Cl}$$

Mannich base

Scheme 2.280 Mechanism of mannich reaction

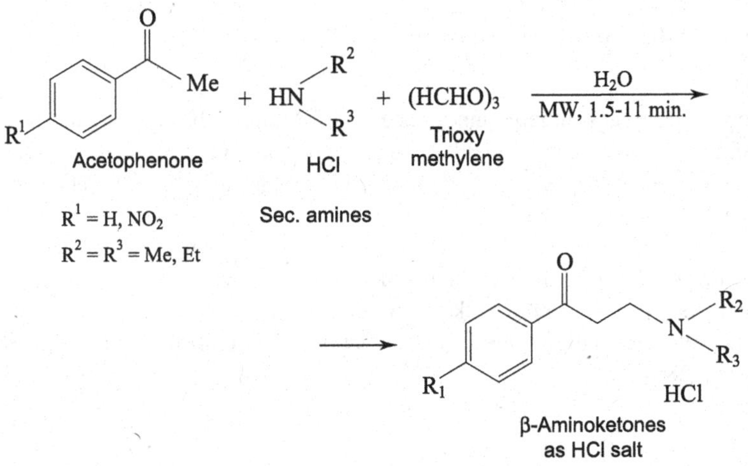

R[1] = H, NO$_2$
R[2] = R[3] = Me, Et

Sec. amines

β-Aminoketones
as HCl salt

Scheme 2.281 Mannich reaction in water

which promotes activation of the C–H bond of alkyne was reported in Scheme 2.282 [298].

Using (S) proline methyl ester as chiral amine, an asymmetric synthesis of propargyl amines was developed with high distereoselectivity (95:5 for R [1] = R [4] = Ph).

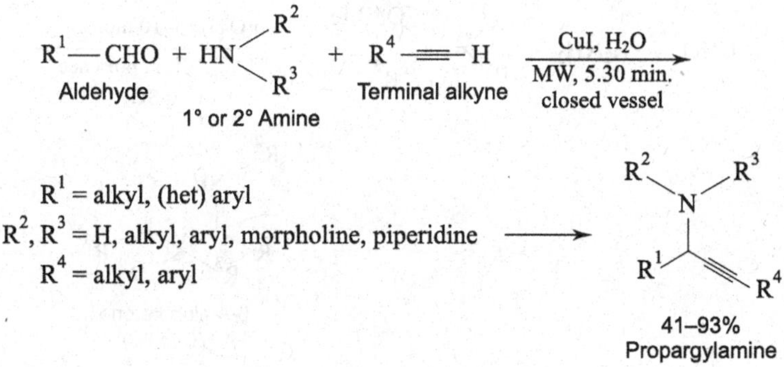

$$R^1-CHO + HN\underset{R^3}{\overset{R^2}{\diagdown}} + R^4-\!\!\!\equiv\!\!\!-H \xrightarrow[\substack{MW, 5.30\ min.\\ closed\ vessel}]{CuI,\ H_2O}$$

Aldehyde   1° or 2° Amine   Terminal alkyne

R^1 = alkyl, (het) aryl

R^2, R^3 = H, alkyl, aryl, morpholine, piperidine $\longrightarrow$

R^4 = alkyl, aryl

41–93%
Propargylamine

Scheme 2.282 A novel type of Mannich reaction

2.36.2 *Mannich-Type Reactions*

A Mannich-type reaction has been developed. In this procedure, the imines react with enolate (especially trimethylsilyl ethers) to give β-amino ketones [299]. The general scheme for the synthesis of β-aminoketones is given in Scheme 2.283.

Imines also react with vinyl ethers in the presence of catalytic amount of Yb(OTf)$_3$ to give the corresponding β-aminoketones (Scheme 2.284) [300].

The above reaction was also used for the synthesis of β-amino esters from aldehydes using Yb(OTf)$_3$ as catalyst (Scheme 2.285) [301].

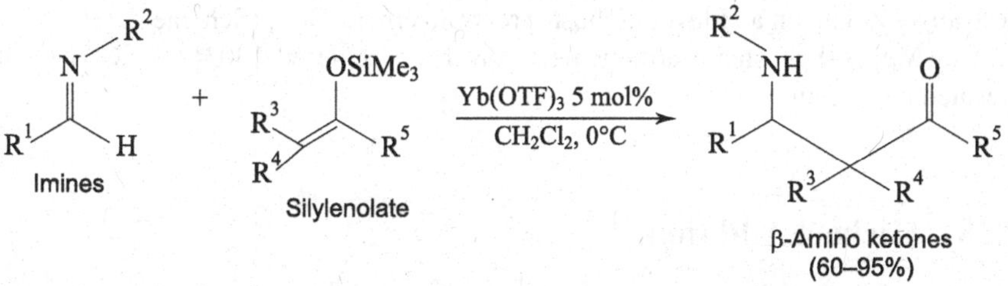

Imines   +   Silylenolate $\xrightarrow[CH_2Cl_2,\ 0°C]{Yb(OTF)_3\ 5\ mol\%}$ β-Amino ketones (60–95%)

Scheme 2.283 Knoevenagel condensation in the presence of CTABr aqueous phase

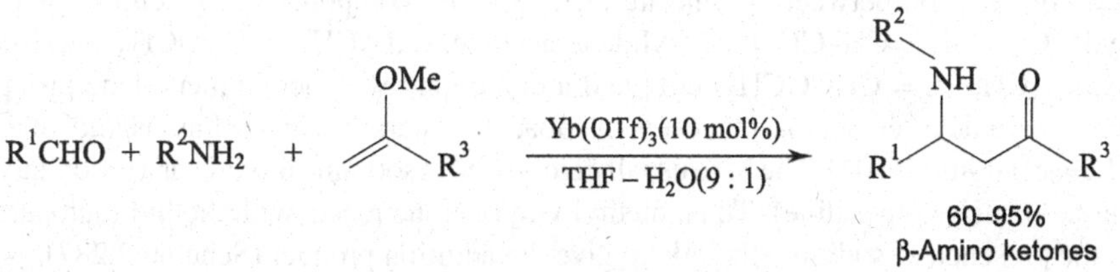

$$R^1CHO + R^2NH_2 + \underset{R^3}{\overset{OMe}{\diagup}} \xrightarrow[THF-H_2O(9:1)]{Yb(OTf)_3(10\ mol\%)}$$

60–95%
β-Amino ketones

Scheme 2.284 Synthesis of β-aminoketones

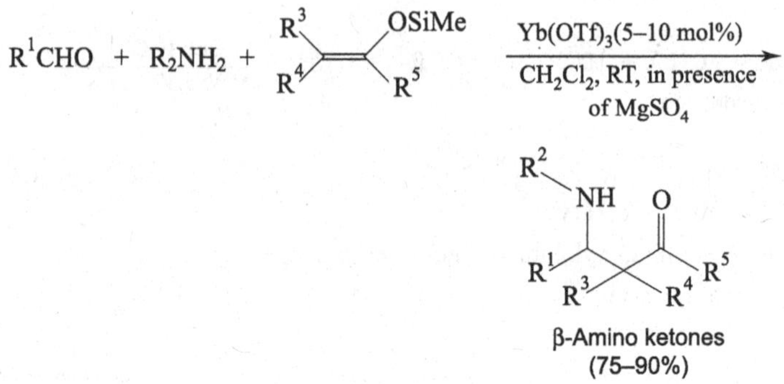

$$R^1CHO + R_2NH_2 + \underset{R^4}{\overset{R^3}{\diagdown}}C=C\underset{R^5}{\overset{OSiMe}{\diagup}} \xrightarrow[\substack{CH_2Cl_2,\ RT,\ in\ presence \\ of\ MgSO_4}]{Yb(OTf)_3(5\text{–}10\ mol\%)}$$

$$\underset{\substack{\beta\text{-Amino ketones} \\ (75\text{–}90\%)}}{\overset{R^2}{\underset{R^3}{\overset{NH}{\diagdown}}}\underset{R^1}{\overset{}{\diagup}}\underset{R^4}{\overset{O}{\diagup}}R^5}$$

Scheme 2.285 Synthesis of a typical β-aminoketones

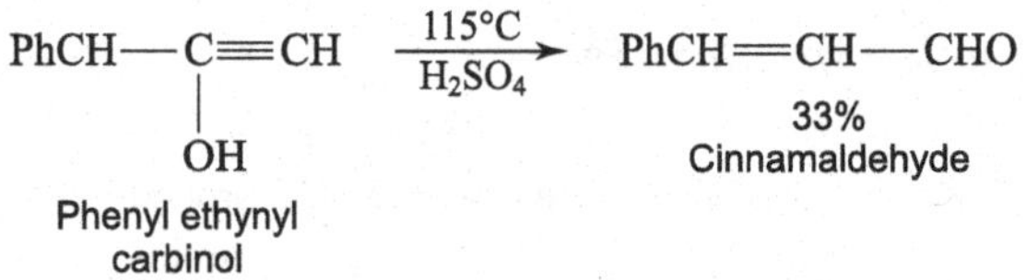

$$PhCH-C\equiv CH \xrightarrow[H_2SO_4]{115°C} PhCH=CH-CHO$$
$$\underset{\substack{OH \\ Phenyl\ ethynyl \\ carbinol}}{|} \qquad \underset{\substack{33\% \\ Cinnamaldehyde}}{}$$

Scheme 2.286 Meyer–Schuster rearrangement

2.37 Meyer–Schuster Rearrangement [302]

Acid-catalysed rearrangement of acetylenic alcohols into α,β-unsaturated carbonyl derivatives is known as Meyer–Schuster rearrangement [302] (Scheme 2.286).

The Meyer–Schuster rearrangement has been reported [303] in SC-H$_2$O at elevated temperature.

2.38 Michael Addition

The Michael addition reaction is one of the most useful C–C bond-forming reactions and has wide synthetic applications in organic synthesis. The base-catalysed addition reaction between α,β-unsaturated carbonyl compounds [e.g., cinnamaldehyde, $C_6H_5CH{=}{=}CH{-}CHO$; benzylidene acetone, $C_6H_5CH{=}{=}CHCOCH_3$; mesityl oxide, $(CH_3)_2C{=}{=}CHCOCH_3$) etc.] and a compound with active methylene group (e.g., malonic ester, acetoacetic ester, cyanoacetic esters, nitroparaffins) is known as Michael addition [304]. The base usually employed is sodium ethoxide or a secondary amine (usually piperidine). Thus, methyl vinyl ketone reacts with diethyl malonate in the presence of sodium ethoxide to give the addition product (Scheme 2.287).

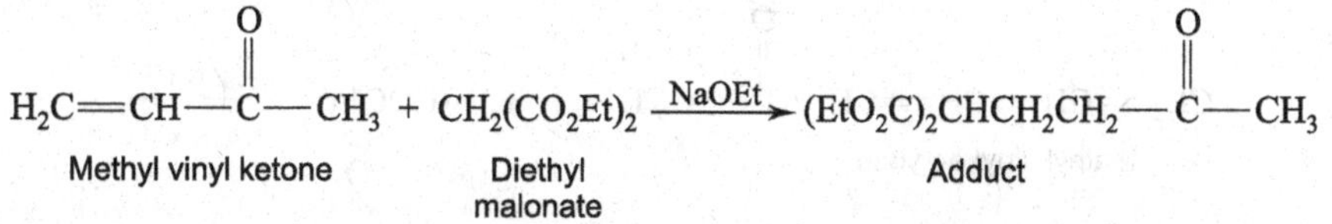

Scheme 2.287 Michael addition

Mechanism

Michael addition is regarded as nucleophilic addition of carbanions to α,β-unsaturated compounds. The base generates a carbanion (1) from active methylene compound, which then adds to the β-carbon of the α,β-unsaturated compound to give another anion (2), which in turn takes a proton from alcohol to produce an enol (3). The enol tautomerizes to give the stable product, ketone. The reaction between diethylmalonate and benzylidene acetone is represented, as shown in Scheme 2.288.

Michael additions take place with various other reagents, e.g., acetylenic esters and α,β-unsaturated nitriles (Scheme 2.289).

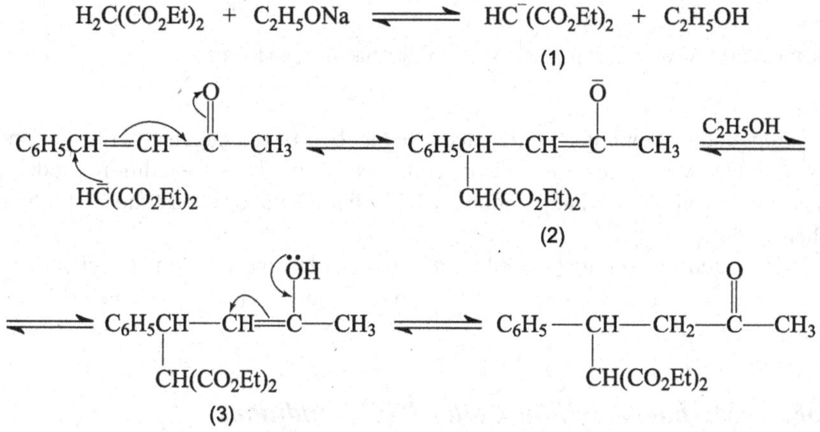

Scheme 2.288 Mechanism of Michael addition

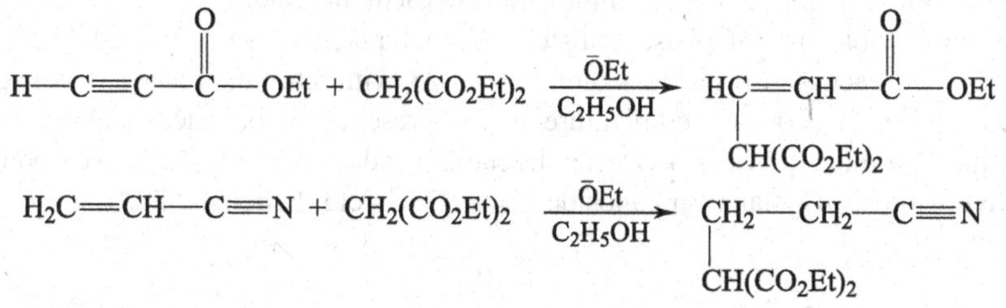

Scheme 2.289 Michael reaction using a various reagents

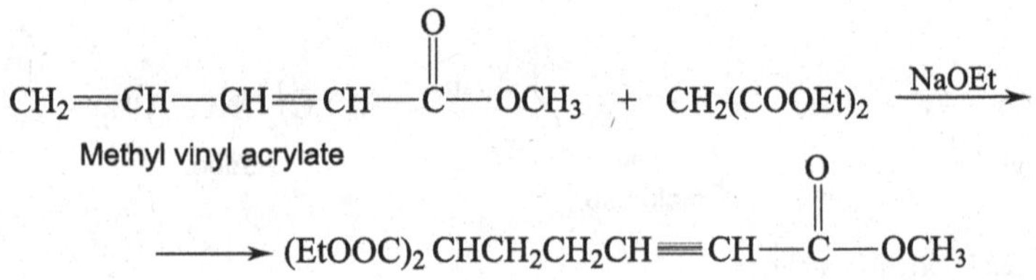

$$CH_2{=}CH{-}CH{=}CH{-}\overset{\overset{\textstyle O}{\|}}{C}{-}OCH_3 \;+\; CH_2(COOEt)_2 \;\xrightarrow{\;NaOEt\;}$$

Methyl vinyl acrylate

$$\longrightarrow (EtOOC)_2\, CHCH_2CH_2CH{=}CH{-}\overset{\overset{\textstyle O}{\|}}{C}{-}OCH_3$$

Scheme 2.290 Reaction of conjugated carbonyl compounds with active methylene compounds

$$Ph{-}\overset{\overset{\textstyle OSiMe_3}{|}}{C}{=}CH_2 + (CH_3)_2C{=}CH{-}\overset{\overset{\textstyle O}{\|}}{C}{-}CH_3 \xrightarrow{\;TiCl_4\;}$$

$$Ph{-}\overset{\overset{\textstyle O}{\|}}{C}{-}CH_2{-}\overset{\overset{\textstyle CH_3}{|}}{\underset{\underset{\textstyle CH_3}{|}}{C}}{-}CH_2{-}\overset{\overset{\textstyle O}{\|}}{C}{-}CH_3$$

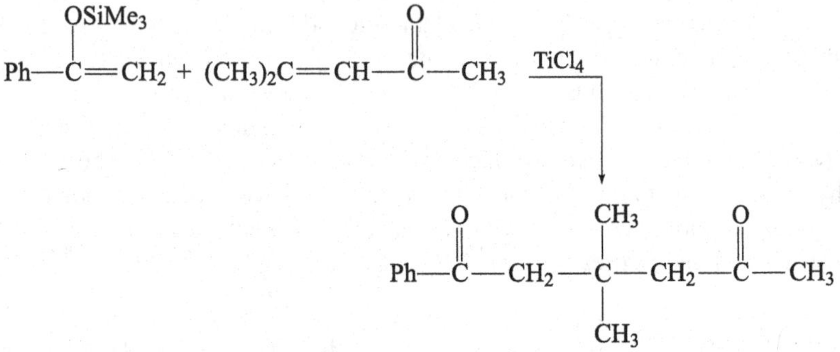

Scheme 2.291 A convenient preparation of Michael addition products

Compounds containing conjugated double bonds (conjugated with carbonyl group) react with active methylene compounds to give 1,6-addition products. Thus, methyl vinyl acrylate condenses with diethyl malonate as shown [305] in Scheme 2.290.

Michael addition products [306] are also obtained by the addition of silyl ethers of enols to α,β-unsaturated ketones and esters when catalysed by TiCl₄ (Scheme 2.291).

2.38.1 Michael Addition Under PTC Conditions

The Michael addition of active nitriles to acetylenes can be catalysed [307, 308] by the addition of quaternary ammonium chloride (Scheme 2.292).

A remarkable use of phase transfer Michael reaction was reported [309] in 1975. The reaction of nitro sugar (1) or (2) with ethylmalonate in benzene in 0.2 N NaOH at room temperature in the presence of hexadecyltributylphosphonium bromide gives 1,6-O-benzylidene-2,3-dideoxy-2-C-bis(ethoxycarbonyl) methyl-3-nitro-α-D-manno-pyranoside (3) in 92% yield (Scheme 2.293).

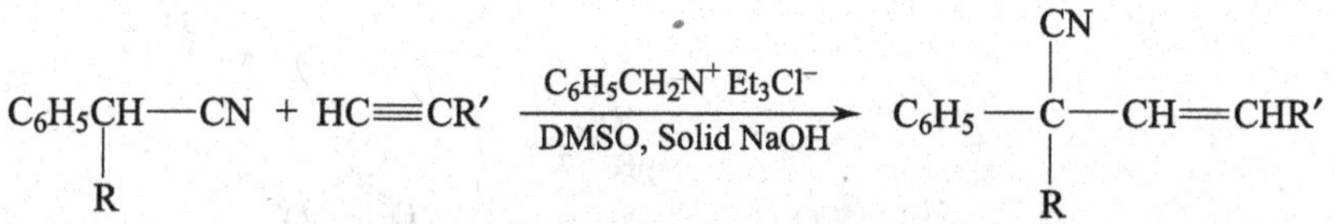

Scheme 2.292 Michael addition under PTC conditions

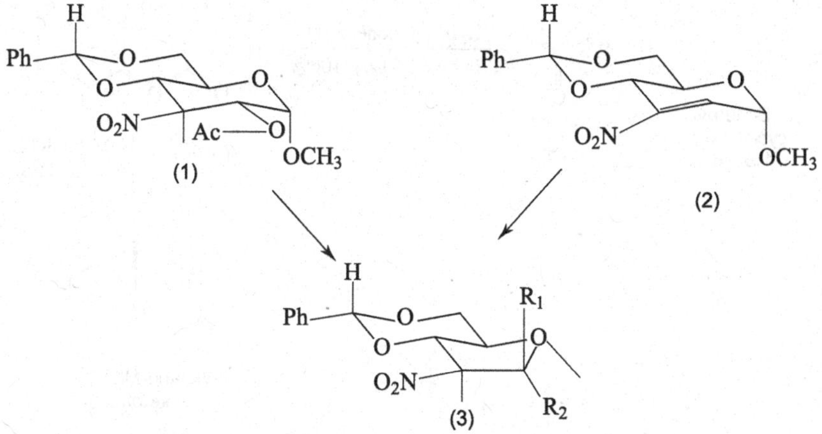

R–	R'	% Yield
Me	H	83
Et	H	80
iso Pr	H	82
C_5H_{11}	H	88
Et	Ph	94
iso Pr	Ph	83
$PhCH_2$	Ph	98

Scheme 2.293 A remarkable use of phase transfer Michael Reaction

2.38.2 Michael Addition in Aqueous Medium

Michael addition in aqueous phase was first reported in 1970s. 2-Methylcyclopentane-1,3-dione on reaction with methyl vinyl ketone in water gave an adduct with the use of a basic catalyst (pH > 7). The adduct further cyclizes to give fused ring systems [310] (Scheme 2.294).

Michael reaction of 2-methyl cyclohexane-1,3-dione with methyl vinyl ketone gave optically pure Wieland-Miescher ketone [311] (Scheme 2.295).

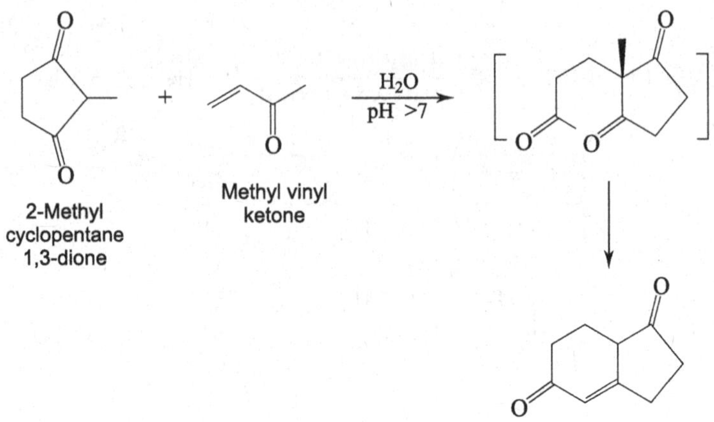

Scheme 2.294 Michael reaction in aqueous phase

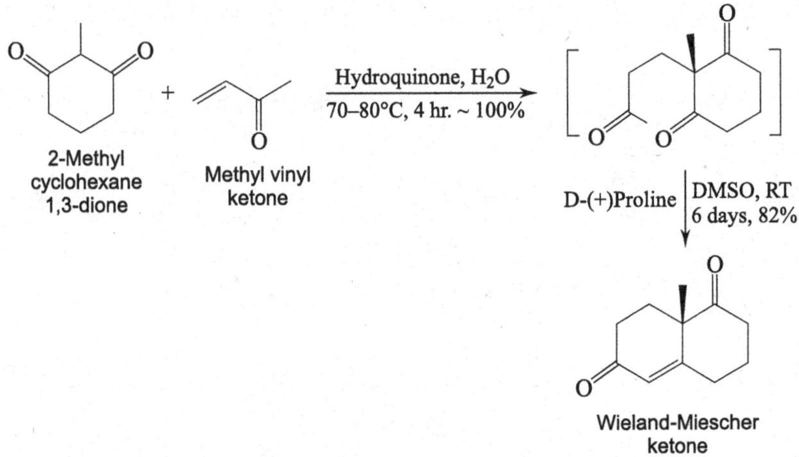

Scheme 2.295 Synthesis of optically pure Wieland-Miescher ketone

The Michael addition of 2-methylcyclopentane-1,3-dione to acrolein in water gave an adduct, which was used for the synthesis of 13-α-methyl-14-α-hydroxysteroid [312] (Scheme 2.296).

The rate of the above Michael addition (Scheme 2.10) was enhanced by the addition of ytterbium triflate [Yb(OTf)₃].

The Michael addition of nitromethane to methyl vinyl ketone in water (in absence of a catalyst) gives [313] 4:1 mixture of adducts (A and B) (Scheme 2.297).

Use of methyl alcohol as solvent (in place of H₂O) gave 1:1 mixture of A and B. The above reaction does not occur in neat conditions or in solvents like THF, PhMe etc., in the absence of a catalyst.

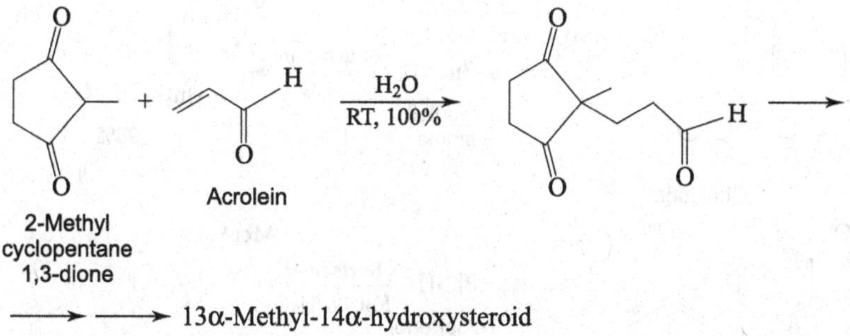

13α-Methyl-14α-hydroxysteroid

Scheme 2.296 Synthesis of an intermediate used for the synthesis of 13-α-methyl-14-α-hydroxysteroid

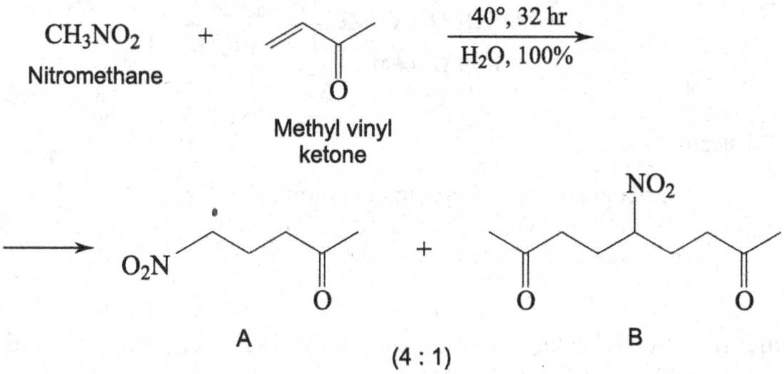

Scheme 2.297 The Michael addition of CH_3NO_2 with methyl vinyl in water

The Michael addition of cyclohexanone to ascorbic acid was carried out in water in the presence of an inorganic acid [314] (rather than a base) (Scheme 2.298).

Effective Michael reactions of amines, thiophenol and methylacetoacetate to chalcone have been developed [315] (Scheme 2.299).

Surfactant is hexadecyltrimethylammonium bromide

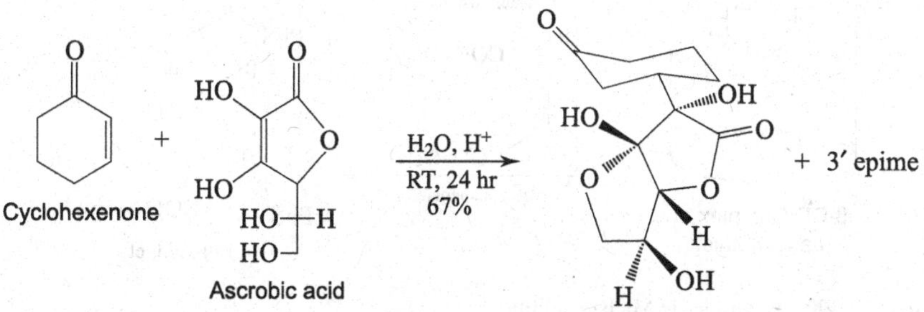

Scheme 2.298 Michael addition of cyclohexanone to ascorbic acid in water in the presence of an inorganic acid

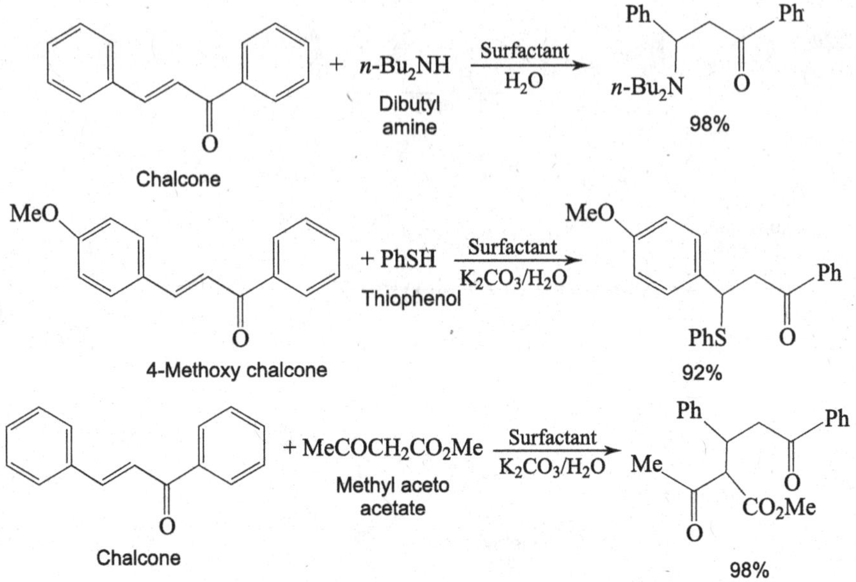

Surfactant is hexadecyltrimethylammonium bromide

Scheme 2.299 Some effective Michael additions

Asymmetric Michael addition of benzenethiol to 2-cyclohexenone and maleic acid esters proceeds enantioselectively in their crystalline cyclodextrin complexes. The adducts were obtained in 38 and 30% ee, respectively. In both cases, the reaction was carried out [316] in water suspension (Scheme 2.300).

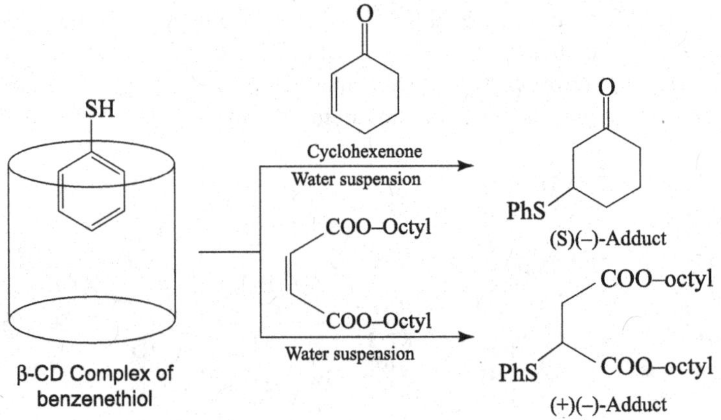

Scheme 2.300 A symmetric Michael addition

2.38.3   Michael Addition in Solid State

A number of 2'-hydroxy-4', 6'-dimethylchalcones undergo a solid-state intramolecular Michael type addition to yield [317] the corresponding flavanones
(Scheme 2.301).

The Michael addition of chalcone to 2-phenylcyclohexanone under PTC conditions gave [318] 2,6-disubstituted cyclohexanone derivatives in high distereoselectivity (90% ee) (Scheme 2.302).

The enantioselective Michael addition of mercapto compounds with an optically
active host compound derived from 1:1 inclusion complex of cyclohexanone with
(–)–A derived from tartaric acid [319] and a catalytic amount of benzyltrimethyl
ammonium hydroxide on irradiation with ultrasound for 1 h at room temperature
gave the adduct in 50–78% yield with 75–80% (Scheme 2.303).

The Michael addition of thiols to 3-methyl-3-buten-2-one in its inclusion crystal
with (–)–A also occurred enantioselectively (Scheme 2.304).

Michael addition of diethyl (acetylamido) malonate to chalcone using asymmetric
phase transfer catalyst (ephedrine salt) in the presence of KOH in the solid state gave
[320] the adduct in 56% yield with ee of 60% (Scheme 2.305).

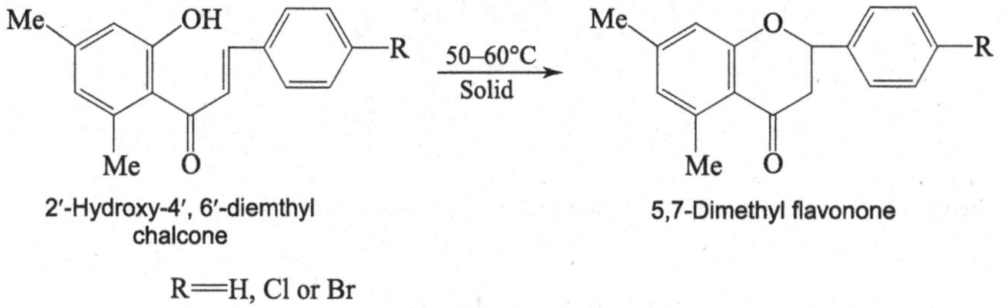

2'-Hydroxy-4', 6'-diemthyl
chalcone

5,7-Dimethyl flavonone

R=H, Cl or Br

Scheme 2.301 Michael addition in solid state

2-Phenyl
cyclohexanone

Chalcone

99%

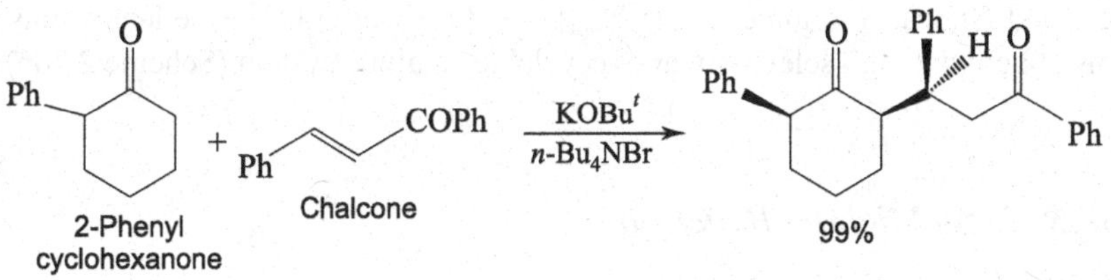

Scheme 2.302 Synthesis of Michael adduct in high distereoselectivity

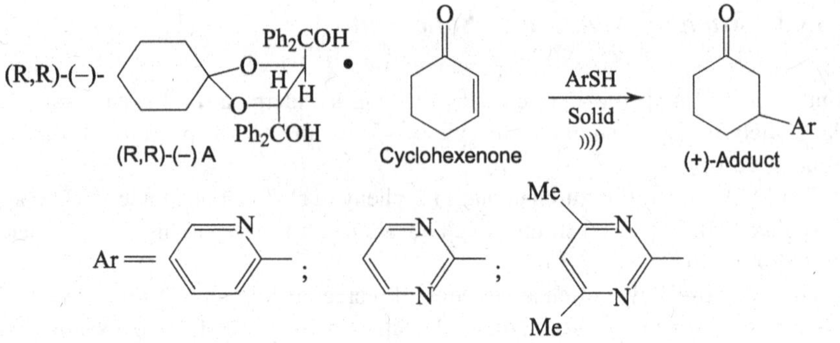

Scheme 2.303 The enantioselective Michael addition

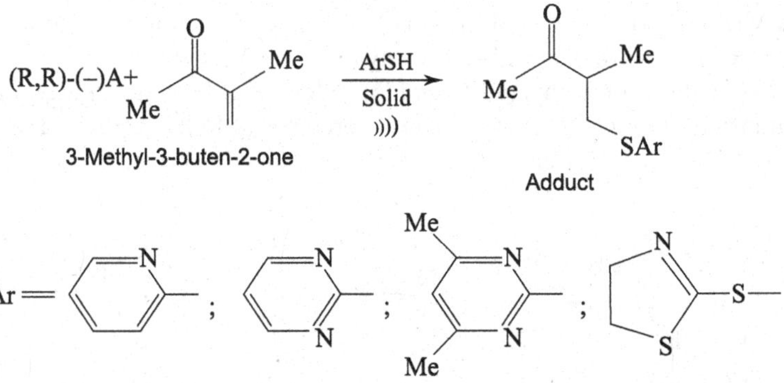

Scheme 2.304 Another example of enantioselective Michael addition

2.38.4   Michael Addition in Ionic Liquids

Michael addition reactions of acetyl acetone to methyl vinyl ketone in the presence of catalyst Ni(acac)$_2$ in ionic liquid [b$_{min}$] [BF$_4$] provides [321] excellent results in terms of activity, high selectivity and recyclable catalytic system (Scheme 2.306).

2.38.5   Aza-Michael Reaction

It is an important class of carbon–nitrogen bond-forming reaction and is an important tool in organic chemistry. A typical and efficient aza-Michael addition of amines catalysed by PSSA in aqueous medium (Scheme 2.307) has been developed [322].

Bis-aza-Michael reaction of alkyldiamines and methyl acrylate and acrylonitrile (Scheme 309) have also been reported [322] (Scheme 2.308).

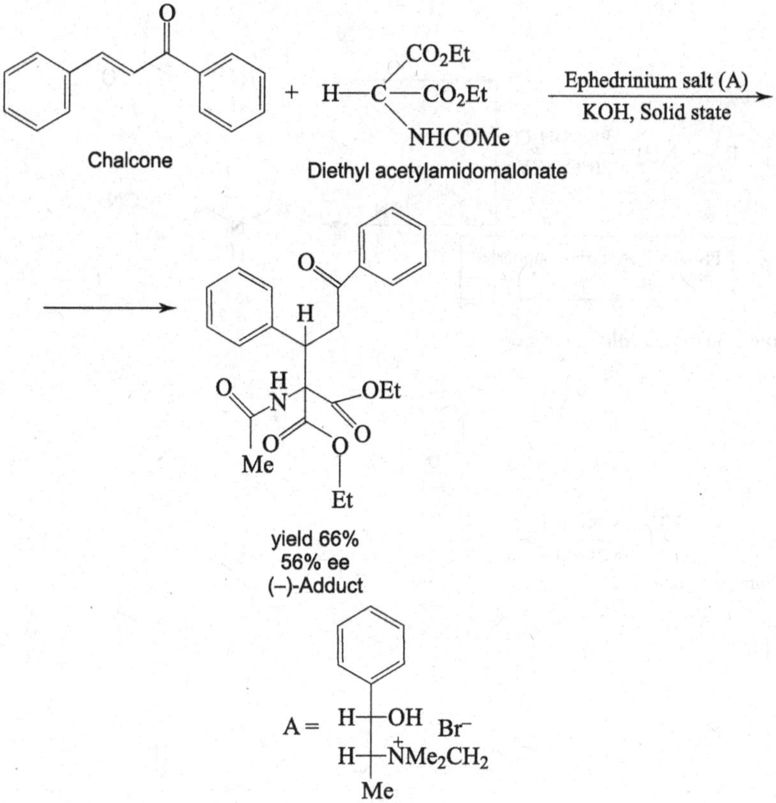

Scheme 2.305 Michael Addition using asymmetric PIC Catalyst

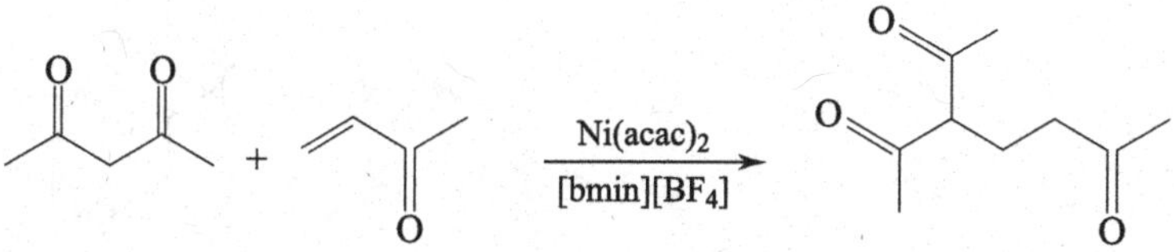

Scheme 2.306 Michael addition in ionic liquids

2.38.6 Applications

In Michael addition, a new carbon–carbon bond is produced. This procedure is of great synthetic importance since a variety of organic compounds can be synthesized. Some of the important applications are given in Scheme 2.309.

(*i*) Synthesis of dimidone

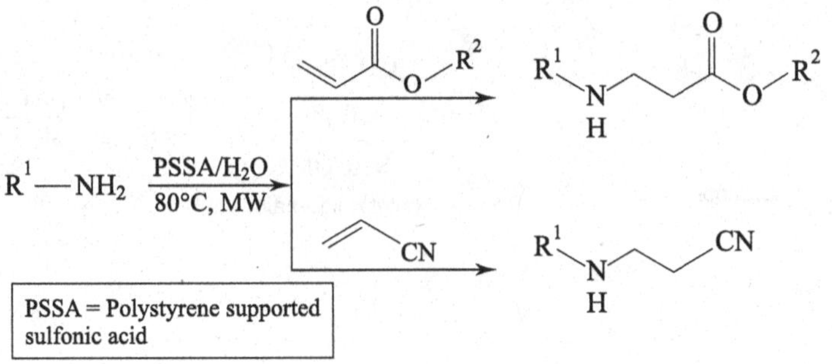

Scheme 2.307 Aza-Michael reaction

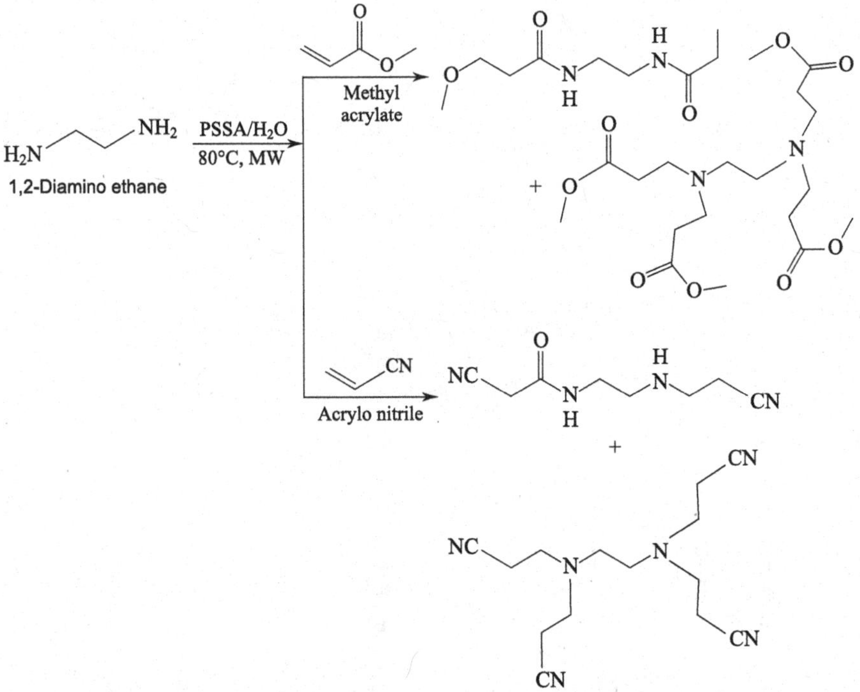

Scheme 2.308 BIS-aza-Michael reaction

The Michael adduct obtained by the condensation of diethyl malonate with mesityl oxide in the presence of sodium ethoxide undergoes **internal Claisen condensation** to give dimidone.

(*ii*) Synthesis of bicyclic ketones

Use of Michael addition followed by aldol condensation is an important route for the synthesis of bicyclic ketones and is known as **Robinson annulation**. Thus,

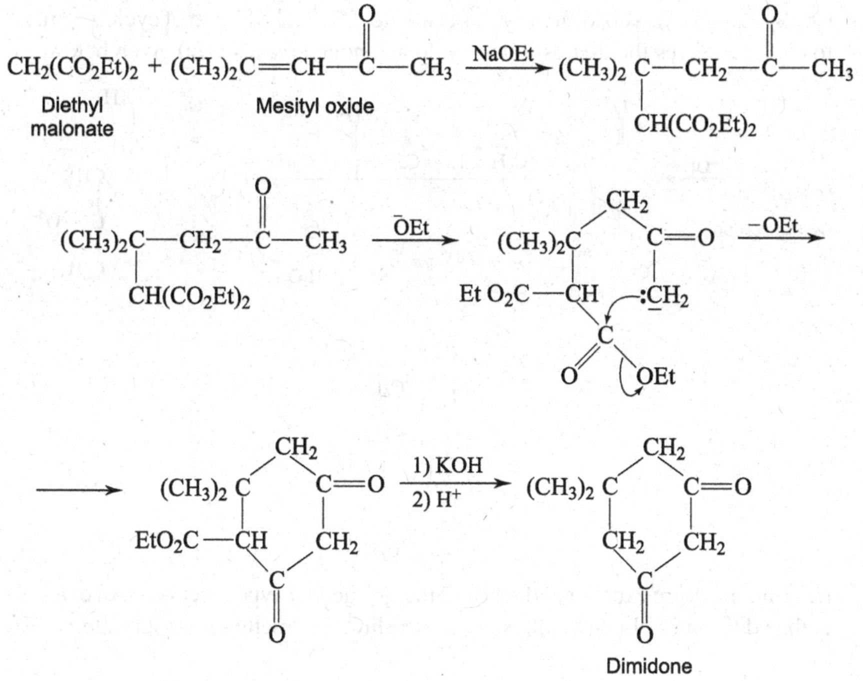

$$CH_2(CO_2Et)_2 + (CH_3)_2C{=}CH{-}\overset{\overset{\displaystyle O}{\|}}{C}{-}CH_3 \xrightarrow{\text{NaOEt}} (CH_3)_2 \overset{\overset{\displaystyle }{|}}{\underset{\displaystyle CH(CO_2Et)_2}{C}}{-}CH_2{-}\overset{\overset{\displaystyle O}{\|}}{C}{-}CH_3$$

Diethyl malonate   Mesityl oxide

$$(CH_3)_2\overset{\overset{\displaystyle }{|}}{\underset{\displaystyle CH(CO_2Et)_2}{C}}{-}CH_2{-}\overset{\overset{\displaystyle O}{\|}}{C}{-}CH_3 \xrightarrow{\bar{O}Et} (CH_3)_2C \qquad \xrightarrow{-OEt}$$

Scheme 2.309 Some applications of Michael reaction

Michael condensation of 2-methyl-1,3-cyclohexanedione with methyl vinyl ketone followed by aldol condensation gives bicyclic ketone.

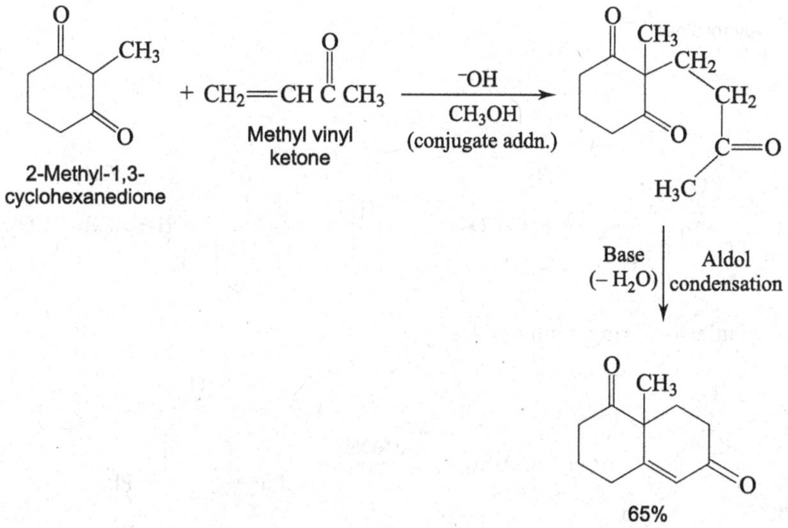

2-Methyl-1,3-cyclohexanedione

Methyl vinyl ketone

$^-$OH
CH_3OH
(conjugate addn.)

Base
$(-H_2O)$ Aldol condensation

65%

(*iii*) Synthesis of *o*-substituted cyclohexanone. Michael addition of cyclohexanone to chalcone gives the 2-substituted cyclohexanone. *See* also (*iv*) given below.

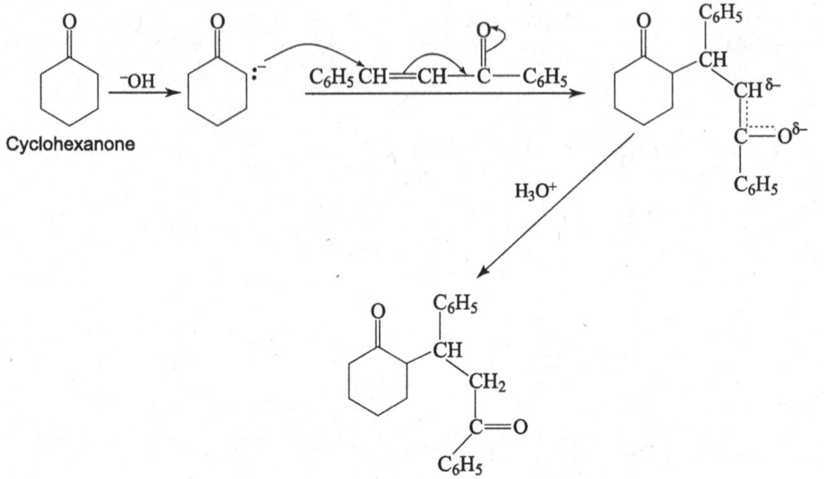

Cyclohexanone

(*iv*) Enamines are excellent addends in many Michael-type reactions. An example is the addition of N-(1-cyclohexenyl)-pyrrolidine to methyl methacrylate.

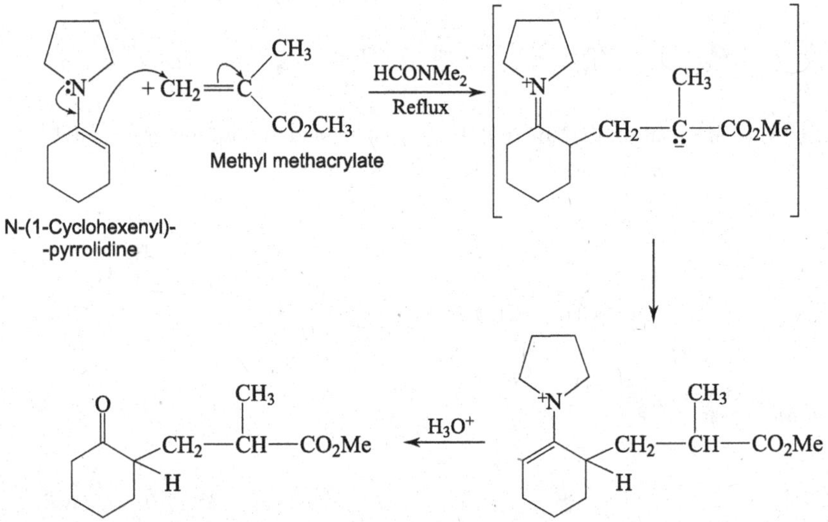

Methyl methacrylate

N-(1-Cyclohexenyl)-
-pyrrolidine

(*v*) Synthesis of ring compounds.

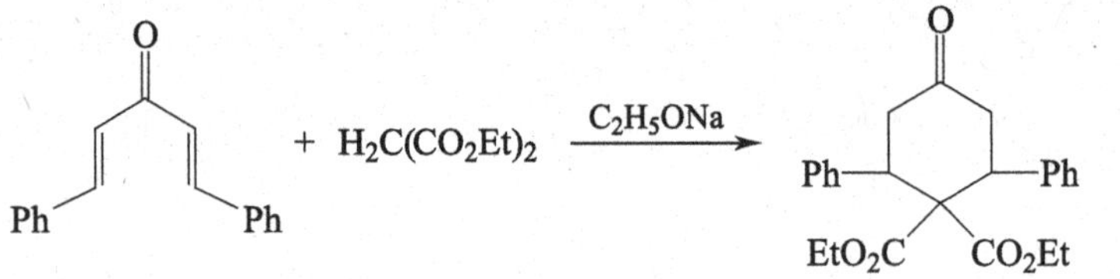

Double Michael additions are often employed for synthesizing ring compounds.
(*vi*) Synthesis of carbonic acid, a cyclopropane derivative.

$(CH_3)_2C{=}CH{-}CO_2Et$ + $CNCH_2CO_2Et$ $\xrightarrow{\text{NaOEt}}$

Ethyl 3-methyl crotonate   Ethyl cyanoacetele

$$\underset{H_3C}{\overset{CH_3}{>}}C\underset{CH{\cdot}CO_2Et}{\overset{CH_2CO_2Et}{<}}$$
$$\overset{|}{CN}$$

$\xrightarrow[\text{2) }\Delta,\,-CO_2]{\text{1) }H_3O^+}$

$$\underset{CH_3}{\overset{CH_3}{>}}C\underset{CH_2COOH}{\overset{CH_2COOH}{<}}$$

$\xrightarrow[\text{2) Alcohol}]{\text{1) Redn. P + Br}}$

$$\underset{H_3C}{\overset{CH_3}{>}}C\underset{CHBr{\cdot}CO_2Et}{\overset{CHBr{\cdot}CO_2Et}{<}}$$

$\Big\downarrow$ 1) Na 2) H$_2$O

$$\underset{CH_3}{\overset{CH_3}{>}}C\underset{CH{\cdot}CO_2H}{\overset{CH{\cdot}CO_2H}{<}}$$

Caronic acid

(*vii*) Synthesis of nitro and cyano compounds.

HCN + $(CH_3)_2C{=}CH{-}NO_2$ $\xrightarrow[(CH_3OCH_3)]{KOH}$ $(CH_3)_2C{-}CH_2NO_2$

2-Methyl-1-nitropropene

$$\overset{|}{CN}$$

2,2-Dimethyl-3-
nitropropanenitrile

CH_3NO_2 + $CH_3CH{=}CHCO_2Et$ $\xrightarrow{C_2H_5ONa}$ $CH_3{-}CH{-}CH_2{-}CO_2Et$

Ethyl crotonate

$$\overset{CH_2NO_2}{\overset{|}{}}$$

Ethyl 3-methyl-4-nitrobuytrate

(*viii*) Synthesis of aconitic acid.

$$\underset{C{-}CO_2Et}{\overset{C{-}CO_2Et}{|||}}$$ + $H_2C(CO_2Et)_2$ $\xrightarrow[\text{2) }H_3O^+,\,\Delta]{\text{1) }C_2H_5ONa}$

Ethyl acetylene
dicarboxylate

$$\overset{CH{\cdot}CO_2H}{\underset{CH_2CO_2H}{\overset{||}{C{\cdot}CO_2H}}}$$

Aconitic acid

(*ix*) Some typical applications of Michael additions are given below:

$$CH_2(CO_2Et)_2 \ + \ CH_2{=}CH{-}\overset{\overset{\displaystyle O}{\|}}{C}{-}CH_3 \ \xrightarrow{\text{NaOEt}} \ \underset{\overset{|}{CH(CO_2Et)_2}}{CH_2CH_2COCH_3}$$

$$CH_2(CO_2Et)_2 \ + \ CH_2{=}CH{-}CHO \ \xrightarrow{\text{NaOEt}} \ \underset{\overset{|}{CH(CO_2Et)_2}}{CH_2CH_2CHO}$$

$$\underset{\overset{|}{CH_3}}{CH_3{-}C}{=}CH{-}CO_2Et \ \xrightarrow[\text{NaOEt}]{CH_2(CO_2Et)_2} \ \underset{\overset{|}{CH(CO_2Et)_2}}{\overset{\overset{\displaystyle CH_3}{|}}{H_3C{-}C}{-}CH_2CO_2Et}$$

(x) Synthesis of allylrethrone.

A typical synthesis of allylrethrone [323], an important component of an insecticidal pyrethroid, has been carried out by a combination of Michael reaction of 5-nitro-1-pentene and methyl vinyl ketone in the presence of Al_2O_3 followed by an intramolecular aldol-type condensation.

5-Nitro-1-pentene   Methyl vinyl ketone

Allylrethrone

2.39   Mukaiyama Reaction

It is a stereoselective aldol condensation [324] of silyl enol ethers of ketones with aldehydes in the presence of titanium tetrachloride. For example, condensation of silyl enol ether of 3-pentanone with 2-methylbutyraldehyde in the presence of $TiCl_4$ gives the aldolate product, which on hydrolysis yield aldol product Manicone, an alarm pheromone (Scheme 2.310).

In Mukaiyama reaction, other Lewis acids, such as tin tetrachloride ($SnCl_4$) and boron trifluoride etherate ($BF_3 \cdot OEt_2$) can also be used.

2.39.1   Mukaiyama Reaction in Aqueous Phase

The Mukaiyama reactions are carried out under non-aqueous conditions. The first water-promoted Mukaiyama reaction of silyl enol ethers with aldehydes was reported in 1986 (Scheme 2.311) [325].

The above reactions were carried out in aqueous medium without any acid catalyst, but the reaction, however, took several days for completion, since water serves as a

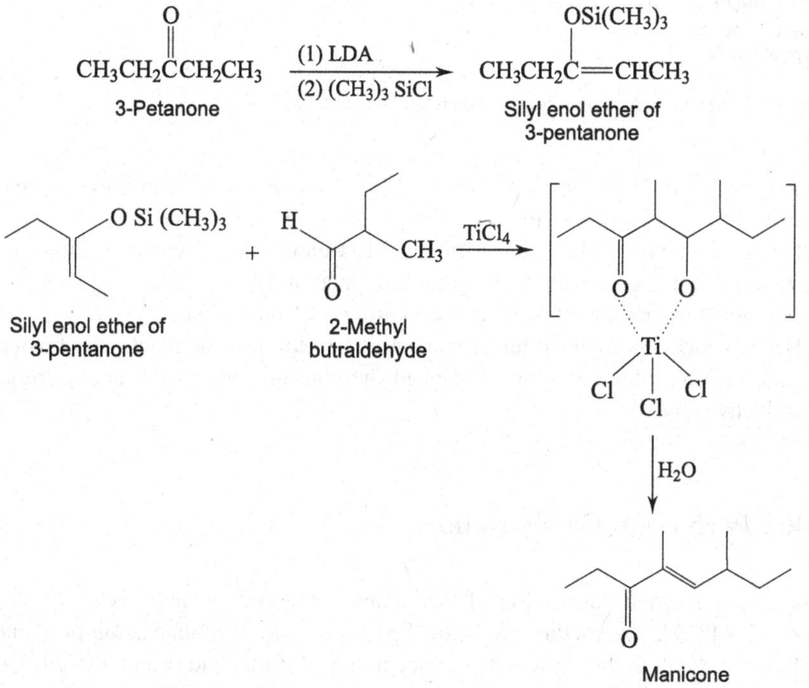

Scheme 2.310 Mukaiyama reaction

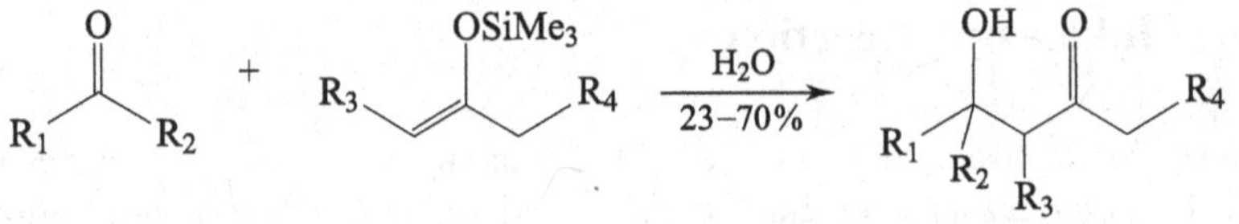

Scheme 2.311 Mukaiyama reaction in water

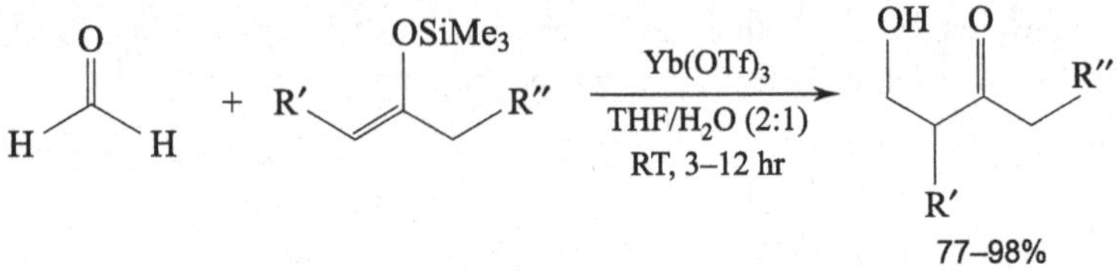

Scheme 2.312 Mukaiyama reaction in aqueous phase in the presence of Lewis acid

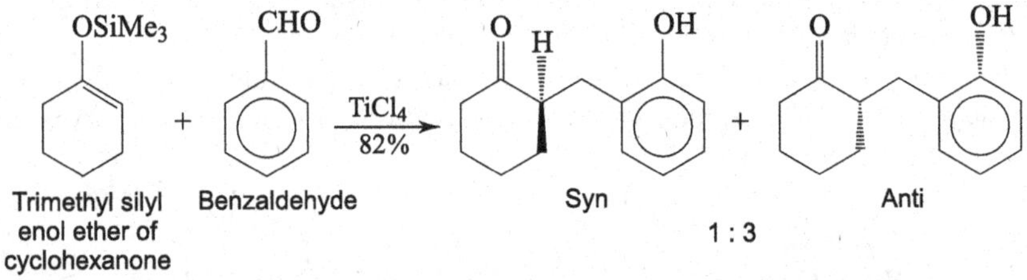

Scheme 2.313 Mukaiyama reaction in heterogeneous phases

weak Lewis acid. The addition of stronger Lewis acid (*e.g.*, ytterbium triflate) greatly improved the yield and also the rate of the reaction [326] (Scheme 2.312).

It has been shown [325] that trimethyl silyl enol ether of cyclohexanone with benzaldehyde that occurs in water in the presence of TiCl$_4$ is heterogeneous phase at room temperature and atmospheric pressure (Scheme 2.313).

Better yields are obtained under sonication conditions. The reaction is favoured by an electron-withdrawing substituent in the para position of the phenyl ring in benzaldehyde.

2.40 Pechmann Condensation

The most important application of Pechmann condensation involves synthesis of coumarins [327]. The method (Scheme 2.314) consists of condensation of phenols with β-ketoesters in the presence of concentrated sulfuric acid or other condensing agents like P$_2$O$_5$, POCl$_3$, AlCl$_3$ etc.

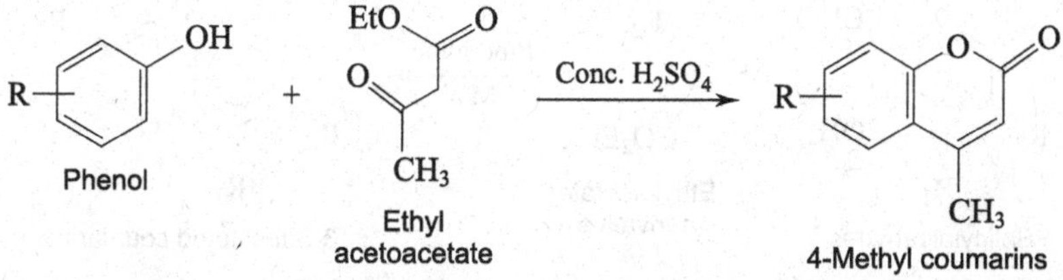

Scheme 2.314 Pechmann Condensation

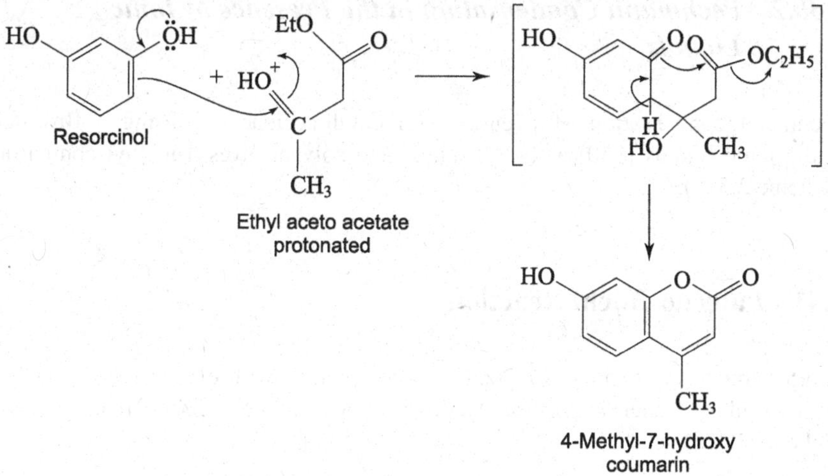

Scheme 2.315 Mechanism of Pechmann condensation

The reaction proceeds by the formation of β-hydroxy ester intermediate, which then cyclizes and dehydrates to give coumarin (Scheme 2.315).

2.40.1 Microwave-Promoted Pechmann Reaction

The Pechmann condensation [328] under microwave irradiation of salicylalde-hydes with alkyl-substituted ethyl acetates under basic conditions (piperidine) give coumarins [328, 329] (Scheme 2.316).

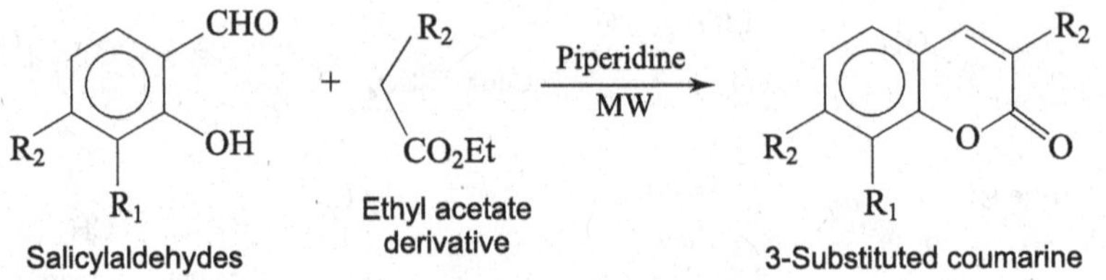

Scheme 2.316 Microwave-promoted Pechmann condensation

2.40.2 Pechmann Condensation in the Presence of Ionic Liquids

Pechmann condensation of phenols with ethyl acetoacetate using a Brönsted acidic ionic liquids [330] as both catalyst and solvent gives 4-methyl coumarins (Scheme 2.317).

2.41 Paterno-Büchi Reaction

Photochemical cycloaddition of carbonyl compounds with olefins gives oxetanes (four-membered ether rings). The reaction is known as Paterno-Büchi reaction [331] (Scheme 2.318).

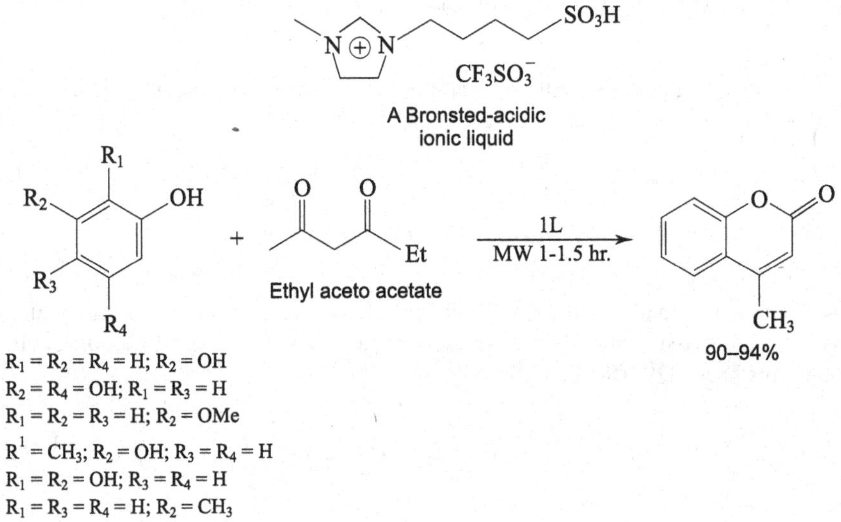

$R_1 = R_2 = R_4 = H; R_2 = OH$
$R_2 = R_4 = OH; R_1 = R_3 = H$
$R_1 = R_2 = R_3 = H; R_2 = OMe$
$R^1 = CH_3; R_2 = OH; R_3 = R_4 = H$
$R_1 = R_2 = OH; R_3 = R_4 = H$
$R_1 = R_3 = R_4 = H; R_2 = CH_3$

Scheme 2.317 Pechmann condensation in the presence of ionic liquids

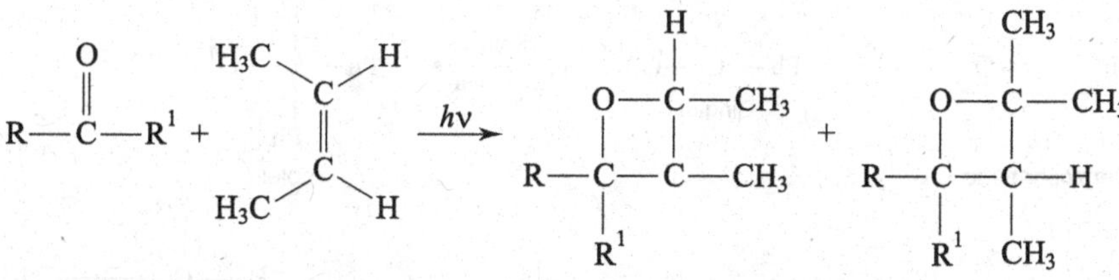

Scheme 2.318 Paterno-Büchi reaction

The Paterno-Büchi reaction usually occurs by the cycloaddition of the triplet state of the carbonyl compound with the ground state of an alkene. An interesting reaction is the photo addition of butyraldehyde with 2-methyl-2-butene to yield a mixture of 2,3,3-trimethyl-4-propyloxetane and 2,2,3-trimethyl-4-propyloxetane (Scheme 2.319).

The oxetane ring is formed in two steps. The carbonyl compound (triplet state) adds through its oxygen atom to give the more stable diradicals. In the second step, the spin inversion occurs with simultaneous bond formation to give oxetane (Scheme 2.320).

The photocycloaddition of benzophenone with *cis*- and *trans*-2-butene gives the same mixture of *cis*- and *trans*-oxetanes (Scheme 2.321) showing that the reaction is not stereospecific. The lack of stereochemical discrimination shows that the reaction is not concerted and that the ring is formed in two stages as shown in Scheme 2.321.

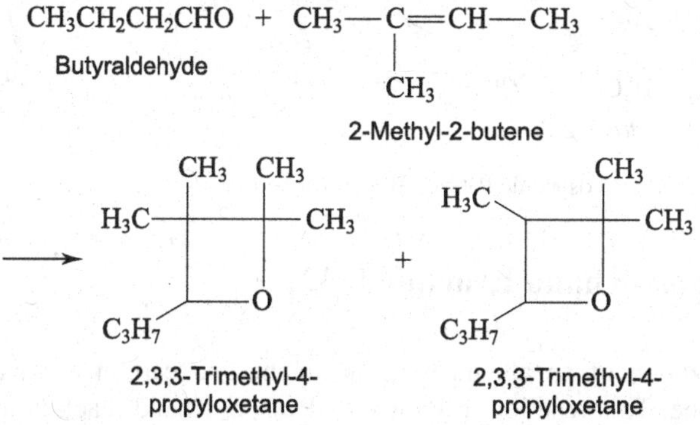

Scheme 2.319 Photochemical Paterno-Büchi reaction

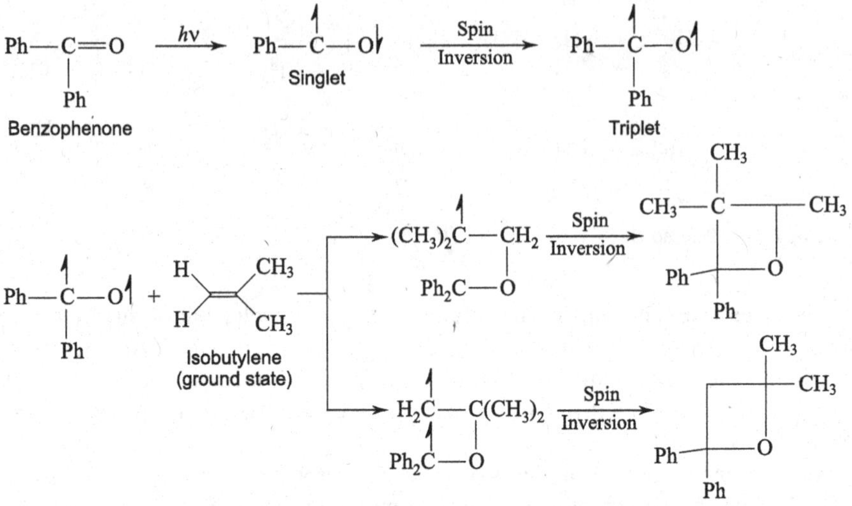

Scheme 2.320 Mechanism of Paterno-Büchi reaction

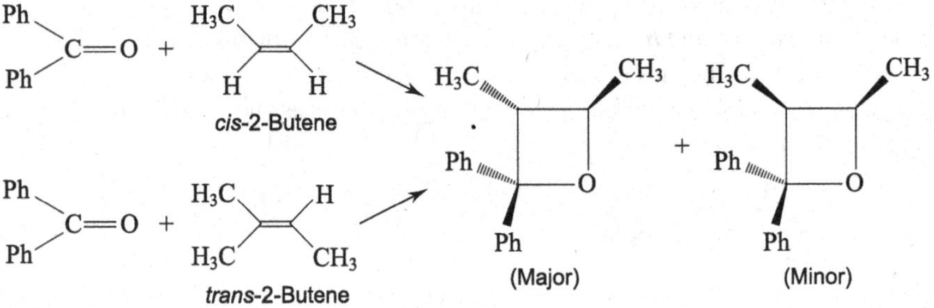

Scheme 2.321 Non-stereospecific Paterno-Büchi reaction

2.42   Pauson–Khand Reaction [332]

The co-cyclization of an alkyne with an alkene and carbon monoxide yields a cyclopentanone. This reaction in known as Pauson–Khand reaction and this reaction has been carried out in SC-CO$_2$ in the presence of dicobalt octacarbonyl as catalyst [332] (Scheme 2.322).

The reaction was also successful in a number of substituted enynes. An intramolecular Pauson–Khand reaction between phenyl acetylene and norbornadiene gave the exo-product with 87% yield (Scheme 2.323).

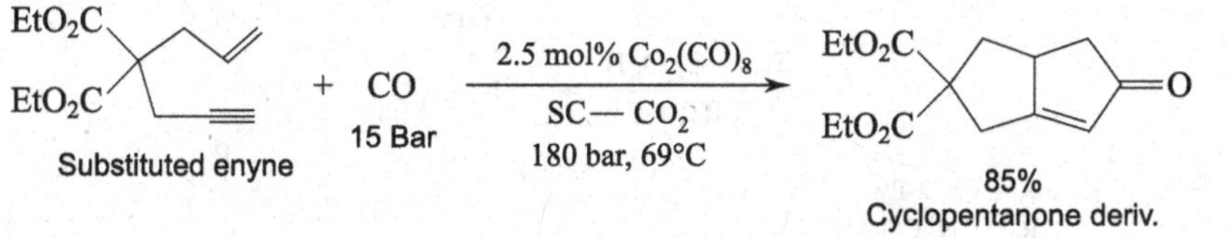

Scheme 2.322 Pauson–Khand reaction

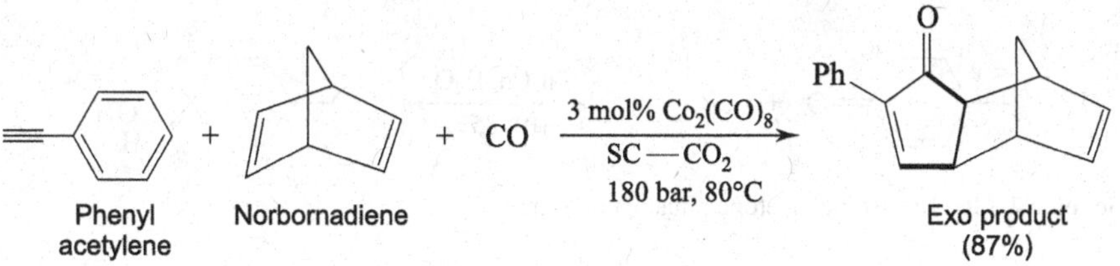

Scheme 2.323 An intramolecular Pauson–Khand reaction

2.43 Pinacol Coupling [333]

The reaction of ketones with magnesium in benzene give 1,2-diols. Thus, acetone under these conditions give pinacol (Scheme 2.324).

This reaction is known as pinacol coupling. The use of Zn–Cu couple-to-couple unsaturated aldehydes to pinacols was known as early in 1982 [333]. Subsequently, chromium and vanadium [334], and some ammonical TiCl$_3$ [335]-based reducing agents were used. It has now been found [336] that pinacol coupling takes place in aromatic aldehydes and ketones in the presence of Ti(III), under alkaline conditions. However, in the presence of acids only, the substrates (aromatic aldehydes and ketones) having electron-withdrawing groups like CN, CHO, COMe, COOH, COOMe, pyridyl as activating groups only underwent pinacol coupling [337] (Scheme 2.325). In case of non-activating carbonyl compounds, it was necessary to use excess of the substrate as a solvent [338].

The coupling reaction of α,β-unsaturated ketones and acetone using a Zn–Cu couple and ultrasound in an aqueous-acetone suspension (Scheme 2.326) gave the corresponding pinacols [339].

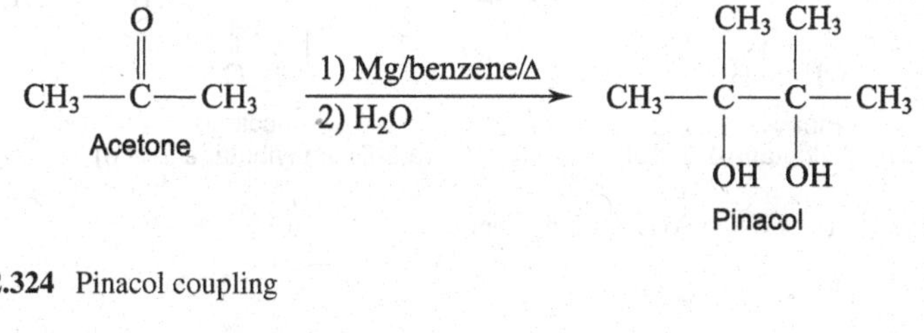

Scheme 2.324 Pinacol coupling

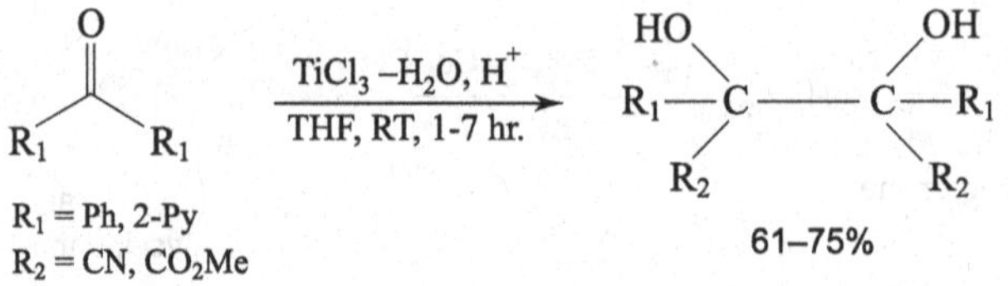

Scheme 2.325 Pinacol coupling of aromatic carbonyl compounds

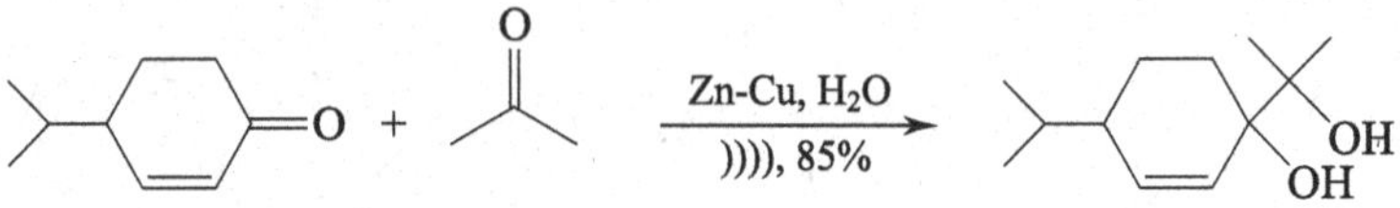

Scheme 2.326 Ultrasound acetone pinacol coupling

2.44   Pinacol–Pinacolone Rearrangement

Acid-catalysed rearrangement of periacols to pincolones is known as pinacol-pinacolone rearrangement [340]. Thus, 2,3-dimethylbutane-2,3-diol (pinacol) on treatment with hot 30% sulphuric acid gives 3,3-dimethylbutan-2-one (pinacolone) (Scheme 2.327).

The mechanism of pinacol-pinacolone rearrangement is given in Scheme 2.328.

2.44.1   Pinacol-Pinacolone Rearrangement in Water Using Microwave Irradiation

Pinacol on heating [186, 341] with water in microwave oven at 270 °C give pinacolone in 76% yield. Pinacolone was isolated by converting it into the 2,4-dinitrophenyl-hydrazone. It is believed that near critical water generated by heating water in microwave oven at 270° acts as an acid catalyst itself (Scheme 2.329).

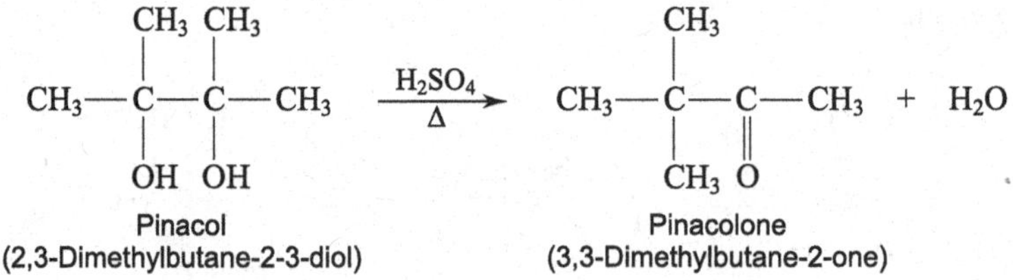

Scheme 2.327 Pinacol-pinacolone rearrangement

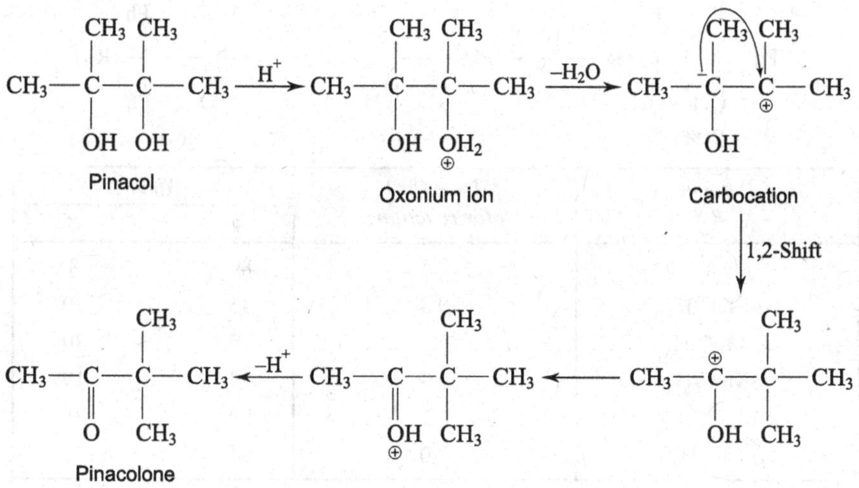

Scheme 2.328 Mechanism of Pinacol-pinacolone rearrangement

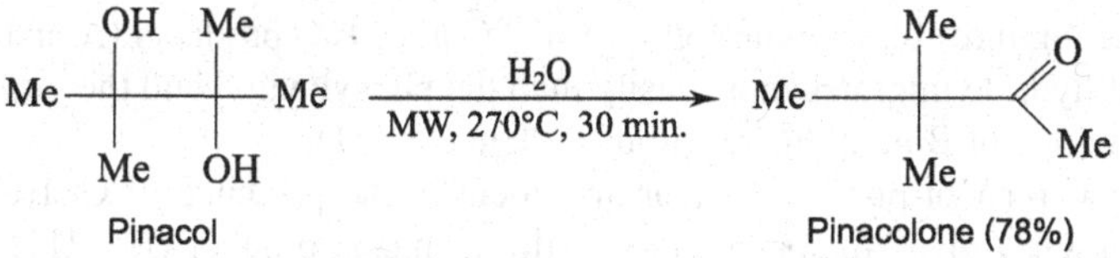

Scheme 2.329 Microwave-assistant pinacol-pinacolone rearrangement in water

2.44.2 Pinacol-Pinacolone Rearrangement on Irradiation with Microwaves in Solid State

Solventless pinacol-pinacolone rearrangement using microwave irradiation [342] has been reported. The process involves irradiation of pinacols with Al^{3+} montmorillonite K10 clay for 15 min to give the rearranged product in good yield (Scheme 2.330). The results are comparable to conventional heating in an oil bath wherein the reaction takes too long (15 h).

Another example of pinacol-pinacolone rearrangement in solid state is given below. Thus, a 1:3 molar ratio of pinacol and p-toluene sulphonic acid (p-TSOH)

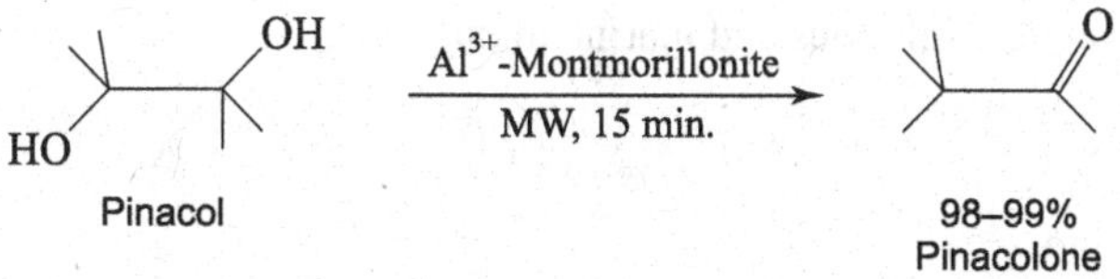

Scheme 2.330 Microwave-assisted solid-state pinacol-pinacolone rearrangement

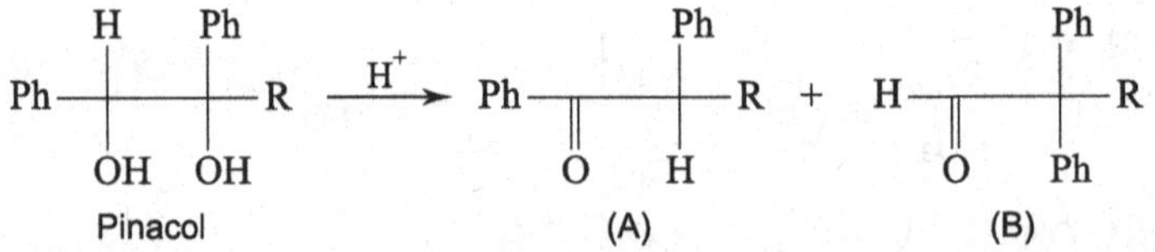

Pinacol R	Time (hr.) for reaction	Yield	
		A	B
Ph	2.5	89	8
o-MeC$_6$H$_4$	0.5	45	29
m-MeC$_6$H$_4$	0.3	70	30
p-MeC$_6$H$_4$	0.7	39	19
p-MeOC$_6$H$_4$	0.7	89	0
p-ClC$_6$H$_4$	1.0	54	41

Scheme 2.331 Pinacol-pinacolone rearrangement in the presence of p-TSOH

(powdered mixture) on keeping [343] at 60 °C gave two products (A and B). It is found that hydride migrates more easily than the phenyl anion and the yield of A is higher than that of B in all the reactions (Scheme 2.331).

However, pinacol-pinacolone rearrangement in the presence of CCl$_3$CO$_2$H (in place of TsOH) gives major amount of the isomeric product (B). This reaction is considerably enhanced if the water formed during the reaction is continuously removed under reduced pressure.

2.45   Prins Reaction [344]

It is a useful C–C bond-forming reaction and is notable in the formation of tetrahydropyran derivatives. It consists of the condensation of olefins with aldehydes under strongly acidic conditions and high reaction temperature; this limits its potential as a useful synthetic methodology.

Using a simple homoallylic alcohol and an aldehyde in the presence of a catalytic amount of cerium triflate, the direct stereoselective formation of tetrahydropyranol derivatives in ionic liquid was achieved [344] (Scheme 2.332).

Tetrahydropyrans are prevalent in number of natural products, including carbohydrates, polyether antibiotics and marine toxins.

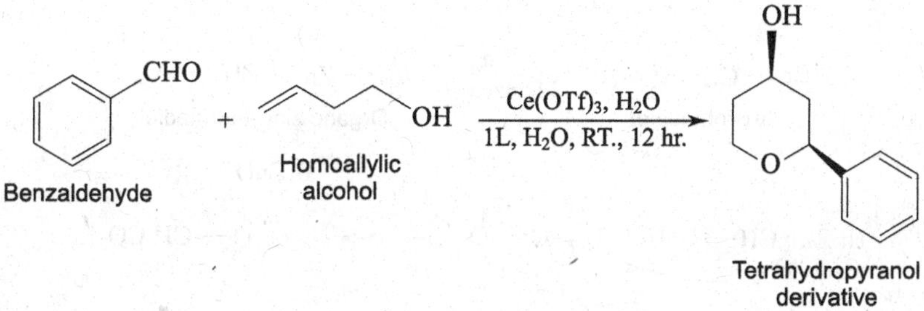

Scheme 2.332 Prins reaction

2.46 Reformatsky Reaction

The reaction of an α-halo-ester (usually an α-bromoester) with an aldehyde or ketone in the presence of zinc metal in an inert solvent (ether-benzene) to produce β-hydroxy ester is known as Reformatsky reaction [345] (Scheme 2.333).

The Reformatsky reaction extends the carbon skeleton of an aldehyde or ketone and yields β-hydroxy esters. The initial product is a zinc alkoxide, which must be hydrolysed to yield the β-hydroxy ester.

Mechanism

Zinc first reacts with the α-bromoester to form an organozinc intermediate (its formation is similar to that of the formation of Grignard reagent). The formed organic zinc intermediate then adds to the carbonyl group of the aldehyde or ketone. Final hydrolysis gives β-hydroxy ester (Scheme 2.334).

Since the organo zinc reagent is less reactive than the organo magnesium reagent (Grignard reagent), it does not add to the ester group. The β-hydroxy esters produced in the Reformatsky reaction are easily dehydrated to α,β-unsaturated esters since dehydration yields a system in which carbon–carbon double bond is in conjugation with the carbon–oxygen double bond of the ester (Scheme 2.335).

In Reformatsky reaction, it is sometimes necessary to activate the zinc by addition of a small crystal of iodine, mercuric bromide or copper.

A modification of the Reformatsky reaction using nitriles (in place of aldehydes or ketones) is called the **Blaise reaction** (Scheme 2.336).

The Reformatsky reaction can also be applied to Schiff's bases to yield β-lactam (Scheme 2.337).

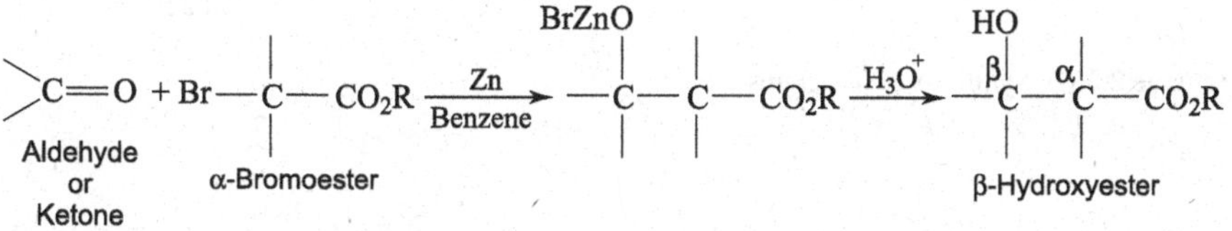

Scheme 2.333 Reformatsky reaction

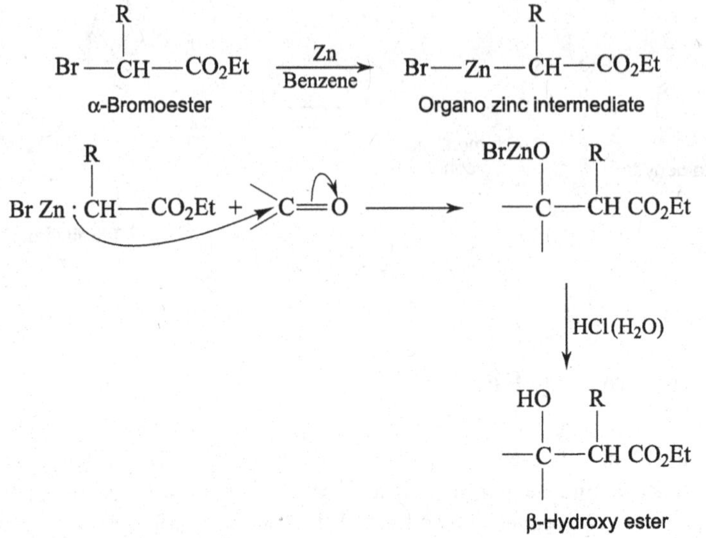

Scheme 2.334 Mechanism of Reformatsky reaction

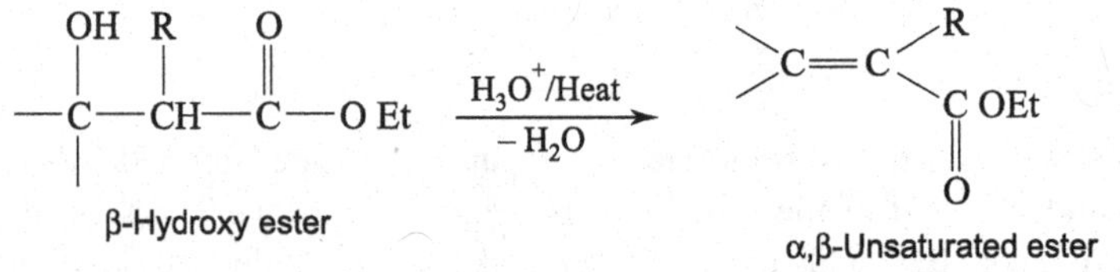

Scheme 2.335 Synthesis of a, b unsaturated esters

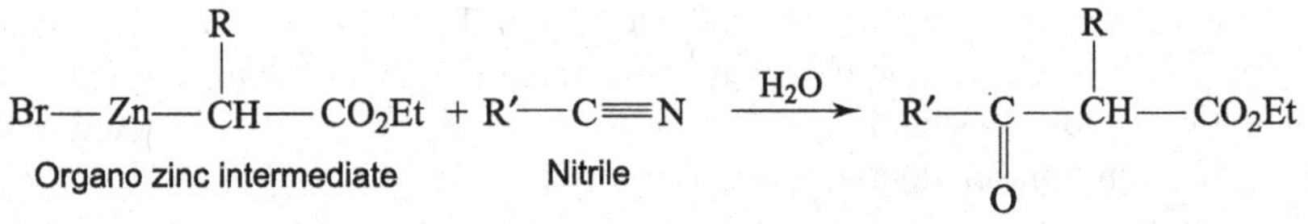

Scheme 2.336 Blaise reaction

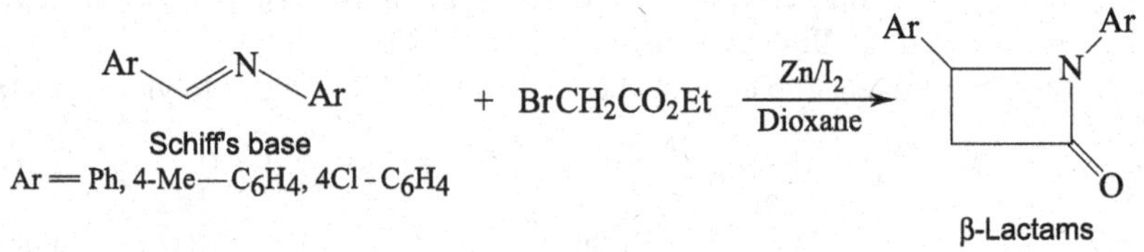

Scheme 2.337 Synthesis of β-lactams

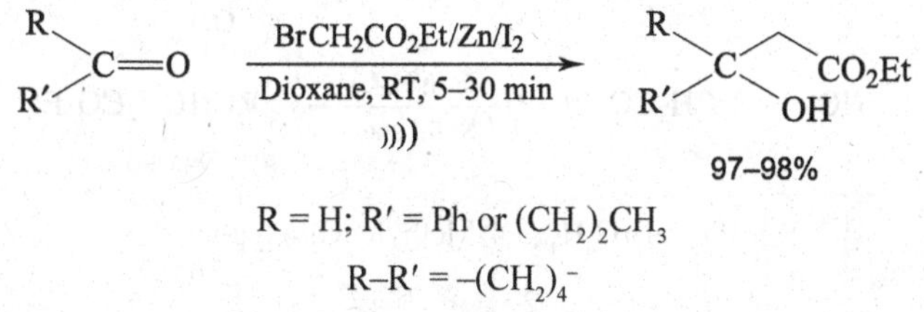

Scheme 2.338 Reformatsky reaction under sonication

R= CH₃, Ph.;
R'= H, CH₃

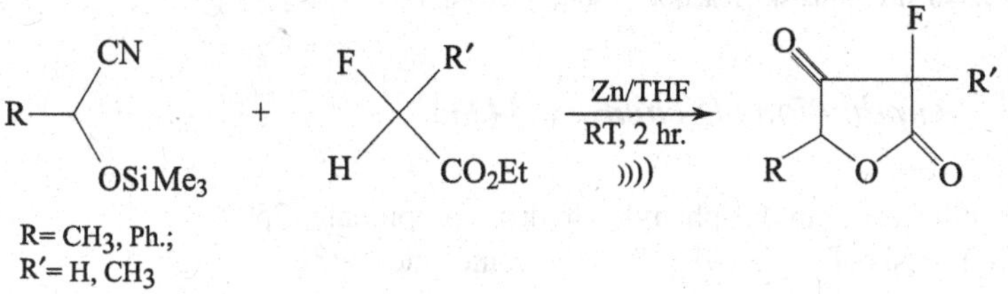

Scheme 2.339 Synthesis of keto-γ-butyrolactone

2.46.1 Reformatsky Reaction Using Sonication

Excellent yields are obtained on sonication compared to more traditional methods, such as those employing activated zinc or trimethylborane as cosolvent [346]. Under optimal conditions, quantitative yields of β-hydroxyester is obtained [347] (Scheme 2.338).

In the sonication procedure, it is necessary to activate zinc with iodine and to carry out the reaction in dioxane.

Even with nitriles, the Reformatsky reaction (known as **Blaise reaction**) under sonication leads to amines which hydrolyse to give ketones. Also, the application of Reformatsky reaction to Schiff's bases provides better yield of β-lactams, but this modification is not of general applicability [348].

Using appropriate nitrile, keto-γ-butyrolactone is obtained in good yield [348] (Scheme 2.339).

2.46.2 Reformatsky Reaction in Solid State

Treatment of aromatic aldehydes with ethylbromoacetate and Zn-NH₄Cl in the solid state give the corresponding Reformatsky products [349] (Scheme 2.340).

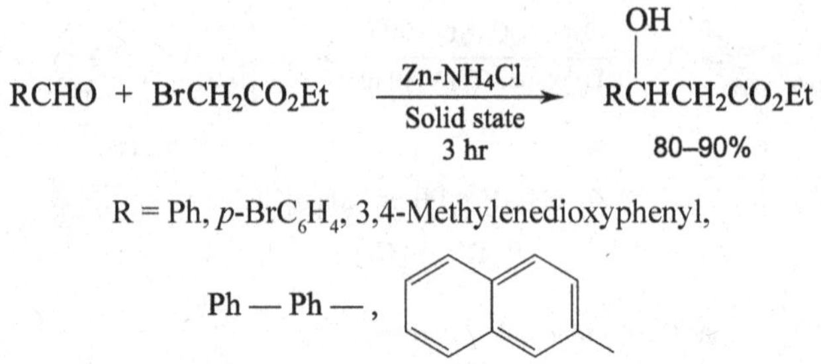

$$\text{RCHO} + \text{BrCH}_2\text{CO}_2\text{Et} \xrightarrow[\substack{\text{Solid state} \\ \text{3 hr}}]{\text{Zn-NH}_4\text{Cl}} \overset{\overset{\text{OH}}{|}}{\text{RCHCH}_2\text{CO}_2\text{Et}}$$

80–90%

R = Ph, *p*-BrC$_6$H$_4$, 3,4-Methylenedioxyphenyl,

Ph — Ph — ,

Scheme 2.340 Reformatsky reaction in solid state

2.46.3 *Applications (Scheme 2.341)*

(*i*) Synthesis of ethyl 3-phenyl-3-hydroxypropionate [350].
(*ii*) Synthesis of α-phenyl-γ-fluorotetronic acid [348].

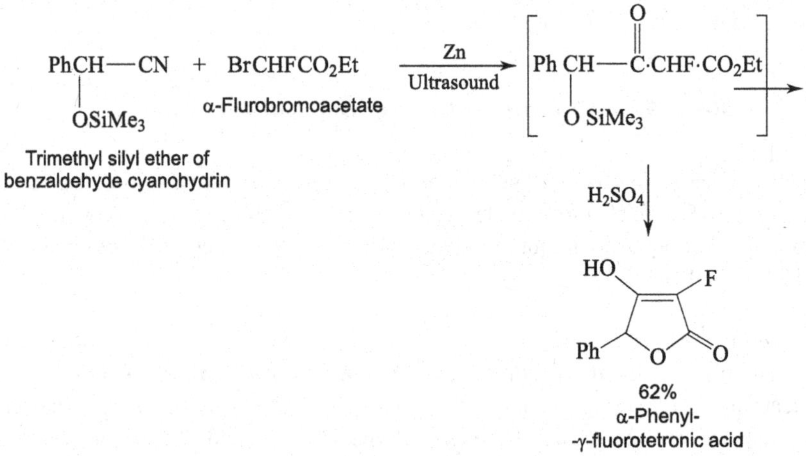

Trimethyl silyl ether of
benzaldehyde cyanohydrin

α-Flurobromoacetate

62%
α-Phenyl-
-γ-fluorotetronic acid

(*iii*) Synthesis of ethyl 2-methyl-3-*p*-tolyl-2-butenoate.

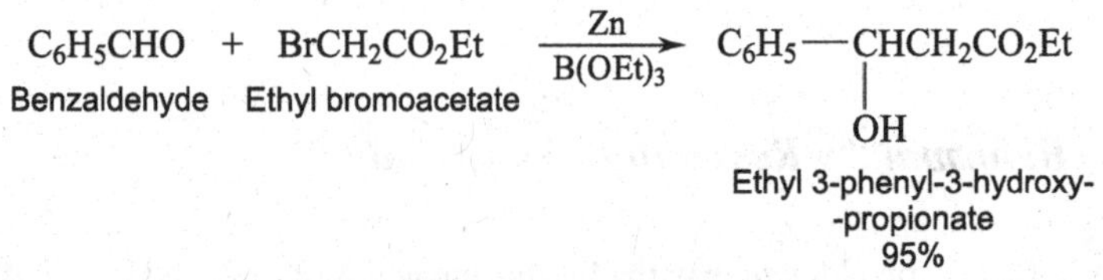

$$\text{C}_6\text{H}_5\text{CHO} + \text{BrCH}_2\text{CO}_2\text{Et} \xrightarrow[\text{B(OEt)}_3]{\text{Zn}} \overset{}{\text{C}_6\text{H}_5-\underset{\underset{\text{OH}}{|}}{\text{CHCH}_2\text{CO}_2\text{Et}}}$$

Benzaldehyde   Ethyl bromoacetate

Ethyl 3-phenyl-3-hydroxy-
-propionate
95%

Scheme 2.341 Some applications of Reformatsky reaction

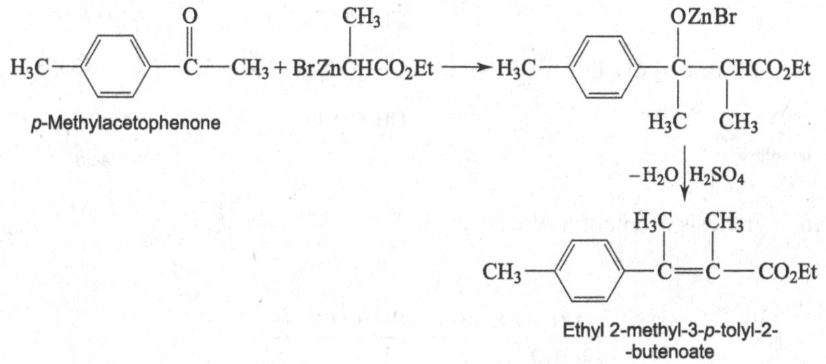

p-Methylacetophenone

Ethyl 2-methyl-3-p-tolyl-2-
-butenoate

(*iv*) Synthesis of citral.

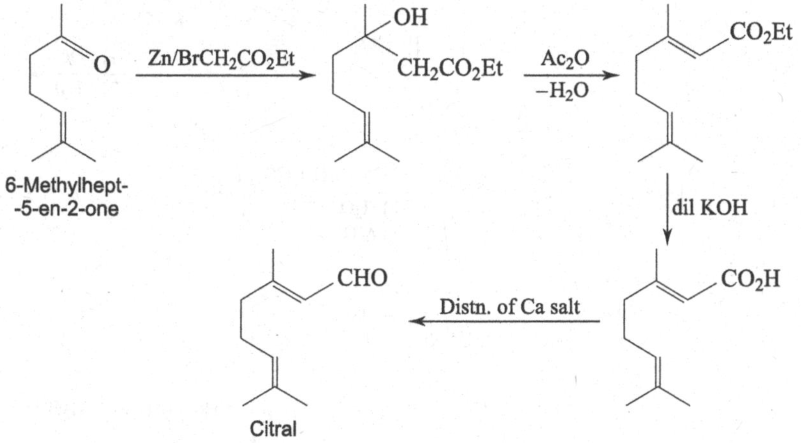

6-Methylhept-
-5-en-2-one

Citral

(*v*) Reformatsky reaction of 6-methyl-2-heptanone [351].

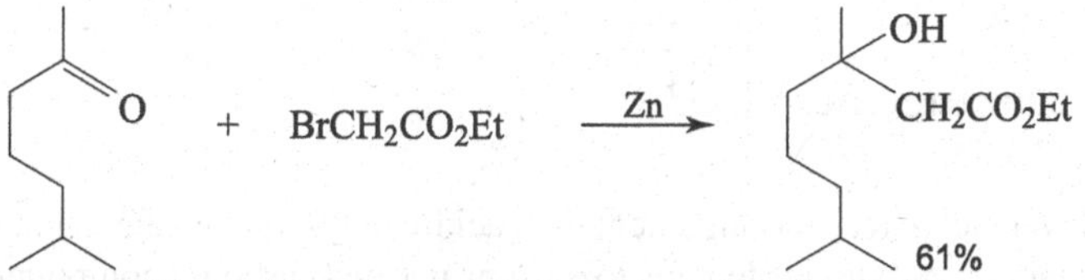

61%

(*vi*) Synthesis of spiro compounds [352].

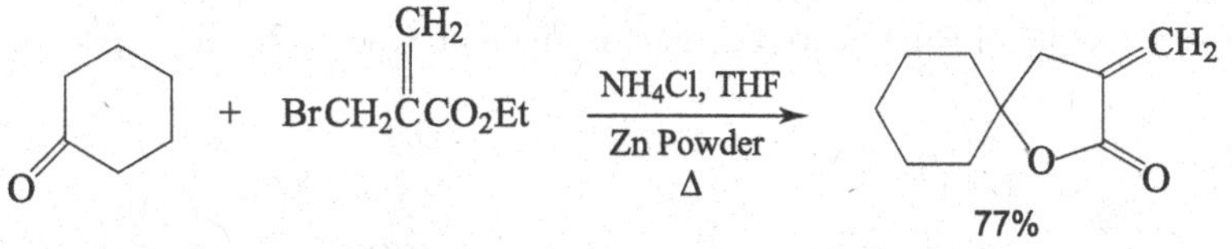

77%

(vii) Synthesis of citric acid.

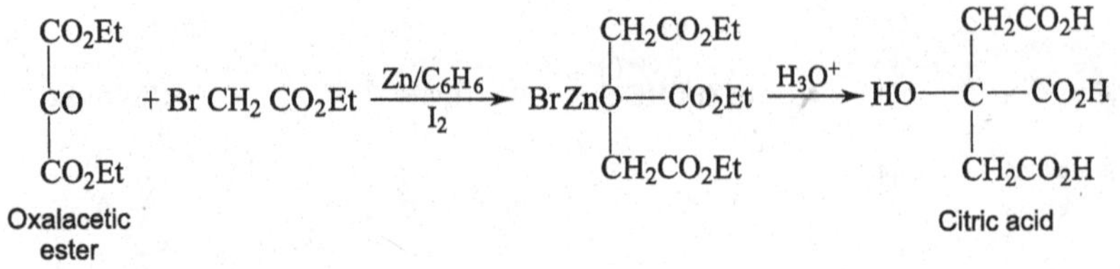

$$\begin{array}{c} CO_2Et \\ | \\ CO \\ | \\ CO_2Et \end{array} + Br\,CH_2\,CO_2Et \xrightarrow[I_2]{Zn/C_6H_6} \begin{array}{c} CH_2CO_2Et \\ | \\ BrZnO-CO_2Et \\ | \\ CH_2CO_2Et \end{array} \xrightarrow{H_3O^+} \begin{array}{c} CH_2CO_2H \\ | \\ HO-C-CO_2H \\ | \\ CH_2CO_2H \end{array}$$

Oxalacetic   Citric acid
ester

(*viii*) Synthesis of vitamin A_1.

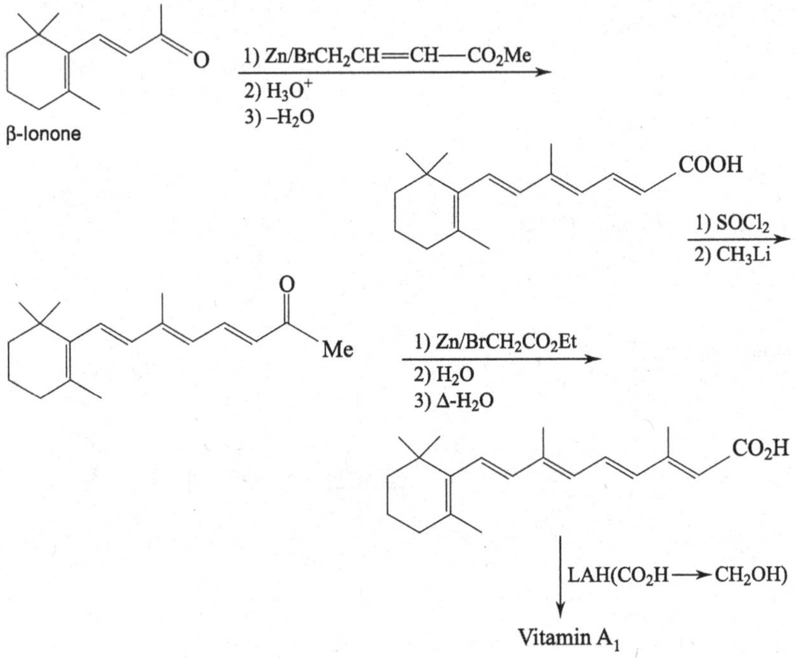

β-Ionone

1) Zn/BrCH$_2$CH=CH—CO$_2$Me

2) H$_3$O$^+$

3) –H$_2$O

COOH

1) SOCl$_2$

2) CH$_3$Li

Me

1) Zn/BrCH$_2$CO$_2$Et

2) H$_2$O

3) Δ-H$_2$O

CO$_2$H

LAH(CO$_2$H ⟶ CH$_2$OH)

Vitamin A_1

2.47   Rupe Rearrangement [353]

Also known as Meyer-Schuster rearrangement, it is acid-catalysed rearrangement of secondary and tertiary α-acetylenic alcohols to α,β-unsaturated carbonyl compounds (Scheme 2.342); aldehydes are formed when the acetylenic group is terminal and ketones are formed when it is internal.

Another example of Rupe rearrangement is given in Scheme 2.343.

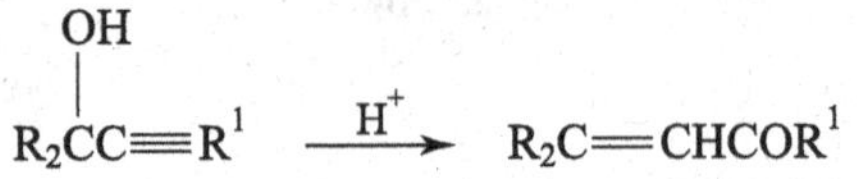

$$\begin{array}{c} OH \\ | \\ R_2CC\equiv R^1 \end{array} \xrightarrow{H^+} R_2C=CHCOR^1$$

Scheme 2.342 Rupe rearrangement

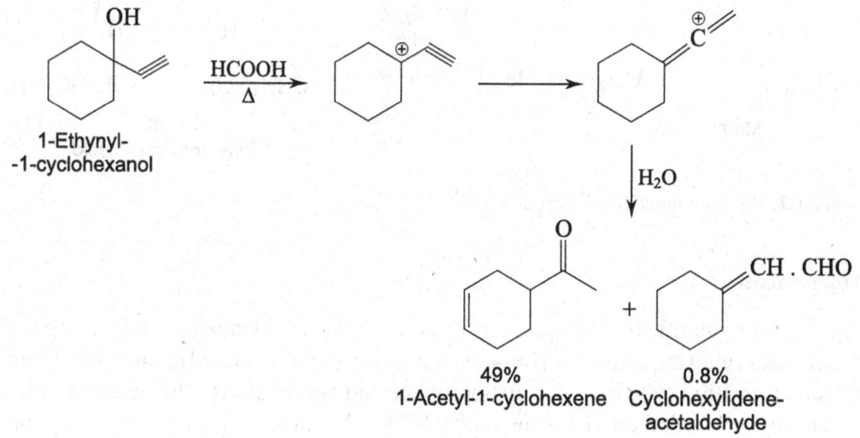

Scheme 2.343 Another example of Rupe rearrangement

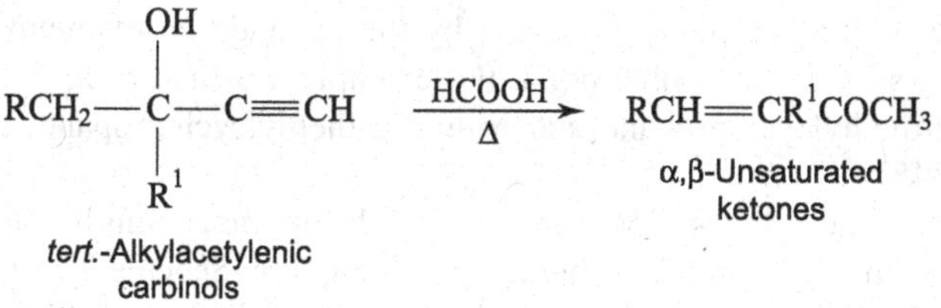

Scheme 2.344 Synthesis of α,β-unsaturated ketones

The Rupe reaction involves the rearrangement of tertiary alkyl acetylenic carbinols with a terminal acetylenic group predominantly to α,β-unsaturated ketones and not the expected aldehydes (Scheme 2.344).

It has now been found [354] that Rupe or Meyer-Schuster rearrangements have been reported in SC-H$_2$O at elevated temperatures. At elevated temperatures, SC-H$_2$O behaves as an acid.

2.48   Simmons–Smith Reaction [355]

This reaction is widely used for the synthesis of cyclopropane derivatives from alkenes by reaction with methylene iodide and zinc-copper or better zinc-silver couple. This is a versatile reaction and has been used with success to a variety of alkenes. Many functional groups are unaffected, enabling one to synthesize a variety of cyclopropane derivatives. As an example, dihydrosterculic acid can be obtained with 51% yield from methyl oleate (Scheme 2.345).

The above reaction is stereospecific and takes place by *cis* addition of methylene to the less hindered side of the double bond.

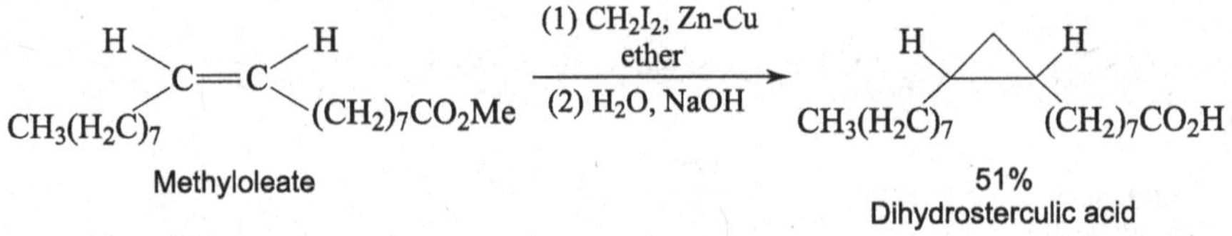

Scheme 2.345 Simmons–Smith reaction

Mechanism

The reactive intermediate is believed to be an iodomethylenezinciodide complex, which reacts with the alkene in a bimolecular process in a concreted manner involving a cyclic transition state to give a cyclopropane and zinc iodide (Scheme 2.346).

The reagent, iodomethylenezinciodide, ICH_2ZnI is obtained in situ by the reaction of diiodomethane and Zn (in the form of zinc copper couple) (Scheme 2.347).

Generally, substituted alkenes react somewhat faster than unsubstituted alkenes. Thus, 1-methylcyclohexene reacts faster than cyclohexene. The reaction, as in the case of other olefins (Scheme 1) occurs by the *cis* addition of methylene to the less hindered side of the double bond. For example, *cis*- and *trans*-3-hexene give pure *cis*-1,2-diethylcyclopropane and *trans*-1,2-diethylcyclopropane, respectively (Scheme 2.348).

Non-terminal acetylenes [356] also react with Simmons–Smith reagent to give the corresponding cyclopropenes, but the yields are low (Scheme 2.349).

Corey et al. [357] observed that there is a pronounced effect of the neighbouring hydroxyl substituents. The oxygen atom of the starting alcohol coordinates with zinc followed by transfer of methylene to the nearer face of the adjacent double bond, thus increasing the rate and control of the stereochemistry of the adduct (Scheme 2.350).

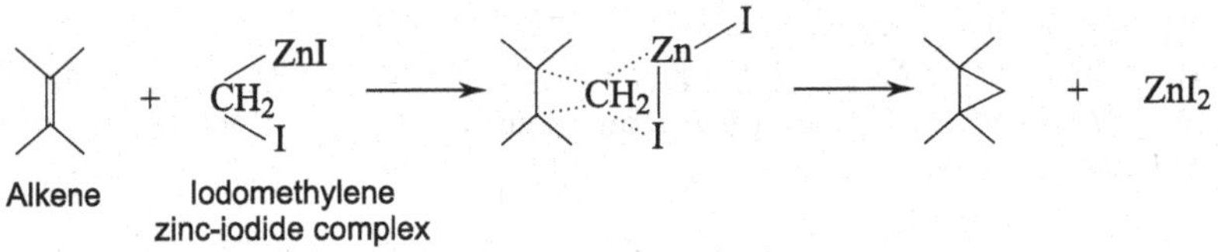

Scheme 2.346 Mechanism of Simmons–Smith reaction

$$CH_2I_2 + Zn \longrightarrow ICH_2ZnI$$

Scheme 2.347 Preparation of ICH_2ZnI

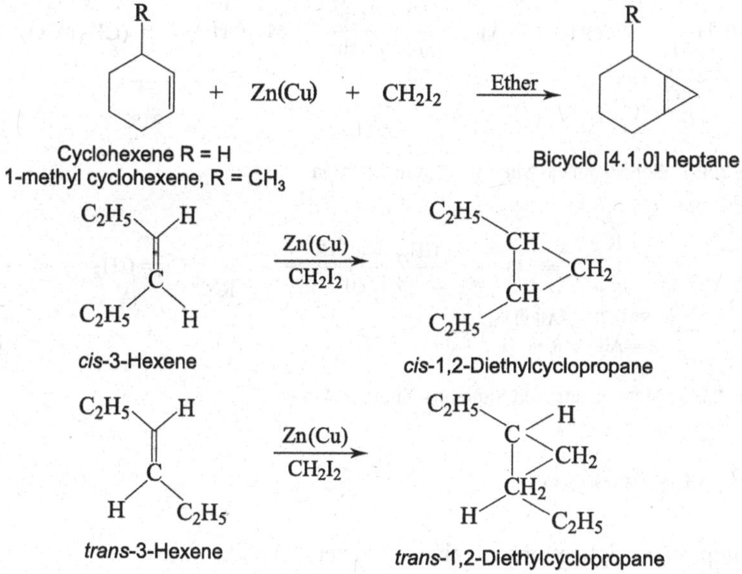

Scheme 2.348 Reactivity in Simmons–Smith reaction

$$R—C\equiv C—R' \xrightarrow[\text{CH}_2\text{I}_2]{\text{Zn(Cu)}} R—C\!=\!C—R'$$

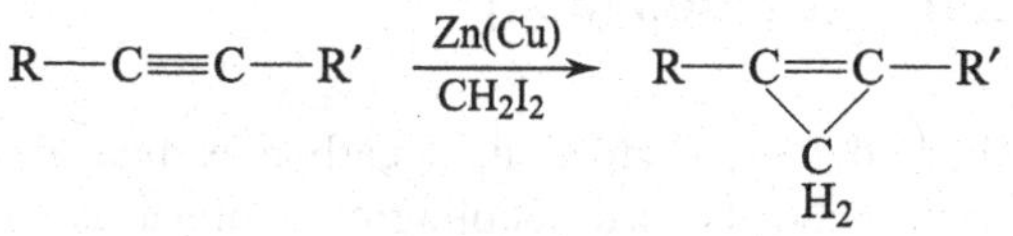

Scheme 2.349 Synthesis of cyclopropenes

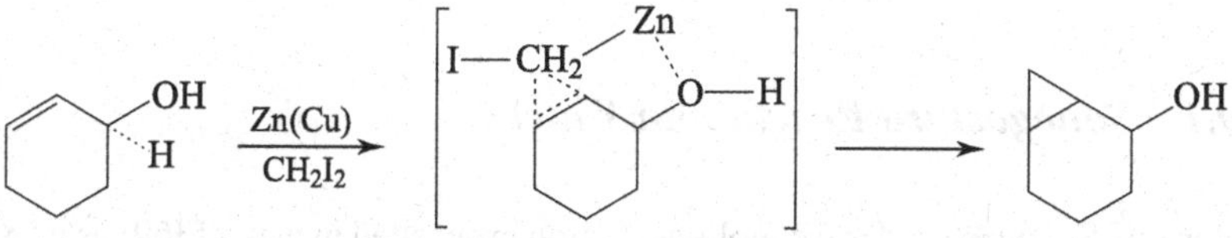

Scheme 2.350 In Simmons–Smith reaction there may be control of stereochemistry

2.48.1 Simmons–Smith Reaction Under Sonication

In this procedure, sonochemically activated zinc and methylene iodide are used [358]. The generated carbene adds on to the olefinic bond to give 91% yield of the cyclopropane derivative compared to 51% yield by the normal route (Scheme 2.351).

The above procedure can be scaled up [359] and has several advantages. Ketones on reaction with Simmons–Smith reagent result in methylenation [360] of the carbonyl group (Scheme 2.352). Normally, such methylenation of carbonyl group requires complex reagents; this can now be accomplished by sonication.

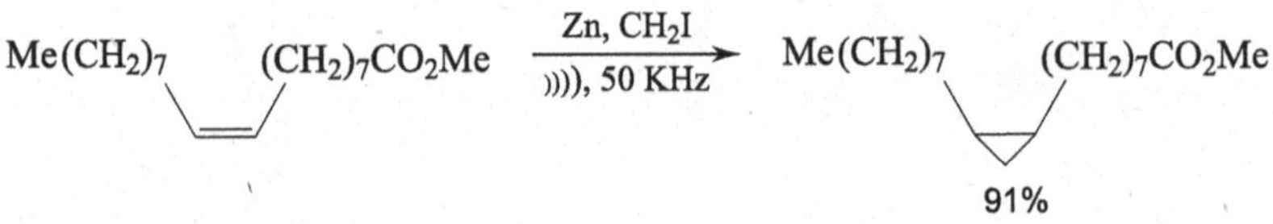

Me(CH$_2$)$_7$ (CH$_2$)$_7$CO$_2$Me $\xrightarrow[\text{)))), 50 KHz}]{\text{Zn, CH}_2\text{I}}$ Me(CH$_2$)$_7$ (CH$_2$)$_7$CO$_2$Me

91%

Scheme 2.351 Sonochemical Simmons–Smith reaction

$$\underset{R'}{\overset{R}{\diagdown}}C=O \xrightarrow[\text{RT,))))}]{\text{CH}_2\text{I}_2/\text{Zn/THF}} \underset{R'}{\overset{R}{\diagdown}}C=CH_2$$

R = R′ = Alkyl
R = Alkyl; R′ = H

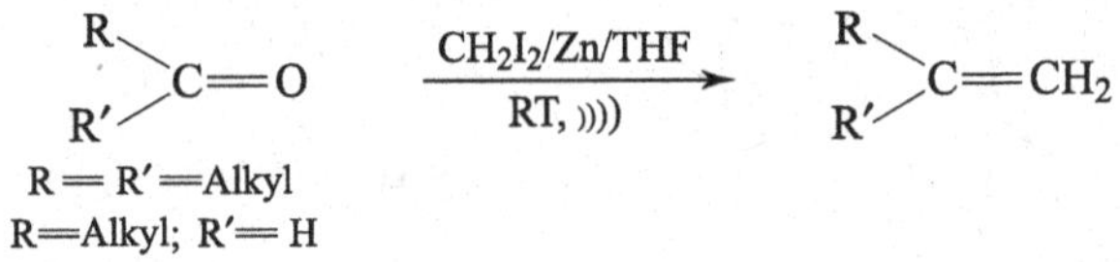

Scheme 2.352 Methylenation in Simmons–Smith reaction

2.48.2 Applications

Some important applications are given in Scheme 2.353.

2.49 Sonogashira Reaction [368]

It is a powerful method for the creation of carbon–carbon bonds [368] and is a palladium and copper co-catalysed coupling of terminal alkynes with aryl and vinyl halides. It is a general method for the preparation of unsymmetrical alkynes (Scheme 2.354).

2.49.1 Sonogashira Reaction in Water

An aqueous Sonogashira-type coupling reaction proceeded in water [369] as the sole solvent, without the need for copper (I) or any transition-metal phosphine complex, which overcome the problem of intrinsic toxicity and air-sensitivity of transition-metal complexes and the use of costly phosphane ligands (Scheme 2.355).

Another example of Sonogashira coupling of aryl bromides and iodides with phenylacetylene using the polymeric complex (A) or the monomeric (B) as catalyst, pyrrolidine as base and TBAB as additive [370] has been performed under microwave heating. Use of catalyst (B) gave better yield (Scheme 2.356).

Sonogashira coupling of aryl and heteroaryl bromides and iodides with phenyl acetylene occurred in the presence of heterogeneous Pd catalyst [371] (C) under microwave heating to give the condensed product (Scheme 2.357).

Sonogashira coupling under transition-metal-free conditions [372] involving ultra-low Pd concentrations as contaminations [373] and using NaOH as base,

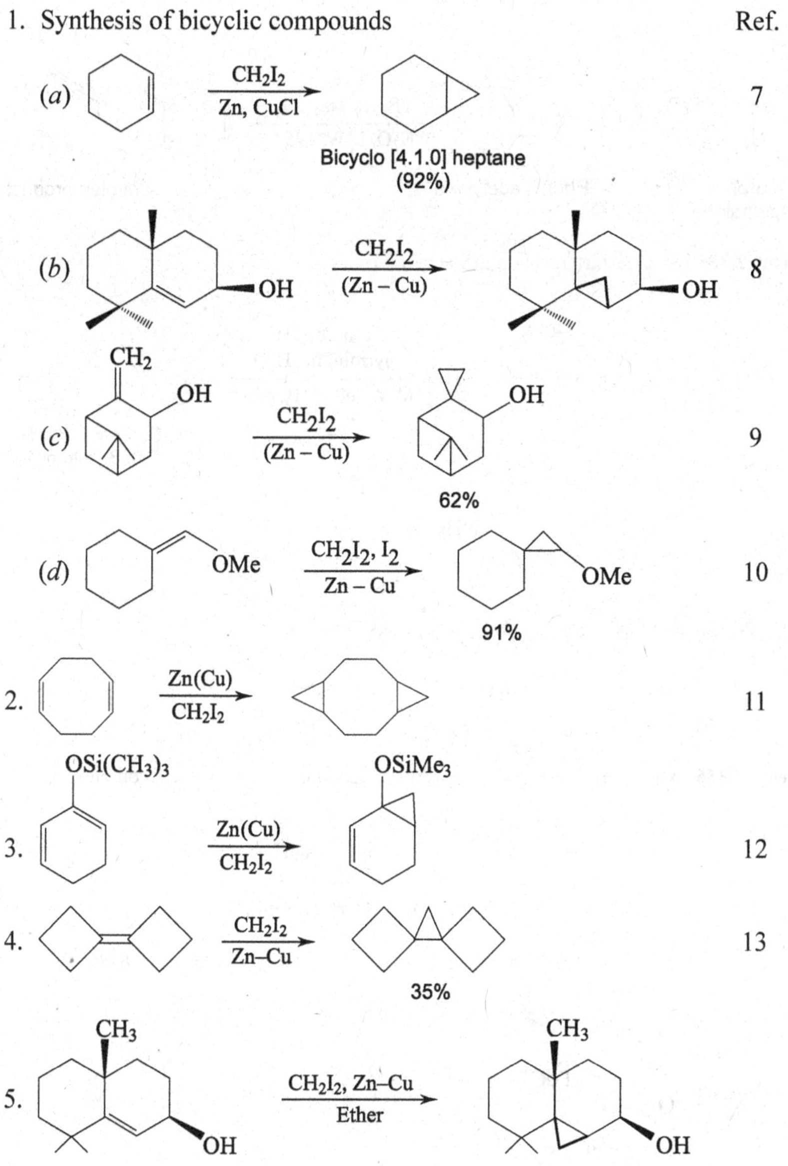

1. Synthesis of bicyclic compounds

Ref.

(a)

$$\xrightarrow[\text{Zn, CuCl}]{\text{CH}_2\text{I}_2}$$

Bicyclo [4.1.0] heptane
(92%)

7

(b)

$$\xrightarrow[\text{(Zn – Cu)}]{\text{CH}_2\text{I}_2}$$

OH

8

(c)

CH$_2$

OH

$$\xrightarrow[\text{(Zn – Cu)}]{\text{CH}_2\text{I}_2}$$

OH

62%

9

(d)

OMe

$$\xrightarrow[\text{Zn – Cu}]{\text{CH}_2\text{I}_2, \text{I}_2}$$

OMe

91%

10

2.

$$\xrightarrow[\text{CH}_2\text{I}_2]{\text{Zn(Cu)}}$$

11

3.

OSi(CH$_3$)$_3$

$$\xrightarrow[\text{CH}_2\text{I}_2]{\text{Zn(Cu)}}$$

OSiMe$_3$

12

4.

$$\xrightarrow[\text{Zn–Cu}]{\text{CH}_2\text{I}_2}$$

35%

13

5.

CH$_3$

OH

$$\xrightarrow[\text{Ether}]{\text{CH}_2\text{I}_2, \text{Zn–Cu}}$$

CH$_3$

OH

Scheme 2.353 Some applications of Simmons–Smith reaction

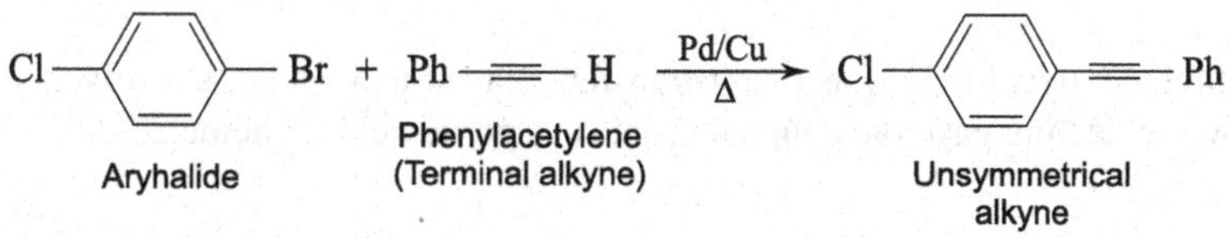

$$\text{Cl} - \underset{\text{Aryhalide}}{\boxed{}} - \text{Br} + \underset{\substack{\text{Phenylacetylene} \\ \text{(Terminal alkyne)}}}{\text{Ph} == \text{H}} \xrightarrow[\Delta]{\text{Pd/Cu}} \text{Cl} - \underset{\substack{\text{Unsymmetrical} \\ \text{alkyne}}}{\boxed{}} - == \text{Ph}$$

Scheme 2.354 Sonogashira reaction

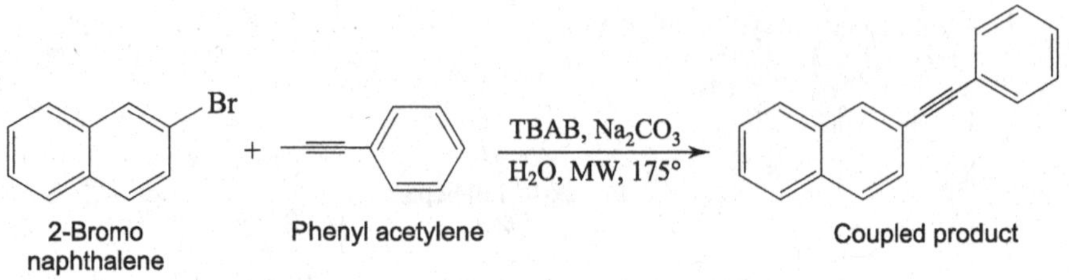

Scheme 2.355 Sonogashira reaction in water

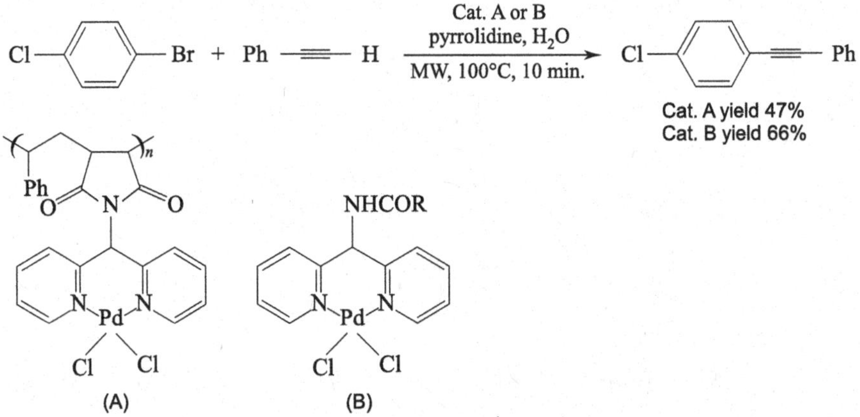

Scheme 2.356 Microwave-assisted Sonogashira reaction using a polymeric complex

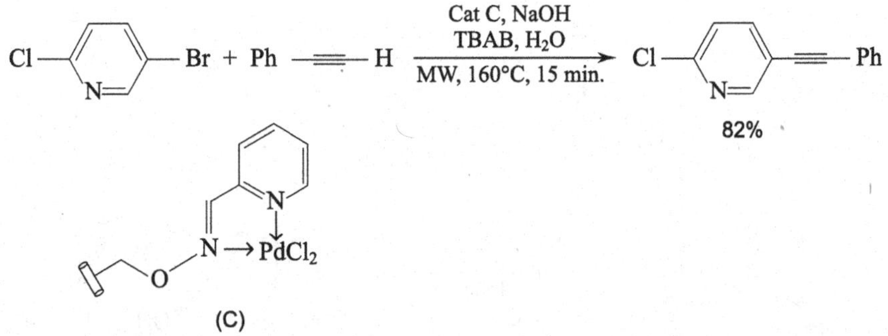

Scheme 2.357 Microwave-assisted Sonogashira coupling in the presence of heteroaryl Pd catalyst

polyethylene glycol (PEG) as phase-transfer catalyst and water as a solvent under microwave heating gave the coupled product in good yield (Scheme 2.358).

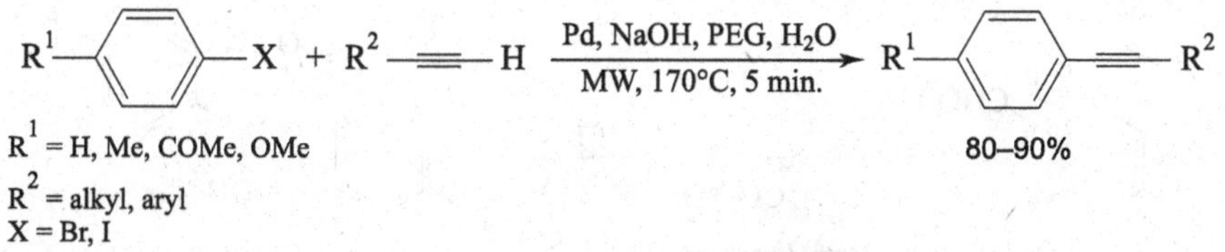

$$R^1\!\!-\!\!\langle\text{---}\rangle\!\!-\!\!X \ + \ R^2\text{---}\!\!\equiv\!\!\text{---}H \xrightarrow[\text{MW, 170°C, 5 min.}]{\text{Pd, NaOH, PEG, H}_2\text{O}} R^1\!\!-\!\!\langle\text{---}\rangle\!\!-\!\!\equiv\!\!\text{---}R^2$$

R^1 = H, Me, COMe, OMe

R^2 = alkyl, aryl

X = Br, I

80–90%

Scheme 2.358 Microwave-assisted Sonogashira coupling in NaOH using PEG as PTC

2.49.2 Sonogashira Reaction in Ionic Liquids

Copper and ligand-free Sonogashira reaction catalysed by Pd(O) nanoparticles proceeds under ultrasound irradiation in ionic liquid, $[bb_{min}][BF_4]$ (Scheme 2.360) [7].

2.50 Stetter Reaction [374]

The reaction of aldehydes with olefins to give 1,4-dicarbonyls is known as Stetter reaction. It is possible to carry out the reaction in ionic liquid using Et_3N as catalyst [374] (Scheme 2.361).

2.51 Stille Coupling Reaction [375]

Coupling of organotin reagents with aryl or vinyl halides or triflates in the presence of palladium give the corresponding coupled product [375] (Scheme 2.361).

The Stille reaction is one of the most widely used step in the preparation of a wide variety of materials including polyarenes and diaryl, and aromatic carbonyl compounds [376].

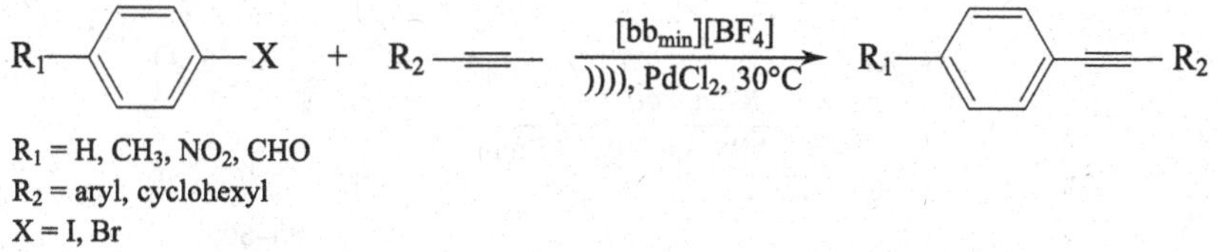

$$R_1\!\!-\!\!\langle\text{---}\rangle\!\!-\!\!X \ + \ R_2\text{---}\!\!\equiv\!\!\text{---} \xrightarrow[\text{)))), PdCl}_2\text{, 30°C}]{[bb_{min}][BF_4]} R_1\!\!-\!\!\langle\text{---}\rangle\!\!-\!\!\equiv\!\!\text{---}R_2$$

R_1 = H, CH_3, NO_2, CHO

R_2 = aryl, cyclohexyl

X = I, Br

Scheme 2.359 Sonogashira reaction in ionic liquids

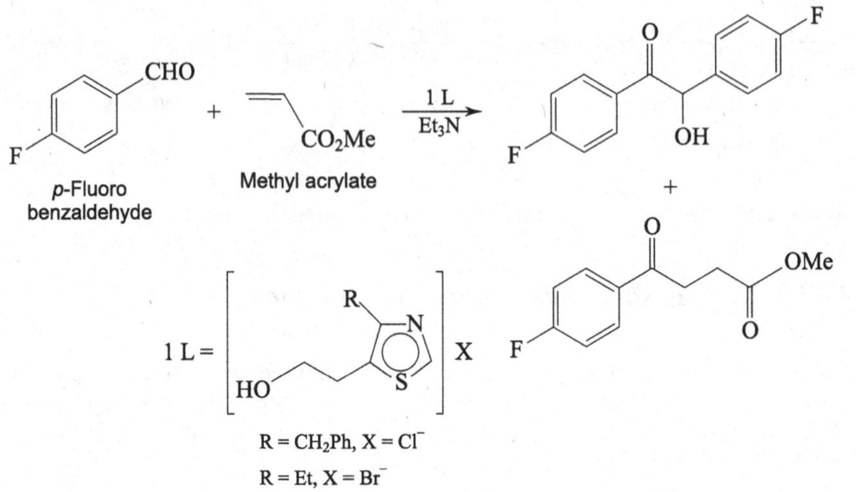

Scheme 2.360 Stetter reaction

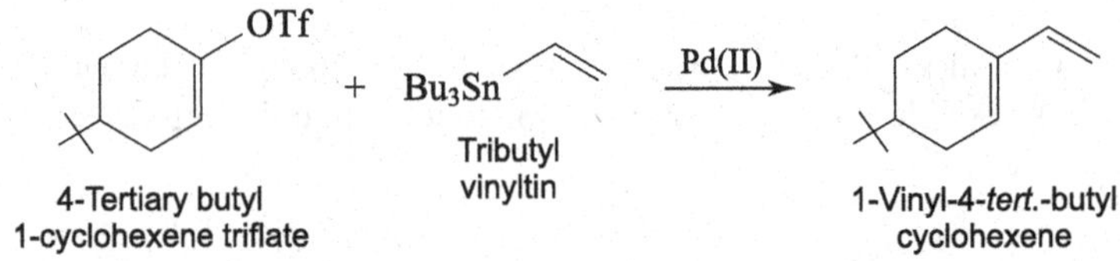

Scheme 2.361 Stille coupling reaction

2.51.1 Stille Coupling Reaction in Water

An aqueous microwave-assisted Stille reaction has been reported [377] in the 2(H) pyrazinone derivatives at the C–3 position (Scheme 2.362).

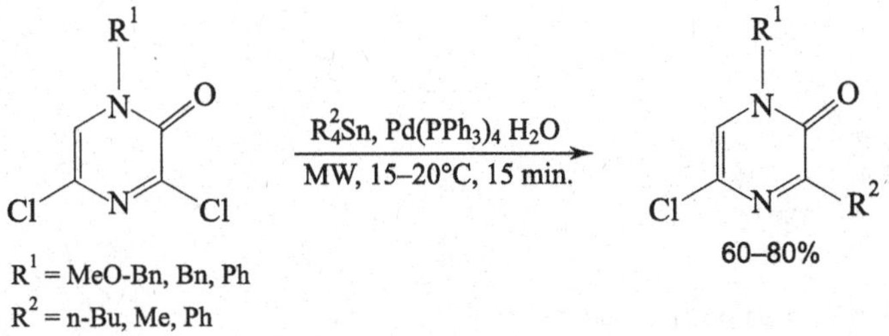

R^1 = MeO-Bn, Bn, Ph
R^2 = n-Bu, Me, Ph

Scheme 2.362 Stille coupling reaction in water

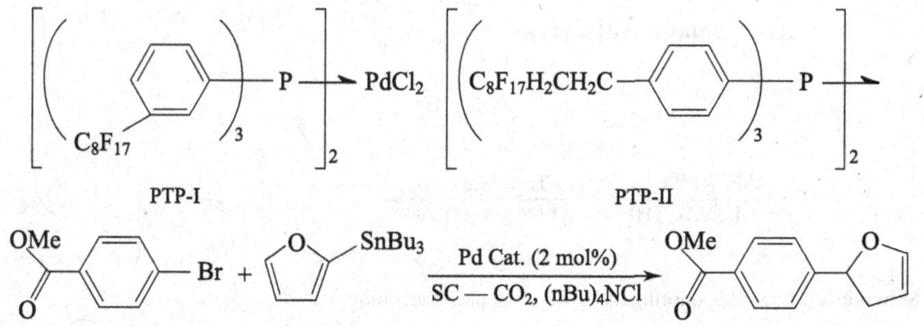

PTP-I   PTP-II

Scheme 2.363 Stille coupling reaction in SC-CO_2

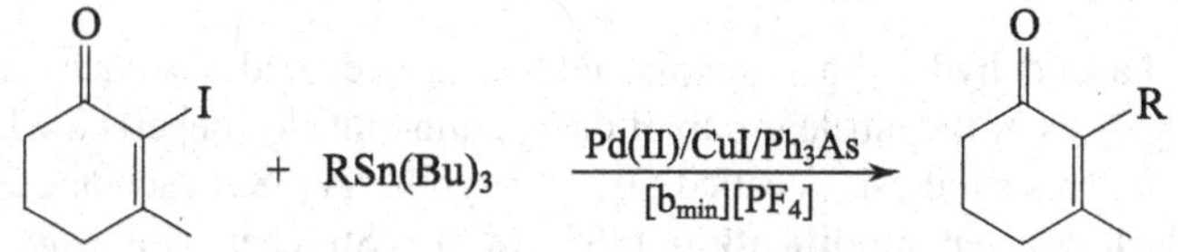

Scheme 2.364 Stille coupling reaction in ionic liquids

2.51.2 Stille Coupling Reaction in SC-CO_2

Stille coupling mediated by fluorous-tagged phosphine (PTP-1 and PTP-2) has been achieved [378] in SC-CO_2 using $(nBu)_4$ NCl as catalyst (Scheme 2.363).

2.51.3 Stille Coupling Reaction in Ionic Liquids

It has been found [379] that the use of palladium complexes immobilized in ionic liquid offers great advantage over the classical organic solvent used for Stille coupling reactions. A large number of Stille coupling reactions with Pd(O) or Pd(II) catalyst associated with Ph_3As in the presence of CuI have been developed in $[b_{min}][BF_4]$ (Scheme 2.364) [379].

2.51.4 Stille Coupling Using Fluorous Phase Technique

Stille coupling has been achieved (Scheme 2.365) with perfluoro-tagged tin compounds [380, 381].

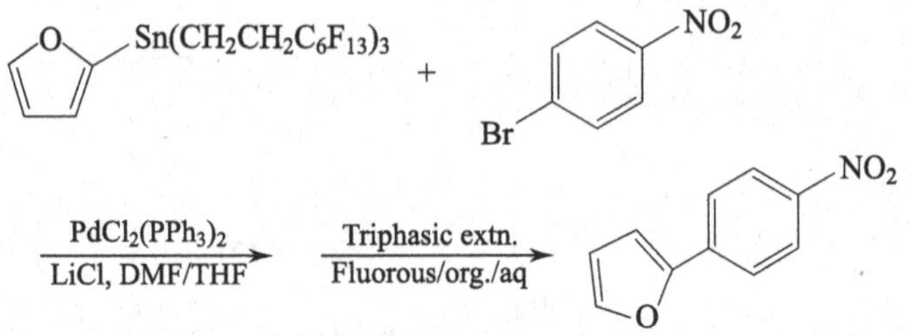

Scheme 2.365 Stille coupling using fluorous phase technique

2.52   Strecker Synthesis [382]

Treatment of an aldehyde with ammonia and hydrogen cyanide produces an α-amino nitrile. Hydrolysis of the nitrile group of the α-amino nitrile converts the latter to an α-amino acid. This synthesis is called Strecker synthesis [382] (Scheme 2.366).

In the **Erlenmeyer modification** [383] of the Strecker synthesis, the aldehyde is treated with HCN and the formed cyanohydrin is reacted with ammonia (Scheme 2.367); the final step is the same as given above.

A more convenient procedure is to treat the aldehyde in one step with ammonium chloride and sodium cyanide (this mixture is equivalent to ammonium cyanide, which in turn decomposes into ammonia and HCN). This procedure is referred to as the **Zelinsky–Stadnikoff modification** [384]. The final step is the hydrolysis of the intermediate α-amino nitrile under basic or acidic conditions to give the corresponding α-amino acid. The synthesis of phenylalanine from phenylacetaldehyde is given in Scheme 2.368.

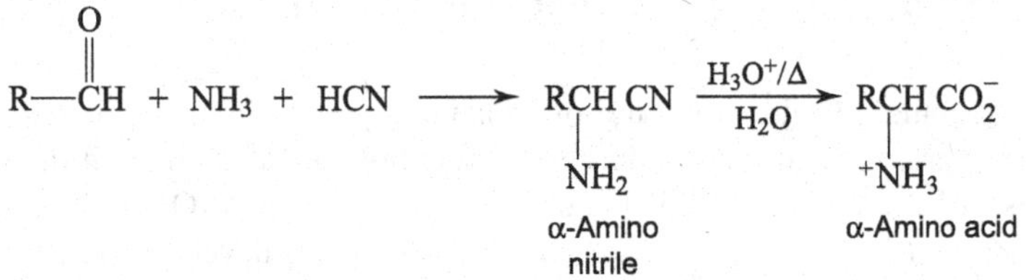

Scheme 2.366 Strecker synthesis

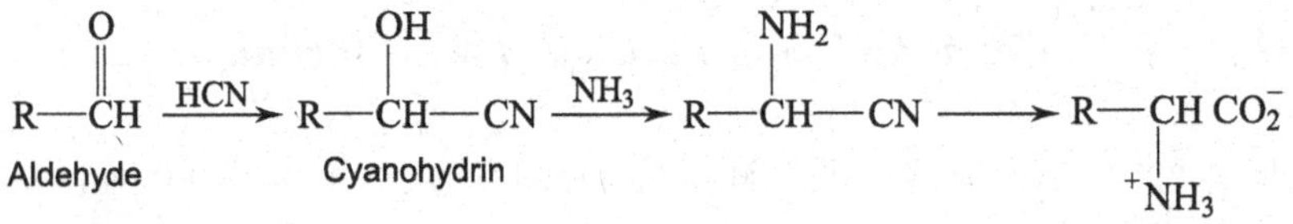

Scheme 2.367 Erlenmeyer modification of Strecker synthesis

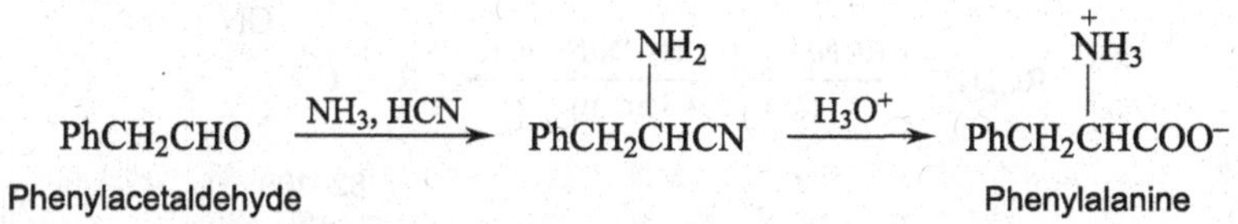

$$\text{PhCH}_2\text{CHO} \xrightarrow{\text{NH}_3, \text{HCN}} \overset{\overset{\displaystyle NH_2}{\displaystyle |}}{\text{PhCH}_2\text{CHCN}} \xrightarrow{\text{H}_3\text{O}^+} \overset{\overset{\displaystyle \overset{+}{N}H_3}{\displaystyle |}}{\text{PhCH}_2\text{CHCOO}^-}$$

Phenylacetaldehyde   Phenylalanine

Scheme 2.368 Zelinsky–Stadnikoff modification of Strecker Synthesis

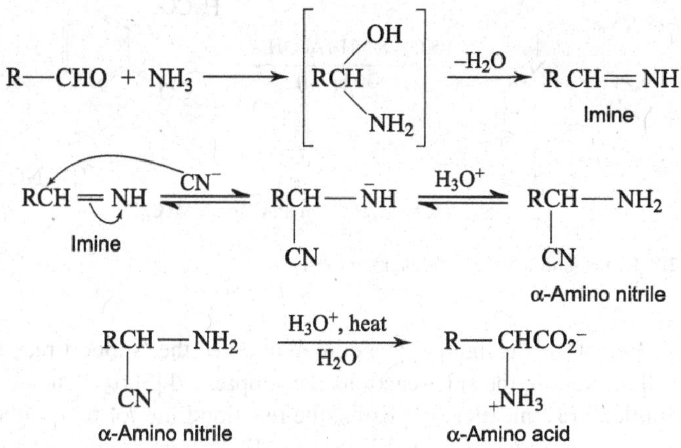

Scheme 2.369 Mechanism of Strecker synthesis

Mechanism

The first step is the formation of an imine from the aldehyde and ammonia. Subsequent addition of HCN gives α-amino nitrile, which on hydrolysis gives α-amino acid (Scheme 2.369).

2.52.1 Strecker Synthesis Under Sonication

Strecker synthesis of amino nitriles with much better yield is possible using ultrasonic acceleration [385] (Scheme 2.370).

A modified Strecker synthesis for the preparation of α-amino nitriles in excellent yields involves the adsorption of the reagent on the surface of a catalyst before

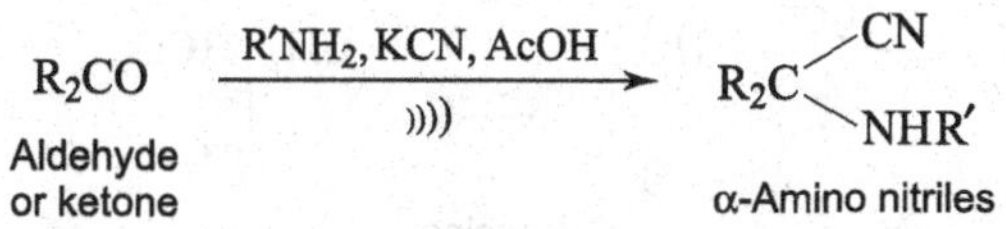

$$\underset{\substack{\text{Aldehyde} \\ \text{or ketone}}}{\text{R}_2\text{CO}} \xrightarrow[\text{)))}]{\text{R}'\text{NH}_2, \text{KCN}, \text{AcOH}} \underset{\text{α-Amino nitriles}}{\text{R}_2\text{C}\overset{\displaystyle \diagup \text{CN}}{\diagdown \text{NHR}'}}$$

Scheme 2.370 Strecker synthesis under sonication

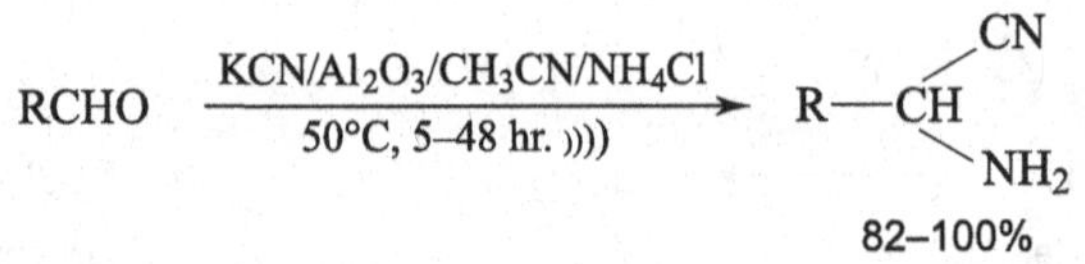

$$RCHO \xrightarrow[\text{50°C, 5–48 hr.)))}]{\text{KCN/Al}_2\text{O}_3\text{/CH}_3\text{CN/NH}_4\text{Cl}} R\text{—CH} \begin{smallmatrix} \text{CN} \\ \\ \text{NH}_2 \end{smallmatrix}$$

82–100%

Scheme 2.371 A modified Strecker synthesis

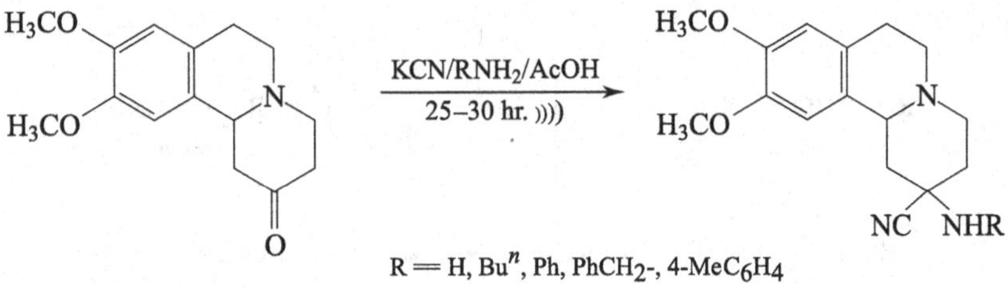

$$\xrightarrow[\text{25–30 hr.)))}]{\text{KCN/RNH}_2\text{/AcOH}}$$

R = H, Bu^n, Ph, PhCH$_2$-, 4-MeC$_6$H$_4$

Scheme 2.372 Ultrasound-assisted Strecker synthesis

the reaction. In fact, this technique is a combination of the 'support reagents' with sonochemical activation; the side reactions are suppressed [386] (Scheme 2.371).

Using simple ultrasonic cleaning baths, the reaction time for the synthesis of α-amino nitriles can be reduced from 12 days to 20–25 h and yields up to 60% are obtained [387] (Scheme 2.372).

2.52.2 Applications (Scheme 2.373)

1. Strecker synthesis is mostly used for the synthesis of α-amino acids. Thus, by using formaldehyde, acetaldehyde, 2-methylpropionaldehyde and phenylacetaldehyde, glycine, alanine, valine and phenylalanine, respectively, can be obtained. The synthesis of dl-Tyrosine is given below.

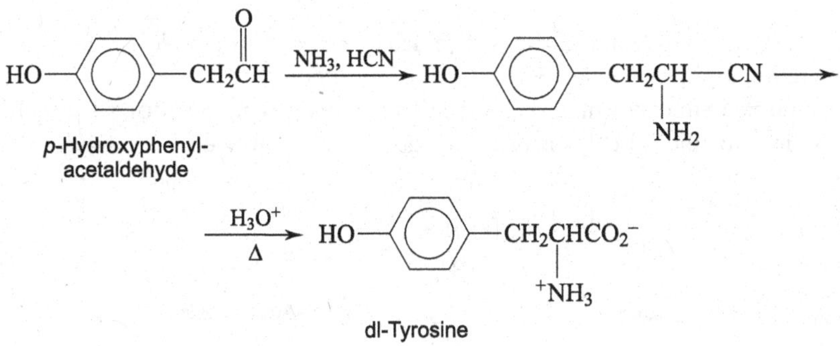

p-Hydroxyphenyl-
acetaldehyde

dl-Tyrosine

Scheme 2.373 Some applications of Strecker synthesis

2. Using Strecker synthesis, disodium iminodiacetate (DSIDA), an intermediate for the Monsanto's Roundup (herbicide), was synthesized.

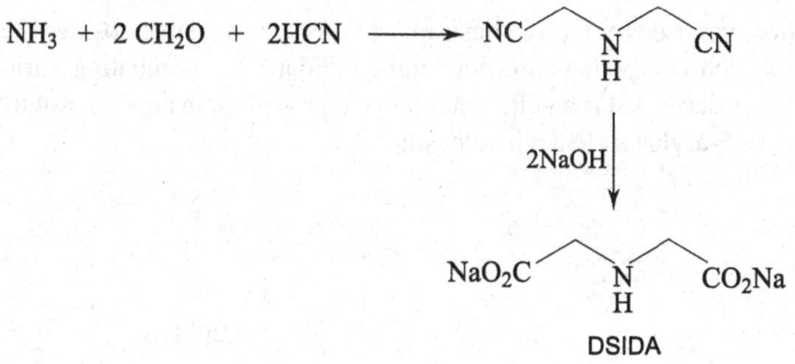

$$NH_3 \ + \ 2\,CH_2O \ + \ 2HCN \ \longrightarrow \ NC \overset{}{\diagdown} \underset{H}{N} \diagup CN$$

$$2NaOH \downarrow$$

$$NaO_2C \overset{}{\diagdown} \underset{H}{N} \diagup CO_2Na$$

DSIDA

2.53 Suzuki Coupling Reaction

The palladium-catalysed cross-coupling of aryl halides with boronic acid, known as Suzuki coupling reaction, is one of the most frequently used C–C cross-coupling reaction. It is a convenient method for the synthesis of biaryls [388, 389].

2.53.1 Suzuki Coupling Reaction in Aqueous Medium

A number of biaryl derivatives were prepared [390, 391] by the reaction of various aryl halides with phenyl boronic acid in aqueous medium using MW irradiation as shown in Scheme 2.374.

The above reaction gave good yields in case of aryl bromides and iodides. However, in case of aryl chloride, the yield was poor. This problem was overcome by carrying out the reaction of aryl chloride and phenyl boronic acid catalysed by Pd/C in aqueous medium using simultaneous cooling technique in conjugation of MW heating [392].

Suzuki reaction also finds application in the synthesis of natural products and heterocyclic synthesis. Thus, the reaction of imidazo [1,2-a] pyridines with aryl

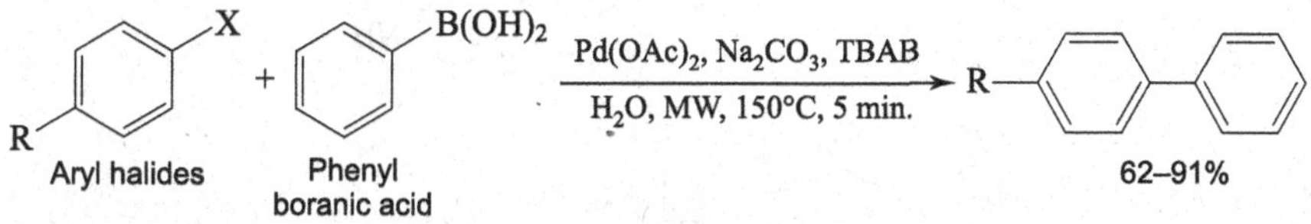

Aryl halides   Phenyl
                boranic acid

R = Me, OMe, COMe
X = Cl, Br, I

$$\xrightarrow[\text{H}_2\text{O, MW, 150°C, 5 min.}]{\text{Pd(OAc)}_2,\ \text{Na}_2\text{CO}_3,\ \text{TBAB}}$$

62–91%

Scheme 2.374 Suzuki coupling reaction in aqueous medium

boronic acid in aqueous medium under MW irradiation gave a convenient synthesis of heterocyclic compounds (Scheme 2.375). This method is more efficient than the conventional conditions [393].

Suzuki reaction was used for the synthesis of 5-aryltriazole acyl nucleosides [394] (Scheme 2.376). Such compounds are potential candidates for combating various viruses. The process described is an efficient one-step procedure in aqueous solution for the synthesis of 5-aryltriazole acylnucleosides.

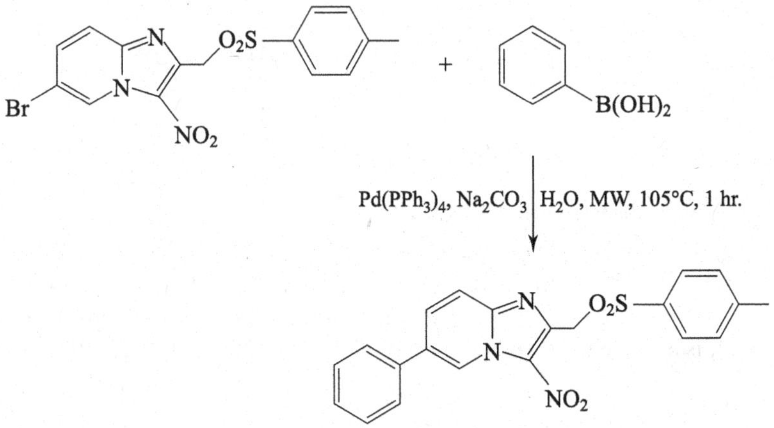

Scheme 2.375 Synthesis of heterocyclic compounds

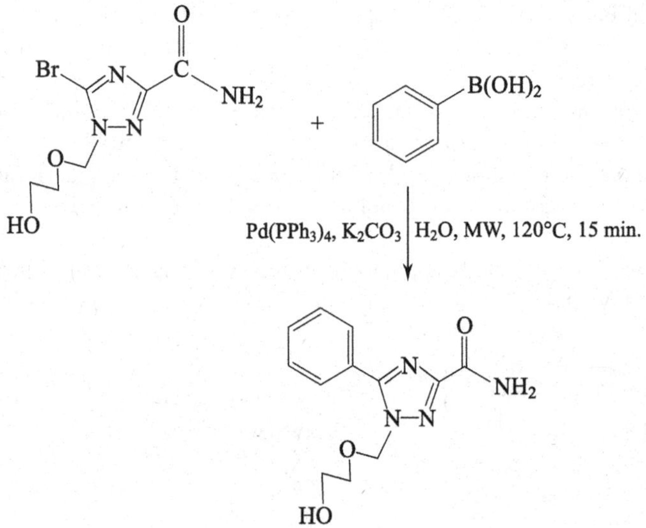

Scheme 2.376 Synthesis of 5-aryltriazole acylnucleosides

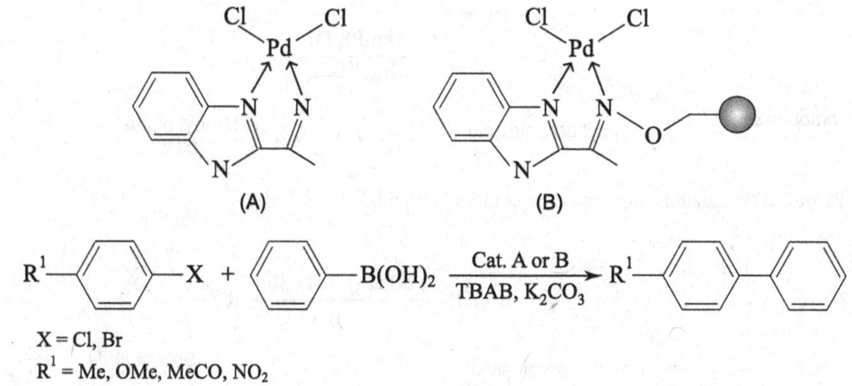

Scheme 2.377 Suzuki coupling reaction under microwave irradiation in water

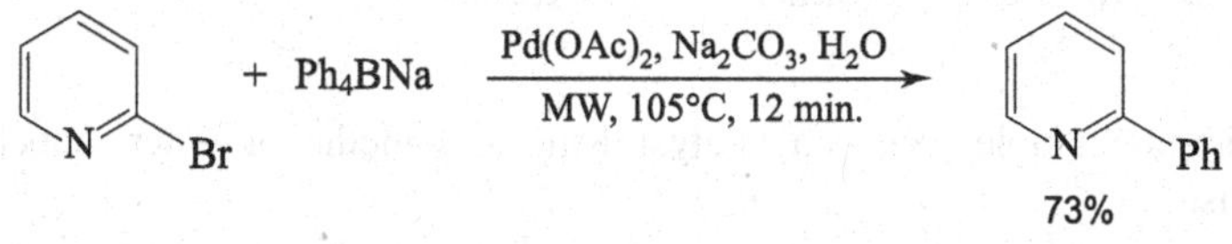

$X = Cl, Br$

$R^1 = Me, OMe, MeCO, NO_2$

Scheme 2.378 Suzuki coupling reaction continued

In the Suzuki coupling reaction of aryl chlorides and bromides with aryl boronic acids under MW irradiation in water, the benzothiazole-based Pd(II)-complexes (A) and (B) were found to be very efficient and active catalyst (the immobilized catalyst B was found to have high durability compared with the mobilized catalyst A). A high turnover number associated with the catalytic activity of these catalysts is very important for mass production (Scheme 2.377) [395].

Another substitute for boronic acid in Suzuki couplings is sodium tetraphenyl borate (Ph_4BNa) (Scheme 2.378) [396].

2.53.2 Suzuki Coupling Reaction in Ionic Liquids

The Suzuki coupling using a Pd catalyst in an ionic liquid as the solvent has been reported [397, 398] to give excellent yield and turnover numbers at room temperature. Thus, the coupling of bromobenzene and tolyl boranic acid under the above conditions give p-methyl biphenyl in good yield (Scheme 2.379).

The Suzuki coupling reaction has also been carried out under mild conditions in an ionic liquid with methanol as a cosolvent (necessary to solubilize the phenyl boronic acid) using ultrasound (Scheme 2.380) [399].

In the above reaction (Scheme 6), due to the formation of inactive Pd black, the recycling of the catalyst was not possible. This problem was overcome by using a

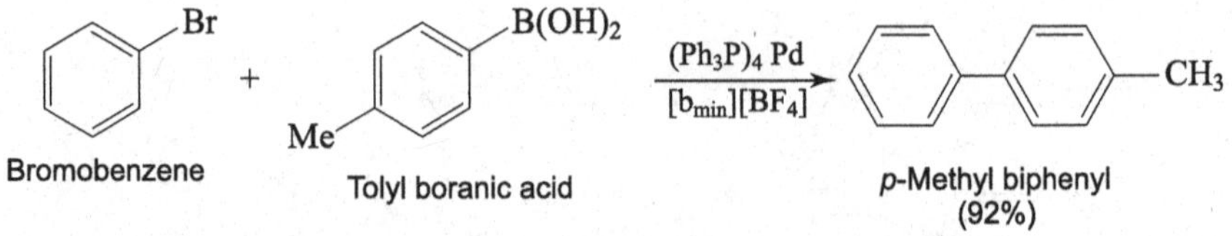

Scheme 2.379 Suzuki coupling reaction in ionic liquids

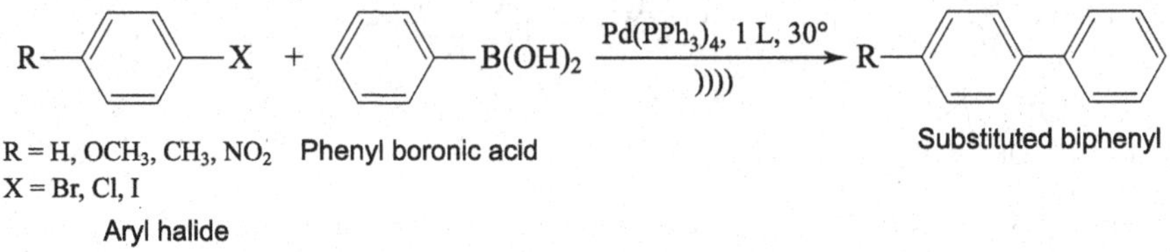

R = H, OCH₃, CH₃, NO₂   Phenyl boronic acid
X = Br, Cl, I

Aryl halide

Scheme 2.380 Suzuki coupling reaction under mild conditions

Pd–biscarbene complex(X), as a catalyst using only methanol under sonochemical conditions.

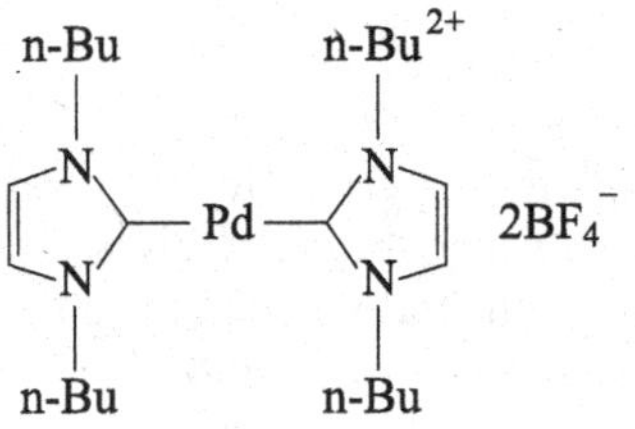

2.53.3   *Suzuki Coupling Reaction in Polyethylene Glycol (PEG)*

In this procedure, substituted aromatic bromides/iodides and aromatic boronic acids in PEG-400 as reaction medium in the presence of potassium fluoride as a base are used. The reaction was conducted in the presence of $PdCl_2$ using microwave irradiation (Scheme 2.381) [400].

Suzuki coupling reactions have also been performed using fluorous phase technique [401].

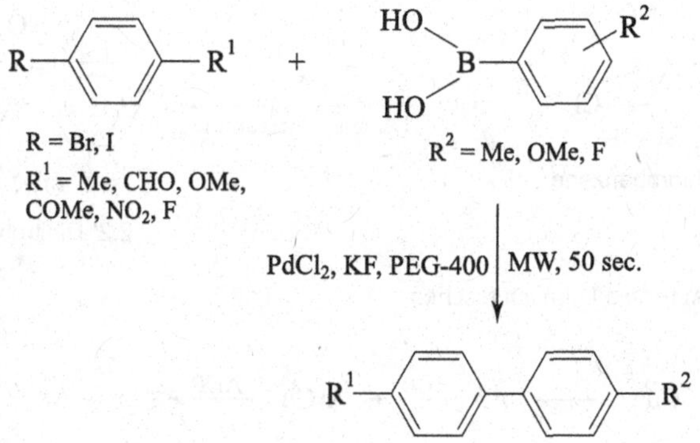

Scheme 2.381 Suzuki coupling reaction in PEG

2.54 Ullmann Reaction [402]

Ullmann reaction is used for the synthesis of diphenylamines, diphenyl ethers and diphenyls. In all these procedures, the reactants are heated with Cu (Scheme 2.382).

The last reaction used for the preparation of diaryls is called **Ullmann coupling reaction**.

Aryl chlorides and bromides usually do not undergo this coupling reactions unless the halogen is activated by a suitable substituent (*e.g.*, NO$_2$) in the *ortho* or *para* position. Thus *o*-nitrochlorobenzene on heating with copper powder gives 2,2'-dinitrobiphenyl (Scheme 2.383).

Mechanism

The mechanism of Ullmann reaction is uncertain. It is believed that the reaction proceeds via radical mechanism (Scheme 2.384).

(*i*) $C_6H_5NHCOCH_3 + C_6H_5Br + K_2CO_3 \xrightarrow[\text{Reflux}]{\text{Cu}} C_6H_5NHC_6H_5 + CH_3COOK + KBr$

    Acetanilide   Bromobenzene   Diphenylamine

(*ii*) $C_6H_5OH + C_6H_5Br + KOH \xrightarrow[\text{Reflux}]{\text{Cu}} (C_6H_5)_2O + KBr + H_2O$

    Phenol   Bromobenzene   Diphenyl ether

(*iii*) $2C_6H_5I + Cu \xrightarrow[\Delta]{C_6H_5NO_2} C_6H_5C_6H_5 + CuI_2$

    Iodobenzene

Scheme 2.382 Ullmann reaction

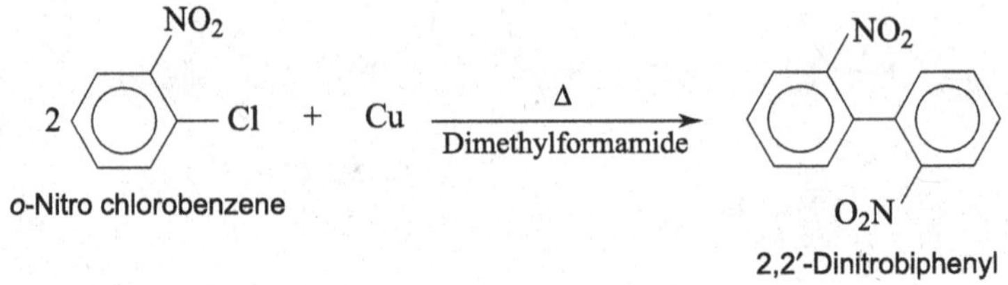

Scheme 2.383 Activated Ullmann reaction

$$ \text{ArX} + \text{Cu} \longrightarrow \text{Ar}^{\cdot} \xrightarrow{\text{Cu}} \text{ArCu} \xrightarrow{\text{ArX}} \text{Ar}-\text{Ar} + \text{CuX}_2 $$

Scheme 2.384 Mechanism of Ullmann reaction

2.54.1 Ullmann Coupling Under Sonication

Under sonication, the size of the copper powder used in Ullmann coupling is considerably reduced [403]. Breaking of the particles brings in contact with reactive solution fresh surface; the reactivity is not even hindered by the usual oxide layer on copper powder. The coupling is carried out in dimethylformamide (Scheme 2.385).

The yield is much lower in decalin (20%) and toluene (5%).

Sonication of arylsulphonates in the presence of in situ generated nickel (O) complex is an interesting Ullmann-type coupling [404]. This method best works for triflates (R = CF$_3$). However, the yields are low for tosylates (R = 4 – CH$_3$C$_6$H$_4$) (Scheme 2.386).

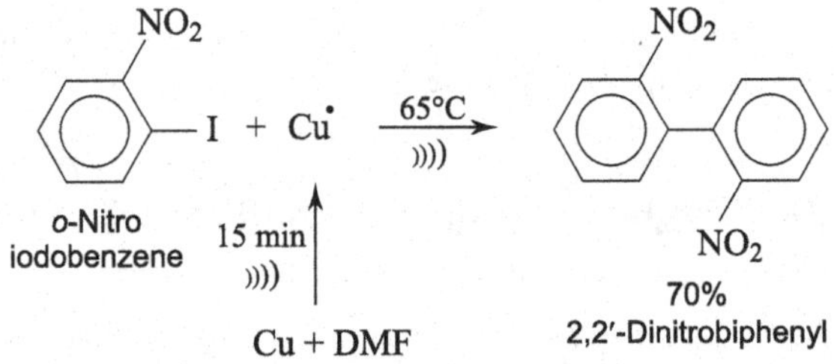

Scheme 2.385 Ullmann coupling under sonication

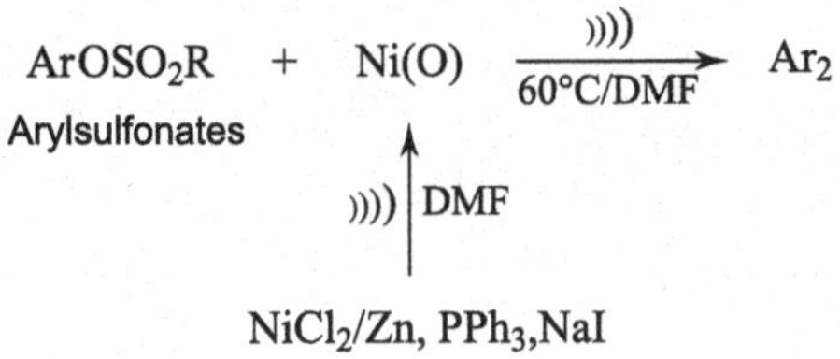

Scheme 2.386 An interesting type Ullmann coupling

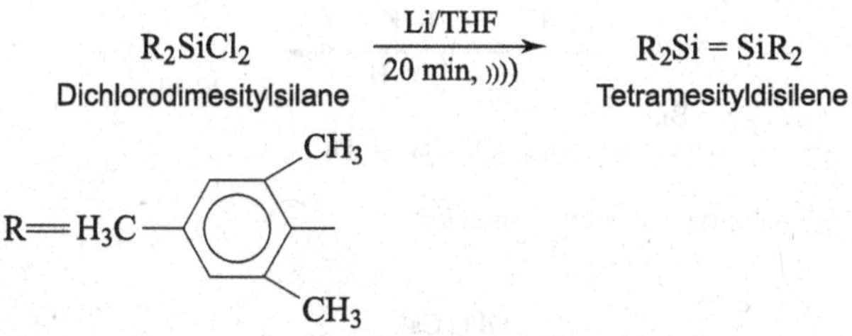

$$R_2SiCl_2 \xrightarrow[\text{20 min, }))))]{\text{Li/THF}} R_2Si = SiR_2$$

Dichlorodimesitylsilane   Tetramesityldisilene

$$R = H_3C-\underset{CH_3}{\overset{CH_3}{\bigcirc}}-$$

Scheme 2.387 Synthesis of hindered tetramesityldisilene

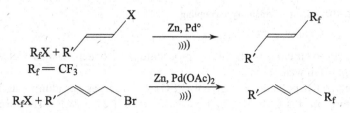

$R_fX + R''$
$R_f = CF_3$

$R_fX + R'$ ⟶ Br

Scheme 2.388 Cross-coupling in Ullmann-type coupling

The above Ullmann type coupling is useful in the formation of silicon bonds. Thus, a number of chlorosilanes can be coupled by sonication in the presence of lithium [405]. Using this method, highly hindered tetramesityldisilene can be prepared in good yield under sonication [406] (Scheme 2.387).

Another Ullmann type coupling is the cross-coupling reactions [407–409] of perfluoroalkylzinc reagents with vinyl, alkyl or aryl halides can be achieved by using a cleaning bath (35–45 kHz) (Scheme 2.388).

2.54.2 Ullmann-Type Coupling in Water

The homocoupling of arylsulfinic acids in the presence of Pd(II) in aqueous solvent gave biaryls (Scheme 2.389) [410].

The above coupling requires stoichiometric amount of palladium. In the presence of hydrogen gas, aryl halides homocoupled to give biaryl compounds in moderate yields [411] (Scheme 2.390).

A facile coupling of aryl halides via a palladium-catalysed reductive coupling using zinc in air and aqueous acetone at room temperature using Pd/C as a catalyst was reported (Scheme 2.391) [412].

$$2ArSO_2Na + Na_2PdCl_4 \xrightarrow{H_2O} Ar\text{—}Ar + 2SO_2 + Pd + 4NaCl$$

Scheme 2.389 Ullmann-Type coupling in water

$$2\text{ArX} \xrightarrow[\text{H}_2\text{O- BuOH}]{\text{H}_2,\ \text{Cat. PdCl}_2,\ \text{K}_2\text{CO}_3} \begin{array}{c} \text{Ar—Ar} \\ \text{30–50\%} \end{array}$$

X = Br, I,
Ar = Ph, p-MePh, p-ClPh, m-CF$_3$Ph

Scheme 2.390 Ullmann-type coupling continued

$$2\ \text{ArX} \xrightarrow[\substack{\text{Zn, H}_2\text{O/18-Crown-6} \\ \text{air atmosphere, rt}}]{\text{Pd/C Cat.}} \text{Ar—Ar}$$

Scheme 2.391 A facile Ullmann-type coupling

Using a crown ether in the above procedure gave better isolated yields of the product [413] in water alone.

Subsequently, polyethylene glycol (PEG) was used as an additive at elevated temperature. In this case, aryl chloride also worked effectively [414]. Reductive homocoupling of chlorobenzene to biphenyls gives high yields (93–95%) in the presence of catalytic PEG-400 and 0.4 mol% of a recyclable heterogeneous trimetallic catalyst (4% Pd, 1% Pt and 5% Bi on carbon) [415].

It has been found that CO_2 promotes the palladium-catalysed zinc-mediated reductive Ullmann coupling of aryl halides. It is believed that in the presence of CO_2, Pd/C, and Zn, various aromatic halides including less reactive aromatic chlorides coupled to give the corresponding homocoupling products in good yields [416].

2.54.3 Applications (Scheme 2.392)

1. Ullmann coupling reaction is helpful for the preparation of different types of biaryls. Some of these are given below.

 (a) Synthesis 4,4'-diphenic acid
 (b) Synthesis 2,2',4,4'-tetramethylbiphenyl

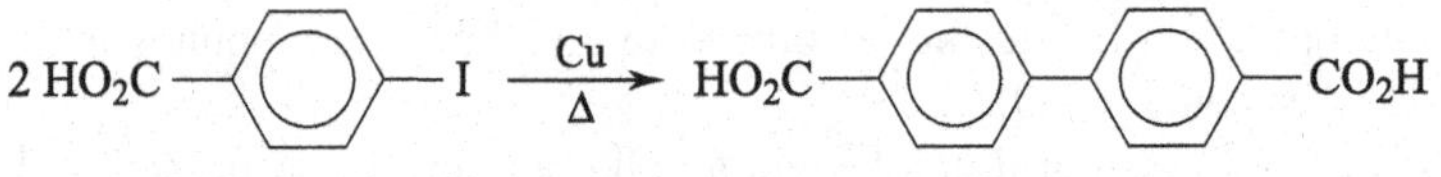

p-Iodobenzoic acid   4,4'-Diphenic acid

Scheme 2.392 Applications of Ullmann coupling reactions

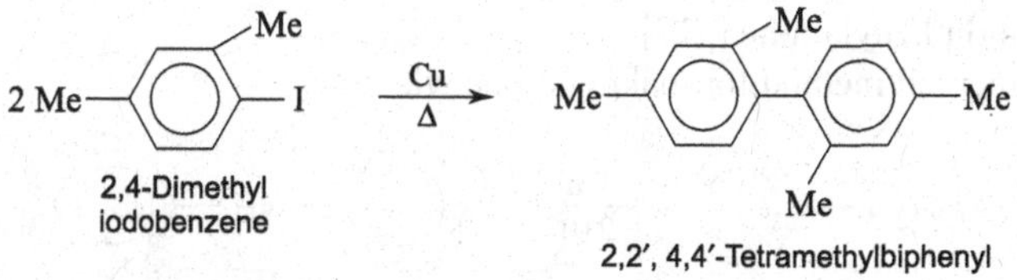

2 Me—⟨ ⟩—I (Me) $\xrightarrow[\Delta]{Cu}$ **Me—⟨ ⟩—⟨ ⟩—Me**

2,4-Dimethyl
iodobenzene

2,2′, 4,4′-Tetramethylbiphenyl

(c) Synthesis of *p,p′*-diaminobiphenyl [410].

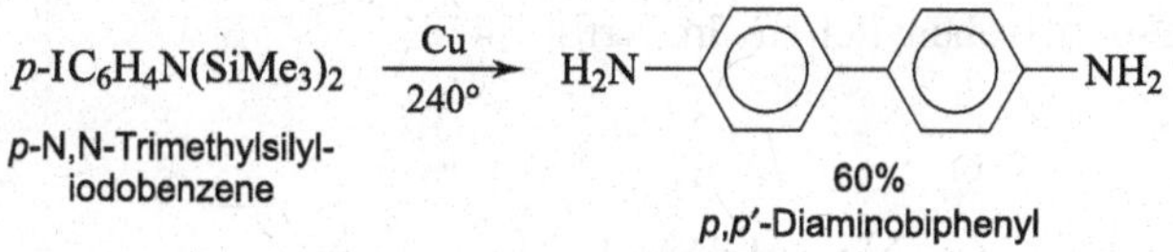

$p\text{-}IC_6H_4N(SiMe_3)_2$ $\xrightarrow[240°]{Cu}$ H_2N—⟨ ⟩—⟨ ⟩—NH_2

p-N,N-Trimethylsilyl-
iodobenzene

60%
p,p′-Diaminobiphenyl

(d) Synthesis of 2,4,6-trinitrobiphenyl

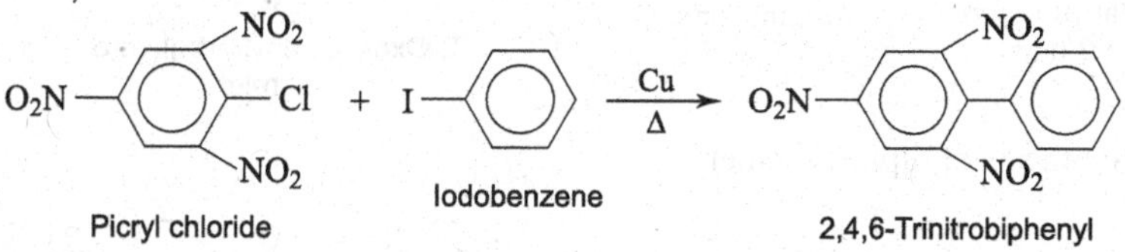

O_2N—⟨ ⟩—Cl (NO$_2$) + I—⟨ ⟩ $\xrightarrow[\Delta]{Cu}$ O_2N—⟨ ⟩—⟨ ⟩ (NO$_2$, NO$_2$)

Picryl chloride   Iodobenzene   2,4,6-Trinitrobiphenyl

2. Synthesis of cyclic hydrocarbons

(a) Synthesis of 2,7-dimethoxy-9,10-dihydrophenanthrene

MeO—⟨ ⟩—I I—⟨ ⟩—OMe $\xrightarrow[\Delta]{Cu}$ MeO—⟨ ⟩—⟨ ⟩—OMe

2,2′-Diiodo-5,5′-dimethoxy
dibenzyl

2,7-Dimethoxy-9,10-
dihydrophenanthrene

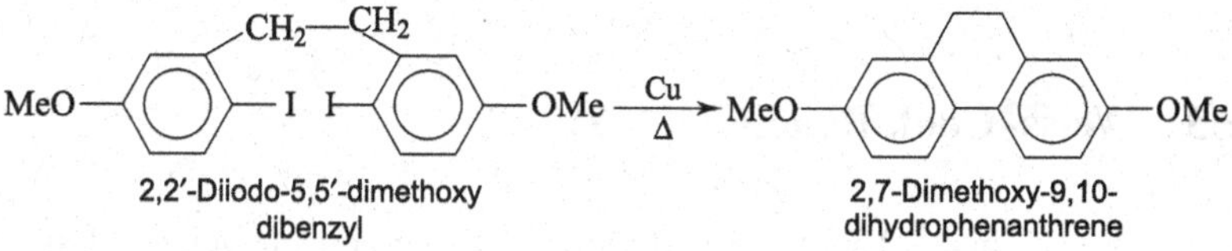

(b) Synthesis of perylene.

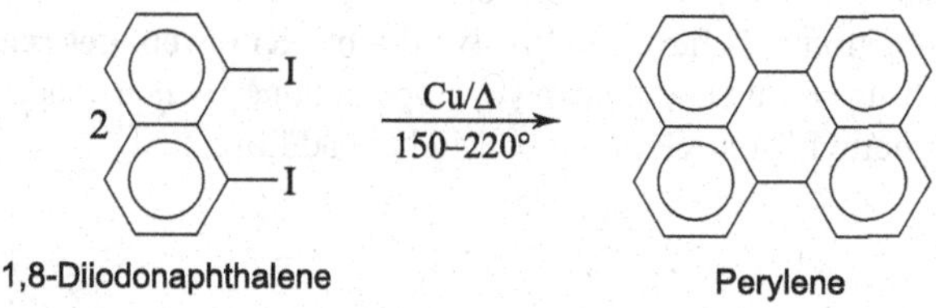

2 ⟨ ⟩ (I, I) $\xrightarrow[150–220°]{Cu/\Delta}$ ⟨ ⟩

1,8-Diiodonaphthalene   Perylene

(c) Synthesis of anthanthrone.

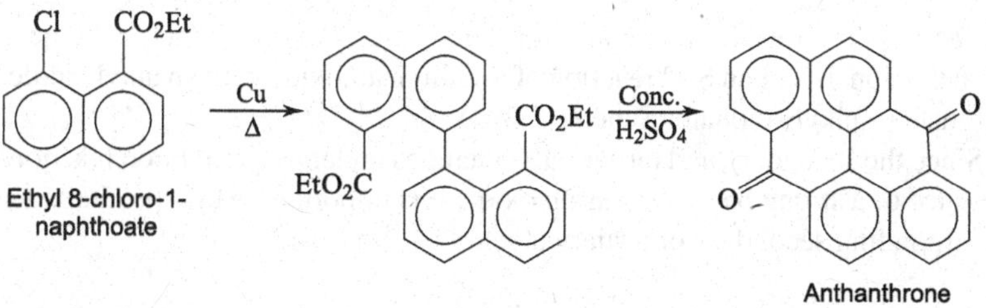

Ethyl 8-chloro-1-
naphthoate $\xrightarrow[\Delta]{Cu}$ (CO$_2$Et, EtO$_2$C) $\xrightarrow[H_2SO_4]{Conc.}$ Anthanthrone

3. Synthesis of diphenylamine [417].
This is a commercial method for making diphenylamine.

Iodobenzene   Aniline   Diphenyl amine
                                                                          85%

4. Synthesis of 2-oxo-1-phenyltetrahydropyrrole [418].

Tetrahydropyrole   Bromobenzene   92%
2-one   2-Oxo-1-phenyltertrahydro
                                                                        pyrrole

5. Synthesis of diphenyl ether.

Iodobenzene   Phenol   Diphenyl ether
                                                                          60%

2.55   Weiss–Cook Reaction [419]

The reaction of dimethyl 3-oxyglutarate with glyoxal in aqueous acidic solution gives
methyl [3.3.0] octane-3,7-dione-2,4,6–8-tetracarboxylate, which on acid hydrolysis followed by decarboxylation gives *cis*-bicyclo [3.3.0] octane-3–7-dione [419]
(Scheme 2.393). The reaction is believed to involve double **Knoevenagel reaction**
that gives [420] an α,β-unsaturated-γ-hydroxycyclopentenone, which reacts with
another molecule of dimethyl-3-oxoglutarate by Michael addition.

2.56   Williamsons Ether Synthesis [421]

It is an important useful procedure for the synthesis of unsymmetrical ethers. This
synthesis consists of an S_N2 reaction of a sodium alkoxide with an alkyl halide, alkyl
sulfonate or alkyl sulphate (Scheme 2.394).

Since the secondary and tertiary alkyl halides undergo elimination reaction in the
presence of a strong base, such as alkoxide, it is important to use the alkoxide of the
corresponding secondary or tertiary alcohol.

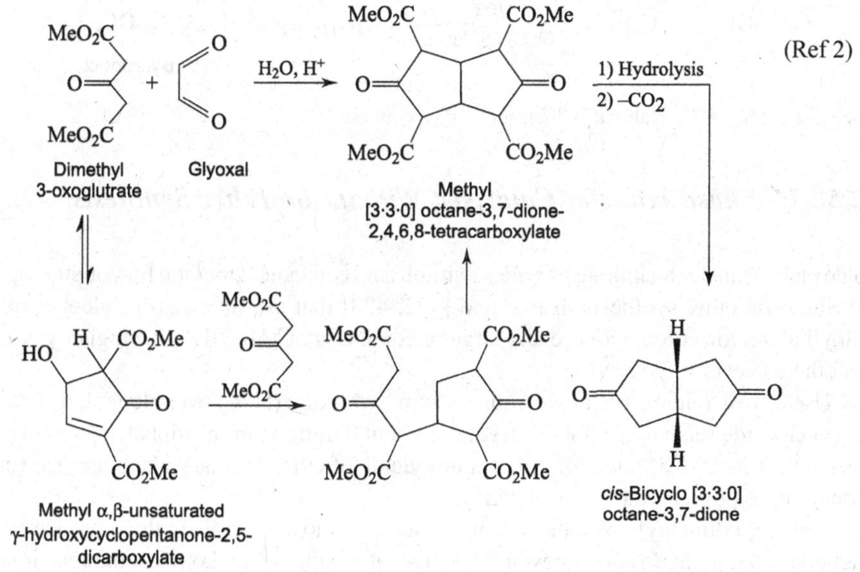

(Ref 2)

Dimethyl
3-oxoglutrate — Glyoxal

Methyl
[3·3·0] octane-3,7-dione-
2,4,6,8-tetracarboxylate

Methyl α,β-unsaturated
γ-hydroxycyclopentanone-2,5-
dicarboxylate

cis-Bicyclo [3·3·0]
octane-3,7-dione

Scheme 2.393 Weiss–Cook reaction

$$R' - X + {}^-OR \longrightarrow R' - O - R' + X^-$$

$$X = I, -Br, -OSO_2R' \text{ or } OSO_2OR'$$

Scheme 2.395: Williamsons Ether Synthesis

Scheme 2.394 Williamson's ether synthesis

Mechanism

The alkoxide ion reacts with the substrate in an S_N2 reaction resulting in the formation of an ether (Scheme 2.395). The substrate must have a good leaving group (as indicated in Scheme 1).

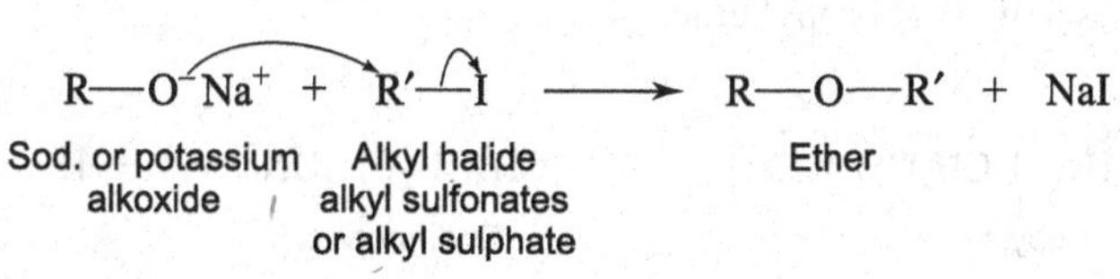

$$R - O^- Na^+ + R' - I \longrightarrow R - O - R' + NaI$$

Sod. or potassium
alkoxide — Alkyl halide
alkyl sulfonates
or alkyl sulphate — Ether

Scheme 2.395 Mechanism of Williamson's ether synthesis

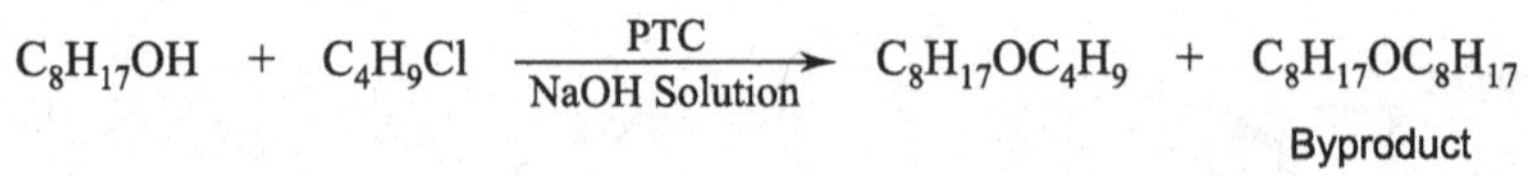

$$C_8H_{17}OH \ + \ C_4H_9Cl \ \xrightarrow[\text{NaOH Solution}]{\text{PTC}} \ C_8H_{17}OC_4H_9 \ + \ C_8H_{17}OC_8H_{17}$$

Byproduct

Scheme 2.396 PTC-catalysed Williamson's ether synthesis

2.56.1   Phase Transfer Catalysed Williamson Ether Synthesis

The phase transfer technique provides a simple and convenient method for conducting Williamson ether synthesis. It is found [422, 423] that use of excess of alcohol or alkylhalide, lower temperature and larger alcohol (e.g., $C_8H_{17}OH$) give higher yield of ethers (Scheme 2.396).

Use of five-fold excess of aqueous sodium hydroxide (50%) over alcohol, excess alkyl chloride (also used as solvent) and tetrabutylammonium bisulphate (1–5 mol) as catalyst at 25–70° gave [424] optimum yields of ether. Primary alcohols require longer time or greater amount of catalyst.

Although dimethyl sulphate does not react with most alcohols in the presence of aqueous sodium hydroxide or even by the use of alkali metal alkoxides, the reaction proceeds easily [424] with tetrabutylammonium salts as catalyst. Activated alcohols and primary alcohols give high yield of ethers, but secondary alcohols react very slowly and tertiary alcohols do not at all react.

In case of phenols, potassium carbonate is mostly used to scavenge protons. It is found that crown ethers can enhance their solubility as well as reactivity. A good example is the conversion of phenol into benzyl phenyl ether in quantitative yield by using K_2CO_3 and a catalytic amount of 18-Crown 6.

The aromatic ethers are obtained by using a phenolate derivative with an alkyl halide. The Williamson's ether synthesis is used for diaryl ethers (*see* Ullmann reaction, Sect. 2.54).

2.56.2   Applications (Scheme 2.397)

(*i*) Synthesis of ethyl propyl ether [425].

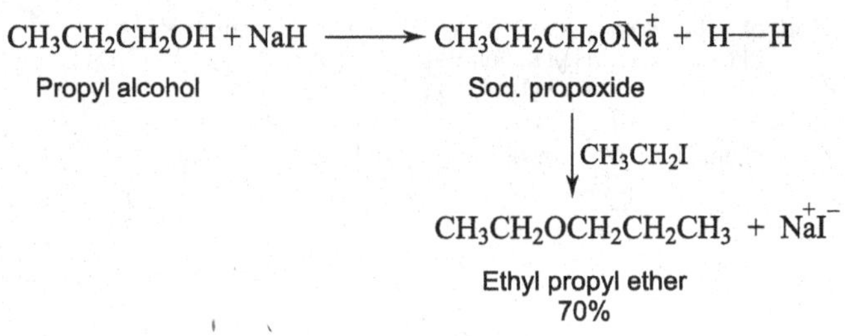

$$CH_3CH_2CH_2OH + NaH \longrightarrow CH_3CH_2CH_2O\overset{+}{N}a + H\text{—}H$$

Propyl alcohol   Sod. propoxide

$$\Big\downarrow CH_3CH_2I$$

$$CH_3CH_2OCH_2CH_2CH_3 + Na\overset{+}{I}^{-}$$

Ethyl propyl ether
70%

Scheme 2.397 Applications of Williamson ether synthesis

(*ii*) Synthesis of p-nitrophenyl butyl ether [426].

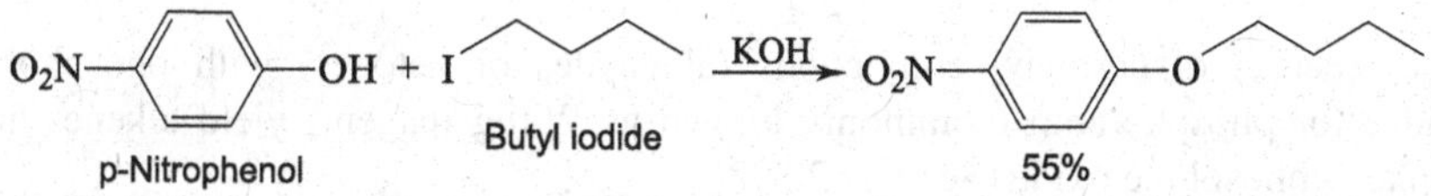

p-Nitrophenol   Butyl iodide   55%

(*iii*) Synthesis of methylphenyl ether [427].

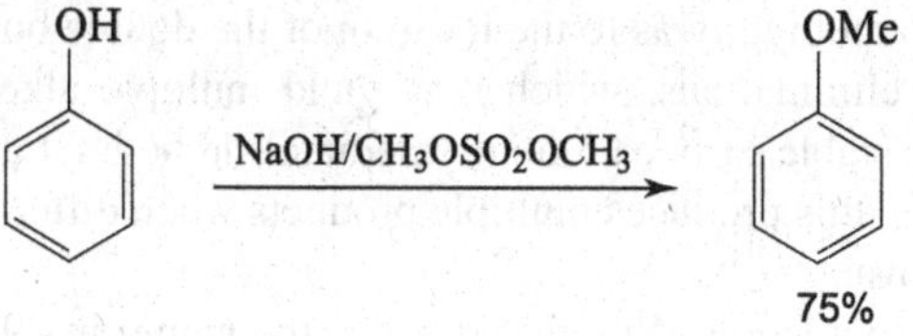

75%

(*iv*) Synthesis of isobutyl ethyl ether [428].

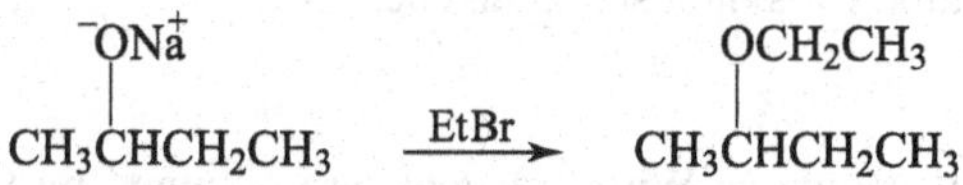

(*v*) A variation of Williamson's [421] ether synthesis is by using thallium (I)
ethoxide. The method [426] is best for substrates containing an additional oxygen
function, such as —OH, —COOR, —CO—NR$_2$.

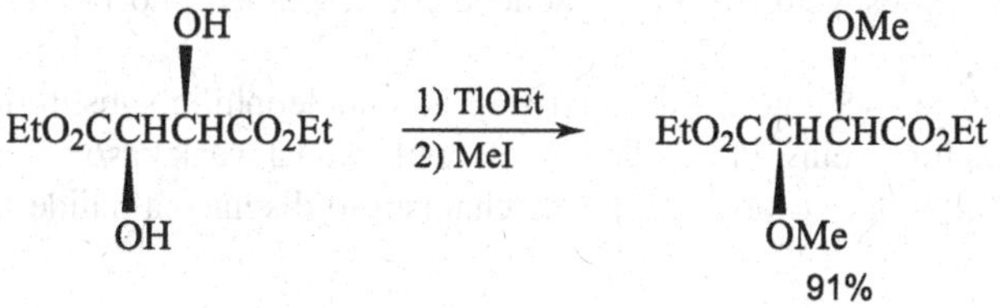

91%

(*vi*) The use of benzyl ether [429] as protecting group is very common, since the
protecting group can be easily removed by catalytic reduction or by heating with
conc. HCl and glacial acetic acid.

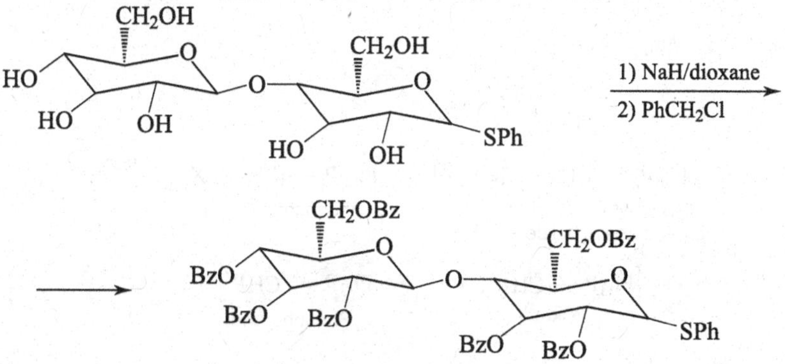

2.57 Wittig Reaction [430]

The reaction of carbonyl compounds (aldehydes or ketones) with phosphorus ylides (or phosphorane) (commonly known as Wittig reagent) yield alkenes and triphenylphosphine oxide (Scheme 2.398).

This reaction is known as the Wittig reaction [430] and is a very convenient method for the synthesis of alkenes (a mixture of (E) and (Z) isomers result). In this reaction, there is absolutely no ambiguity as to the location of the double bond in the product, in contrast to E1 eliminations, which may yield multiple alkene products by rearrangement to more stable carbocation intermediate and both E1 and E2 elimination reactions may occur; this produces multiple products when different β-hydrogens are available for removal.

Wittig reaction was discovered by George Wittig (and hence the name) in 1954 and was awarded Nobel Prize in Chemistry in 1979 due to its tremendous synthetic potentialities, being a valuable method for synthesizing alkenes.

The Phosphorus Ylides

The phosphorus ylides (commonly known as Wittig reagent) are obtained by the reaction of an alkyl halide with triphenylphosphine. The formed phosphonium salt is treated with a strong base like sodium hydride or phenyl lithium to give phosphorus ylide. These phosphorus ylides carry a positive and negative charge on adjacent atoms and can be represented by doubly bonded species called phosphoranes (Scheme 2.399).

In the above procedure (Scheme 2), the first step is a nucleophilic substitution reaction, triphenylphosphine being an excellent nucleophile and a weak base readily reacts with 1° and 2° alkyl halides by an S_N2 mechanism to displace a halide ion

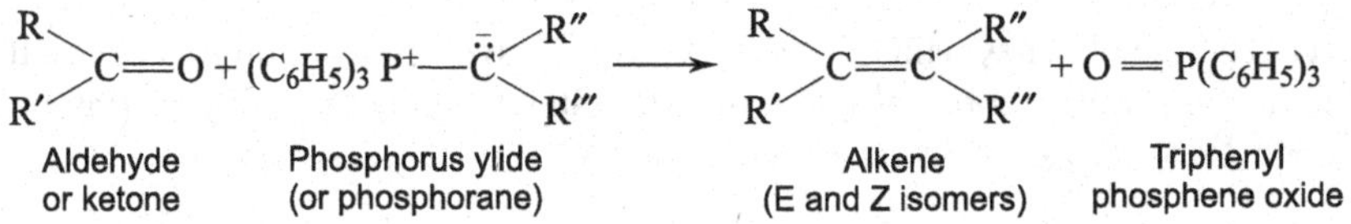

Scheme 2.398 Wittig reaction

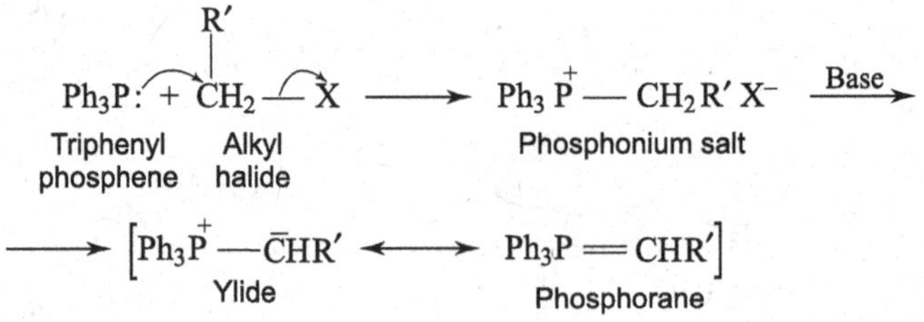

Scheme 2.399 Wittig reagent

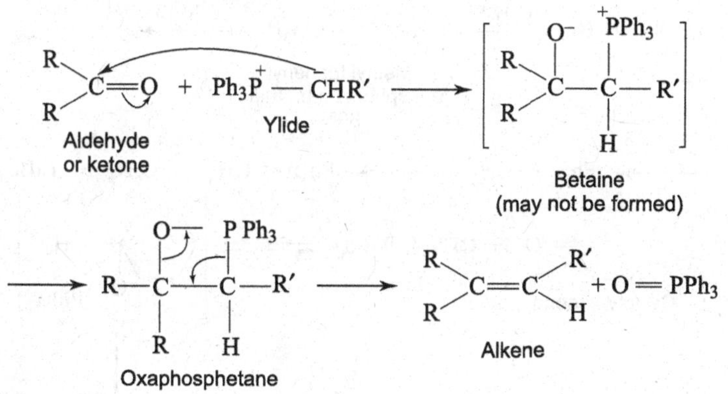

Scheme 2.400 Mechanism of Wittig reaction

from the alkyl halide to give an alkyl triphenyl phosphonium salt. In the second step (which is an acid/base reaction), a strong base removes a proton from the carbon atom that is attached to phosphorus to give the ylide.

Mechanism

The mechanism of Wittig reaction has been the subject of considerable study. It was earlier suggested that the ylide, acting as a carbanion, attacks the carbonyl carbon of the aldehyde or ketone to form an unstable intermediate with separated charge called betaine. In the subsequent step, the betaine being unstable gets converted into a four-membered cyclic system called an oxaphosphetane, which spontaneously loses a molecule of triphenyl phosphine oxide to form an alkene. Subsequent studies have shown that betanine is not an intermediate and that the oxaphosphetane is formed directly by a cycloaddition reaction. The driving force for the Wittig reaction is the formation of very strong phosphorus–oxygen bond in triphenylphosphine oxide (Scheme 2.400).

A typical example is given in Scheme 2.401.

Though Wittig synthesis appears to be complicated, in fact, these are easy to carry out as one pot reaction.

2.57.1 The Wittig Reaction with Aqueous Sodium Hydroxide

It has been shown [431–433] that phase transfer catalyst. viz., alkyl-triphenylphosphonium salts react with aqueous sodium hydroxide to generate ylides which combine with organic phase aldehydes to produce olefins (Scheme 2.402).

In the above synthesis, the yield of the olefin increases [433] with the increase in concentration of alkali up to a maximum and then decreases. The yields depend on the

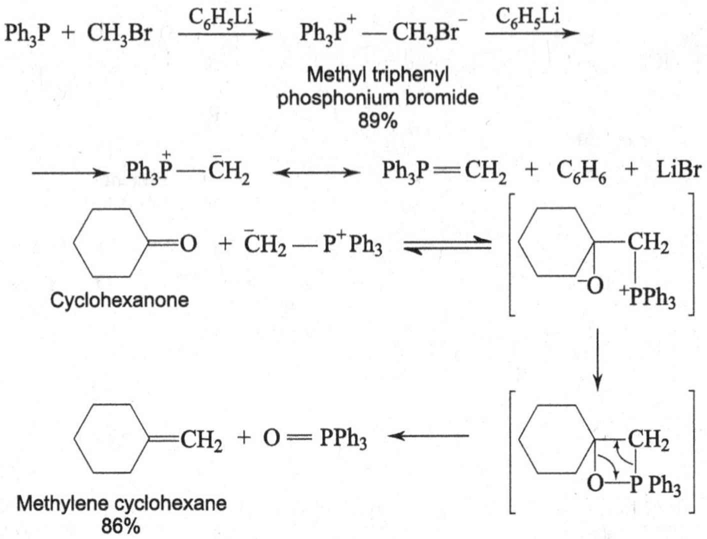

$$Ph_3P + CH_3Br \xrightarrow{C_6H_5Li} Ph_3P^+ - CH_3Br^- \xrightarrow{C_6H_5Li}$$

Methyl triphenyl
phosphonium bromide
89%

$$\longrightarrow Ph_3\overset{+}{P} - \overset{-}{C}H_2 \longleftrightarrow Ph_3P = CH_2 + C_6H_6 + LiBr$$

Cyclohexanone

Methylene cyclohexane
86%

Scheme 2.401 Synthesis of methylene cyclohexane

$$Ph_3P^+ - CH_2C_6H_5Cl^- + NaOH(org) \xrightarrow[\text{Solution}]{CH_2Cl_2} [Ph_3P = CHPh]$$

$$\downarrow RCHO$$

$$RCH = CHPh + Ph_3PO$$

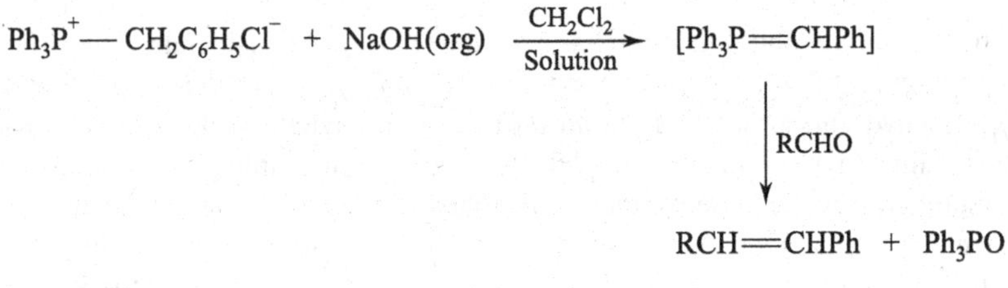

Scheme 2.402 Wittig reaction with aqueous sodium hydroxide

alkyl group attached to the triphenylphosphonium salt. The quaternary phosphonium salts are better than the quaternary ammonium salts.

The PTC-catalysed reaction is limited to only aldehydes and so is useful for preparation of olefins of the type $RCH = CHR'$; the reaction is unsuccessful in case of ketones.

In certain cases, it is possible to isolate [434] crystalline phosphonium ylides directly by treatment of the phosphonium halides with aqueous sodium hydroxide (Scheme 2.403).

Stable crystalline derivatives are also obtained when the phosphonium salts contain a –CO₂Me [435] or a –CHO [436] group instead of a cyano group.

$$X^- Ph_3P^+CH_2CH = CHCN + NaOH \longrightarrow Ph_3 \overset{+}{P} \overset{-}{C}H - CH = CHCN + NaCl$$
$$(aq)$$

Scheme 2.403 Isolation of crystalline phosphonium

It has been found [437] that triphenyl alkylphosphonium fluoride reacts with aldehydes to give olefins in good yield (Scheme 2.404).

Wittig reaction can be conveniently performed [438] in ionic liquid [b$_{min}$]BF$_4$. The advantage is easy separation of alkene from Ph$_3$PO and also recycling of the solvent.

Modifications of Wittig Reagent

Several modifications of the Wittig reagent have been made to improve the reactivity of ylides:

(i) **Horner-Wadsworth-Emmons Modification**

In this modification, the ylides are obtained from phosphonate ester (instead of a triphenylphosphonium salt), which in turn are readily available from alkyl halide and triethylphosphite via an **Arbuzov rearrangement** [439]. These ylides are more reactive than the corresponding phosphoranes and often react with ketones that are inert to phosphoranes. Thus, the reaction of ethyl bromoacetate with triethylphosphite gives phosphonate ester, which on treatment with base (NaH) and reaction with cyclohexanone gives α,β-unsaturated ester, ethyl cyclohexylidene acetate with 70% yield (with triphenylphosphorane only 20% yield is obtained). This method, commonly known as **Horner–Wadsworth–Emmons modification** of the Wittig reaction (Scheme 2.405).

The major product obtained in Horner–Wadsworth–Emmons modification is usually the (E) alkene isomer compared to (E) and (Z) isomeric mixture that is obtained in the usual Wittig reaction.

The above reaction is sometimes performed in an organic/water biphase systems [440, 441] and in place of strong base like NaH, a PTC can be used in aq. NaOH with good results. Even much weak base like K$_2$CO$_3$ or KHCO$_3$ can be used. Substrates with base and acid-sensitive functional groups can be used directly.

Water has been shown to be an efficient medium for the Wittig reaction employing stabilized ylides and aldehydes yields ranging from 80 to 98% and high E-selectivities (up to 99%) are achieved. Typical examples are given in Scheme 2.406.

(ii) **The Wittig–Horner Reaction**

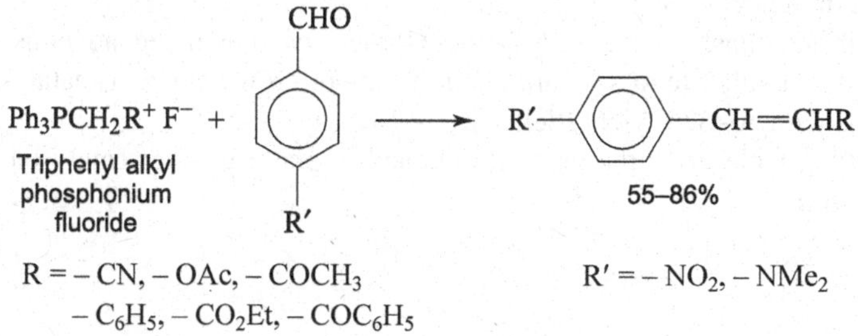

$$Ph_3PCH_2R^+ F^- +$$

Triphenyl alkyl
phosphonium
fluoride

R' ——⟨○⟩—— CH=CHR

55–86%

R = – CN, – OAc, – COCH$_3$
   – C$_6$H$_5$, – CO$_2$Et, – COC$_6$H$_5$

R' = – NO$_2$, – NMe$_2$

Scheme 2.404 Convenient synthesis of olefins

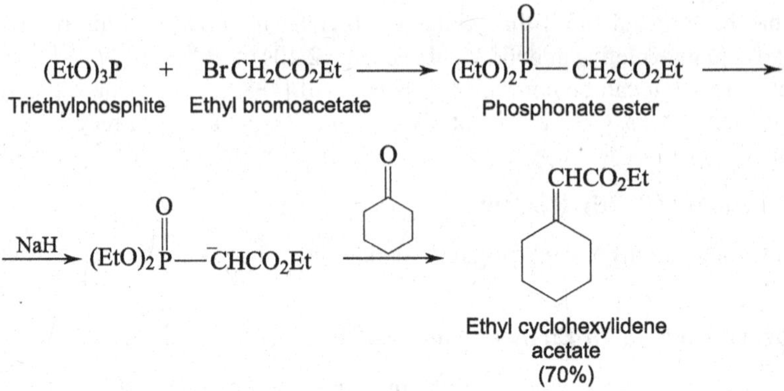

Scheme 2.405   Horner–Wadsworth–Emmons modification of Wittig reaction

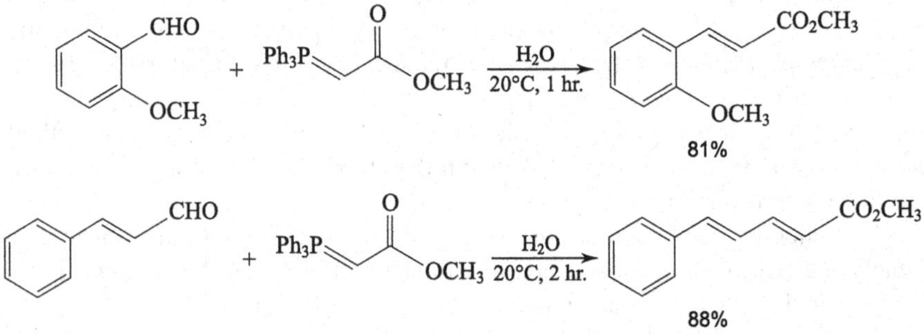

Scheme 2.406   Wittig reaction in aqueous phase

In this reaction, which is a modification of Wittig reaction, the readily available phosphine oxide Ph_2 is used. Its lithio derivative is made to react with aldehydes and ketones to yield β-hydroxyphosphine oxides, which on treatment with sodium hydride smoothly eliminates water to give the corresponding alkene. This step is stereospecific, erythro hydroxy phosphine oxide gives the Z-alkene and the threo compound gives the E-alkene by preferential syn elimination. Various steps involved are given in Scheme 2.407.

The phase transfer catalysed Wittig–Horner reaction using aqueous sodium hydroxide and either tetra-alkylammonium salts or crown ethers as catalysts gives the olefins with 50–87% yield (Scheme 2.408).

Following table gives the product obtained in phase transfer catalysed Wittig–Horner reaction.

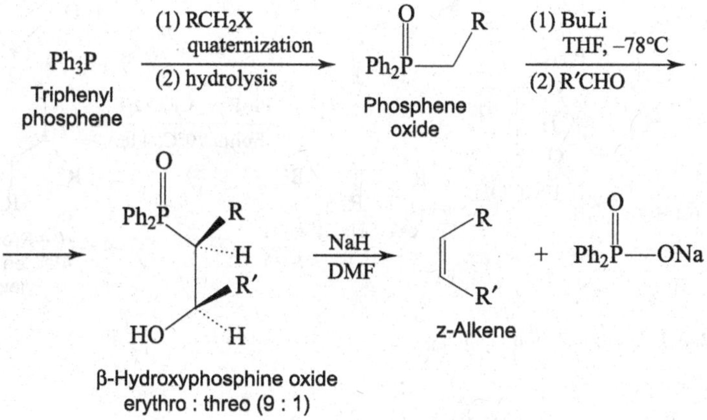

Scheme 2.407 Wittig–Horner reaction

$$(EtO)_2P(O)CH_2R + \begin{array}{c} R' \\ C=O \\ R'' \end{array} + NaOH \xrightarrow{PTC} RCH=C\begin{array}{c} R' \\ R'' \end{array} + (EtO)_2PO_2Na$$

Org.   Org.   Aq.   Olefin

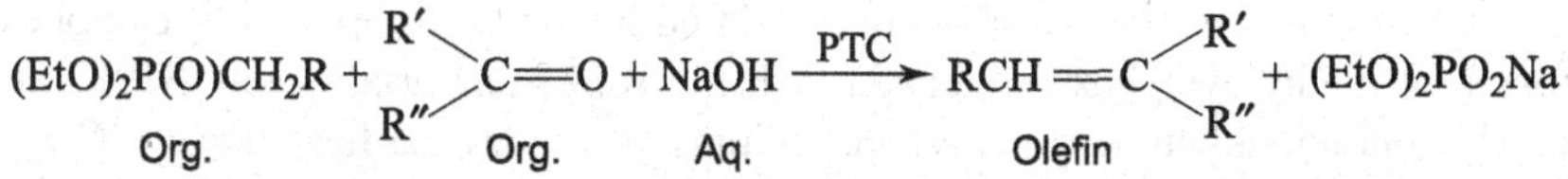

Scheme 2.408 PTC-catalysed Wittig–Horner reaction

R	R'	R'	Product	% Yield	Ref.
–CN	C_6H_5	H	cinnamaldehyde	77	12–14
–CN	CH_3	H	crotonaldehyde	51	13
–CN	CH_3	CH_3	3-methyl-2-butenitrile	62	13
$–CO_2Et$	Ph	H	ethyl transcinnamate	56	13
$–CO_2Et$	Me	H	ethyl trans-crotonate	54	13
$–CO_2Et$	Ph	H	cinnamic acid	95	8
$–COC_6H_5$	Ph	H	$C_6H_5CH{=}CHCOC_6H_5$	55	12
$–COC_6H_5$	P–Cl C_6H_4	H	P-Cl $C_6H_4CH{=}CHCOC_6H_5$	65	12
$–COC_6H_5$	p–BrC_6H_4	H	p-Br $C_6H_4CH{=}CHCOC_6H_5$	63	12
$C_6H_5CH{=}CH–$	2–pyridyl	H	$C_6H_5 CH = CH — CH = CH-$ 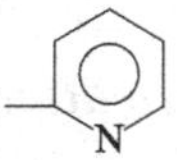	57	12–15
$C_6H_5CH{=}CH$	2–Furyl	H	$C_6H_5 CH = CH —CH -$ 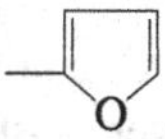	84	16
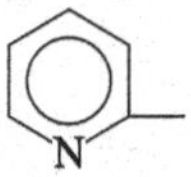	C_6H_5	H	$C_6H_5 CH = CH$ 	71	13

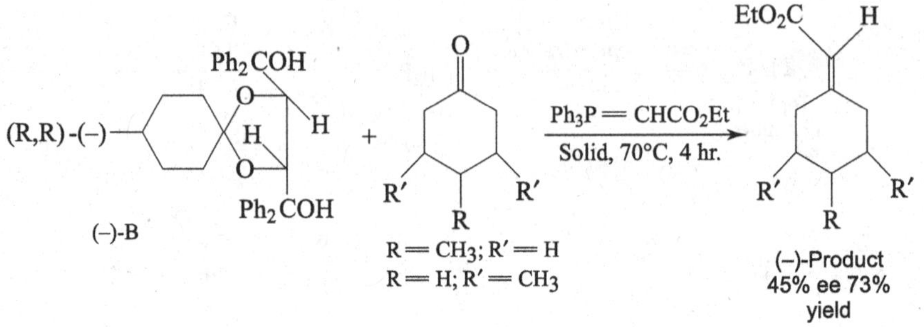

Scheme 2.409 Wittig reaction in solid phase

2.57.2   Wittig Reaction in Solid Phase

The well-known Wittig reaction has been reported [447] to occur in solid phase. In this procedure a 1:1 mixture of finely powdered inclusion compound of cyclohexanone or 4-methylcyclohexanone and (–)–B (derived from tartaric acid [319] and a catalytic amount of benzyltrimethylammonium hydroxide) was heated to 70 °C with Wittig reagent, carbethoxymethylene triphenylphosphorane to give optically active 1-(carbethoxymethylene) cyclohexane or the corresponding 4-methyl or 3,5-dimethyl compound (Scheme 2.409).

2.57.3   Wittig Reaction in Ionic Liquids

In the Wittig reaction, the separation of the alkene from the by-product (Ph_3PO) is a classical problem which is usually done by crystallization or chromatography. The ionic liquid [b_{min}] [BF_4] can be used as a medium to perform the reaction [448]. The advantage is easy separation of alkene from Ph_3PO and also recycling of the solvent. Also, the E-stereoselectivity was observed in ionic liquid solvents similar to that observed in organic solvents.

2.57.4   Applications

Wittig reaction has tremendous applications for the synthesis of alkenes. It gives a great advantage over most other synthesis as there is no ambiguity as to the location of the double bond in the product. Some of the important applications are given in Scheme 2.410.

(*i*) Synthesis of E-stilbene.
(*ii*) Synthesis of exocyclic methylene group compounds.

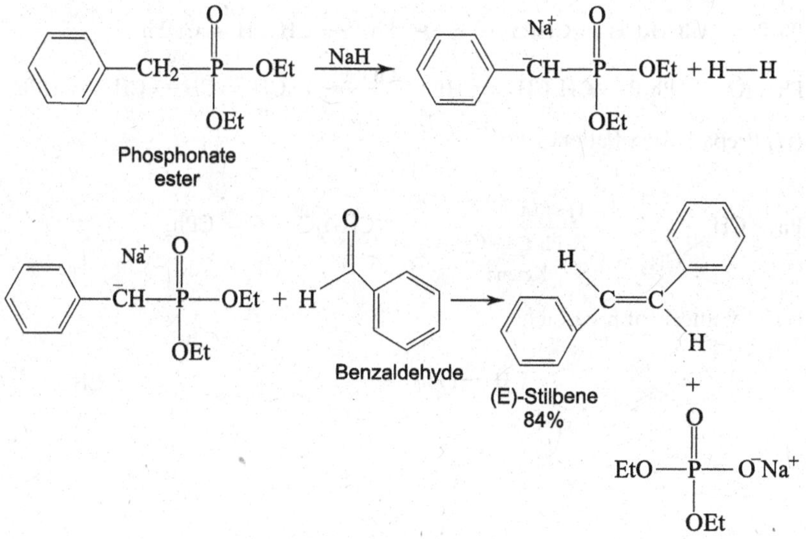

Scheme 2.410 Some applications of wittig reaction

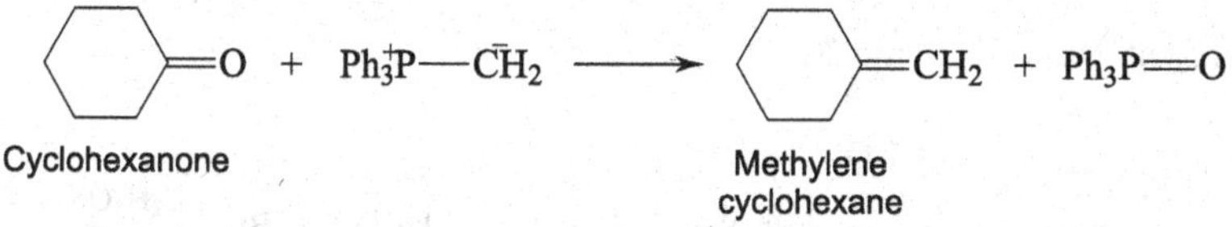

Cyclohexanone   Methylene cyclohexane

(*iii*) Synthesis of α,β-unsaturated esters.

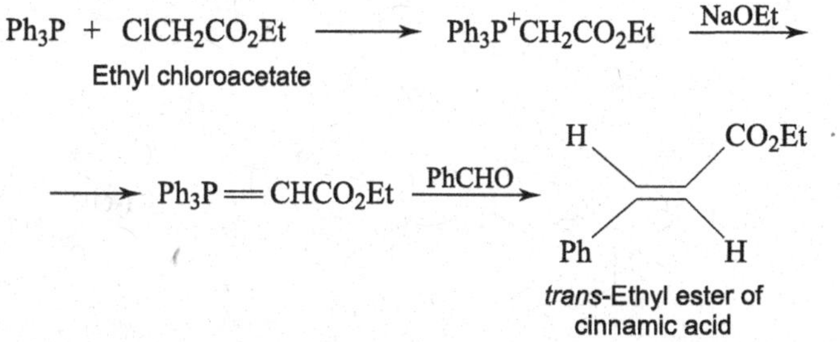

$$Ph_3P + ClCH_2CO_2Et \longrightarrow Ph_3P^+CH_2CO_2Et \xrightarrow{NaOEt}$$

Ethyl chloroacetate

trans-Ethyl ester of cinnamic acid

(*iv*) Synthesis of polyzonimine, a natural insect repellent produced by millipedes.

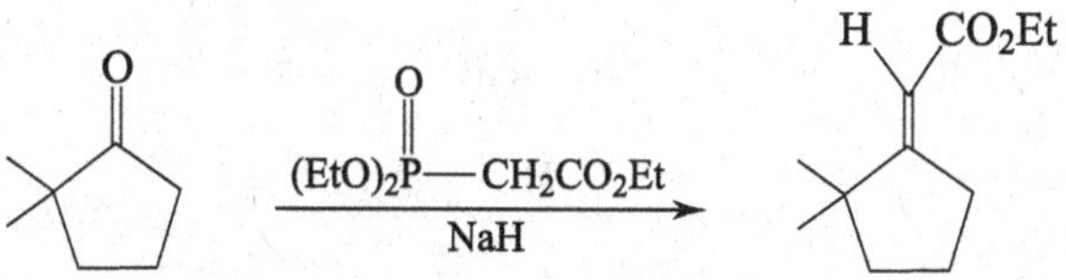

(*v*) Synthesis of dienes.

$$Ph_3P + ClCH_2CH{=}CHPh \longrightarrow Ph_3\overset{+}{P}{-}CH_2CH{=}CHPh$$

$$PhCHO + Ph_3\overset{+}{P}{-}CH_2CH{=}CHPh \xrightarrow{\text{LiOEt}} PhCH{=}CH{-}CH{=}CHPh$$

(*vi*) Preparation of allenes.

$$Ph_3\overset{+}{P}CH\overset{\text{CH}_3}{\underset{\text{CH}_3}{\diagdown}} \xrightarrow[\substack{2)~Ph_2C{=}C{=}O \\ \text{Ketene}}]{1)~PhLi} (CH_3)_2C{=}C{=}CPh_2$$

(*vii*) Synthesis of β-carotene.

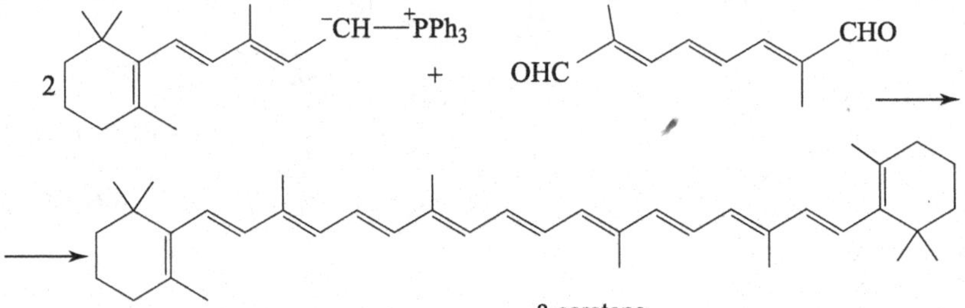

β-carotene

(*viii*) Synthesis of Vitamin A.

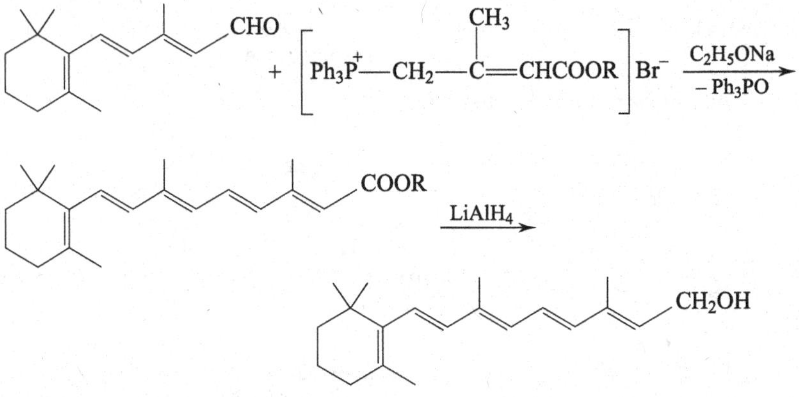

Vitamin A

(*ix*) Synthesis of heterocyclic compounds [434].

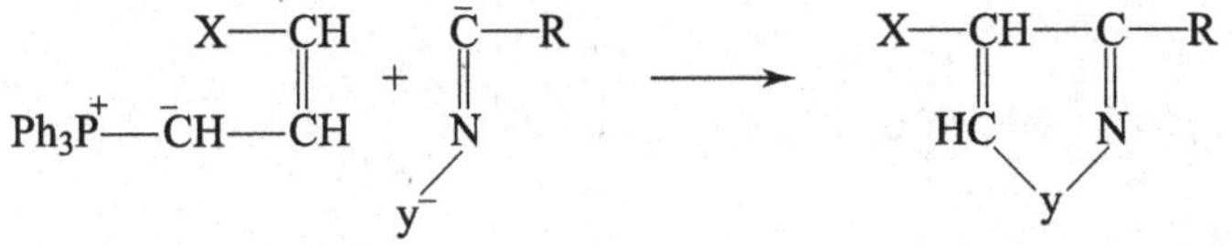

x = – CN , – CHO, – CO$_2$Me   R = Ph, – CO$_2$CH$_3$, p-NO$_2$C$_6$H$_4$

y = O, – NPh

(*x*) Synthesis of bicyclic compounds [449].

31%

(*xi*) The Wittig reaction in aqueous alkaline medium provides a convenient synthesis of olefins as per equation.

$$Ph_3P^+CH_2C_6H_5^-Cl + NaOH\ (aq) \longrightarrow \left[Ph_3P\!=\!CHPh \right]$$

$$\downarrow RCHO$$

$$RCH\!=\!CHPh + (C_6H_5)_2CHO$$

Aldehyde RCHO	R' in phosphonium salt $Ph_3\,P^+\,R'\,X^-$	Yield^a % R—CH=CH—R'	Ref.
OHC—CHO		48^b	3
OHC—CHO		23–29^b	3
OHC—CHO		13^b	3
C$_6$H$_5$CHO	CH$_3$	99	4
C$_6$H$_5$CHO		0	3
—CHO	CH$_3$	63	4

a Ratio of *cis* and *trans* is approx. 1:1, when R' ≠ H; *b* Product is R'(CH = CH)$_2$ R'

2.58 Wurtz Reaction

The coupling of alkyl halides with sodium in dry ether to give hydrocarbons is known as Wurtz reaction [450] (Scheme 2.411).

Using two different alkyl halides (like RX and R′ X) will give a mixture of different alkanes, viz., R–R + R′–R′ + R–R′ and so this method known as crossed Wurtz reaction is of not much importance. So this reaction is useful for the preparation of alkanes containing even number of carbon atoms (symmetrical). However, the hydrocarbons prepared by this method contain small amounts of olefins as by-products.

Coupling of alkyl halides or sulfonates with Grignard reagents or RLi in the presence of Cu(I) salts is also a modified form of Wurtz reaction. Thus, alkyl halides react with Grignard reagent in the presence of catalytic amount of cobaltous chloride to give alkanes. Coupling products are also obtained by simply adding cobaltous chloride to a solution of Grignard reagent (Scheme 2.412).

Another variation of Wurtz reaction is known as **Wurtz–Fittig reaction** [451]. In this procedure, an alkyl halide and an aryl halide couple to form alkylated aromatic ring. For example, bromobenzene and butyl bromide react with sodium to give *n*-butylbenzene (Scheme 2.413).

The by-products in this reaction are R-R and Ar–Ar, which can be separated easily.

Mechanism

Two mechanisms have been proposed:

$$CH_3CH_2CH_2Br \xrightarrow[\text{Ether}]{Na} CH_3(CH_2)_4CH_3$$

n-Propyl bromide   *n*-Hexane

Scheme 2.411 Wurtz reaction

$$RMgX + R'X \xrightarrow{CoCl_2} R-R' + MgX_2$$

$$2\,RMgX \xrightarrow{CoCl_2} R-R$$

Scheme 2.412 Ullmann coupling under sonication

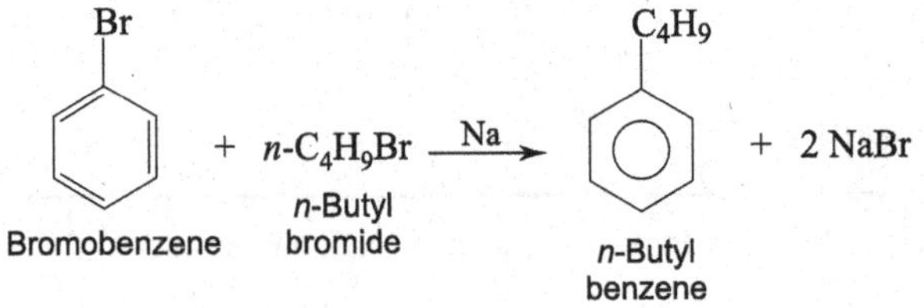

Scheme 2.413 Wurtz–Fittig reaction

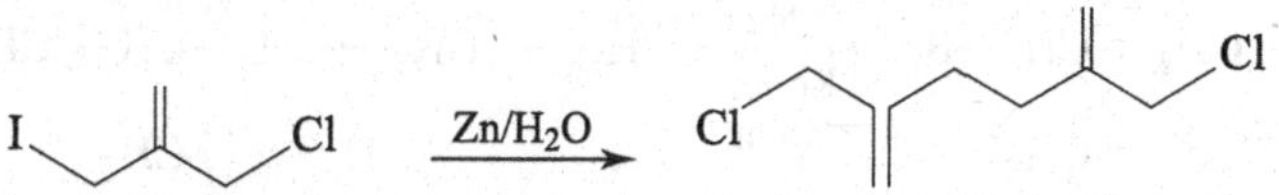

Scheme 2.414 Ullmann coupling under sonication

(*i*) Free radical mechanism. It operates in vapour state.

$$R-X + \overset{\cdot}{Na} \longrightarrow \overset{\cdot}{R} + NaX$$
$$\overset{\cdot}{R} + \overset{\cdot}{R} \longrightarrow R-R$$

(*ii*) Ionic mechanism. It operates in solutions. Initially sodium atom reacts with alkyl halide to form alkyl sodium having carbanion character. Final step is an S_N2 type displacement on another molecule of alkyl halide.

$$RX + 2\overset{\cdot}{Na} \longrightarrow \overset{-}{R}\overset{+}{Na} + NaX$$
$$R + R-X \xrightarrow{S_N2} R-R + X^-$$

2.58.1 Wurtz Reaction Under Sonication

Wurtz reaction carried out under sonication gives much better yields.

2.58.2 Wurtz Reaction in Water

It has been shown [452] that Wurtz coupling can also be carried out by Zn/H_2O (Scheme 2.414).

Ullmann coupling reaction is also considered to be Wurtz-type coupling (see Ullmann reaction, Sect. 2.54).

2.58.3 Applications

Some important applications are given in Scheme 2.415:
   1. Synthesis of 3,4-dimethylhexane [453].

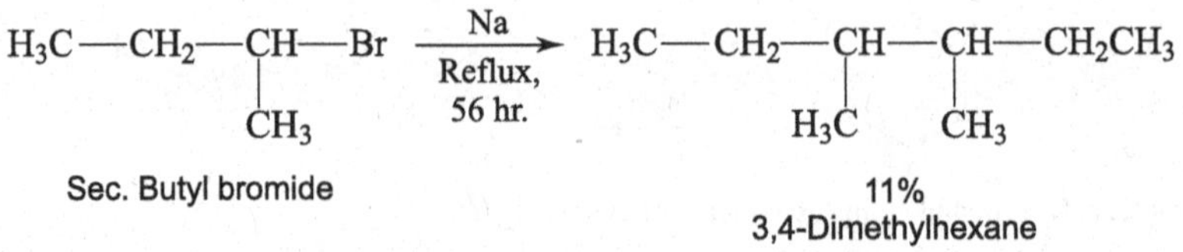

$$H_3C—CH_2—CH—Br \xrightarrow[\text{Reflux,}\atop\text{56 hr.}]{Na} H_3C—CH_2—CH—CH—CH_2CH_3$$

with CH_3 branch on left reactant; H_3C and CH_3 branches on right product.

Sec. Butyl bromide

11%
3,4-Dimethylhexane

Scheme 2.415 Some applications of Wurtz reaction

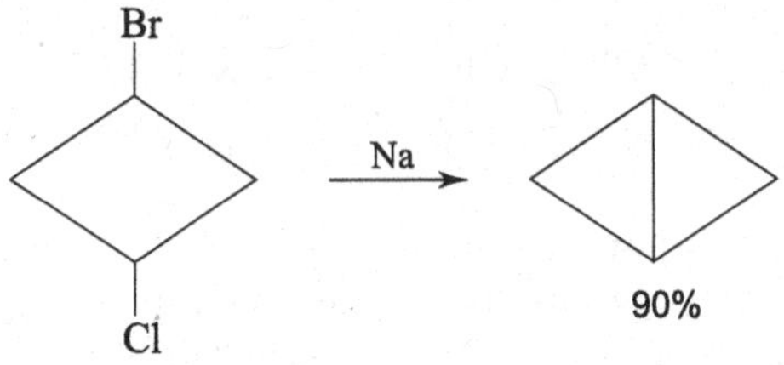

90%

3. Synthesis of bicyclic compounds.

References

1. P.G. Gassman, J. Seter and F.J. Williams, J.Am. Chem. Soc., 1971, *93*, 1673. J.M. Bloomfield and D.C. Owsley, J. Org. Chem., 1975, *40*, 393.
2. K. Ruhlann, H. Seefluth and H. Becker, Chem. Ber., 1967, *100*, 3820., J.J. Bloomfield, Tetrahedron Lett., 1968, 587.
3. G.D. Gutsche, I.Y.C. Tao and J. Kozma, J. Org. Chem., 1967, *32*, 1782.
4. A. Krebs, Tetrahedron Lett., 1968, 4511.
5. R.C. Cookson and S.A. Smith, J. Chem. Soc. Perkin I, 1979, 2447.
6. A.T. Nielsen, W.J. Houlihan, Organic Reactions, 1968, *16*, 1; W. Forest Ed., Newer Method of Preparative Organic Chemistry, 1971, *6*, 48, H.O. House, Modern Synthetic Reactions (W.A. Benzman, California, 2nd. ed.), 1972, pp. 629–682.
7. R.S. Verma and G.W. Kabalka, Heterocycles, 1985, *23*, 139.
8. F. Fringuelli, G. Pani, O. Piermalti and F. Pizzo, Tetrahedron, 1994, *50*, 11499; F. Fringuelli, G. Pani, O. Piermatti and F. Pizzo, Life Chem. Rep. 1995, *13*, 133.
9. T. Mukaiyama, K. Narasaka and T. Banno, Chem. Lett., 1973, 1011; T. Mukaiyama, K. Banno and K. Narasaka, J. Am. Chem. Soc., 1974, *96*, 7503. T. Mukaiyama, Org. React., 1982 *28*, 2303.
10. K. Takai, C.M. Heathcock, J. Org. Chem., 1985, *50*, 3247; A.E. Vougioukas and H.B. Kagan Tetrahedron Lett., 1987, *28*, 5513.
11. .A. Lubineau, J. Org. Chem., 1986, *51*, 2142., A. Lubineau, E. Mayer, Tetrahedron, 1988, *44*, 6065.
12. (a) S. Kobayashi and I. Hachiya, J. Org. Chem., 1994, 59, 3590, (b) For a review on lanthanide catalysed organic reaction in aqueous media, S. Kobayashi, Synlett., 1994, 589.
13. S. Murata, M. Suzuki and R. Nojori, Tetrahedron Lett., 1980, *21*, 2527.
14. S. Kobayashi and I. Hachiya, Tetrahedron Lett., 1992, 1625.
15. Y. Wef and R. Bakthavatechalan, Tetrahedron Lett., 1991, *32*, 1535.
16. S.C. Welch, J.M. Assercq and J.P. Loh, Tetrahedron Lett., 1986, 1115.
17. O. Kajmoto, Chem., Rev., 1999, *99*, 355; P.E. Savage, Chem., Rev., 1999, *99*, 603; D. Broli, C. Kaul, A. Krammer, P. Krammer, T. Richtu, M. Jung, H. Vogal and P. Zehner, Angew. Chem., Int. Ed., 1999, *38*, 2998; M. Siskin and M. Katrinzky, Chem., Rev., 2001, *101*, 825.

18. J. An, L. Bagnell, T. Cablewski, C.K. Strauss and R.W. Trainer, J. Org., Chem., 1997, 62(8), 2505.
19. C.P. Mchnett, N.C. Dispenziere and R.A. Crok, Chem., Commun., 2006, 1610.
20. P. Kotrusz, I Kmentova, S. Gotov, S. Toma and E. Solacaniova, Chem., Cummum., 2002, 2510.
21. S. Chandrasekher, N. Ramakrishna Reddy, S. Shmeem Sultana, Ch. Narsihmulu and K. Venkatram Reddy, Tetrahedron Lett., 2006, 338.
22. W. Hoffmann and H. Siegel, Tetrahedron Lett., 1975, 533.
23. F. Arndt and B. Eistert, Ber., 1935, 68, 200.
24. T. Aoyama and T. Shioiri, Chem., Pharm. Bull. 1981, 29, 3249.
25. A.B. Smith, III, Chem. Commum., 1974, 695.
26. E.J. Walsh, Jr. and G.B. Stone, Tetrahedron Lett., 1986, 1127.
27. A.B. Smith, III, B.D. Dorsey, M. Visnick, T. Maeda and M.S. Malamas, J. Am. Chem. Soc., 1986, 108, 3110.
28. J. Meinwald and Y.C. Meinwald. In advances in Alicyclic chemistry, ed. H. Hart and G.J. Karabatsos, Vol. 1, p. 1 (New York, Academic Press).
29. A.V. Baeyer and V. Villiger, Ber., 1899, 32, 3625.
30. H.O. House, Modern Synthetic Reactions, 2nd edn. Benjamine; Menlo Park, New York, 1972, p. 321; W.D. Emmons and G.B. Lucas, J. Am. Chem. Soc., 1955, 77, 2287.
31. M. Suzuki, H. Takada and R. Noyoro, J. Org. Chem., 1982, 47, 902.
32. F. Fringulli, R. Germani, E. Pizzo and G. Savelli, Gazz. Chem. Ital., 1989, 119, 249.
33. A.E. Thomas and F. Ray, Tetrahedron, 1992, 48, 1927.
34. K. Tanka and F. Toda, Chem. Rev., 2000, 100, 1028-29.
35. C.C. Ryerson, D.P. Ballon and C. Walsh, Biochemistry, 1982, 21, 2644.
36. N.A. Donoghue, D.B. Norris and P.W. Trudgill, Eur. J. Biochem., 1976, 63, 175.
37. B.P. Branchaud and C.T. Walsh, J. Am. Chem. Soc., 1985, 107, 2153.
38. G. Magnusson, Tetrahedron Lett., 1977, 2713.
39. J.T. McCurdy and R.D. Garrett, J. Org. Chem., 1968, 33, 660.
40. F.J. Fried, R.W. Thoma and A. Klingsberg, J. Am. Chem. Soc., 1953, 75, 5764.
41. R.L. Prairie and P. Talalay, Biochemistry, 1963, 2, 203.
42. A. Hassner. J. Org. Chem., 1978, 43, 1774.
43. W. Baker, J. Chem. Soc., 1933, 1381; H.S. Mahal and K. Venkataraman, J. Chem. Soc., 1934, 1767; K. Venkataraman in Zechmeister Progress in the Chemistry of Natural Products, 1959, 17, 2; E. Levmi, Chem., Rev., 1954, 54, 493.
44. V.K. Ahluwalia et al. unpublished results.
45. P. Barbier, C.R. Acad. Sci., 1898, 128, 110.
46. C. Blomberg, Synthesis, 1977, 18.
47. P.J. Pearce, D.H. Richards and N.F. Scilly, J. Chem. Soc. Perkin Trans. I, 1972, 1655.
48. J.-L. Luche and J.C. Damanio, J. Am. Chem. Soc., 1980, 102, 7926.
49. T. Uychara, J. Yamada, K. Ogata and T. Kato, Bull. Chem. Soc. Japan, 1985, 58, 211.
50. M. Ihara, M. Katogi, K. Fukumoto and T. Kametani, J. Chem. Soc., Chem. Commun., 1987, 721.
51. I.C. Burkow, L.K. Sydnes and D.C.N. Ubeda, Acta. Chem. Secand. Ser. B. 1987, B41, 235
52. S.B. Singh and G.R. Pettit, Syn. Commun., 1987, 17, 877.
53. S. Araki and Y. Butsugan, Chemistry Lett., 1988, 457.
54. R.L. Snowden, P. Sonnay, J. Org. Chem. 1984, 49, 1465.
55. C. Petrier, A.L. Gemal, J.-L. Luche, Tetrahedron Lett., 1982, 23, 3361.
56. J. Einhorn, J.-L. Luchi, Tetrahedron Lett., 1986, 27, 1791.
57. .J. Einhorn, J.-L. Luchi, Tetrahedron Lett., 1986, 27, 501.
58. C. Einhron, J. Einhorn and J.L. Luche, Synthesis, Review, 1989, 784.
59. D.H.R. Barton, J.M. Reaton, L.E. Geller, M.M. Pechet, J. Am. Chem. Soc., 1960, 82, 2640; 1961, 83, 4076; For review, see Barton, Pure Appl. Chem., 1968, 16, 1–15; M. Akhtar, Adv. Phytochem., 1964, 2, 263–304.
60. D.H.R. Barton et al. J. Chem. Soc. Perkin. Trans I, 1979, 1159.

61. A.L. Nussbaum and C.H. Robinson, Tetrahedron, 1962, *17*, 35.
62. K. Heusler and J. Kalvoda, Angew. Chem. Internat. edn., 1964, *3*, 525.
63. Kalvoda and K. Heusler, Synthesis, 1971, 501.
64. E.J. Corey, J.F. Arnett, and G.N. Widiger, J. Am. Chem., Soc., 1975, *97*, 430.
65. P.D. Hobbs and P.D. Magnus, J. Am. Chem. Soc., 1976, *98*, 4594.
66. S.W. Baldwin and H.R. Blomquist, J. Am. Chem., Soc., 1982, *104*, 4990.
67. D.H.R. Barton, N.K. Basu, R.H. Day, N.M. Pechet and A.N. Starrat, J. Chem., Soc. Perkin Trans. I, 1975, 2243.
68. K.R. Laumas and M. Gut, J. Org. Chem., 1962, *22*, 314; D.H.R. Barton and J.M. Beaton, J.Am. Chem. Soc., 1961, *83*, 4083.
69. A.B. Baylis and M.E.D. Hillman, Ger. Pat. 2155113, CA 1972, *77*, 3417.
70. D. Basavaiah, Tetrahedron Lett., 1986, *27*, 2031; 1987, *28*, 4591 and 4351; 1990, *31*, 1621.
71. M.K. Kundu, S.B. Mukherjee, N. Balu, R. Padmakumar and S.V. Bhatt, Synlett., 1994, 444.
72. P.M. Rose, A.A. Clifford and C.M. Rayner, Chem. Commun, 2002, 968.
73. D. Basavaiah, P.D. Rao and R.S. Huma, Tetrahedron, 1996, *52*, 8001.
74. V.K. Aggarwal, A. Mereu, G.J. Tarver and R. McCague, J. Org. Chem., 1998, *63*, 7183.
75. J–C. HSu, Y–H. Yen and Y–H. Chu, Tetrahedron Lett., 2004, *45*, 4673.
76. W. Jurćik and R. Wilheum, Green Chem., 2002, *7*, 844.
77. A Kumar and S.S. Pawar, J. Mol. Catal. A. Chem., 2004, *211*, 43.
78. S. Chandrasekher, Ch. Narsihmulu, B. Saritha and S.S. Sultana, Tetrahedron Lett., 2004, *45*, 5865.
79. D. Basavaiah, A.J. Rao and T. Satyanarayana, Chem., Rev., 2003, *103*, 811
80. E. Beckmann, Chem., Ber., 1886, *19*, 988.
81. I. Almena, A. Diaz-Ortiz, E. Diez-Barra, A. Hos and A. Loupy, Chem. Lett., 1996, 333; S. Caddick, Tetrahedron, 1995, 10400.
82. A.L. Bosch, P. dela Cruiz, E. Diez-Barra, A. Luppy and F. Langa, Synlett., 1995, 1259.
83. S. Guo, Z. Du, S. Zhang, D. Li, Z.li and Y. Deng, Green Chem., 2006, *8*, 296.
84. J.C. Eck and C.S. Marrel, OS, II, 1943, *76*, 371.
85. S. Chowdhury, R.S. Mohan and J.L. Scott., Tetrahedron, 2007, *63*, 2363–2369 and the references cited therein.
86. S. Selman and J.F. Eastham, Quart. Rev., 1960, *14*, 221.
87. F. Toda, Y. Tamaka, Y. Kagawa and Y. Sakaino, Chem., Lett., 1990, 373.
88. H.–M. yu, S.–T. Chen, M.–J. Tseng and K.T. Wang, J. Chem. Res (S), 1999, 62.
89. M. Furukawa, T. Yoshida and S. Hayashi, Chem., Pharm. Bull., 1975, *23*, 580; Chem., Abs., 1975, *83*, 43241.
90. A.J. Lapworth, J. Chem. Soc., 1903, *83*, 955; 1904, *85*, 1206; Bulk, Organic Reaction Vol. (IV) 269 (1948).
91. Y. Yano, Y. Tamura and W. Tagaki, Bull. Chem. Soc. Japan, 1980, *53,* 740.
92. C.M. Starks and C. Liotta, Phase Transfer Catalysts, Principles and Techniques, Academic Press Inc. NY, 1978, p. 343.
93. J. Solodar, Tetrahedron Lett., 1971, 287.
94. W. Tagaki and H. Hara, J. Chem. Soc., Chem. Commun, 1973, 891.
95. E.T. Kool and R. Breslow, J. Am. Chem. Soc., 1988, *110*, 1596.
96. H. Stetter and G. Dambkes, Synthesis, 1977, 403; H. Sletter and H. Kuhlmann (to Bayer A.G.) Ger. Patent 2, 437, 319 (1974); CA, 1976, *84*, 164172.
97. G. Sumrell, J.I. Stevens and G.E. Goheen, J. Org. Chem., 1957, *22*, 39.
98. G.H. Gholamhosein, H. Hakimelahic, C.B. Boyle and H.S. Kasmai, Helv., 1977, *60*, 342.
99. D.P. Macaione and S.E. Wentworth, Synthesis, 1974, 716.
100. P. Biginelli, Ber., 1891, *24*, 1317, 2962; 1893, *26*, 447.
101. C.O. Kappe, D. Kumar and R.S. Verma, Synthesis, 1999, 1799.
102. J. Penj and Y. Deng, Tetrahedron Lett., 2001, *42*, 5917.
103. L. Bouveault, Bull. Soc., Chem. France, 1904, *31*, 1306, 1322. N. Smith, J. Org. Chem., 1941, *6*, 489; Sice, J. Am. Chem. Soc., 1953, *75*, 3697; Jones *et al.*, J. Chem. Soc., 1958, 1054.

104. Evans, Chem. and Ind. (London), 1957, 1956.
105. J.L. Luche, Ultrasonics, 1987, 25, 40.
106. S. Cannizzaro, Ann., 1853, 88, 129; K. List and H. Limpricht, Ann., 1854, 90, 180.
107. A. Fuentes and V.S. Sinistera, Tetrahedron Lett., 1986, 27, 2967.
108. R.S. Varma, G.W. Kabalka, L.T. Evans and R.M. Pagni, Synth. Commun., 1985, 15, 279.
109. R.A. Bruce, Org. Prep. Proced Int., 1987, 19.
110. T.A. Geissmann, Org. Reactions Vol. II, 1944, 94.
111. T.A. Geissmann, Org. React, 1944, 2, 95.
112. R.S. McDonald and C.E. Sibley, Can. Jour., Chem., 1981, 59, 1061.
113. (a) L. Claisen, Ber., 1912, 45, 3157; (b) L. Claisen and O. Eisleb, Annalen 1913, 401, 21; (c) L. Claisen and E. Tietze, Ber., 1925, 58, 275; (d) D.S. Tarbell, Org. React., 1944, 2, 1.
114. M.R. Saidi, Heterocycles, 1982, 19, 1473.
115. P. Vitorelli, T. Winkler, H.J. Hansen and H. Schmidt, Helv. Chim. Acta, 1968, 51, 1457; R.K. Hill, A.G. Edwards, Tetrahedron Lett., 1964, 3239.
116. E.N. Marvel and J.L. Stephenson, J. Org. Chem., 1960, 25, 676; H. Hart, J. Am. Chem. Soc., 1954, 76, 4033.
117. W.N. White and E.F. Wolfartt, J. Org. Chem., 1970, 35, 2196.
118. E. Brandes, P.A. Grieco and J.J. Gajewski, J. Org. Chem., 1989, 54, 515.
119. D. Dallinger and C.O. Kappe, Chem. Rev., 2007, 107. Page 2587 and the references cited therein.
120. E.J. Corey, R.L. Danheiser, S. Chandrasekaran, P. Siret, G.E. Keck and J.-L., Gras, J. Am. Chem. Soc., 1978, 100, 8031.
121. J.W.S. Stevenson and T.A. Bryson, Tetrahedron Lett., 1982, 3143.
122. S.J. Rhoades, J. Am. Chem. Soc., 1955, 73, 5060.
123. E. Brandes, P.A. Grieco and J.J. Gajewski, J. Org., Chem., 1989, 54, 515.
124. J.J. Gajewski, J. Jurayj, D.K. Kimbrough, M.E. Gande, B. Ganem and B.K. Carpenter, J. Am. Chem. Soc., 1987, 109, 1170.
125. P.A. Grieco, E.B. Brandes, S. McCann and J.D. Clark, J. Org. Chem., 1989, 54, 5849.
126. Mc Murry, A. Andrus, G.M. Ksander, J.J. Musser and M.A. Johnson, Tetrahedron, 1981, 37 (Suppl. No. 1), 319.
127. P.A. Grieco, E.B. Brandes, S. McCann and J.D. Clark, J. Org. Chem., 1989, 54, 5849.
128. A. Lubineau, J. Auge, N. Bellanger and S. Caillebourdin, J. Chem. Soc. Perkin Trans. I, 1992, 13, 1631.
129. P. Wipf, Comprehensive Organic Synthesis, B.M. Trost, I. Fleming, L.A. Paquette Eds., Pergamon Press, New York, 1991, Vol. 5, p. 827; F.E. Zieglar Chem. Rev., 1988, 88, 1423; S.J. Rhoads and N.R. Raulins, Org. React., 1975, 22, 1. Claisen Rearrangements in aqueous medium in organic synthesis in water. Edited by Paul A. Grieco, Blackie Academic and Professional, London, 1998, pp. 47–101 and the reference cited therein.
130. L. Claisen and A. Claperede, Ber. 1881, 14, 2460; J.G. Schmidt, Ber., 1881, 14, 1459; H.O. House, Modern Synthetic Reactions (W.A. Benjamin, California, 2nd. ed., 1972), 632–639.
131. T. Mukaiyama, K. Banno and K. Narasaka, J. Am. Chem. Soc., 1974, 96, 7503.
132. Y. Yamamoto, K. Maruyama and K. Matsumoto, J. Am. Chem. Soc., 1983, 105, 6963.
133. A. Lubinean, J. Org. Chem. 1986, 51, 2142; A. Lubineau and E. Meyer. Tetrahedron, 1988, 44, 6065.
134. K.R. Nivalkar, C.D. Mudaliar and S.H. Mashraqui, J. Chem. Res. (s), 1992, 98.
135. F. Fringulli, G. Pani, O. Piermatti, and F. Pizzo, Tetrahedron, 1994, 50, 11499. F. Fringulli, G. Pani, O. Piermatti and F. Pizzo, Life Chem. Rep., 1955, 13, 133.
136. P. Formentin, H. Garcia and A.J. Mol., Catal A: Chem. 2004, 214, 137.
137. V. Jurćik and R. Wilhelm, Green Chem., 2002, 7, 844.
138. E. Clemmensen, Ber., 1913, 46, 1838; 1914, 51, 681; E.L. Martin in Organic Reactions I (New York, 1942), 155; J. G. St. C. Buchamam, P.G Woodgate, Quart. Rev., 1961, 23, 522; V. Vedejs, Org. Reactions, 1975, 22, 401.
139. W.P. Reeves, J.A. Murry, D.W. Willoughby, and W.I. Friedrich, Synthetic Communication, 1988, 18, 1961.

140. E. Clemmensen, Ber., 1913, *46*, 1838.
141. B. Bannister and B.B. Elsner, J. Chem. Soc., 1951, 1055.
142. E.L. Martin, J. Am. Chem., Soc., 1936, *58*, 143.
143. Reported in review by Vedejs (reference 1).
144. A.K. Banerjee, J. Alvarez, M. Santan and M.C. Carrasco, Tetrahedron, 1986, *42*, 6615.
145. W.G. Dauben, J. Am. Chem. Soc., 1954, *76*, 3864.
146. T. Curtius, J. Prakt. Chem., 1894, *50*, 275; J.H. Saunders, R. Slocombe, J. Chem. Rev., 1948, *43*, 203.
147. W. Lwowski, S. Linke and G.T. Tisue, J. Am. Chem., Soc., 1967, *89*, 6308.
148. C.W. Porter and L. Young, J. Am. Chem. Soc., 1938, *60*, 1497.
149. H.D. Dakin, OS, 1941, *1*, 149; J.E. Lettler, Chem. Rev., 1949, *45*, 385; H.D. Dakin, J. Am. Chem. Soc., 1909, *42*, 477.
150. H.D. Dakin, Org., Synth., Coll. Vol. 1, 1941.
151. C.A. Bunton in Peroxide Reaction Mechanism, edited by J.O. Edwards, Interscience.
152. G.W. Kabalka, N.K. Reddy and C. Narayana, Tetrahedron Lett., 1992, *33*, 865.
153. R.S. Varma and K.P. Naicker, Org. Letters, 1999, *1*, 189.
154. M.B. Hocking and J.H. Ong., Can. J. Chem., 1977, *55*, 102. M.B. Hocking, M.K.O. and T.A. Smyth, Can. J. Chem., 1978, *56*, 2646.
155. Y. Agasimundin and S. Siddappa, J. Chem. Soc., Perkin I, 1973, 503.
156. J. Baker, J. Chem. Soc., 1953, 1615.
157. C. Darzens, Compt. Rend., 1904, *139*, 1214; 1905, *141*, 766; 1906, 142, 214; M. Ballester, Chem. Rev., 1955, *55*, 283.
158. A. Jonczyk, M. Fedorynski and M. Makosza, Tetrahedron Lett., 1972, 2395.
159. C. Kimura, K. Kashiwaya and K. Murai, Asahi Garasa Kogyo Gijutsu Shoreika Kenkyu Hokoku, 1975, *26*, 163.
160. M. Makosza and M. Ludwikow, Angew. Chem., Int. Ed. Engl, 1974, *13*, 655.
161. J.M. Melntosh and H. Khalili, J. Org. Chem., 1977, *42*, 2123.
162. H. Achenbach and J. Witzke, Tetrahedron Lett., 1979, 1579.
163. A. Knorr, E. Large and A. Weissenborn, C.A., 1939, *28*, 2367.
164. T.T. Tung, J. Org. Chem., 1963, *28*, 1514.
165. W. Dieckmann, Ber., 1894, *27*, 102, 965; 1900, *33*, 2670; Ann., 1901, 317, 53, 93; C.R. Housen and B.E. Hudson, Organic Reactions, 1942, I, 274.
166. F. Toda, T. Suzuki and S. Higa, J. Chem. Soc., Perkin Trans. I. 1998, 3521.
167. Cited in a review by J.M. Khurana, Chemistry Education, 1990 (Oct, Dec.) p. 27.
168. J.L. Luche, C. Petrier and C. Duputy, Tetrahedron Lett., 1985, *26*, 753.
169. J.L. Crowley and H. Rapoport, J. Chem., Soc., 1970, *92*, 6363.
170. J. Schalffer, J.J. Bloomfield, Org. React., 1967, *15*.
171. H.L. Lochte and A.G. Pinman, J. Org. Chem., 1960, *25*, 1462.
172. O.S. Bhanot, Indian J. Chem., 1967, *5*, 127.
173. C.A. Broun, Synthesis, 1975, *5*, 326.
174. G. Nee and B. Tehboubar, Tetrahedron Lett., 1979, 3717.
175. H.-J. Liu and H.K. Lai, Tetrahedron Lett., 1979, 1193.
176. O. Diels, K. Alder, Ann., 1928, *460*, 98; 1929, *62*, 470; Ber., 1929, *62*, 2081, 2087; J.A. Norton, Chem. Res., 1943, *31*, 319; D.A. Oppolzer, Angew Chem. Int. Ed., 1977. *16*, 10.
177. B. Perio, M.J. Dozias, P. Jacquault and J. Hamilin, Tetrahedron Lett., 1997, *38*, 7867.
178. D. Bogdal, J. Pielichowski and A. Boron, Synlett, 1996, 873.
179. D. Ruhua, W. Yuliang and Y. Yaozhong, Synth. Commun., 1994, *24(13)*, 1917.
180. A. Vass, J. Toth and E. Pallai Varsanyi, Effect of Inorganic solid support for microwave assisted organic reactions, presented at the International Conference on Microwave Chemistry Prague, Czech. Republic, Sept. 6–11, 1998.
181. D.C. Rideout and R. Breslow, J. Am. Chem., Soc., 1980, *102*, 7816; R.B. Woodward and Harold Baer, J.Am. Chem. Soc., 1948, *70*, 1161.
182. R. Breslow, U. Maitra and D. Rideout, Tetrahedron Lett., 1983, *24*, 1901; R. Breslow and U. Matra, Tetrahedron Lett., 1984, *25*, 1239.

183. J.A. Berson, Z. Hamlet and W.A. Muller, J.Am. Chem. Soc., 1962, *84*, 297; A.A.Z.Samil, A. Desavignac, I. Rico and A. Latters, Tetrahedron, 1985, *41*, 3683.

184. Chao-Jun Li and Tak-Hang Chan, Organic Reactions in Aqueous Media, John Wiley & Sons Inc., Page 16.

185. P.E. Savage, Chem., Rev., 1999, *99*, 603–621 and the references cited therein.

186. J.M. Kremsner and C.O. Kappe, Eur. J. Org. Chem., 2005, 3672.

187. V.K. Ahluwalia and M. Kidwai, New Trends in Green Chemistry, Anamaya Publishers, 2006, Page 82.

188. J. Lee and J.K. Snyder, J. Am. Chem., Soc., 1989, *111*, 1522.

189. T. Fisher, T. Sethi, T. Welton and J. Woolf, Tetrahedron Lett., 1999, *40*, 793.

190. M.J. Earle, P.B. McMormac and K.R. Seddon, Green Chem., 1999, *1*, 23.

191. A.R. Renslo, R.D. Weinstein, J.W. Tester and R.L. Danherisel, J. Org. Chem., 1997, *62*, 4530.

192. R.S. Oakes, T.J. Happenstall, N. Shezad, A.A. Clifford and C.M. Rayner, Chem., Commun., 1999, 1459–1460.

193. S. Fukuzawa, K. Metoki, Y. Komuro and T. Funazukuri, Synlett, 2002, *28*, 134–136.

194. K. Ishihara and H. Yamamoto, J. Am. Chem., Soc., 1994, *116*, 1561; K. Ishihara, H. Kurihara and H. Yamamoto, J. Am. Chem. Soc., 1996, *118*, 3049; K. Ishihara, H. Kurihara, M. Mausumoto and H.J. Yamamoto, J. Am. Chem., Soc., 1998, *120*, 6920.

195. D. Boger and S. Weireb, Hetero-Diels–Alder Methodology in Organic Synthesis, Academic Press, Orlando, 1987.

196. P.A. Grieco and S.D. Larsen, J.Am. Chem. Soc., 1985, *107*, 1768.

197. W. Opolzer, Angew. Chem., Int. Ed. Engl., 1972, *11*, 1031.

198. P.A. Grieco, D.T. Parker, W.F. Fobare and R. Ruckle, J. Am. Chem., Soc., 1987, *109*, 5859.

199. G. Briger and J.N. Bennett, Chem. Rev., 1980, *80*, 63.

200. W. Oppolzer, Pure and Appl. Chem., 1981, *53*, 1181.

201. T. Kametani and H. Nemoto., Tetrahedron, 1981, *37*, 3.

202. A.G. Fallis, Canad. J. Chem., 1984, *62*, 183.

203. M.P. Edwards, S.V. Ley, S.G. Lister, B.D. Palmer and D.J. Williams, J. Org. Chem., 1984, *49*, 3503.

204. K.C. Nicolaou and R.L. Magolda, J. Org. Chem., 1981, *46*, 1506.

205. W.K. Roush and A.G. Meyers, J. Org. Chem., 1981, 1506, 1509.

206. E. Fischer and F. Jourdan, Ber, 1883, *16*, 2241; E. Fischer and O. Hes, Ber., 1884, *17*, 559.

207. G.B. Jones and J. Chapman, J. Org. Chem., 1993, *58*, 5558.

208. D. Villemin, B. Labiad and Y. Ouhilal, Chem. and Ind., 1989, 607.

209. C.R. Strauss and R.W. Trainer, Aust. J. Chem., 1998, *51*, 703.

210. C. Friedel and J.M. Crafts, Compt. Rend., 1887, 84, 1392, 1450; P. Gore. Chem. Revs., 1955, 55, 229.

211. A.K. Boon, J.A. Levisky, J.L. Pflug and J.S. Wilkis, J. Org. Chem., 1986, *51*, 480.

212. A. Stark, B.L. McClean and R.D. Singer, J. Chem. Soc., Dalton Trans, 1999, 63.

213. C.W. Lee, Tetrahedron Lett., 1999, *40*, 2461.

214. E. Ota, J. Electrochem. Soc., 1987, *134*, 512.

215. J.A. Boon, S.W. Lander, J.A. Levusky, J.L. Pflug, L.M. Skrzynecki-cooke and J.S. Wilkers, J. Electrochemical Soc., 1987, *134*, 501.

216. S. Kobayashi, J. Synth. Org. Chem., 1955, *53*, 370.

217. C.J. Adams, M.J. Earle, G. Roberts and K. Seddon, Chem., Commun., 1998, 2097.

218. D. Clarke, M.A. Ali, A.A. Clifford, A Parrat, P. Rose, D. Schwinn, W. Bannwarth and C.M. Rayner, Current Topics in Medicinal Chemistry, 2004, *4*, 731 and the references cited therein.

219. C.E. Song, W.H. Shim, E.J. Roh and J.H. Chori, Chem., Commun., 2000, 1695.

220. R. Hitzler, F.R. Smail, S.K. Ross and M. Poleakoff, J. Chem., Soc., Chem. Commun., 1998, 359.

221. A. Kawada, S. Mitamura and S. Kobayashi, Synlett., 1994, 545.

222. A.R. Katritzky, S.M. Allin and M. Siskin., Acc. Chem., Res., 1996, *29(8)*, 399–406.

223. K. Chandler, F. Deng, A.K. Dillow, C.L. Liotta, and C.A.Eckert, Ind. Eng. Chem. Res., 1997, *36(12)*, 5175.

224. P. Boadjout, W.H. Ohrbm and J.B. Woell, Synth. Commun., 1986, *16*, 401.
225. D.M. Trost and B.P. Coppola, J. Am. Chem. Soc., 1982, *104*, 6879.
226. M.J. Burk, S. Feng, M.G. Gross and W. Tumas, J.Am. Chem. Soc., 1995, *117*, 8277.
227. C.J. Adams, M.J. Earle, G. Roberts and K.R. Seddan, Chem. Cummun., 1998, 2097.
228. M.J. Earle and K.K. Seddon, Pure Appl. Chem., 2000, *72*, 1391 and the references cited therein.
229. M.R. Bell, T.E. Dambra, V. Kumar, M.A.Eissenstat, J.L. Herrmann, J.R. Wetzel, D. Rosi, R.E. Philion, S.J. Daum, D.J. Hlasta, R.K. Kullnig, J.H. Ackerman, D.R. Haubrich, D.A. Luttinger, E.R. Baiziman, M.S.Miller and I. Ward, J. Med. Chem., 1991, *34*, 1099.
230. D. Clarke, M.A. Ali, A.A. Clifford, A Parratt, P. Rose, D. Schwinn, W. Bannwarth and C.M. Rayner, Current Topics in Medicinal Chemistry, 2004, *4*, pages 736–745.
231. A.G.M. Barrett, D.C. Braddock, D. Catterick, J.P. Henschke and R.M. McKinnell, Synlett., 2000, 847.
232. K. Mikami, Y. Mikami, Y Matsumolo, J. Nishikedo, F. Yamamoto and H. N akajima, Tetrahedron Lett., 2001, *42*, 289.
233. T. Kitazume, J. Fluorine Chem., 2000, *105*, 265.
234. Friedlander, Ber., 1882, 15, 2572; P. Friedlander and C.F. Gohring, Ber., 1883, 16, 1833; Manske, Chem. Rev., 1942, 30, 124.
235. G. Sabitha, R.S. Babu, B.V.S. Reddy and J.S. Yadav, Synth. Commun., 1999, *29(4)*, 4403.
236. K. Fries and G. Fink, Ber., 1908, *41*, 4271; K. Fries and W. Pfaffendorf, Ber., 1910, *43*, 1910; A.H. Bhatt, Chem. Rev., 1940, *27*, 429.
237. N.M. Cullinane, A.G. Evans and E.J. Lloyd, J. Chem., Soc., 1956, 2222.
238. D. Bellus et al., Chem., Rev., 1967, 599.
239. J.C. Anderson and C.B. Reese, Proc. Chem. Soc., 1960, 217.
240. V. Sridar and V.S. Sundara Rao, Indian J. Chem., 1994, *33B*, 184.
241. C. Graebe and F. Ullmann, Ann. 1896, 291, 16; F. Ullmann, Ann., 1904, 332, 82.
242. A. Mohna, J.I. Vaguero, J.I. Garefa and J. Alvarez-Builla, Tetrahedron Lett., 1993, *34*, 2673.
243. V. Grignard, Compt. Rend., 1900, *130*, 1322.
244. J.L. Luche and J.C. Damino, J. Am. Chem. Soc., 1980, 102, 7964; J.D. Sprinch and G.S. Lewandos, Inorg. Chem. Acta, 1982, 76, 1241; W. Oppolzer and A, Nakao, Tetrahedtron Lett., 1986, 27, 5471.
245. W. Oppolzer and A. Nakao, Tetrahedron Lett., 1986, *27*, 5471.
246. F. Todd, H. Takumi and H. Yamaguchi. Chem. Exp., 1980, *4*, 507.
247. D. Holt, Tetrahedron Lett., 1981, 2243.
248. R.F. Heck, Acc. Chem. Res., 1979, *12*, 146; Organic Reactions, 1982, *27*, 345; H.A. Dieck, J. Organometallic Chem., 1975, *93*, 259.
249. T. Jeffery, Chem., Commun., 1984, 1287.
250. N.A. Bumagin, P.G. More and I.P. Beletskaya, J. Organometallic Chemistry, 1989, *371*, 397.
251. N.A. Bumagin, N.P. Andryukhova, and I.P. Beletskaya, Akad. Nauk. SSSR, 1990, *313*, 107.
252. W.A. Herrann and V.P.W. Bohn. J. Organomet. Chem., 1999, *572*, 141.
253. N.A. Bumagin, L.I. Sukhomlinova, A.N. Vanchikov, T.P. Tolstaya and I.P. Beletskaya, Bull. Russ. Acad. Sec., Div. Chem. Sec. 1992, *41*, 2130; N.N. Denik, M.M. Rabachnik, M.M. Novikova and I.P. Beletskaya, Zh. Org. Khim, 1995, *31*, 64; T. Jeffery, Tetrahedron Lett., 1994, *35*, 3051; Organic Synthesis in Water. Ed. Paul A. Grieco, Blackie Academic and Professionals, London (1998) pp. 181–188 and the references cited therein.
254. R.K. Arvela and N.E. Leadbeater, J. Org. Chem., 2005, *70*, 1786.
255. R.K. Avela, S. Pasquini and M. Larhed, J. Org. Chem. 2007, *72*, 6390.
256. D.K. Morita, D.R. Pesiri, S.A. David, W.H. Glaze and W. Tumas, J. Chem. Soc., Chem. Commun; 1998, 1397–1398.
257. B.M. Bhanage, Y. Ikushima, M. Shirai, and M. Arai, Tetrahedron Lett., 1999, *40*, 6427.
258. T.R. Early, R.S. Gordin, M.A. Carroll, A.B. Holmes, R.E. Shute, I.F. McConvey, I.F. Early, R. Tessa, S.R. Richard, Michael A. Carroll, Andrew B. Holmes, E. Richard and I.F. McConvey, Chem. Commun., 2001, *19*, 1966.
259. Y. Kayaki, Y Noguchi and T. Ikarya, Tetrahedron Lett., 1999, *40*, 6427.

260. L.K. Yeng, Jr. K.P. Johnstone and R.M. Cooks, Chem. Commun., 2001, 2290.
261. A. Carmichacl, M.J. Earle, I.D. Holbrey, P.B. McCormac and K.R. Seddon, Org. Lett., 1999, *1*, 997.
262. L. Xu, W. Chen and J. Xiao, Organometallics, 2000, *19*, 1123.
263. M.T. Rutz, R. Breinbauer and K. Wanninger, Tetrahedron Lett., 1996, *37*, 4499.
264. S. Chandrasekhar, Ch. Narsihmulu, S.S. Sultana and N.R. Reddy, Org. Lett., 2002, 4390.
265. L. Xu, W. Chen, I. Ross and J. Xiao, Org. Lett., 2001, *3*, 293.
266. W. Cabri, F. Candiani, A. Bedeschi and S. Penco, J. Org. Chem., 1992, *57*, 1481; W. Cabri, I. Candiane and A. Bedischi, J. Org. Chem., 1993, *58*, 7421; W. Cabri and I. Candiani, Acc. Chem. Res., 1995, *28*, 2; M. Larhed and H. Hallberg, J. Org. Chem., 1997, *62*, 7858.
267. A. Hantzsch, Ann., 1882, *215*, 1, 72; Ber, 1885, *18*, 1744; 1886, *19*, 289.
268. C. Cottrill, A.Y. Usyatinsky, J.M. Arnold, D.S. Clark, J.S. Dornick, P.C. Michels and Y.L. Khmelnitsky, Tetrahedron Lett., 1998, *39*, 1117.
269. L. Henry, Compt. Rend., 1895, *120*, 1265.
270. R.S. Varma, R. Dahiya and S. Kumar, Tetrahedron Lett., 1997, *38*, 5131.
271. A Kumar and S.S. Pawar, J. Mol. Catal. A., 2005, *235*, 244.
272. T. Jiang, H. Gao, B. Han, G. Zhao, Y. Chang, W. Wu, L. Gao and G. Yang, Tetrahedron Lett., 2004, *45*, 2699.
273. T. Hiyama in Metal-catalysed cross coupling Reactions; F. Diederich and P.J. Stand Eds.; Wiley-VCH, New York, 1998, Chapter 10, p. 421; S.E. Denmark and M.H. Ober, Aldrichim. Acta, 2003, *36*, 75; C.J. Handy, A.S. Manoso, W.T. McEloy, W.M. Seganish and P. Deshong, Tetrahedron, 2005, *61*, 12201.
274. D. Dallinger and C.O. Kappe, Chem. Rev., 2007, *107*, 2263–2591 and the references cited therein.
275. A.W. Hoffmann, Chem. Ber., 1881, *14*, 659.
276. C.R. Strauss, K.D. Raner, R.W. Trainor and J.S. Thorn, Aust. Pat. 677876 (1977).
277. F. Knoevenagel, Ber., 1896, *29*, 172; J.R. Johnson, Org. Reactions, 1942, *1*, 210.
278. O. Doebner, Ber., 1900, *33*, 2140.
279. F. Texier-Boullet and A. Foucand, Tetrahedron Lett., 1982, 4927.
280. H. Stobbe, Ber., 1893, *26*, 2312.
281. K. Takai, C.H. Heathcock, J. Org. Chem., 1985, *50*, 3247; A.E. Vougioakas and H.B. Kagan, Tetrahedron Lett., 1987, *28*, 5513; K. Mikami, M. Terada and T. Nakai, J. Chem. Soc., Chem. Commun., 1993, 343 and references cited therein.
282. Paul A. Grieco Ed. in Organic synthesis in water, Blackie Academic and Professional, London, 1998, p 255.
283. J. Aug., M. Lubin and A. Lubineau, Tetrahedron Lett., 1994, *35*, 7947.
284. S.A. Ayoubi, F. Texier-Boullet and J. Hamelin, Synthesis, 1994, 258.
285. V. Singh, J. Singh, P. Kaur and G.L. Kad, J. Chem. Res. (S), 1997, 58; D. Bogdal, J. Chem. Res. (S), 1998, 468.
286. M. Kidwai, P. Sapra and K.R. Bhushan, J. Indian Chem. Soc., 2002, *79*, 596.
287. G. Bram, A. Loupy and M. Majoub, Tetrahedron Lett., 1990, *46*, 5167.
288. D. Villemins and B. Martin, Synth. Commun., 1995, *25*, 3135.
289. P. Formentin, H. Garcia and A. Leyva, J. Mol. Catal. A: Chem., 2004, *214*, 137.
290. J.R. Harjani, S.J. Nara and M.M. Salunkhe, Tetrahedron Lett., 2002, *43*, 1127.
291. C.G. Butler, R.K. Callow and N.C. Johnson, Proc. Royal Soc., 1961, *B*, *155*, 417.
292. H. Kolbe, Ann., 1860, 113, 125; R. Schmitt, J. Prak. Chem., 1885, *(2) 31*, 397; A.S. Lindsey and H. Leskey, Chem. Rev., 1957, *57*, 583.
293. A.R. Renslo, R.D. Weinstein, J.W. Tester and R.L. Danheiser, J. Org. Chem., 1997, *62*, 4530.
294. .R.D. Weinstein, A.R. Renslo, R.L. Danheiser, J.G. Harris and J.W. Tester, J. Phys. Chem., 1966, *100*, 12337.
295. C. Mannich and W. Krosche, Arch. Pharm, 1912, *250*, 647; F.F. Blicke, Org. Reactions, 1942, *1*, 303.
296. Y. Peng, R. Dou, G. Song and J. Jiang, Synlett, 2005, 2245.
297. P. Peng and G. Song, Green Chem., 2001, *3*, 302.

298. L. Shi, Y.Q. Tu, M. Wang, F.M. Zhang and C.A. Fan, Org. Lett., 2004, *6*, 1001.
299. I. Ojima, S – I. Inaba and K. Yoshoda, Tetrahedron Lett., 1977, 3643.
300. S. Kobayashi and H. Ishitani, J. Chem. Soc. Chem. Commun., 1995, 1379.
301. S. Kobayashi, M. Araki and M. Yasuda, Tetrahedron Lett., 1995, *36*, 5773.
302. K.H. Meyer and K. Schuster, Chem. Ber., 1922, *55*, 819.
303. J. An., L. Baycell, T. Cablewski, C.R. Strauss and R.W. Trainer, J. Org. Chem., 1997, *62(8)*, 2505.
304. A. Michael, J. Prakt. Chem. 1987 (2), 35, 349; E.D. Bergmann, D. Ginsburg and R. Pappo, Org., Reactions, 1959, *10*, 179.
305. Kohler and Butler. J. Am. Chem., Soc., 1926, *48*, 1040.
306. W.S. Wadsworth and W.D. Emmons, J. Am. Chem. Soc., 1961, *83*, 1733.
307. M. Makosza, Tetrahedron Lett., 1966, 5489; Polish Patent 55113 (1968); CA, 1969, *70*, 106006.
308. M. Makosza, J. Czyzewski and M. Jawdosiak, Org., Synth., 1976, *55*, 99.
309. T. Sakakibara and R. Sudoh, J. Org. Chem., 1975, *40*, 2823; T. Sakakibara, M. Yamada and R. Sudoh, J. Org. Chem., 1976, *41*, 736.
310. Z.G. Hajos and D.R. Parrish, J. Org. Chem., 1974, *39*, 1612; U. Elder, G. Sauer and R. Wiechert, Angew. Chem. Int. Edn. Engl., 1971, *10*, 496.
311. N. Harada, T. Sugioka, U. Uda and T. Kuriki, Synthesis, 1990, 53.
312. J.F. Lavelle and P. Deslongchamps, Tetrahedron Lett., 1988, *29*, 6033.
313. A. Lubineau and J. Auge, Tetrahedron Lett., 1992, *33*, 8073.
314. K. Sussang Karn, G. Fodor, I. Karle and C. George, Tetrahedron, 1988, *44*, 7047.
315. F. Toda, M. Takumi, M. Nagami and K. Tanaka, Heterocycles, 1988, *47*, 469.
316. H. Sakuraba, Y. Tanaka and F. Toda, J. Incl. Phenon, 1991, *11*, 195.
317. B. Satish, K. Panneesel-Vam, D. Zacharids and G.R. Desivaju, J. Chem. Soc. Perkin Trans., 1995, *2*, 325.
318. E. Diez-Barra, A. de la Hoz, S. Merino and P. Sanchez-Verdu, Tetrahedron Lett., 1997, *38*, 2359.
319. F. Toda, K. Tanka and J. Sato, Tetrahedron Asymmetry, 1993, *4*, 1771.
320. A. Loupy, J. Sansoulet, A. Zaparucha and C. Merinne, Tetrahedron Lett., 1989, *30*, 333.
321. M.M. Dell' Anna, V. Gallo, P. Mastrorilli, C. Francesco, G. Rommanazzi G.P. Suranna, Chem., Commun., 2002, 434.
322. V. Polshettiwar and R.S. Varma, Tetrahedron Lett., 2007, *48*, 8735.
323. R.Ballini, Synthesis, 1993, 687.
324. T. Mukaiyama, Chem., Lett., 1982, 353; J. Am. Chem. Soc., 1973, 95, 967; Chem. Lett., 1986, 187; T. Mukaiyama, Organic Reactions, 1982, 28, 187.
325. A. Lubineau, J. Org. Chem., 1986, *51*, 2142; A. Lubineau, E. Meyer, Tetrahedron, 1988, *44*, 6065.
326. S. Kobayashi and I. Hachiya, J. Org., Chem., 1994, *59*, 3590. For a review on lanthanides catalysed organic reactions in aqueous media, *see* S. Kobayashi, Synlett, 1994, 589.
327. H.V. Pechmann and C. Duisberg, Ber., 1883, 16, 2119; S. Sethna, Chem. Rev., 1945, 36, 10; K. D.Kaufmann, J. Org. Chem., 1967, 32, 504.
328. V. Singh, P. Singh, P. Kaur and G.L. Kad, J. Chem. Res., (S), 1997, 58.
329. D. Bogdal, J. Chem. Res., (S), 1998, 468.
330. Y. Gu, J.Zhang, Z. Duan and Y. Deng, Adv. Synth. Catal., 2005, 347, 512.
331. E. Paterno and G. Cheffi Gazz. Chin. Ital., 1909, *39*, 341; G. Büchi, C.G. Inmand and E.S. Lipinsky, J. Am. Chem. Soc., 1954, *76*, 4327; D.R. Arnold, Advan. Photochem., 1968, *6*, 301.
332. N. Jeong, S.H. Swang, Y.W. Lee and J.S. Lim., J. Am. Chem. Soc., 1977, *119*, 10549.
333. Grinder, Ann. Chem. Phys., 1892, *26*, 369.
334. J.B. Conant and H.B. Cutter, J.Am. Chem. Soc., 1926, *48*, 1016.
335. P. Karrer, Y. Yen and I. Reichstein, Helv. Chem. Acta, 1993, *13*, 1308.
336. A Clerici and O. Porta, Tetrahedron Lett., 1982, *23*, 3517.

337. A Clerici and O. Porta, J. Org. Chem., 1982, *47*, 2852; A Clerici, O. Porta and M. Riva, Tetrahedron Lett., 1981, *22*, 1043.

338. A Clercci and O. Porta J. Org. Chem., 1983, *48*, 1690; Tetrahedron, 1983, *39*, 1239; A. Clerici, O. Porta and P. Zago, Tetrahedron, 1986, *42*, 561; A. Clerici and O. Porta, J. Org. Chem., 1989, *54*, 3872.

339. P. Delair and J. L. Luche, J. Chem. Soc. Chem. Commun., 1989, 398.

340. R. Fitting, Ann., 1859, *110*, 17; 1860, *114*, 54.

341. .B. Kuhlman, E.M. Arnett and M. Siskin, J. Org. Chem., 1994, *59*, 5377.

342. E. Gutirrez, A. Loupy, G. Bram and E. Ruiz-Hitzky, Tetrahedron Lett., 1989, *30*, 945.

343. F. Toda and T. Shigemasa, J. Chem. Soc. Perkin Trans I, 1989, 209.

344. C.C.K. Keh, V.V. Namboodri and R.S. Varma, Tetrahedron Lett., 2002, *43*, 4993.

345. S. Reformatsky, Ber., 1887, 20, 1210; J. Russ. Chem. Soc., 1890, 22, 44.

346. B. Han, P. Boudjouk, J. Org. Chem., 1982, 47, 5030.

347. A.K. Bose, K. Gupta and M.S. Manhas, J. Chem. Soc., Chem. Commun., 1984, 86; C. Petrier, L. Gemai and J. L. Luchi, Tetrahedron Lett., 1982, 3361.

348. T. Kitazume, Synthesis, 1986, 855.

349. H. Tanka, S. Kishigami and F. Toda, J. Chem., Soc., 1991, *56*, 4333.

350. M.W. Rathke, J. Org. Chem., 1970, *35*, 3966.

351. R. Heilmann and R. Glenat, Bull. Soc. Chim. Fr., 1955, 1586.

352. H. Mattes and C. Benezra, Tetredron Lett., 1985, 5697.

353. K.H. Meyer and K. Schuster, Ber., 1922, *55*, 819; H. Rupe and E. Kambli, Hel. Chem. Acta, 1926, *9*, 672; S. Swaminathan and K.V. Narayana, Chem. Rev., 1971, *71*, 429.

354. J. An, L. Bagnell, T. Cablewski, C.R. Strauss and R.W. Trainer, J. Org. Chem., 1997, *62(8)*, 2505.

355. H.E. Simmons and R.D. Smith. J. Am. Chem. Soc., 1958, *80*, 5323; H.E. Simmons, T.R. Carins, S.A. Vladuchick, and C.M. Hoiness, Organic Reactions, 1973, *29*, 1.

356. N.T. Castellucci and C.E. Griffin, J. Am., Chem., Soc., 1960, *82*, 4107.

357. E.J. Corey and H. Uda, J. Am. Chem., Soc., 1963, *85*, 1788.

358. O. Repic and S. Vogt, Tetrahedron Lett., 1982, *23*, 2729.

359. H. Tso, T. Chou and H. Hung, J. Chem. Soc. Chem. Commun., 1887, 1552.

360. C. Petrier, A.L. Gemal and J.L. Luck, Tetrahedron Lett., 1982, *23*, 3361.

361. R.J. Rawson and I.T. Harrison, J. Org. Chem., 1970, *35*, 2057.

362. W.G. Douben and A.C. Asheroft, J. Am. Chem. Soc., 1963, *85*, 3673.

363. C. Filliatre and C. Gueraud, C.R. Acad. Sci., *C*, 1971, *273*, 1186.

364. E. Wenkart, R.A. Mueller, E.J. Reardon, Jr., S.S. Sathe, D.J. Scharf and G. Tosi, J. Am. Chem. Soc., 1970, *92*, 7428.

365. S.D. Koch, J. Org. Chem., 1961, *26*, 3122.

366. S.D. Koch, R.M. Kliss, D.V. Lopiekes and R.J Wireman. J. Org. Chem., 1961, *26*, 3122.

367. L.K. Bee, J. Beeby, J.W. Everett and P.J. Garrat. J. Org. Chem., 1975, *40*, 2212.

368. R. Chinchilla and C. Nájera, Chem. Rev., 2007, *107*, 874.

369. P. Appukkutan, W. Dehaen and E.V. der Eycken, Eur. J. Org. Chem., 2003, 4713.

370. J. Gil-Moltó, S. Karlström and C. Nájera, Tetrahedron, 2005, *61*, 12168.

371. K.M. Dawood, W. Solodenko, A. Kirschning, ARKIVOC, 2007, 104; K.M. Dawood and A. Kirshning, Tetrahedron, 2005, *61*, 12121.

372. N.-E. Leadbeater, M. Marco and B.J. Tominack, Org. Lett., 203, *5*, 3919.

373. R.K. Arrela, N.E. Leadbeater, M.S. Sangi, V.A. Williams, P. Granados and R.D. Singer, J. Org. Chem., 2005, *70*, 161.

374. A. Anjaiah, S. Chandrasekhar and R. Gree, Adv. Synth. Catal., 2004, 346, 1329.

375. J.K. Stille, J.Am. Chem. Soc., 1984, 106, 4630; J. Org. Chem., 1990, 55, 3019; Angew. Chem. Int. Ed., 1986, 25, 508.

376. M. Kosugi and K. Fugani, J. Organomet. Chem., 2002, *653*, 50.

377. N. Kaval, K. Bisztray, W. Deharen, C.O. Kappe, and E. Van der Eycken, Mol. Diversity, 2003, *7*, 125.

378. S. Schneider and W. Bannwarth. Angew. Chem. Int. Edn., 2000, *39*, 4142.

379. S.T. Handy and X. Zhang, Org. Lett., 2001, *3*, 233.

380. M. Hoshino, P. Degenbolb and D.P. Curran, J. Org. Chem., 1997, *62*, 8341.

381. J.K. Stille, Angeus Chem. Int. Ed. Engl., 1986, *25*, 508.

382. A. Strecker, Ann., 1850, *75*, 25; 1954, *91*, 349; D.T. Mowry, Chem., Rev., 1948, *42*, 236.

383. D.T. Mowry, Chem. Rev., 1948, *42*, 189.

384. K. Weinges, Chem. Ber., 1971, *104*, 3594.

385. J. Menedez, G.G. Trigo and M.M. Solthuber, Tetrahedron Lett., 1986, *27*, 3285.

386. T. Hanafusa, J. Ichihara and T. Ashida, Chemistry Lett., 1987, 687.

387. D.R. Borthakur and J.S. Sandhu, J. Chem., Soc., Chem. Commun., 1988, 1444.

388. N.E. Leadbeater, Chem., Commun., 2005, 2881.

389. N. Miyaura and A. Suzuki, Chem. Rev., 1995, *95*, 2457.

390. N.E. Leadbeater and M. Marco, J. Org. Chem., 2003, *68*, 888.

391. L. Bai, J. –X. Wang and Y. Zhang, Green Chem., 2003, *5*, 615.

392. R.K. Arvela and N.E. Leadbeater, Org. Lett., 2005, *7*, 2101.

393. M.D. Crozet, C. Castera-Ducros and P. Vanelle, Tetrahedron Lett., 2006, *47*, 7061.

394. R. Zhu, F. Qu, G. Quéléverb and L. Peng, Tetrahedron Lett., 2007, *48*, 2389.

395. K.M. Dawood, Tetrahedron, 2007, *63*, 9642.

396. R.K. Arvela, N.E. Leadbeater, T.L. Mack and C.M. Kormos, Tetrahedron Lett., 2006, *47*, 217.

397. T. Welton, P.J. Smith, C.J. Mathew, 221st American Chemical Society National meeting, 1 EC-311, 2001.

398. C.J. Methew, P.J. Smith and T. Wellon, Chem., Commun., 2000, 1249.

399. R. Rajgopal, D.V. Jarikote and K.V. Srinivasan, Chem., Commun., 2002, 616.

400. V.V. Namboodiri and R.S. Varma, Green Chemistry, 2001, *3*, 146.

401. D. Clarke, M.A. Ali, A.A. Clifford, A. Parratt, P. Rose, D. Schwinn, W. Bannwarth and C.M. Rayner, Current Topics in Medicinal Chemistry, 2004, *4*, 740 and references cited therein.

402. F. Ullmann, Ann., 1904, *332*, 38; F. Ullmann and P. Sponagel, Ber., 1905, *38*, 2211; P.E. Fanta, Chem. Rev., 1946, *38*, 139.

403. J. Lindley, T.J. Manson and J.P. Lorimer, Ultrasonics, 1987, *25*, 45.

404. T. Yamashita. Y. Inouse, T. Konodo and H. Hashimoto, Chem., Lett., 1986, 407.

405. P. Boudjouk and B. Hans. Tetrahedron Lett., 1981, *22*, 3813.

406. P. Boudjouk, B. Hans and K.R. Anderson, J. Am. Chem. Soc., 1982, *104*, 4992.

407. T. Kitazume and N. Ishikawa, Chemistry Lett., 1981, 1679.

408. T. Kitazumi and N. Ishikawa, J. Am. Chem. Soc., 1985, *107*, 5186.

409. N. Ishikawa and T. Kituzume, European Patent, 0082 252 Al, 1982.

410. K. Garres, J. Org. Chem., 1970, *35*, 3273.

411. D.V. Davydov and I.P. Beletskaya, Russ. Chem. Bull., 1955, *44*, 1139.

412. S. Venkataraman and C.-J. Li, Org. Lett., 1999, 1, 1133.

413. S. Venkataraman, C.-J. Li, Tetrahedron Lett., 2000, *41*, 4831; S. Venkataraman, T. Huang and C.-J. Li, Adv. Synth. Catal; 2002, *344*, 399; S. Mukhopadhyay, A. Yaghmur, M. Baidossi, B. Kundi and Y. Sasson, Org. Process. Res. Dev., 2003, *7*, 641.

414. S. Mukhopadhyay, G. Rothenberg, D. Gitis and Y. Sasson, Org. Lett., 2000, *2*, 211.

415. S. Mukhopadhyay, G. Rothenberg and Y. Sasson, Adv. Synth. Catal., 2001, *343*, 274.

416. J. -H. Li, Y. -X. Xie and D. -L. Yin, J. Org. Chem., 2003, *68*, 9867; J. -H. Li, Y. -X. Xie, Chin. J. Chem., 2004, *22*, 966; J.-H. Li, Y.-X. Xie, H. Jiang, and M. Chen., Green Chem., 2002, *4*, 424.

417. T. Yamamoto, Can. J. Chem., 1983, *61*, 86.

418. B. Renger, Synthesis, 1985, 856.

419. W. Weiss and J.M. Edwards, Tetrahedron Lett., 1968, 4885.

420. C. Mannich and W. Krosche, Arch. Pharm. 1912, *250*, 647, F.E. Blike, Organic Reactions, 1942, *1*, 303.

421. A. W. Williamson, J. Chem. Soc., 1852, *4*, 229; O.C. Dermer, Chem. Revs., 1934, *14*, 409.

422. J. Jarrouse, C.R. Hebd, Scances Acad. Sci. Ser. C., 1951, *232*, 1424.

423. H.H. Freeman and R.A. Dubois, Tetrahedron Lett., 1975, 3251.

424. A. Merz, Angew Chem. Int. Ed. Engl, 1973, *12*, 846.
425. O. Gurney, J. Am. Chem. Soc., 1922, *44*, 1742.
426. H. Kaunowki, G. Cross and D. Seebach, Ber., 1981, *114*, 477.
427. G.S. Hiers and F.D. Hager, Org., Synth. 1941, *Coll. Vol. 1*, 58.
428. J.F. Norris and G. W. Rigby, J. Am. Chem. Soc., 1932, *54*, 2088.
429. R.C. Beier, B.P. Mundy and G.A. Strobel, Carbohyd. Res., 1983, *121*, 79.
430. For a general Treatise, *see* Cadogan: Organophosphorus Reagents in Organic Synthesis, Academic Press, N.Y. 1970. For a monograph *see* Johnson; Ylid Chemistry', Academic Press, N.Y. 1966; Reviews Bestmann and Vostrowsky, Top. Curr., Chem., 1983, *109*, 85, 164.
431. G. Markl and A. Merz, Synthesis, 1975, 295.
432. S. Hung and I. Stemmler, Tetrahedron Lett., 1974, 3151.
433. W. Tagaki, I. Inouse, Y. Yano and T. Okonoge; Tetrahedron Lett., 1974, 2587.
434. P.D. Croce, J. Chem. Soc., Perkin Trans, I, 1976, 619.
435. F. Bohlmann and C. Zdero. Chem. Ber., 1973, *106*, 3779.
436. M.J. Borenquer, J. Castells, J. Fernandiz and R.M. Galard, Tetrahedron Lett., 1971, 493.
437. G.P. Schiemenz, J. Becker and J. Stoeckigt, Chem. Ber., 1970, *103*, 2077.
438. V.L. Boulaire, K.N. West, C.L. Liotta, C.A. Eckert., Chem. Commun., 2001, 887.
439. Prbuzov, Pune Applied Chem., 1964, *9*, 307.
440. C. Piechuchi, Synthesis, 1976, 187.
441. M. Milolajczk, W. Grzejsezk, W. Midura and A. Zatoria, Synthesis. 1976, 397.
442. M. Mikolajezyk, S. Grzejszezk, W. Miclura and A. Zatorski, Synthesis, 1976, 396.
443. C. Piechucki, Synthesis, 1974, 869.
444. L.D. Incan and J. Seyden-Penne, Synthesis, 1975, 516.
445. M. Mikokajezuk, S. Grzejszezak, W. Modura and A. Zatorshi, Synthesis, 1975, 278.
446. A. Merz and G. Markl, Angew. Chem. Int. Ed. Engl. 1973, *12*, 845.
447. F. Toda and H. Akai, J. Org. Chem. 1990, *55*, 3446.
448. V. Le Boulaire and R. Gree, Chem. Commun., 2000, 2195.
449. H.J. Altenbach, Angew., Chem. Int., Ed., 1979, *18*, 940.
450. A. Wurtz, Ann. Chem. Phys., 1855, (*3*) *44*, 275; Ann., 1855, *96*, 364; R.E. Buntrock, Chem. Rev., 1968, *68*, 209.
451. B. Tollens and R. Fittig. Ann., 1864, *131*, 303; R. Fittig and J. Koneg., Ann., 1867, *144*, 277.
452. C.J. Li and T.H. Chan., Organometallics, 1991, *10*, 2548.
453. W.J. Balley, J. Org. Chem., 1962, *27*, 3088.

Chapter 3
Green Preparation

In this chapter, some organic preparations which are considered green synthesis and can be performed in a chemical laboratory are described. Few examples of the following types of reactions are as below:

1. Aqueous phase reactions
2. Solid-state (solventless) reactions
3. Photochemical reactions
4. PTC-catalysed reactions
5. Rearrangement reactions
6. Microwave-induced reactions
7. Enzymatic transformations
8. Sonication reactions
9. Esterification
10. Enamine reaction
11. Reactions in ionic liquids.
12. Green preparations using renewable resources.
13. Reactions using the principle of atom economy (avoiding waste).
14. Extraction of d-limonene from orange peels using liquid CO_2.

© The Author(s) 2021
V. K. Ahluwalia, *Green Chemistry*,
https://doi.org/10.1007/978-3-030-58513-6_3

3.1   Aqueous Phase Reactions

3.1.1   Hydrolysis of Methyl Salicylate with Alkali

Alkaline hydrolysis of esters is called *saponification* and is an irreversible process. It is mostly carried out by healthy ester with alkali.

Mechanism (Scheme 3.1)

Methyl salicylate on saponification gives sodium salicylate, which on acidification gives salicylic acid.

Materials

Methyl salicylate   2 ml
Sodium hydroxide 10%,   15 ml

Procedure

Reflux a mixture of methyl salicylate (2 ml) with sodium hydroxide solution (15 ml, 10%) in a round-bottomed flask using a reflux condenser or a sand bath (temp. 90–100 °C) for about 30 min till the ester layer disappears. Cool the solution and acidify with hydrochloric acid. Cool the resultant solution (ice bath). Filter the separated salicylic acid and recrystallize from hot water, it yields 1.2 g, m.p. 158–159 °C.

Notes

1.   Use of a crown ether, viz., [18] crown-6 in small amount for saponification gives quantitative yield. The special feature of using crown ether is that even the sterically hindered esters, which are difficult to saponify with alkali, can be saponified conveniently by using [18] crown 6. (C.J. Pedersen and K.K. Friensdorff, Angew Chem. Int. Engl., 1972, 11, 16; C. J. Pedersen, J. Am. Chem., Soc, 1967, 89, 2485, 7017; 1970, 92, 386, 391).
2.   Saponification of esters can be brought about by using microwave under solid–liquid PTC conditions without solvent. The reaction can be completed in few minutes (Scheme 3.2) (A. Loupy, P. Pigeon, M. Ramdani, P. Jaequault, Synthetic Common., 1994, 24(2), 159).

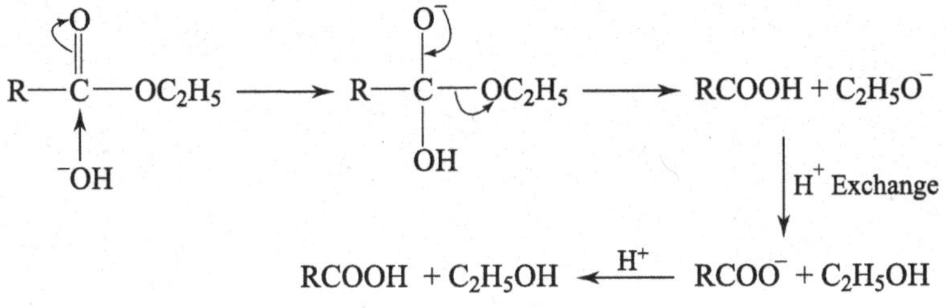

Scheme 3.1 Mechanism of saponification

Scheme 3.2
Microwave-assisted
saponification

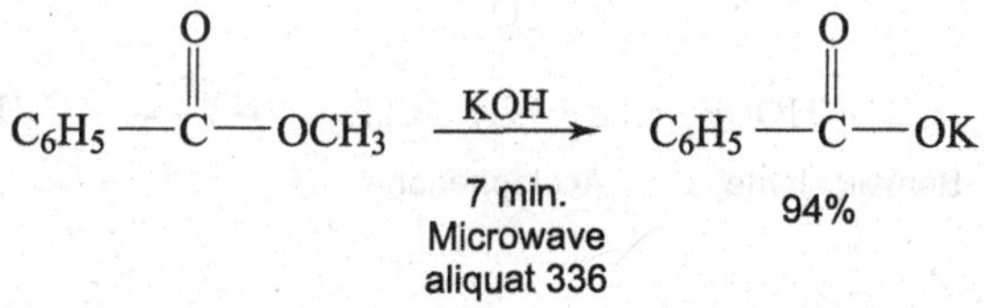

$$C_6H_5-\overset{\overset{\displaystyle O}{\|}}{C}-OCH_3 \quad\xrightarrow[\substack{7\ min.\\Microwave\\aliquat\ 336}]{KOH}\quad C_6H_5-\overset{\overset{\displaystyle O}{\|}}{C}-OK$$

94%

3. Ester hydrolysis or saponification can also be conducted under milder conditions when sonication is used (S. Moon, L. Duclin, J. V. Croney, Tetrahedron Lett., 1979, 3971).

In this case, rate increase is attributed to the emulsifying effect. A typical example is given in Scheme 3.3.

3.1.2 Chalcone

Crossed aldol condensation of benzaldehyde with a ketone (e.g., acetophenone) in the presence of alkali gives chalcone (Scheme 3.4).

Materials

Benzaldehyde	2.2 ml
Acetophenone	2.5 ml
Rectified spirit	7 ml
Sodium hydroxide solution	1.1 g in 10 ml H_2O

Procedure

Place benzaldehyde (2.2 ml) into a conical flask (100 ml capacity) and add acetophenone (2.5 ml) and rectified spirit (7 ml). Stir the mixture, so as to obtain a homogeneous solution. If necessary, the mixture may be warmed to get a clear solution.

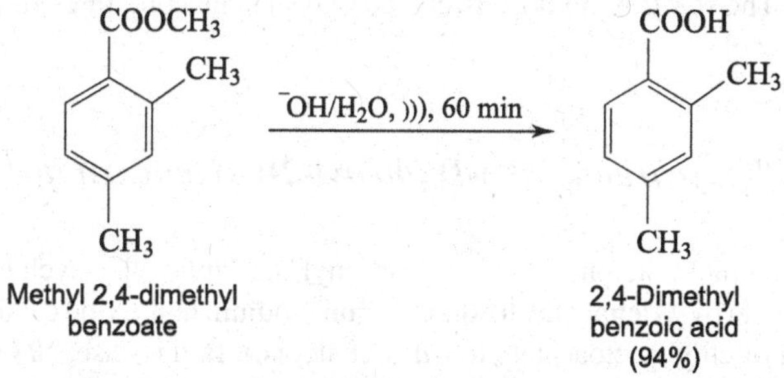

Methyl 2,4-dimethyl benzoate $\xrightarrow[]{^-OH/H_2O,\ ))),\ 60\ min}$ 2,4-Dimethyl benzoic acid (94%)

Scheme 3.3 Saponification using sonication

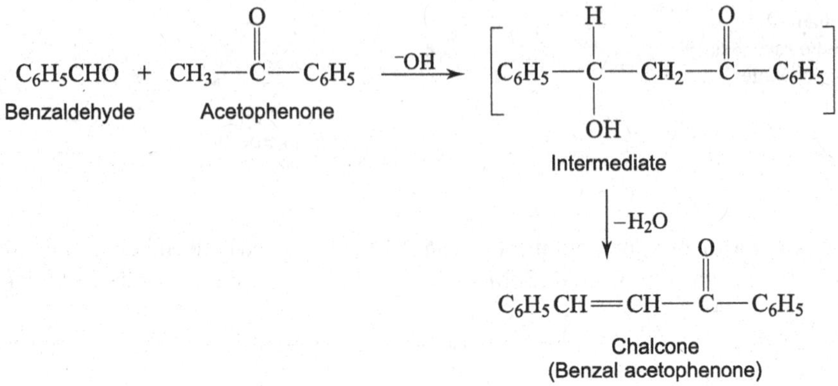

$$C_6H_5CHO + CH_3-\overset{\overset{\displaystyle O}{\|}}{C}-C_6H_5 \xrightarrow{^-OH} \left[C_6H_5-\overset{\overset{\displaystyle H}{|}}{\underset{\underset{\displaystyle OH}{|}}{C}}-CH_2-\overset{\overset{\displaystyle O}{\|}}{C}-C_6H_5 \right]$$

Benzaldehyde   Acetophenone

Intermediate

$$\downarrow -H_2O$$

$$C_6H_5\,CH=CH-\overset{\overset{\displaystyle O}{\|}}{C}-C_6H_5$$

Chalcone
(Benzal acetophenone)

Scheme 3.4 Synthesis of chalcone

To get the clear solution at room temperature, add sodium hydroxide solution (1.1 g NaOH in 10 ml H_2O). Stir the mixture. Keep the mixture overnight in a refrigerator. Filter the separated chalcone, wash with cold water and recrystallize from alcohol. Yield 3.35 g (86.6%), m.p. 56–57 °C.

Notes

1. Follow the same procedure and using veratraldehyde (3, 4-methylenedioxy benzaldehyde), p-anisaldehyde and 3-nitrobenzaldehyde in place of benzaldehyde prepare 3,4-methylenedioxy chalcone (m.p. 112 °C), 4-methoxychalcone (m.p. 74 °C) and 3-nitrochalcone (m.p. 146°), respectively.
2. Chalcones are very conveniently obtained by the use of microwave ovens. There is considerable reaction rate enhancement and the reaction could be completed in 30 s to 2 min time (R. Gupta, A.K. Gupta, S. Paul, P. L. Kachroo, Indian J Chem., 1995, 34 B, 61). In this procedure, a solution of ketone and aromatic aldehyde in dry ethanol with catalytic amount of sodium hydroxide (1–2 pellet) is heated in a microwave oven for 30 s to 2 min (at 210 watts, 30% microwave power). The reaction mixture is cooled and the chalcone filtered, and washed with ethanol.

3.1.3   6-Ethoxycarbonyl-3,5-Diphenyl-2-Cyclohexenone[1]

The α,β-unsaturated ketone, 6-ethoxycarbonyl-3,5-diphenyl-2-cyclohexenone is prepared (note 1) by **Michael addition reaction** (sodium hydroxide catalysed conjugate addition of ethyl acetoacetate to *trans*-chalcone). In this case, NaOH serves as a source of $^-$OH to catalyse the reaction. The Michael adduct obtained above on base-catalysed **aldol condensation reaction** gives (note 2) stable six-membered ring compound (6-ethoxycarbonyl-3,5-diphenyl-3-hydroxy-cyclohexanone).

Finally, the aldol intermediate is dehydrated to the required 6-ethoxycarbonyl-3.5-diphenyl-2-cyclohexenone. Various steps are given below (Scheme 3.5).

Materials Required

trans-chalcone	0.72 g
ethyl acetoacetate	0.45 g
absolute ethanol	15 ml
sodium hydroxide solution (2.2 M)	0.75 ml

Procedure[1]

Add finely powdered *trans*-chalcone (0.72 g) to a round-bottomed flask (50 ml capacity), followed by the addition of ethyl acetoacetate (0.45 g) and absolute ethanol (15 ml). Swirl the flask until the solid almost dissolves. Add sodium hydroxide solution (2.2 m, 0.75 ml) and heat the flask which had been fitted with a reflux condenser on a hot plate. After the mixture starts boiling gently, continue reflux for 1 h when

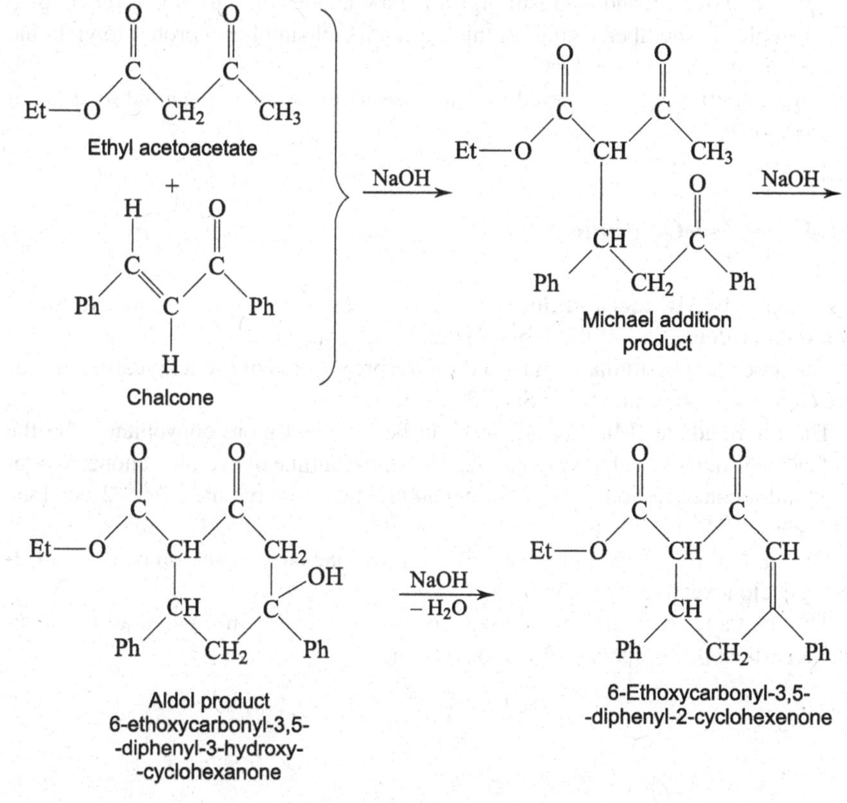

Scheme 3.5 Synthesis of 6-ethoxycarbonyl-3, 5-diphenyl-2-cyclohexenone

the mixture becomes turbid. Cool the reaction mixture to room temperature, scratch the sides of the flask with a stirring rod to induce crystallization. Cool the mixture in an ice-bath for an hour. Filter the separated product using ice-cold water (2–3 ml) for transfer. Dry the crystals in the air (overnight) or at 75–80 °C for 30 min.

To the separated solid in a test tube, add acetone (4–5 ml) and stir the mixture. Most of the solid dissolves leaving behind some alkali. Separate the clear solution by centrifugation or filtration using filter acid. Evaporate the acetone from the clear solution. The oil produced solidifies on scratching with a glass rod. Crystallize from ethanol (4–5 ml). Weigh the formed 6-ethoxycarbonyl-3,5-diphenyl-2-cyclohexenone. m.p. 111–112 °C.

If possible, record 1 R spectra. You should observe absorbance at 1734 and 1660 cm^{-1} for the ester carbonyl and enone group, respectively.

Notes

1. The method is described by A. Garcia Ruso, J. Garcia–Raso, J.V. Sinisterra and R. Mestres, Michael Addition and Aldol condensation. "A simple teaching model for organic laboratory" Journal of Chemical Education, 1986, 63, 443.
2. From the Michael addition products, the methyl group loses a proton in the presence of base and the resulting carbanion attacks the carbonyl group to give a stable six-membered ring; in this case, ethanol supplies a proton to yield the aldol intermediate product.
3. Since the reaction is carried out in aqueous phase, it is regarded as a green synthesis.

3.1.4 $\Delta^{1,9}$-2-Octalone

It is obtained by **Michael addition** of methyl vinyl ketone to cyclohexanone followed by **aldol condensation** of the formed adduct (Scheme 3.6).

The procedure is similar to that used for the preparation of 6-ethoxycarbonyl-3,5-diphenyl-2-cyclohexenone (see Sect. 3.1.3).

The intermediate (Michael adduct) can be obtained more conveniently by the Michael addition of methyl vinyl ketone with the enamine of cyclohexanone. Subsequent aldol condensation of the Michael adduct gives the required $\Delta^{1,9}$–2-octalone (1) (Scheme 3.7).

For method of preparation of enamine at cyclohexanone, see preparation of 2-acetyl cyclohexanone (Sect. 3.10.1).

The overall procedure is similar to that used for the preparation of 6-ethoxycarbonyl-3,5-diphenyl-2-cyclohexenone.

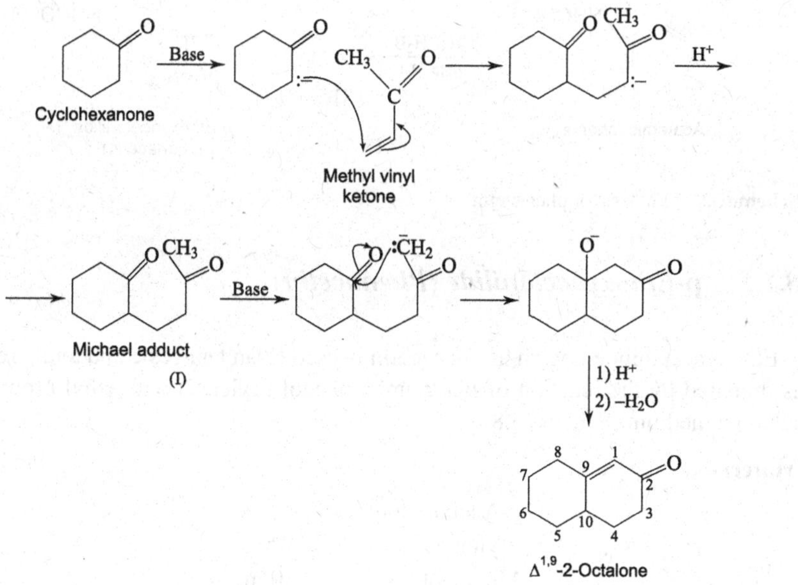

Scheme 3.6 Synthesis of $\Delta^{1,9}$-2-octalone

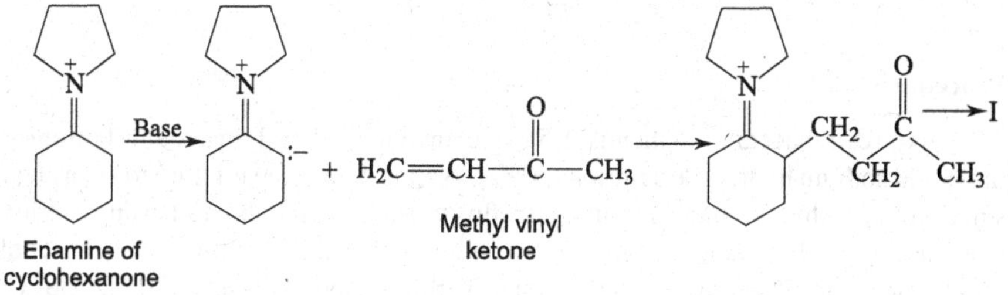

Scheme 3.7 Robinson annulation reaction

Note

1. Reactions that combine the Michael addition reaction and aldol condensation (as in the above case) to give a six-member ring fused on another ring are mostly used in steroid field and are known as **Robinson annulation reactions**.

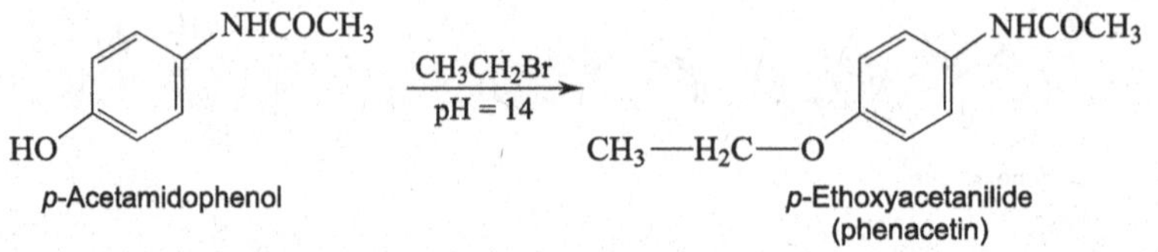

Scheme 3.8   Synthesis of phenacetin

3.1.5   p-*Ethoxyacetanilide (Phenacetin)*

p-Ethoxyacetanilide known as phenacetin is used as an analgesic and antipyretic. It is prepared by the reaction of p-acetamidophenol (Tylenol) with ethyl bromide in alkaline medium (Scheme 3.8).

Materials

p-Acetamidophenol	
(Tylenol)	1.5 g
Methanol	10 ml
Sodium hydroxide	0.63 ml
Solution (50%)	
Ethyl bromide	1.5 ml

Procedure

To a mixture of p-acetamidophenol (1.5 g) and methanol (10 ml) in a round-bottomed flask, add sodium hydroxide solution (50%, 0.63 ml). Shake the mixture to get a clear solution. Add ethyl bromide (1.5 ml) and reflux the mixture for about 2 h using a reflux condenser. Add hot water (20 ml). Leave the reaction mixture for overnight, cool (cold water) and filter the separated product. Yield 3 g (about 80%), m.p. 137–138 °C.

Note

p-Ethoxyacetanilide can also be prepared by ethylation of p-aminophenol followed by acetylation with acetic anhydride at a pH of about 5.

3.1.6   p-*Acetamidophenol (Tylenol)*

p-Acetamidophenol, known as Tylenol, is used as an analgesic. It is prepared by the acetylation of p-aminophenol with acetic anhydride at a pH of about 5.

Synthesis of tylenol

Materials

p-Aminophenol	1.1 g
Acetic anhydride	1.1 ml
Conc. hydrochloric acid	0.9 ml
Sodium acetate (anhydrous)	1 g

Procedure

To a solution of p-aminophenol (1.1 g) in water (10 ml) containing conc. hydrochloric acid (0.9 ml) in a conical flask, add acetic anhydride (1.1 ml). Shake the mixture so that all the acetic anhydride dissolves. Finally add a solution of anhydrous sodium acetate (1 g) in water (6 ml), stir the solution and leave it to stand overnight. Cool the solution in ice-bath, filter the separated product. Crystallize from hot water. Yield 1.1 g (75%), m.p. 169 °C.

Notes

1. p-Aminophenol should be pure. If it is dark-coloured, suspend it in methanol and filter.
2. Anhydrous sodium acetate is used, since it is easily soluble in water than the trihydrate.
3. Tylenol can also be prepared directly by the acetylation of p-aminophenol with acetic anhydride in aqueous medium, i.e., the use of HCl and NaOAc can be avoided.

3.1.7 Vanillideneacetone

It is obtained by the **Claisen–Schmidt reaction** of vanillin with acetone under basic conditions.

Mechanism

Mechanism of the synthesis of vanillidene acetone

Materials

Vanillin	3.04 g
Acetone	12 ml
Sodium hydroxide solution (10%)	9 ml

Procedure

Dissolve vanillin (3.04 g) in acetone (12 ml) in a conical flask. To this solution, add sodium hydroxide solution (10%, 9 ml). The solution which was yellow in the beginning, gradually darkens and becomes deep red in colour. Keep the reaction mixture for 72 h with occasional shaking. Add water (50 ml) while stirring vigorously and then acidify by adding hydrochloric acid (10%, 15 ml). Filter the resulting product, wash with water and dry. Recrystallize from 1:1 ethanol–water (10–12 ml). Yield 3.3 g (85%). Record its m.p.

3.1.8 2,4-Dihydroxybenzoic Acid (β-Resorcylic Acid)[1]

It is prepared from resorcinol, potassium bicarbonate and carbon dioxide gas (Scheme 3.9).

Materials

Resorcinol	15 g
Potassium bicarbonate	75 g
Water	150 ml

Procedure[1]

Heat the mixture of resorcinol (15 g), potassium bicarbonate (75 g) and water (150 ml) in a three-necked round-bottomed flask (250 ml capacity) fitted with a reflux condenser and a gas inlet tube. Heat the mixture gently on a steam bath for 3 h and reflux for 30 min, while passing a rapid stream of carbon dioxide gas through the solution. Acidify the hot solution by adding concentrated hydrochloric acid (45 ml)

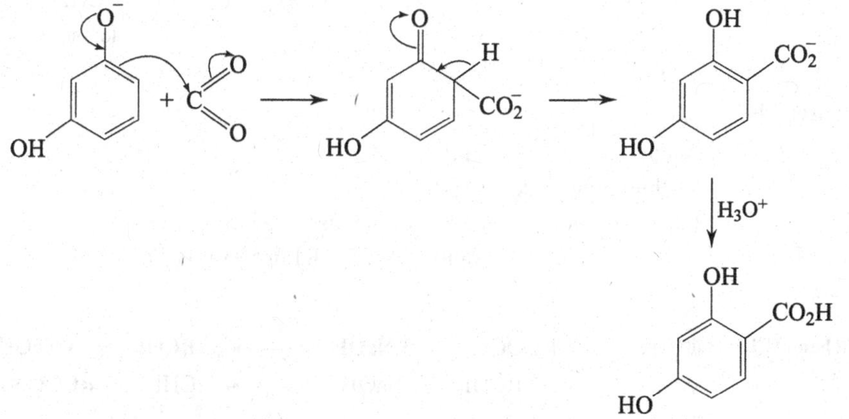

Scheme 3.9 Synthesis of β-resorcylic acid

from a separatory funnel with a long tube delivering the acid to the bottom of the flask. Cool the flask to room temperature and then in ice-bath. Filter the separated β-resorcylic acid. Yield 9 g (64%), m.p. 216–17 °C.

Note

1. Nierensltein, Clibbens, Org. Syn. Coll. Vol II, 1943, 557.

3.1.9 Iodoform

It is prepared by the well-known **haloform reaction**, (A. Lieben, Ann (Suppl.) 1870, 218) which consists of the cleavage of methyl ketones (CH_3–CO–R), acetaldehyde, ethanol and secondary methyl carbinols (CH_3CHOR) with halogens (mostly iodine) and a base to give haloform (iodoform if iodine is used) and carboxylic acids (Scheme 3.10).

From Acetone

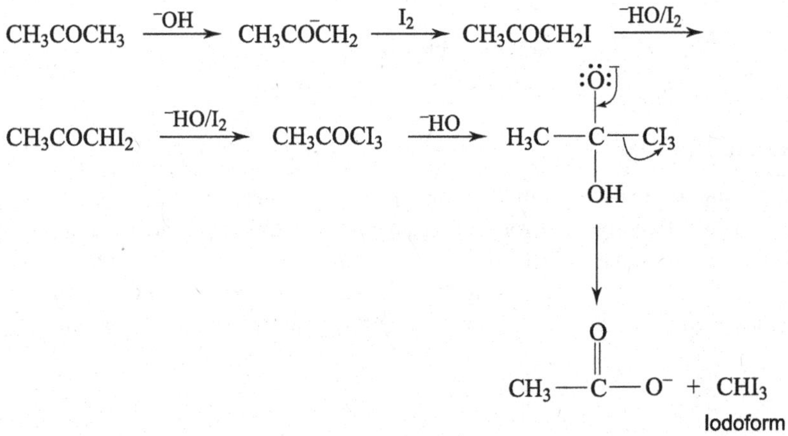

$$CH_3COCH_3 \xrightarrow{\ ^-OH\ } CH_3CO\bar{C}H_2 \xrightarrow{\ I_2\ } CH_3COCH_2I \xrightarrow{\ ^-HO/I_2\ }$$

$$CH_3COCHI_2 \xrightarrow{\ ^-HO/I_2\ } CH_3COCI_3 \xrightarrow{\ ^-HO\ } H_3C-\overset{\overset{\displaystyle :\ddot{O}:^-}{|}}{\underset{\underset{\displaystyle OH}{|}}{C}}-CI_3$$

$$CH_3-\overset{\overset{\displaystyle O}{\|}}{C}-O^- + CHI_3$$

Iodoform

Materials

Acetone	3 ml
Sodium hydroxide	10% 15 ml
Iodine solution	12.5 g iodine dissolved in a solution of 25 g KI in 50 ml H_2O

Scheme 3.10 Halo form reaction

$$RCOCH_3 + 3\,NaOI \longrightarrow RCOI_3 + 3\,NaOH$$
$$RCOI_3 + NaOH \longrightarrow CHI_3 + RCOONa$$

Procedure

To a solution of acetone (3 ml) in water (30 ml) and sodium hydroxide solution (10%, 15 ml), add iodine solution (12.5 g iodine dissolves in a solution of 25 g potassium iodide in 50 ml H_2O). Heat the reaction mixture at 60 °C (water bath). Continue heating till the precipitates of iodoform settles down. Filter the yellow crystals of iodoform and crystallize from dilute methanol. Yield 5 g (31.2%), m.p. 119 °C.

Note

1. Iodoform can also be conveniently obtained from ethyl alcohol.

3.1.10 Endo-cis-1,4-Endoxo–Δ^5-Cyclohexene-2,3-Dicarboxylic Acid

It is obtained by **Diels–Alder reaction** of furan with maleic acid or maleic anhydride in water at room temperature (Scheme 3.11).

Materials

Furan   2.5 g

Maleic acid 5 g

Procedure[1]

Stir furan (2.5 g) with maleic acid (5 g) in water (25 ml) for 2–3 h at room temperature. Filter the separated adduct, and wash with water. Yield 7.2 g. Record its m.p.

Note

1. R.B. Woodward and Harold Baer, J. Am. Chem. Soc., 1948, *70*, 1161, R. Breslow and D. Rideout, J. Am. Chem. Soc., 1980, *102*, 7816.

Scheme 3.11 Synthesis of endo-cis-1,4-endoxo-Δ^5-cyclohexene-2,3-dicarboxylic acid

Maleic anhydride

Hot water

Furan

Maleic acid

H_2O RT

Endo-*cis*-1,4-endoxo-Δ^5-cyclohexene-2,3--dicarboxylic acid

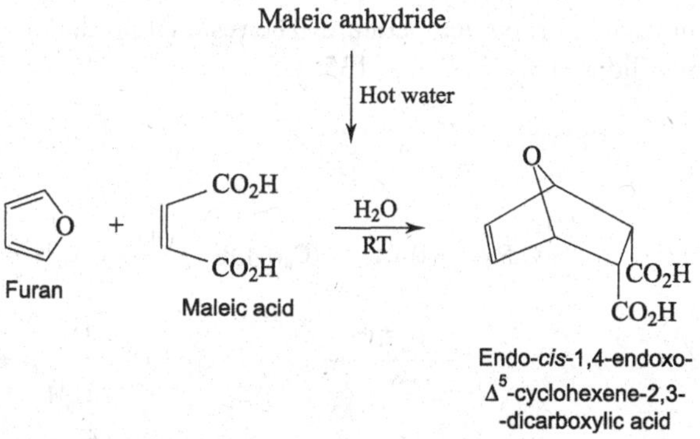

3.1.11 Trans Stilbene

Trans stilbene was earlier prepared (G. Markl and A. Merz, Synthesis, 1973, 295) by the reaction of benzaldehyde with benzyltriphenyl phosphonium chloride (Scheme 3.12).

It has now been possible to carry out the above reaction in the presence of aqueous sodium hydroxide (John C. Warner, Paul T. Anastas and Jean–Pierre Anselme, J. Chem. Education 1985, *62*, 346).

Materials

Benzaldehyde	4.24 g
Benzyltriphenyl phosphonium chloride (see note 1)	15.72 g
Methylene chloride	20 ml
Sodium hydroxide solution	50%, 20 ml

Procedure

Take benzaldehyde (4.24 g) and benzyltriphenyl phosphonium chloride (15.72 g) in a three-neck round-bottom flask (250 ml capacity) filled with a thermometer, a condenser and a separatory funnel. Add methylene chloride (20 ml), stir the mixture (magnetic stirrer) and add aqueous sodium hydroxide solution (50%, 20 ml) slowly (through the separatory funnel). The temperature rises to 50 °C. Keep the reaction mixture at 50 °C for 30 min with vigorous stirring. The mixture turns yellow and cloudy. Separate the organic layer, wash with water (25 ml) and then with sodium bicarbonate solution (40 ml) and finally with water. Dry the extract over sodium sulphate, filter and evaporate to dryness. To the residue, add absolute ethanol (30 ml) and cool the solution in ice-bath for 15 min. Filter the separated *trans*-stilbene (2 g), m.p. 122–123 °C.

Evaporate the filtrate, add petroleum ether (b.p. 40–60 °C, 40 ml). Filter the precipitated triphenyl phosphine oxide (10 g), m.p. 146–47 °C. Evaporate the filtrate *in vacuo* at room temperature. The residual product consisted of *cis*-stilbene (3 g); is solidified at –5 °C, b.p. 135 °C.

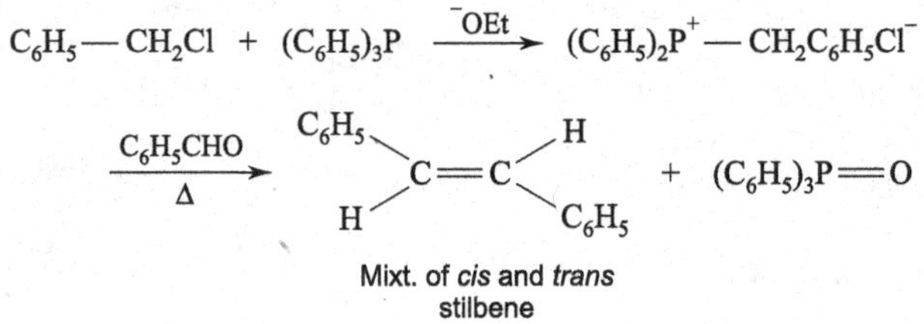

$$C_6H_5-CH_2Cl \ + \ (C_6H_5)_3P \ \xrightarrow{\ ^-OEt\ } \ (C_6H_5)_2\overset{+}{P}-CH_2C_6H_5Cl^-$$

$$\xrightarrow[\Delta]{C_6H_5CHO} \quad \underset{H}{\overset{C_6H_5}{\diagdown}}C=C\underset{C_6H_5}{\overset{H}{\diagup}} \ + \ (C_6H_5)_3P=O$$

Mixt. of *cis* and *trans* stilbene

Scheme 3.12 Synthesis of trans stilbene

Notes

1. Benzyl triphenyl phosphonium chloride is prepared by refluxing a mixture of benzyl chloride (4.4 ml), triphenyl phosphine (14.3 g) and xylene (70 ml). The separated product is filtered, m.p. 310–11 °C. It is washed with xylene and dried in vacuum desiccator.
2. If possible take the NMR spectra of *cis* and *trans* stilbene and characterize both compounds.
3. The above procedure of Wittig condensation can also be performed with *trans*-cinnamaldehyde, *p*-methyl-, *p*-methoxy-, *p*-chloro-, *m*- and *p*-nitrobenzaldehydes and 9-anthracenealdehyde. The product is obtained with 65–85% yield (G. Markl and A. Merz, Synthesis, 1973, 295).

3.1.12 2-Methyl-2-(3-Oxobutyl)-1,3-Cyclopentanedione

It is prepared by the reaction of 2-methylcyclopentane-1,3-dione and methyl vinyl ketone in water, and is 87% yield. The reaction is considered a **Michael addition** under acidic reaction conditions due to the enolic nature of the dione (Z.G. Hojas and D.R. Parrish, J. Org. Chem., 1974, *12*, 39). (Scheme 3.13)

Materials

2-methylcyclopentane-1,3-dione 6.5 g

Methyl vinyl ketone   9.6 ml

Procedure

To a suspension of 2-methylcyclopentane-1,3-dione (6.5 g) in water (14 ml), add methyl vinyl ketone (9.6 ml). Stir the mixture under nitrogen at 20 °C for 5 days. Extract with benzene, treat with Na_2SO_4, charcoal and $MgSO_4$. Filter the solids, extract with hot benzene (20 ml) and evaporate the combined benzene extract. The oily product (10 g) consisted of the required 2-methyl-2-(3-oxobutyl)-1,3-cyclopentanedione. Characterize the product on the basis of IR spectra (1770 and 1725 cm^{-1}) and NMR spectral data [δ 1.12 (s, 3, 2, –CH$_3$), 2.22 (s, 3, CH$_2$CO$^-$) 2.82 (m, 4, –COCH$_2$CH$_2$CO$^-$)].

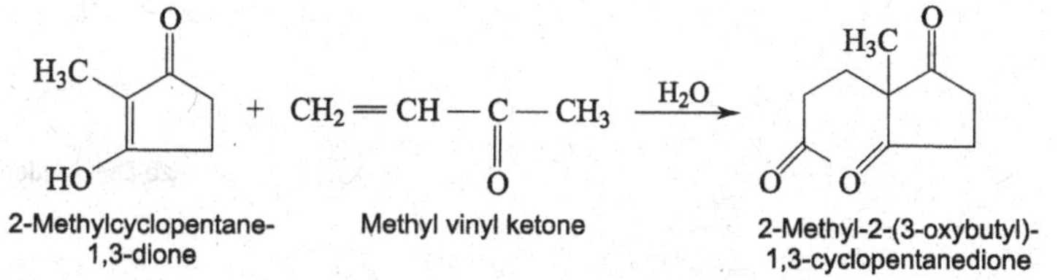

Scheme 3.13 Synthesis of 2-methyl-2-(3-oxobutyl)-1,3-cyclopentanedione

Note

1. 2-methyl-2-(3-oxobutyl)-1,3-cyclopentanedione was earlier reported to be prepared by the reaction of the dione and methyl vinyl ketone as a solid, m.p. 117–118 °C (C.B.C. Boycl and J.S. Whitehurst, J. Chem. Soc., 1959, 2022). However, the compound was not the required product (Z.G. Hajos and D. R. Parrish, J. Org Chem., 1974, 1612).

3.1.13 Hetero Diels–Alder Adduct

Hetero Diels–Alder reaction is useful for the synthesis of heterocyclic compounds with nitrogen or oxygen containing dienophiles. In this reaction (Note 1), an iminium salt (generated in situ under Mannich-like conditions) reacted with dienes in water to give aza-Diels–Alder reaction products (Scheme 3.14).

Materials

Benzylamine hydrochloride	11.7 g
Formalin 40%	25 ml
Cyclopentadiene	13 g

Procedure

Cyclopentadiene (13 g) was added to a stirred mixture of benzylamine hydrochloride (11.7 g) and formalin (25 ml). The heterogeneous reaction mixture was vigorously stirred for 3 h at room temperature. The separated adduct was obtained in quantitative yield (bicyclic amine). Record the NMR spectra of the product.

Notes

1. Scot D. Larnsen and Paul A. Grieco, J. Am. Chem. Soc. 1985, *107*, 1768.
2. This procedure can be used for large number of bicyclic compounds.

Scheme 3.14 Hetero Diels–Alder adduct

$$C_6H_5CH_2NH_2 \cdot HCl \xrightarrow[\text{H}_2\text{O}]{\text{HCHO}} \left[C_6H_5CH_2\overset{+}{N}{=}CH_2Cl^- \right]$$

Aza-Diels-Alder adduct

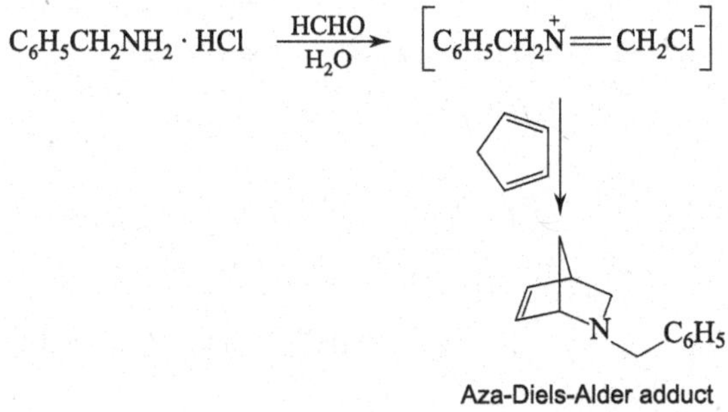

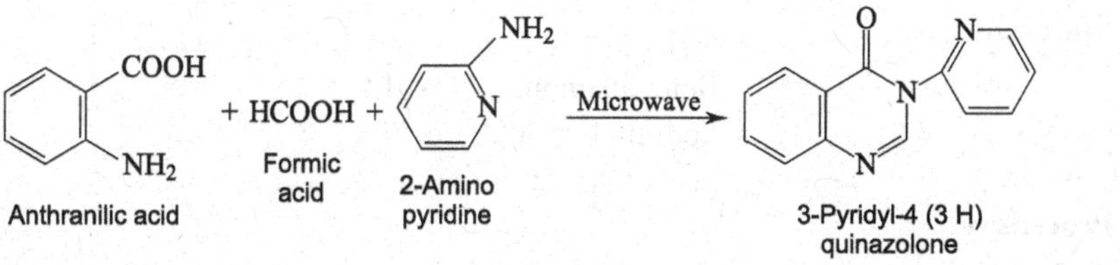

Scheme 3.15 Solid-state synthesis of 3-pyridyl-4 (3H) quinazolone

3.2   Solid State (Solventless) Reactions

3.2.1   3-Pyridyl-4 (3H) Quinazolone[1]

It is obtained by the reaction of anthranilic acid with formic acid and 2-aminopyridine under microwave irradiation (Scheme 3.15).

Materials

Anthranilic acid	1.26 g
Formic acid	5 g
2-Aminopyridinie	0.92 g

Procedure

A mixture of anthranilic acid (1.26 g), formic acid (5 g) and 2-amino pyridine (0.92 g) was heated in a microwave oven for 4 min. The product obtained melted at 156–157 °C (92% yield).

Notes

1. The procedure adopted is the one described by M. Kidwai, S. Rastogi, R. Mohan and Ruby, Croatica, Chemica Acta, 2003, 76 (*4*), 365.
2. The methodology in environmentally benign and completely eliminates the need of solvent from the reaction.

3.2.2   Diphenylcarbinol

It is obtained by reduction of benzophenone with sodium borohydride in solid state (Scheme 3.16).

Scheme 3.16 Solid-state
synthesis of diphenylcarbinol

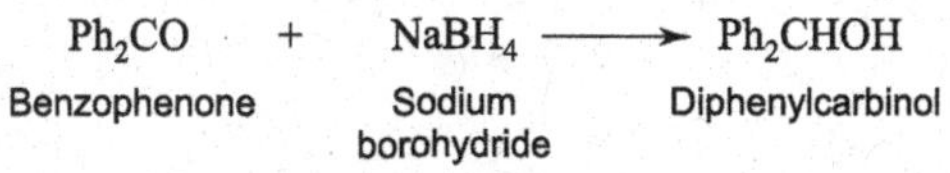

$$Ph_2CO \quad + \quad NaBH_4 \longrightarrow Ph_2CHOH$$

Benzophenone   Sodium   Diphenylcarbinol
borohydride

Materials

Benzophenone   1.8 g
Sodium borohydride   4.8 g

Procedure[1]

A mixture of powdered benzophenone (1.8 g) and sodium borohydride (4.8 g) was kept in a dry box at room temperature with occasional mixing and grinding using an agate mortar and pestle for 5 days. Diphenyl carbinol was obtained with 100% yield.

Note

1. The procedure is that described by K. Tanlka and F. Toda, Chem. Rev., 2000, *100*, 1028.

3.2.3   Phenylbenzoate

It is obtained by **Baeyer–Villiger oxidation** of benzophenone with *m*-chloroperbenzoic acid in solid state (Scheme 3.17).

Materials

Benzophenone   1.8 g
m-Chloroperbenzoic acid   3.5 g

Procedure[1]

Keep a powdered mixture of benzophenone (1.8 g) and *m*-chloroperbenzoic acid (3.5 g) at room temperature for 24 h. Macerate the product with sodium bicarbonate solution and extract with ether. Distillation of the dried ether extract gave phenyl benzoate with 85% yield.

Notes

1. The procedure is that described by K. Tanka and F. Toda, Chem. Rev., 2000, *100*, 1028.
2. Using phenyl benzyl ketone ($PhCOCH_2Ph$), the product, benzylbenzoate, $PhCOOCH_2$ Ph, is obtained with 97% yield.
3. The conventional Baeyer–Villiger oxidation in chloroform gives low yields (40–50%) of the product.

Scheme 3.17 Solid-state synthesis of phenylbenzoate

$$C_6H_5COC_6H_5 \xrightarrow{\text{\textit{m}-Chloroperbenzoic acid}} C_6H_5COOC_6H_5$$

Benzophenone   Phenylbenzoate

3.2.4 Azomethines

These are Schiff's bases and are obtained by the reaction of a carbonyl compound with an substituted amine (Scheme 3.18).

Generally, azomethines are prepared by condensing together equimolar quantities of aromatic aldehydes and substituted aniline using toluene as solvent in the presence of anhyd. $ZnCl_2$ at about 160° C for about 30 min. The product azomethine is isolated by neutralization and crystallization (Scheme 3.19).

Procedure

A mixture of 4-aminotoluce (2.8 m mole) and 4 hydroxy-3-methoxybenzaldehyde is finally ground using a mortar and pestle. The reaction mixture becomes liquid within a minute or so and then solidifies to give a crystalline solid within the next 8–10 min. The water produced is removed at 80 °C under vacuo. Yield at 95 °C, m.p. 117 °C.

Notes

1. In 1R, azomethine showed a characteristic band at about 1625 cm^{-1} due to
   $$-C=N-$$
   $$|$$
2. On TLC examination, it was shown that the reaction is complete.
3. Other aromatic aldehydes also reacted with substituted amines to give the corresponding azomethine.

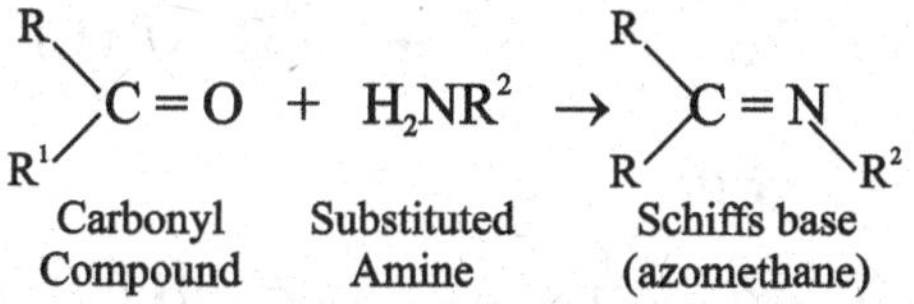

Scheme 3.18 Synthesis of azomethane

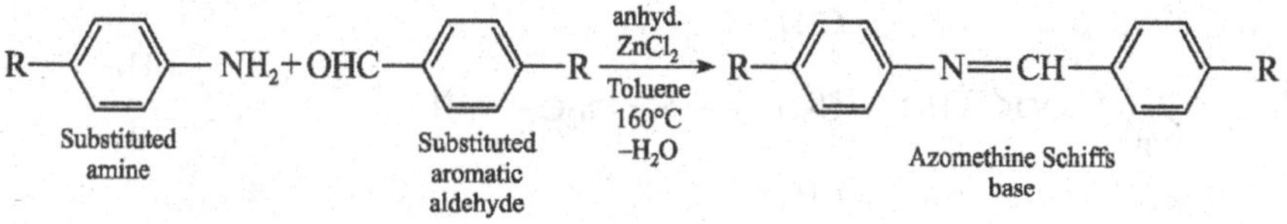

Scheme 3.19 Earlier synthesis of azomethine

3.3   Photochemical Reactions

3.3.1   *Benzopinacol*

Benzopinacol is obtained by photo reduction of benzophenone. It is a dimerization reaction brought about by exposing a solution of benzophenone in isopropyl alcohol to natural light (Scheme 3.20).

Mechanism (Scheme 3.21)

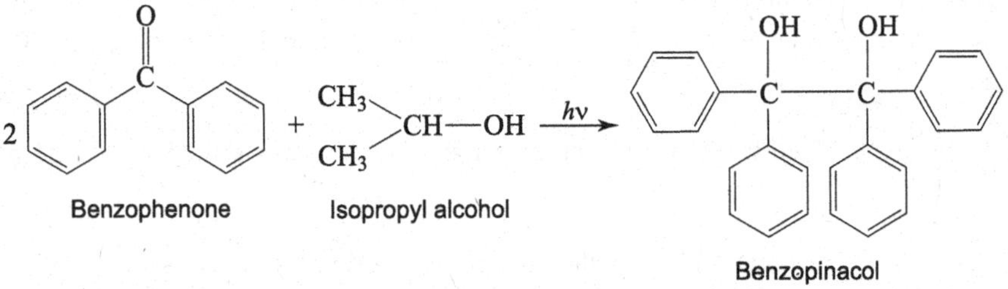

Scheme 3.20 Photochemical synthesis of benzopinacol

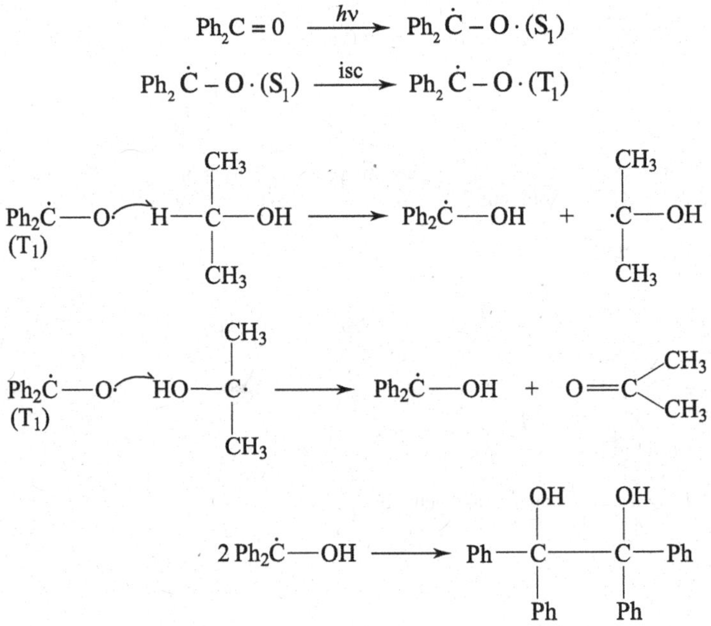

Scheme 3.21 Mechanism of formation of benzophenone

Materials

Benzophenone   2.5 g

Isopropyl alcohol   10 ml

Procedure[1]

Dissolve benzophenone (2.5 g) in isopropyl alcohol (10 ml) in a round-bottomed flask (25 ml capacity). Add glacial acetic acid (0.2 ml) and fill the flask with isopropyl alcohol. Stopper the flask tightly and place the flask in bright sunlight. The reaction takes about a week for completion (3 h with a sun lamp). The product gets separated. It is filtered and crystallized from glacial acetic acid. Yield 98%, m.p. 185–186 °C.

Notes

1. It is a green reaction since it is carried out photochemically and there are no by-products formed.
2. The reaction can be completed in a much shorter time (3–4 h) by using a UV lamp.

3.3.2 Conversion of Trans-Azobenzene to Cis-Azobenzene

The azobenzene as ordinarily obtained is the *trans*-isomer. It can be photo chemically isomerised to the *cis* form. This conversion can be achieved by irradiation with sunlight or with a fluorescent lamp (Scheme 3.22).

Materials

trans-Azobenzene   1 g   (see note 2)

Petroleum ether   50 ml

b.p. 40–60 °C

Procedure

Dissolve *trans*-azobenzene (1 g) in petroleum ether (50 ml) in a round-bottomed flask (100 ml capacity). Place the flask in bright sunlight. The reaction takes 4–5 days for completion (3 h with UV lamp irradiation). Meanwhile, prepare a chromatographic column from activated acid alumina grade I (50 g) as a slurry in petroleum ether. After the column is ready, place a circular filter paper on the top of the column (so that the alumina is not disturbed). Slowly pour the irradiated solution into the column taking care not to disturb the alumina in the column Elute and the column with petroleum

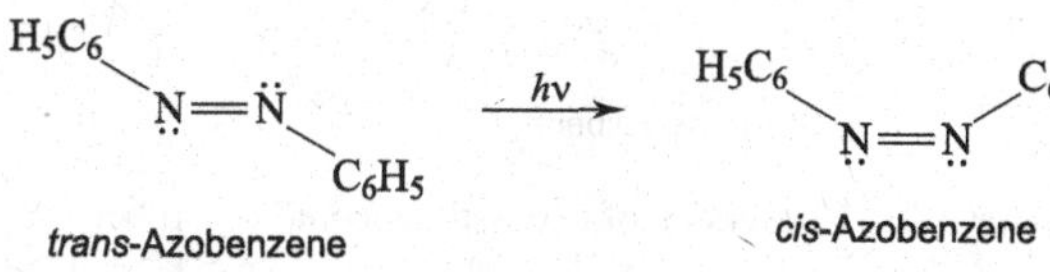

Scheme 3.22 Conversion of *trans*-azobenzene into *cis*-azobenzene

ether (b.p. 40–60 °C). A sharp-coloured band (about 2 cm) of *cis* azobenzene appears at the top of the column, while the diffuse coloured region of the *trans* form moves down the column. Protect the upper band by covering with a carbon paper (so that the *cis* form is not reconverted into *trans*). When the chromatography is complete, extrude the column, cut the upper band and shake with petroleum ether (100 ml) containing methanol (1 ml). Filter the solution, wash the filtrate with water (2 × 15 ml), dry the solution over anhydrous sodium sulphate and evaporate the solution. The residual coloured product (m.p. 71.5 °C) is pure *cis*-azobenzene.

Notes

1. The purity of *cis* azobenzene can be confirmed by recording the UV absorption spectrum in ethanol solution as soon as possible after its isolation. The *cis*-azobenzene has λ_{max} 281 nm (ε 5260); *trans* azo-benzene has λ_{max} 320 nm (ε 21,300) in ethanol solution.

2. The azobenzene (*trans*) required is prepared as follows:
    To a suspension of magnesium turnings (1 g), nitrobenzene (1.8 ml), methanol (35 ml) add a small crystal of iodine in a round-bottomed flask. Fit the flask with a reflux condenser. If the reaction does not commence in 2–3 min, warm the mixture (water-bath) to start the reaction. In case, the reaction becomes too vigorous, use a cold water-bath for few seconds. When most the magnesium has reacted, add more magnesium turnings (1 g). Let the reaction proceed (as above). Finally, heat the reaction mixture on a water-bath. Cool the mixture, pour into water (65 ml), acidify with glacial acetic acid (until the mixture is acidic to congo red). Cool the mixture (ice-bath), filter the separated azobenzene and crytallize from ethanol (90%). Yield 1 g (34%), m.p. 67–68 °C.

3.3.3 *Conversion of* **trans** *Stilbene into* **cis** *Stilbene*

The stilbene as ordinarily obtained is the *trans* isomer. It can be photochemically isomerized to the *cis* form. This can be achieved by irradiation with sunlight or with a UV fluorescent lamp (Scheme 3.23).

The *trans* form is readily available by a variety of reactions and is much more stable than the *cis* isomer because it is less sterically hindered. However, it is possible to produce a mixture containing mostly the *cis* isomer by irradiation of a solution of

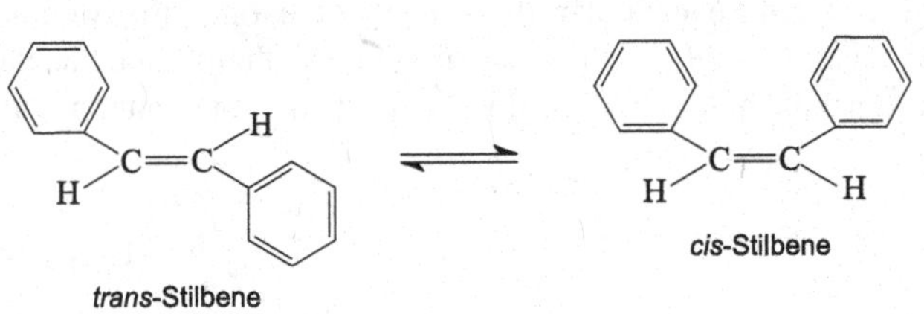

Scheme 3.23 Conversion of *trans*-stilbene into *cis*-stilbene

the *trans* isomer in the presence of a suitable photosensitizer (e.g., benzophenone). The stilbene required is prepared by Wittig reaction of benzaldehyde with triphenyl phosphonium salt (Sect. 3.1.11).

3.4 PTC-Catalysed Reactions

3.4.1 *Phenylisocyanide (C_6H_5N≡≡C)*

It is prepared by the reaction of dichlorocarbene (generated in situ) by the PTC method from aniline (1° amine) (Scheme 3.24).

Materials

Aniline	4 ml
Chloroform	12 ml
aq. NaOH (50% solution)	20 ml
Benzyltriethyl ammonium chloride	0.25 g

Procedure

To the mixture of aniline (4 ml), chloroform (12 ml) and benzyltriethyl ammonium chloride (0.25 g), add a vigorously stirred solution of sodium hydroxide solution (50%, 20 ml). Use a reflux condenser, since the mixture starts refluxing. The reaction subsides in 4–5 min, and the stirring continued for 1 h. After the reaction is over, add cold water. Extract the mixture with methylene chloride, wash the organic layer with aqueous sodium chloride (5%, 30 ml) and dry over anhydrous sodium sulphate. Distil the clear solution under nitrogen. Phenyl isocyanide is collected at 50–52 °C/11 mm. Yield 57%.

Notes

1. The required PTC, viz., benzyl triethylammonium chloride is obtained by refluxing a solution of triethylamine (3.3 g) and benzylchloride (5 g) in absolute ethanol for about 50 h. Cool the solution to room temperature and add ether. Filter the separated benzyltriethyl ammonium chloride. Purify it by dissolving in hot acetone and reprecipitation with ether. Yield 8.5 g (90%).
2. The usual method of preparation of phenyl isocyanide without the PTC gives very poor yield (5–10%).

$$C_6H_5NH_2 + CHCl_3 + NaOH \xrightarrow{\quad C_6H_5CH_2\overset{+}{N}Et_3Cl^- \quad} C_6H_5N≡≡C$$
$$\text{aq} \qquad\qquad\qquad\qquad\qquad\qquad\qquad 55\%$$

Scheme 3.24 Synthesis of C_6H_5N≡≡C in the presence of PTC

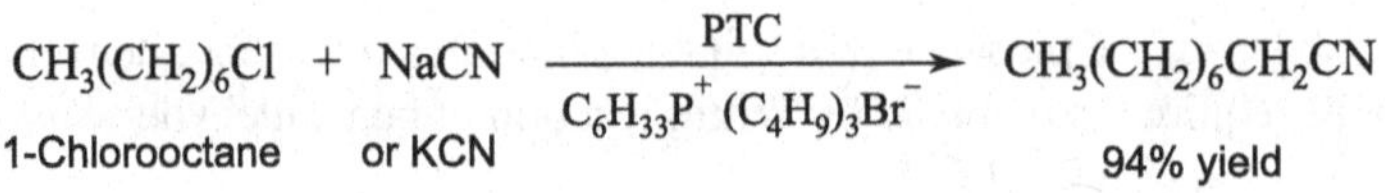

$$CH_3(CH_2)_6Cl \ + \ NaCN \ \xrightarrow[\underset{C_6H_{33}\overset{+}{P}(C_4H_9)_3Br^-}{}]{PTC} \ CH_3(CH_2)_6CH_2CN$$

1-Chlorooctane   or KCN   94% yield

Scheme 3.25 Synthesis of 1-cyanooctane using PTC

3.4.2 1-Cyano Octane ($CH_3(CH_2)_6CH_2CN$)

We know that alkyl halides do not react with sodium cyanide under a variety of conditions, e.g., stirring and heating for long time. However, if a small quantity of a phase transfer catalyst is used, the reaction goes to completion in about 2 h time. Thus, 1-cyanooctane is obtained as follows (Scheme 3.25).

Materials

1-Chlorooctane	5 g
Sodium cyanide	5 g
Water	5 ml
Hexadecyl tributyl phosphonium bromide	0.5 g (see note 1)

Procedure[2]

Heat a mixture of 1-chlorooctane (5 g), sodium cyanide (5 g), water (5 ml) and hexadecyl tributyl phosphonium bromide (0.5 g) with stirring at 105 °C for 2 h in a round-bottomed flask (250 ml capacity), cool the reaction mixture, add water (100 ml) and extract with dichloromethane. Wash the organic extract with water, dry over anhydrous sodium sulphate and distil. 1-Cyanooctane gets collected at 140°/1 mm. Yield 94% (97% pure).

Notes

1. The PTC, hexadecyl tributyl phosphonium bromide [n–$C_{16}H_{33}$ $P^+(n$–$C_4H_9)_3$ Br^-] is prepared (CM. Starks J. Am. Chem. Soc., 1971, 93, 195) by heating a mixture of 1-bromohexadecane (1.5 g) and tributyl phosphine (1 g) at 70–90 °C for 3–4 days. Cool the mixture, filter the separated PTC and crystallize from hexane. Yield 1.7 g (60%), m.p. 54 °C.
2. C.M. starks, J. Am. Chem Soc, 1971, 93, 195
3. PTC methodology can also be used for the preparation of following alkyl or acyl halides.

(a)

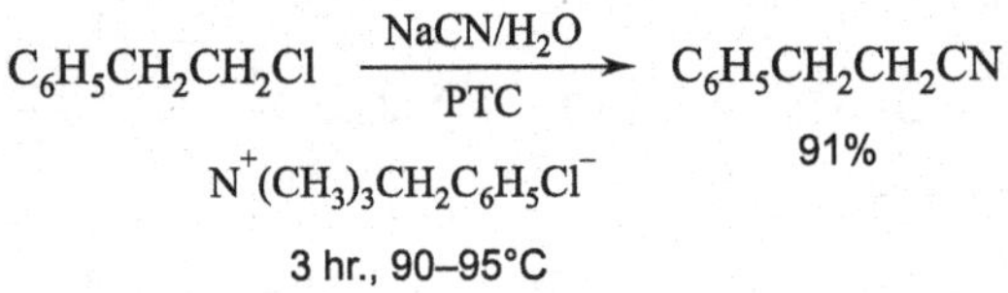

$$C_6H_5CH_2CH_2Cl \ \xrightarrow[\underset{N^+(CH_3)_3CH_2C_6H_5Cl^-}{PTC}]{NaCN/H_2O} \ C_6H_5CH_2CH_2CN$$

91%

3 hr., 90–95°C

Ref. N. Sugemol, T. Fujita, N. Shigematsu and A. Ayadha, Chem. Pharm. Bull., 1962, *10*, 427; Japanese Patent 1961/63.

(b) $C_6H_5COCl \xrightarrow[\text{PTC Bu}_4\text{N}^+\text{X}^-]{\text{NaCN/H}_2\text{O}} C_6H_5COCN$

                                                                    60–70%

Ref. K.E. Koening and W.P. Weber, Tetrahedron Lett., 1974, 2274.

3. Benzonitrite can be conveniently prepared from benzamide by the reaction with dichlorocarbene [generated in situ].

3.4.3 1-Oxaspiro-[2,5]-Octane-2-Carbonitrile

It is obtained by the PTC-catalysed Darzen condensation of cyclohexanone with chloroacetonitrile (A. Jonczyk, M. Fedorynski and M. Makosza, Tetrdhedron Lett., 1972, 2395) (Scheme 3.26).

Materials

Cyclohexanone	5.4 g
Sodium hydroxide solution (50%)	10 ml
Benzyltriethyl ammonium chloride	0.2 g
Chloroacetonitrile	3.8 g

Procedure

Chloroacetonitrile (3.8 g) added dropwise to a stirred mixture of cyclohexanone (5.4 g), sodium hydroxide solution (50%, 10 ml) and benzyltriethyl-ammonium chloride (0.2 g) at 15–20 °C. Stir the mixture at 15–20 °C for 30 min. Extract the mixture with dichloro methane, wash the organic phase with water, dry over anhydrous sodium sulphate and distil. 1-Oxaspiro-[2,5]-octane-2-carbonitrile is obtained at 87 °C/5 mm. Yield 70%.

Note

1. Darzen reaction was earlier performed under anhydrous conditions. It has now been possible in aqueous phase using PTC.

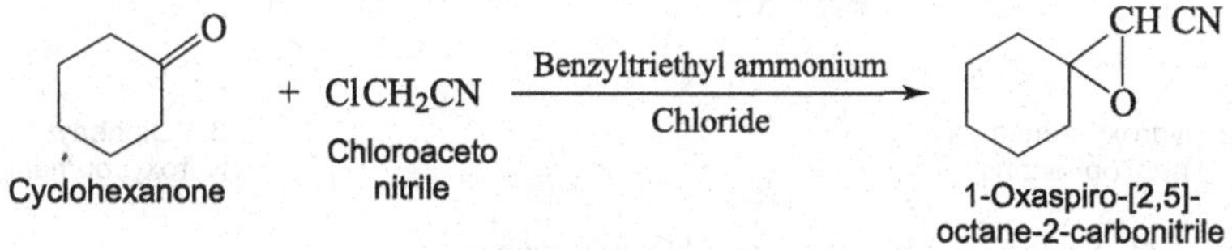

Scheme 3.26 Synthesis of 1-oxaspiro-[2,5]-octane-2-carbonitrile

3.4.4   3,4-Diphenyl-7-Hydroxycoumarin

3,4-Diphenyl coumarins, known for their antifertility activity, were prepared earlier in low yields and required anhydrous conditions. These are now obtained in excellent yield and purity by the use of a phase transfer catalyst in the presence of aqueous potassium carbonate by the reaction of *o*-hydroxy benzophenones with phenyl acetyl chloride (Scheme 3.27).

Materials

2-hydroxy-4-methoxy benzophenone	2.45 g
Phenyl acetyl chloride	1.5 ml
Tetrabutyl ammonium hydrogen sulphate	0.1 g
Aqueous potassium carbonate (20%)	50 ml

Procedure[1]

Add dropwise a solution of phenyl acetyl chloride (1.5 ml) in benzene (10 ml) to a stirred mixture of 2-hydroxy-4-methoxybenzophenone (2.45 g) in benzene (50 ml), tetrabutylammonium hydrogen sulphate (100 mg) and aqueous potassium carbonate (20%, 50 ml). Stir the mixture for 5 h, separate the organic layer, wash with water, dry (anhyd. Na_2SO_4) and distil the solvent. Crystallize the residue from ethanol to give 3,4-diphenyl-7-hydroxycoumarin, m.p. 168–69 °C. Yield 80%.

Notes

1.   V.K. Ahluwalia and C.H. Khanduri, Indian J. Clem., 1989, *28B*, 599.
2.   The PTC, tetrabutyl ammonium hydrogen sulphate is obtained (W.T. Ford and R.J. Haurt, J. Am. Chem. Soc., 1973, *95*, 7381) as follows.

Add dimethyl sulphate (4.6 g) to a stirred mixture of tetrabutyl ammonium bromide (9.8 g) and chlorobenzene (15 ml) at 80–85 °C in a two-neck round-bottomed

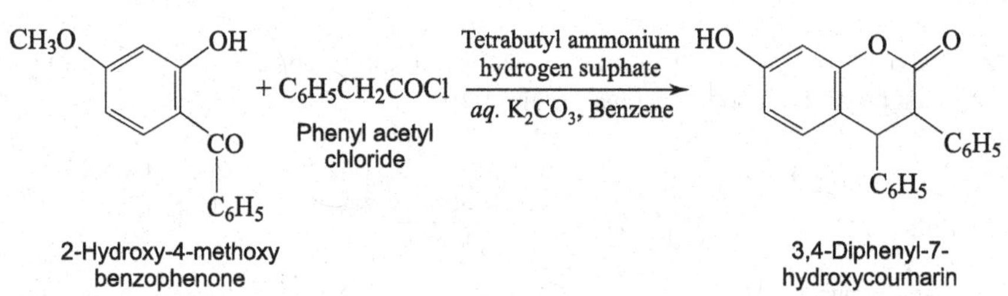

2-Hydroxy-4-methoxy benzophenone   3,4-Diphenyl-7-hydroxycoumarin

Scheme 3.27   Synthesis of 3,4-Diphenyl-7-hydroxycoumarin

flask fitted with a short distillation column and a dropping funnel. Collect the formed methyl bromide as a distillate using a trap cooled in acetone–dry ice mixture. After the distillation of methyl bromide ceases, increase the heating until the temperature at the top of the distillation column starts to rise rapidly. Add cautiously a solution of concentrated sulphuric acid (0.75 ml) in water (300 ml). Reflux the mixture for 48 h. Evaporate the solution to almost dryness under reduced pressure. Dissolve the residue in dichloromethane (250 ml), wash the solution with water (2 × 30 ml), dry (anhydrous sodium sulphate) and distil the solvent. The PTC, tetrabutylammonium hydrogen sulphate (10 g) separates. It is almost pure and can be crystallized from isobutyl methyl ketone.

3.4.5 Flavone

Flavones, a class of natural products were synthesized by a number of methods, and in most of the methods the yield is low and the work-up procedure is difficult. These can be obtained in excellent yield (90%) by the reaction of appropriate o-hydroxyacetophenone with appropriately substituted benzoyl chloride in benzene solution with a phase transfer catalyst in the presence of sodium hydroxide or sodium carbonate followed by cyclization of the formed o-hydroxydibenzoyl methane with p-toluene sulphonic acid. The last step is known as **Baker–Venkataraman synthesis**. Simple flavone is synthesized as shown in Scheme 3.28.

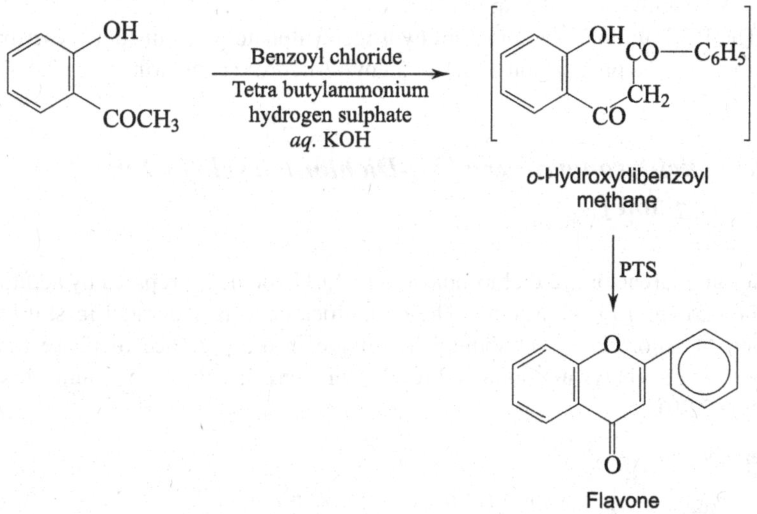

o-Hydroxydibenzoyl
methane

Flavone

Scheme 3.28 Synthesis of flavone on using PTC methodology

Materials

o-Hydroxyacetophenone	0.43 g
Benzoyl chloride	0.42 g
Tetrabutyl ammonium hydrogen sulphate	0.2 g
Aq. potassium hydroxide (10%)	20 ml
p-toluene sulphonic acid	0.1 g

Procedure[1]

Add a solution of benzoyl chloride (0.42 g) in benzene (20 ml) to a stirred mixture of o-hydroxyacetophenone (0.43 g), tetrabutyl ammonium hydrogen sulphate (0.2 g) and *aq.* potassium hydroxide (10%, 20 ml). Continue stirring for 2–3 h until the starting ketone disappeared (TLC). Separate the benzene solution (separatory funnel), wash with water (2 × 15 ml), dry (anhydrous sodium sulphate) and distil. The residual product consisting of o-hydroxydibenzoylmethane was dissolved in benzene (50 ml) and refluxed with p-toluene sulphonic acid (0.1 g). Remove the water formed by distillation using a Dean–Stark apparatus. Reflux the mixture for 45–60 min. Extract the benzene solution with sodium bicarbonate solution (5%, 25 ml) to remove p-toluene sulphonic acid. Remove the solvent by distillation and crystallize the residue of flavone from ethyl acetate-petroleum ether. Yield 95%, m.p. 117 °C.

Notes

1.   V. K. Ahluwalia et al. unpublished results.
2.   The PTC, tetrabutylammonium hydrogen sulphate is obtained as described in note 2 in the preparation at 3,4-diphenyl-7-hydroxycoumarin (Sect. 3.4.4).

3.4.6   Dichloronorcarane [2,2-Dichlorobicyclo (4.1.0) Heptane]

Dichloronorcarene or 2,2-dichlorobicyclo [4.1.0] heptane is prepared by addition of dichlorocarbene to cyclohexene. The dichlorocarbene is generated in situ by the reaction of chloroform and sodium hydroxide in the presence of phase transfer catalyst (tetra-n-butyl ammonium bromide or benzyltriethylammonium chloride) (Scheme 3.29).

Materials

Chloroform	12 ml
Cyclohexene	4.1 g (5.1 ml)
Tetra-n-butyl ammonium bromide	0.25 g
Sodium hydroxide	20 ml (50%)

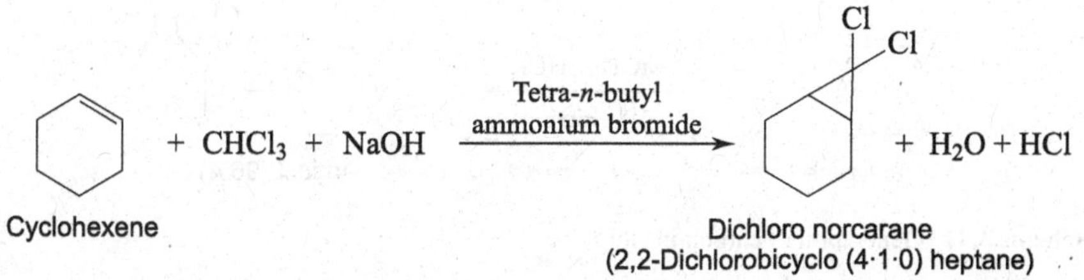

Scheme 3.29 Synthesis of dichloronorcarane [2,2-dichlorobicyclo (4.1.0) heptane]

Procedure

Take chloroform (12 ml) in a round-bottomed flask (100 ml capacity) followed by addition of cyclohexene (4.1 g or 5.1 ml) and tetra-*n*-butyl ammonium bromide (0.25 g). Fit the flask with a reflux condenser, continue stirring with the help of a magnetic needle. Pour in sodium hydroxide solution (50%, 20 ml) and water (15 ml). Heat the flask gently and stir the mixture vigorously for 30 min. Heating should be done in such a way that chloroform refluxed slowly as seen by an occasional drop falling from the reflux condenser.

After the heating is complete, add ice to cool the mixture to room temperature. Then extract with ether (2 × 30 ml) by using a separatory funnel. Dry the ether solution over anhydrous magnesium sulphate and distil. Collect the fraction distilling between 192 and 197 °C. Yield 5.7 g (about 65%).

Notes

1. Using styrene (5.2 g) in place of cyclohexene, you can prepare 1,1-dichloro-2-phenyl cyclopropane (b.p. 103 °C/10 Torr).

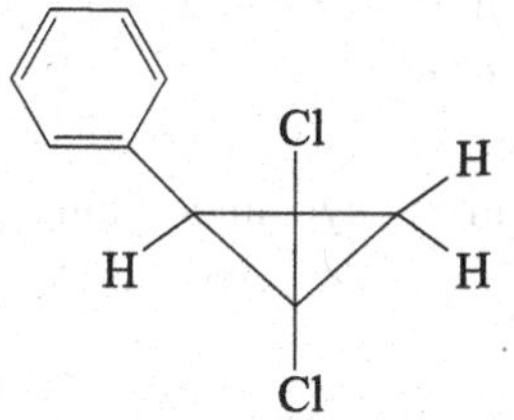

2. Using 1,5-cyclooctadiene (3.1 ml) in place of cyclohexene in the above experiment, you can get the *bis*-adduct. (Scheme 3.30)

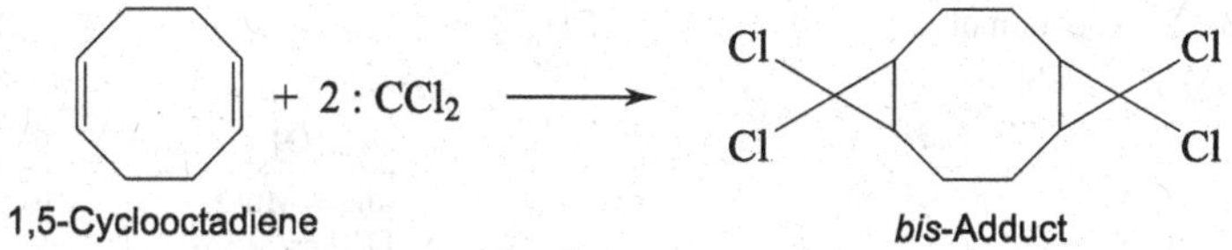

Scheme 3.30 Synthesis of bis-adduct

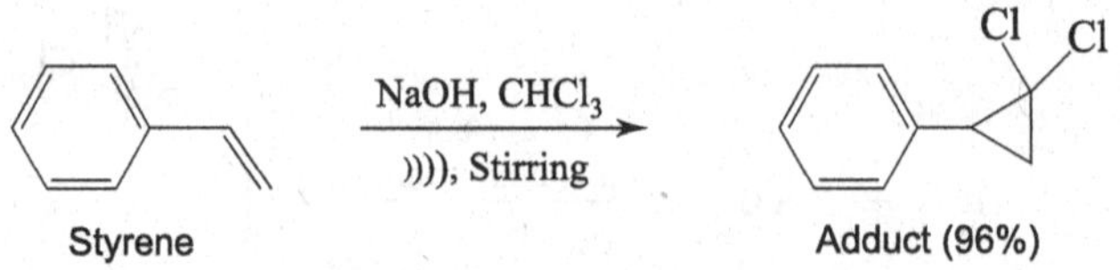

Scheme 3.31 Generation of carbon in situ

3. Dichlorocarbene can also be generated by direct sonication (S.L Regen, A.K. Singh, J. Org. Chem, 1982, *47*, 1587) between powdered sodium hydroxide and chloroform. This procedure is simple and efficient, and avoids the use of phase transfer catalyst. The generated dichlorocarbene in situ undergoes addition reaction to alkenes (Scheme 3.31).

4. Reimer–Tiemann reaction can also be performed using the same procedure (Sect. 3.4.10).

3.4.7 Oxidation of Toluene to Benzoic Acid

Normally toluene is oxidized to benzoic acid by alkaline $KMnO_4$ solution. However, even after prolonged refluxing, the yield is only 40–50%. Use of a phase transfer catalyst (e.g., cetyltrimethyl ammonium chloride) or a crown ether (e.g., [18] crown 6) gives much better yield (80–90%) in shorter time (Scheme 3.32).

Materials

Toluene	2.5 ml
Potassium permanganate	4 g
Sodium carbonate (2 N)	1.6 ml
[18] crown 6.	
or	0.1 g
Cetyltrimethyl ammonium chloride	

Procedure[1]

Place a mixture of toluene (2.5 ml), sodium carbonate solution (2 N, 1.6 ml) and [18]-crown-6 (0.1 g) in a round-bottomed flask (100 ml capacity) filled with a reflux condenser. Add potassium permanganate solution through the reflux condenser while

Scheme 3.32 Oxidation of toluene to benzoic acid

Toluene   Benzoic acid (80–90%)

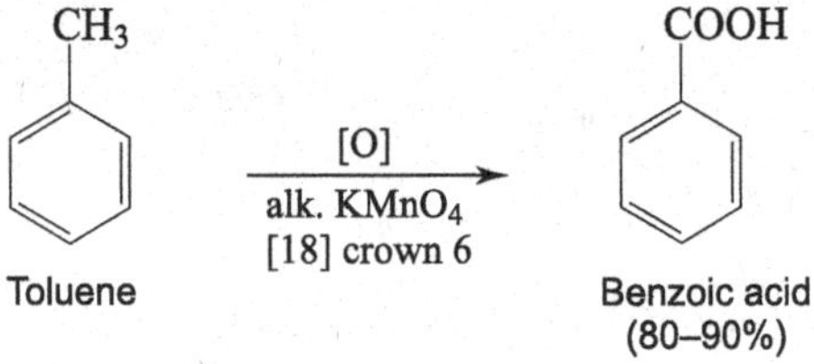

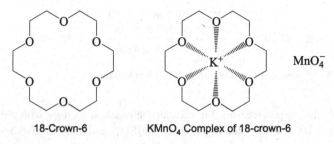

18-Crown-6   KMnO$_4$ Complex of 18-crown-6

Scheme 3.33 18-Crown-6 and its complex with KMnO$_4$

the mixture is kept gently refluxing on a wire gauge. Reflux for a total of 3 h. Cool the reaction mixture. Filter the alkaline solution to remove precipitated MnO$_2$ and pass sulphur dioxide gas or add of saturated solution of sodium sulphite and dilute sulphuric acid till the solution become colourless. Extract the solution with ether (2 × 10 ml) and evaporate the ether (caution). Crystallize the residual product from hot water. Yield 2.2 g. m.p. 122 °C.

Note

1. Crown ether forms a complex with KMnO$_4$ which is soluble in organic phase and thus, KMnO$_4$ becomes more effective for oxidation of toluene (Scheme 3.33).

3.4.8   Benzonitrile from Benzamide

Benzamide on reaction with dichlorocarbene generated in situ by the reaction of chloroform and sodium hydroxide in the presence of a PTC gives benzonitrile (Scheme 3.34).

Materials

Benzamide	6.05 g
Benzyl triethyl ammonium chloride	0.34 g
Chloroform	6 ml
aq. sodium hydroxide (50%)	25 g NaOH in 50 ml H$_2$O

Procedure[1]

Stir a mixture of benzamide (6.05 g), benzyltriethyl ammonium chloride (0.34 g), chloroform (6 ml) and aq. sodium hydroxide (50% solution, 25 g NaOH in 50 ml

$$C_6H_5CONH_2 + CHCl_3 + NaOH \xrightarrow[aq]{C_6H_5CH_2\overset{+}{N}Et_3Cl^-} C_6H_5CN$$

Scheme 3.34 PTC-catalysed conversion of C$_6$H$_5$CONH$_2$ with C$_6$H$_5$CN

$$R\text{—}CHO \xrightarrow[\text{2) } CS_2 \text{ 20°C, 16 or 48 hr.}]{\substack{\text{1) } NH_2OHHCl \text{ on } Al_2O_3\text{–KF} \\ \text{Microwave 350 W, 5 min.}}} RC\equiv N$$

Scheme 3.35 A convenient conversion at R—CHO into RC≡≡N

water) at room temperature for 2 h. Extract the reaction mixture with chloroform, wash organic layer with water, dry (anhyd. Na_2SO_4) and distil to give benzonitrile (84% yield).

Notes

1. T. Saraie, K. Ishiguno, K. Kawashima and K. Morita, Tetrahedron Lett., 1971, 2121.
2. For the preparation of PTC (benzyl triethyl ammonium chloride, see note 1 of preparation of phenyl isocyanide (Sect. 3.4.1).
3. A number of methods are available for the synthesis of nitriles. In most of the cases, the aldoxime is initially prepared and then dehydrated by a wide variety of reagents. The method described for the synthesis of benzonitrile from benzamide using dichlorcarbene in aqueous phase in the presence of PTC is a convenient procedure.
4. In an efficient procedure, the aldehyde is converted into adsorbed oximate by reaction with hydroxylamine hydrochloride and potassium fluoride on alumina under microwave activation and without a solvent. The absorbed oximate is transformed into nitrile by treatment with carbon disulphide at room temperature (D. Villemin, M. Lalaoui, A.B. Alloum, Chem. Ind., 1991, 176) (Scheme 3.35).

Using this procedure, a number of nitriles can be prepared in good yield.

3.4.9 n-Butyl Benzyl Ether

An unsymmetrical ether n-butyl benzyl ether is obtained by an improved **Williamson ether synthesis**[1] using phase transfer catalyst. Thus, the reaction of butyl alcohol with benzyl chloride in the presence of sodium hydroxide and tetrabutylammonium bisulphate (TBAB) as catalyst gives n-butyl benzyl ether (Scheme 3.36).

$$CH_3CH_2CH_2CH_2OH \; + \; C_6H_5CH_2Cl \xrightarrow[\text{TBAB}]{\text{50% NaOH}} CH_3CH_2CH_2CH_2OCH_2C_6H_5$$

n-Butyl alcohol   Benzyl chloride   *n*-Butyl benzyl ether
                                                                    + NaCl + H_2O

Scheme 3.36 Synthesis of *n*-butyl benzyl ether

Materials

n-Butyl alcohol	6.2 g
Benzyl chloride	6.3 g
Tetrabutylammonium bisulphate (TBAB)	1 g
Sodium hydroxide (50%)	10 g in 20 ml H_2O.

Procedure

A mixture of *n*-butyl alcohol (6.2 g), benzyl chloride (6.3 g), sodium hydroxide solution (10 g sodium hydroxide in 20 ml water) and TBAB (1 g) is stirred at 35–40 °C for 1.5 h. Extract the mixture with THF, wash THF solution with NaCl saturated with 50% aqueous sodium hydroxide. Distillation of the dried THF solution gives the required *n*-butyl benzyl ether with 92% yield.

Notes

1. The method adopted is described by H.H. Freedman and R.A. Dubois, Tetrahedron Lett., 1975, 3251.
2. This is the best procedure for the preparation of mixed ethers. The usual Williamson method gives a mixture of products.
3. In this procedure, only a minor amount of symmetrical ether is formed.
4. Primary alcohols are completely alkylated by aliphatic chlorides; secondary alcohols require longer time or greater amount of catalyst.

3.4.10 Salicylaldehyde

It is obtained by the reaction of phenol with dichlorocarbene, which is generated in situ from chloroform and sodium hydroxide solution in the presence of a PTC catalyst[1]. The reaction is known as **Reimer–Tiemann Reaction** (Scheme 3.37).

Materials

Phenol	16.3 g
Chloroform	27.3 ml
Sodium hydroxide (50%)	20 ml
Tetra-*n*-butyl ammonium bromide	0.25 g

Procedure

Add chloroform (16.3 g) to a round-bottomed flask (100 ml capacity) followed by the addition of phenol (16.3 g) and tetra-*n*-butyl ammonium bromide (0.25 g). Fit the flask with a reflux condenser, continue stirring with the help of a magnetic needle.

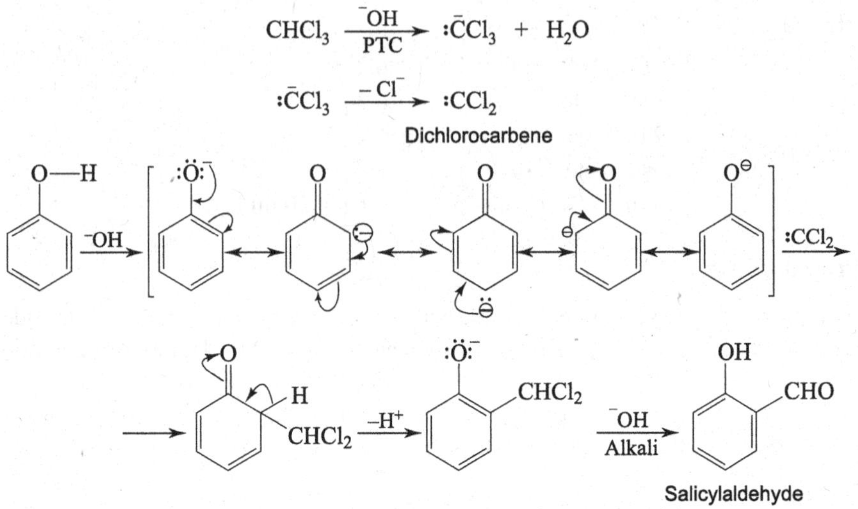

Scheme 3.37 Reimer–Tiemann reaction

Pour sodium hydroxide solution (50%, 20 ml) in one lot and water (15 ml). Heat the flask gently and stir the mixture vigorously for 30 min.

Heating should be done so that chloroform refluxed slowly as seen by an occasional drop falling from the condenser.

After completion of the reaction, remove excess chloroform by steam distillation. Acidify the remaining aqueous alkaline solution cautiously with dilute sulphuric acid and then steam distil till no more oily drops are collected. Extract the distillate containing salicylaldehyde with ether (2 × 15 ml). Remove the ether by distillation, shake vigorously the residual product containing phenol and salicylaldehyde with a saturated solution of sodium bisulphate and allow to stand for 1 h. Filter the bisulphite adduct, wash with water and a little alcohol, and finally with ether. Decompose the adduct by warming on a water bath with dilute sulphuric acid. Extract the cooled mixture with ether, dry the ether solution over anhydrous sodium sulphate and evaporate the ether. Salicylaldehyde is collected by distillation, b.p. 195–197 °C, yield 7.1 g (40%).

The p-hydroxybenzaldehyde obtained as a by-product is isolated from the residue left after steam distillation. Filter the residual solution, cool and extract with ether. Remove the solvent. Crystallize the yellow solid from aqueous sulphurous acid. Yield 1.5 g (7%), m.p. 116–117 °C.

Notes

1.   Salicylaldehyde can also be obtained without using PTC, but the yield is low.
2.   Using the same procedure, β-hydroxy naphthaldehyde can be obtained with 90% yield (m.p. 80–81 °C) starting from β-naphthol.

3.5 Rearrangement Reactions

3.5.1 *Benzopinacolone*

It is obtained by the rearrangement of benzopinacol under the influence of iodine in glacial acetic acid (Scheme 3.38).

The above acid catalyst rearrangement is called the **pinacol rearrangement**.

Mechanism (Scheme 3.39)

Materials

Benzopinacol   2.5 g
Iodine solution   12.5 ml

Iodine solution is 0.015 ml solution of iodine dissolved in glacial acetic acid.

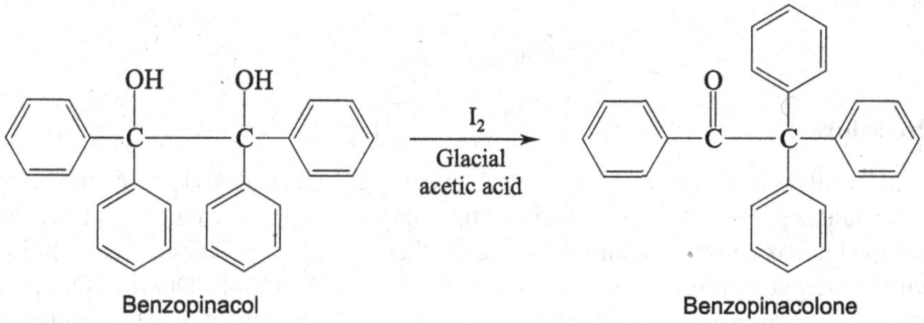

Benzopinacol   Benzopinacolone

Scheme 3.38 Synthesis of benzopinacolone

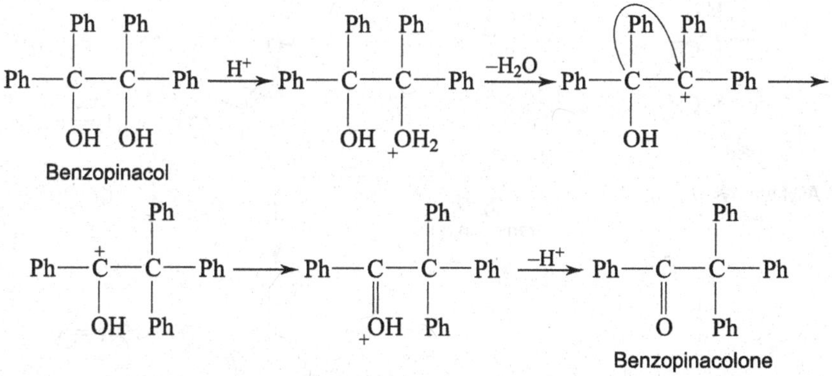

Scheme 3.39 Mechanism for the formation benzopinacolone

Procedure

In a round-bottomed flask, take 12.5 ml solution of iodine dissolved in glacial acetic acid. Add 2.5 g benzopinacol. Reflux the solution for 5 min. Crystals appear from the solution. Cool the solution, filter the separated benzopinacolone, wash with cold glacial acetic acid (2 ml) and dry. Record the yield. m.p. 182° C.

Note

Since it is a rearrangement reaction, so there is 100% atom economy and so is a green reaction.

3.5.2   2-Allyl Phenol

It is obtained by **Claisen rearrangement** of allyl phenyl ether (Scheme 3.40).

Materials

$$\text{Allyl phenyl ether} \quad 2.1 \text{ g}$$

Procedure

Heat the allyl phenyl ether (2.1 g) in a test tube using a sand bath and an air condenser. The liquid is gently refluxed for 4–5 h. After this time, cool the reaction mixture, add sodium hydroxide solution and extract with ether. Acidify the clear alkaline solution with dilute hydrochloric acid and extract with ether (2 × 15 ml). Dry the combined ether extract over anhyd. sodium sulphate and distil the solvent. 2-Allyl phenol is distilled under reduced pressure. Yield 1.5 g, b.p. 103–105 °C/19 mm.

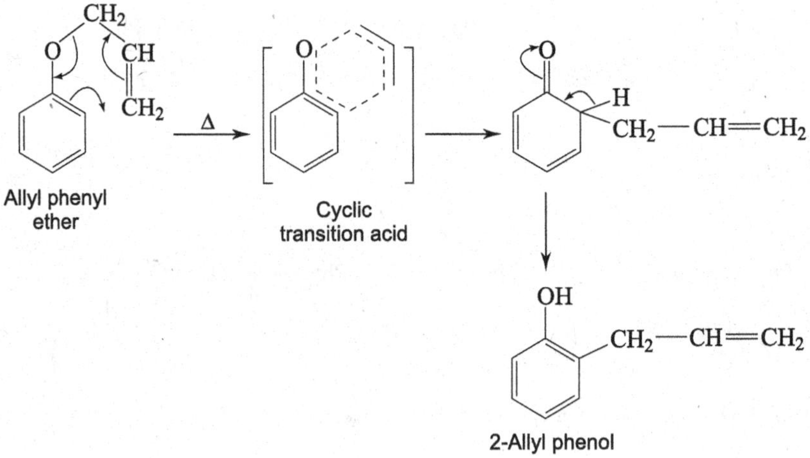

Scheme 3.40 Preparation of 2-allyl phenol

Note

1. The required allyl phenyl ether is obtained as follows:
   Reflux a mixture of phenol (1.9 g), allyl bromide (2.5 g), anhydrous potassium carbonate (3 g) and dry acetone (20 ml) on a water bath using a round-bottomed flask fitted with a reflux condenser and calcium chloride guard tube for 6 h. Distil the acetone (as much as possible) on a boiling water bath and add water (15 ml) to the residue. Extract the mixture with ether (2 × 20 ml). Wash the ether extract with sodium hydroxide solution (10% 2 × 10 ml) and finally with water (2 × 10 ml). Dry the ether solution over anhydrous sodium sulphate and distil. The allyl phenyl ether is obtained by distillation under reduced pressure b.p. 65 °C/19 mm, yield 2.1 g (78%).

3.6 Microwave-Induced Reactions

3.6.1 9,10-Dihydroanthracene-Endo-α,β-Succinic Anhydride (Anthracene-Maleic Anhydride Adduct)

It is obtained by **Diels–Alder reaction** of anthracene with maleic anhydride. Normally, the reaction is carried out by refluxing the mixture in xylene for 15–20 min. However, by using microwave the reaction can be completed in less than one minute (Scheme 3.41).

Materials

Anthracene	3 g
Maleic anhydride	1.15 g
Diglyme	5 ml

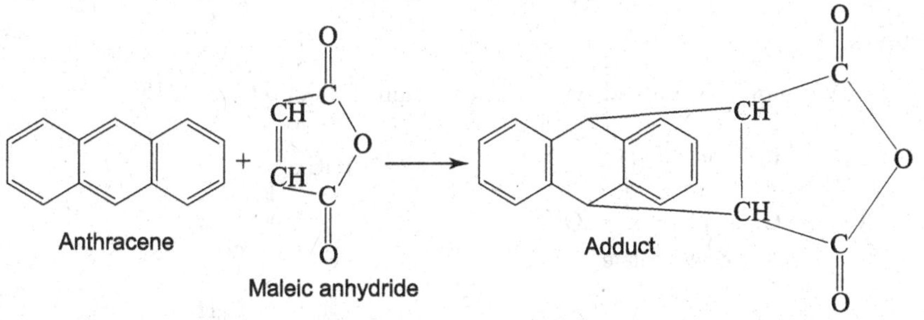

Scheme 3.41 Microwave-assisted synthesis of anthracene-maleic anhydride adduct

Procedure

Place the grinded mixture of anthracene (3 g) and maleic anhydride (1.15 g) in a beaker (250 ml capacity). Add diglyme (5 ml), shake the mixture gently and cover the beaker with a watch glass and irradiate the mixture in a microwave oven for 90 s at a medium power level. Remove the beaker from the oven, allow it cool to room temperature. Filter the separated adduct by suction, wash the product with methanol, yield 3.3 g. m.p. 262–263 °C (decom.).

Note

1. It is possible to conduct Diels–Alder reaction in aqueous phase also (R. Breslow (a review) Acc. Chem. Research, 1991, *24*, 159). Also see Diels–Alder reaction of furan with maleic acid (Sect. 3.1.10).

3.6.2   3-Methyl-1-Phenyl-5-Pyrazolone

It is prepared by the condensation of ethyl acetoacetate with phenyl hydrazine by heating in oil bath at 110–120 °C for 4 h. However, in microwave oven, the reaction takes only 10 min for completion (Scheme 3.42).

Materials

$$\text{Ethyl acetoacetate   2.9 g}$$
$$\text{Phenyl hydrazine   2.7 g}$$

Procedure[1]

Heat a mixture of ethyl acetoacetate (2.9 g) and phenyl hydrazine (2.7 g) in a conical flask (50 ml capacity) in microwave oven (280 W) for 10 min. Cool the reaction mixture, filter the separated product and recrystallize from alcohol–water (1:1) to get quantitative yield of 3-methyl-1 phenyl-5-pyrazolone. m.p. 129 °C.

Note

1. D. Villenmin, B. Labiad, Synthetic commun, 1990, *20 (20)*, 3213.

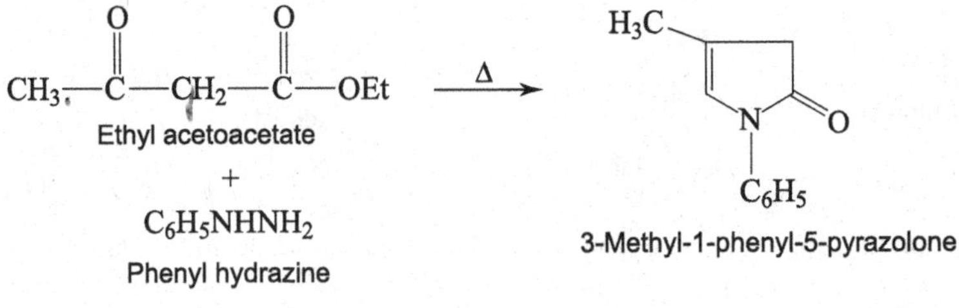

Scheme 3.42 Microwave-assisted synthesis of 3-methyl-1-phenyl-5-pyrazolone

3.6.3 Preparation of Derivatives of Some Organic Compounds

Microwave irradiation has been used for the preparation of following derivatives:
p-Nitrobenzyl esters from carboxylic acids, aryloxyacetic acids from phenols, oximes from aldehydes and ketones.

(a) **p-Nitrobenzyl esters from carboxylic acids.**

A mixture of carboxylic acid (10 m mol) and water (1 ml) is neutralized with sodium hydroxide solution (10%). A solution of p-nitrobenzyl chloride (9 m mol) in ethanol (5 ml) is added and the mixture is heated in microwave oven for 2 min in a sealed 150 ml Teflon bottle. After cooling, water (2 ml) is added. The separated ester is filtered, washed with sodium carbonate solution (5%) and crystallized from alcohol. Using cinnamic acid, corresponding p-nitrobenzyl cinnamate, m.p. 115–16 °C is obtained with 40% yield.

(b) **Aryloxyacetic acids from phenols.**

A mixture of phenol (1.4 m mol) in sodium hydroxide (6 M, 3 ml) and aqueous chloroacetic acid (50%, 0.5 ml) is heated in a sealed Teflon tube (150 ml capacity) in a microwave oven for 2 min. After cooling, water (2 ml) is added and the solution acidified with dilute hydrochloric acid. The mixture is extracted with ether, and the ether extract is washed with water and extracted with sodium carbonate solution (5%). The sodium carbonate extract on acidification gives the aryloxyacetic acid derivative. Using this procedure, β-naphthol gives 2-naphthoxyacetic acid, m.p. 154–56 °C. Conventional methods take 60 min.

(c) **Oximes from aldehydes and ketones.**

A mixture of carbonyl compound (5.5 m mol), hydroxylamine hydrochloride (14.4 m mol), pyridine (5 ml) and absolute ethanol (5 ml) is heated in a sealed Teflon bottle (150 ml capacity) in a microwave oven for 2 min. After cooling, the solvent is removed under reduced pressure. The formed oxime is stirred with cold water and crystallized from alcohol.

Using this procedure, benzophenone gives the oxime, m.p. 140–141 °C in 70% yield.

(d) **Carboxylic acids from ester (saponification).** A mixture of the ester (7.4 m mol) and aqueous sodium hydroxide (25%, 10 ml) is heated in a sealed Teflon tube (150 ml capacity) for 2.5 min in a microwave oven. The solution is cooled and acidified with hydrochloric acid and extracted with ether. The dried ether extract is evaporated to give the acid. Using this procedure, methyl benzoate gives benzoic acid, m.p. 121–122 °C with 84% yield.

Note

The procedure followed is described by R.N. Gedye, F.E. Smith, K.C. Westaway, Can. J. Chem. 1988, *66*, 17.

3.6.4 Copper Phthalocyanine

The phthalocyanines are fast to light, heat, acids and alkalis. These also find its use as photodynamic agents for cancer therapy and other medical applications.

Copper phthalocyanine is prepared by heating a mixture of phthalic anhydride, urea and copper(I) chloride in a high boiling solvent like nitrobenzene or trichlorobenzene in the presence of catalytic amounted of ammonium molybdate.

Procedure

A mixture of phthalic anhydride (592 mg, 4.0 m mol), anhydrous $CuCl_2$ (134 mg, 1.0 m ml), urea (900 mg, 15.0 m mol) is powdered (mortar and paste) and placed in a beaker and covered with a watch glass. The beaker is then placed in a silica jacket and heated in a microwave oven for about 6 min at an interval of 1 min). After the reaction, the mixture is allowed to cool to room temperature, ground and washed with 5% NaOH, H_2O and 2% HCl and again with H_2O. The copper phthalocyanine is obtained with 90% yield based on phthalic anhydride (Scheme 3.43).

Notes

1. R.K. Sharma, C. Sharma and I.T.J. Sindhwani, Chem. Edu., 2011, 88, 86–87.
2. Freshly prepared phthalic anhydride is prepared by sublimation.
3. Anhyd, copper chloride is flame dried as indicated by brown colour.
4. The progress of the reaction can be visually monitored as copper phthalocyanine is of deep shade of blue and the starting mixture is not coloured.
5. The I.R. spectod of copper phthalocyanine (KBr Pallett) shows peak at 726 versus 754 s, 776 m, 874 W, 899 m, 1048 m, 1091 s, 1119 s, 1165 m, 1286.m, 1332 s, 1420 m, 1469 W, 1506 m, 1607 w, 3048 W m^{-1}.

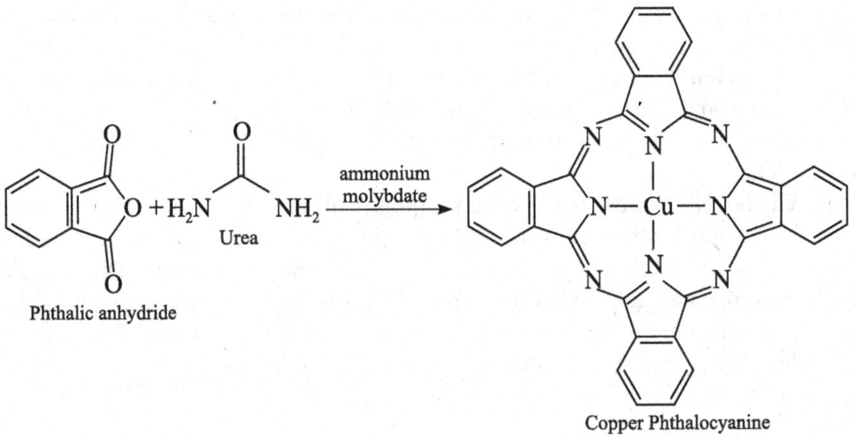

Scheme 3.43 Preparation of copper phthalocyanine

3.7 Enzymatic Transformations

3.7.1 Ethanol

It is obtained by the fermentation of sucrose. It is not possible to obtain more than 10–15% ethanol by this method (since fermentation is inhibited if the concentration of alcohol exceeds 15%). More concentrated alcohol is isolated by fractional distillation. The fermentation of sucrose is represented as shown in Scheme 3.44.

Materials

Sucrose   10 g
Baker's yeast   1 g
Pasteur's salts 10 ml
solution

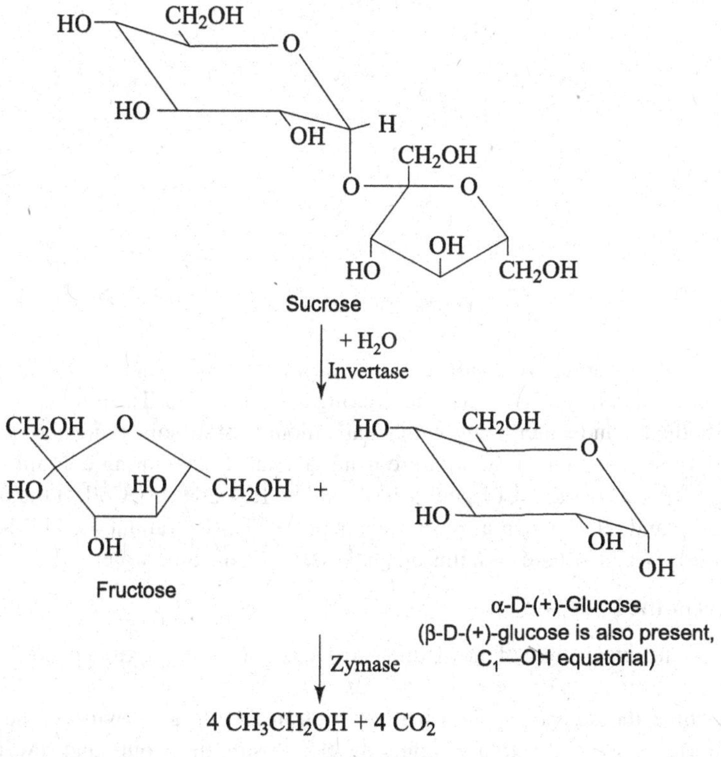

Scheme 3.44 Preparation of ethanol

Note

A solution of Pasteur's salt consists of potassium dihydrogen phosphate (1 g), calcium phosphate (monobasic) (0.1 g), magnesium sulphate (0.1 g) and ammonium tartrate (diammonium salt) (5 g) dissolved in water (430 ml).

Procedure

Place sucrose (10 g) in an Erlenmeyer flask (500 ml capacity). Add water (70 ml) warmed to 25–30 °C, Pasteur's salt solution (10 ml) and dried baker's yeast (1 g). Shake the contents to mix them thoroughly and then attach a balloon directly to the flask (see figure below). The gas will make the balloon expand as fermentation continues. In this way, contact with atmospheric oxygen is excluded, otherwise ethanol formed would be converted into acetic acid. As long as CO_2 continues to be liberated, alcohol is being formed.

Allow the mixture to stand at about 30–35 °C for a week. After this period, remove the flask from the source of the heat and detach the balloon.

Fermentation apparatus

Transfer the clear liquid without disturbing the sediment to another container. This can be done either by centrifugation or filtering using filter aid. The liquid (or filtrate) contains ethanol and water along with small amounts of dissolved metabolites (fuel oils) from the yeast. Distil the mixture using a small fractionating column using a sand bath for heating (temp. of sand bath should be 150–200 °C). Collect the fraction between 77 and 79 °C when most of the ethanol is distilled (about 4 ml of distillate is obtained). The distillate contains ethanol dissolved in some water.

Analysis of the Distillate

Determine the total weight of the distillate to determine the approximate yield. (Water and alcohol form an azeotrope, which boils at 78 °C.)

Determine the density of the distillate by transferring a known volume of the distillate (using a pipette) to a weighted bottle, closing the mouth and reweighing. This procedure gives good values up to two significant figures. Using the following table, determine the percentage composition by weight of alcohol in the distillate from the density as determined.

% Ethanol by weight	Density at 20 °C (g.ml)
75	0.856
80	0.843
85	0.831
90	0.818
95	0.804
100	0.789

This preparation is green synthesis, since no waste is obtained as a by-product which may cause pollution.

3.7.2 (S)-(+)-Ethyl 3-Hydroxybutanoate

It is obtained by chiral reduction of ethyl acetoacetate using baker's yeast (Scheme 3.45).

A small amount (<10%) of the enantiomer with (R) configuration is also obtained.

Materials

Ethyl acetoacetate   6.0 g
Sucrose   90 g
Dry baker's yeast   10.4 g

Procedure

Fit up the apparatus as shown in Fig. 3.1. Place a magnetic stirring bar and a one-hole rubber stopper with a glass tubing leading to a beaker containing a solution of barium hydroxide. Add some mineral oil on the top layer in the beaker to protect the barium hydroxide from air.

Add 150 ml water, 45 g sucrose and 5.2 g baker's yeast to the flask in the order given while stirring. Continue stirring for 1 h (about 20–30 °C). Add 6 g of ethyl acetoacetate, allow the fermenting mixture to stand at room temperature, while stirring for 24 h. Add a solution of sucrose (45 g) in water (150 ml) solution (prepared

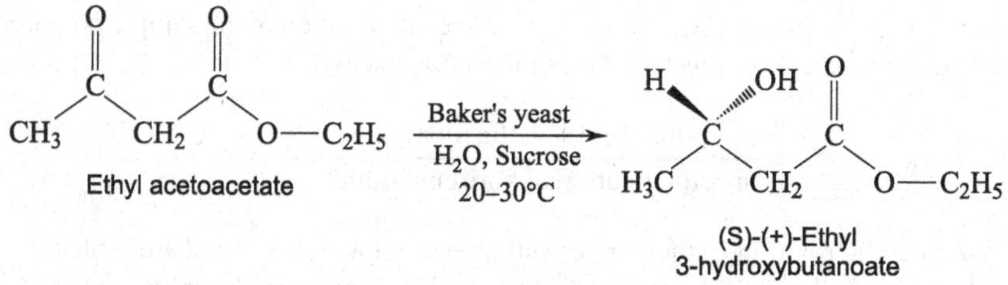

Scheme 3.45 Enzymatic synthesis of S-(+)-ethyl 3-hydroxybutanoate

Fig. 3.1 Setup of apparatus for the reduction of ethyl acetoacetate

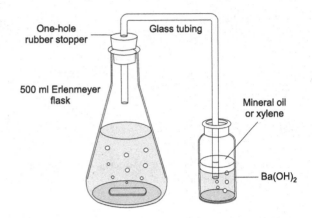

at 40 °C) along with 5.2 g baker's yeast to the fermenting liquid. Allow the mixture to stand for 48 h (with the trap attached at room temperature).

After the reaction is complete, filter it first by decantation through a Buchner funnel (filter paper) containing filter aid (the filter aid should be covered with water) and then transfer the total contents into the Buchner funnel (using mild pressure). At no stage the filter paper will be allowed to dry. Wash the residue with water (20–30 ml). To the filtrate, add sodium chloride (about 60 g), stir vigorously and extract with ether (3 × 40 ml). Dry the combined ether extract over anhydrous magnesium sulphate. Decant the ether and remove by distillation using a boiling stone and evaporate the ether using a warm water bath in a hood in nitrogen atmosphere. About 3–4 ml of the reduced product is obtained.

Purify the crude hydroxy ester by column chromatography on alumina column using methylene chloride (pour the ester in the column) and eluting with methylene chloride. Evaporation of the eluate gives the pure hydroxy ester. Find its weight.

Take the 1R spectrum of the hydroxy ester. Make sure that a stretching peak at 3200–3500 cm^{-1} observed for O—H stretching peak and a stretching peak at about 1715 cm^{-1} has disappeared (there will be still a C==O stretching peak from the ester functional group around 1735 cm^{-1}). In case, C==O stretching peak (at about 1715 cm^{-1}) remains or if the O—H stretching peak (at about 3200–3500 cm^{-1}) remains, the reduction has not taken place and the experiment has to be repeated.

Calculation of Optical Purity or Enantiomeric Excess

With the help of a polarimeter, determine the specific rotation of the sample prepared. Then percentage optical purity = % enantiomeric excess

$$= \frac{\text{Observed specific rotation}}{\text{specific rotation of pure enantiomer}} \times 100$$

The specific rotation of the pure enantiomer is known from standard tables.

The published specific rotation of (+)-ethyl-3-hydroxybutanoate is $[\alpha]_D^{25} = +43.5°$.

Let us suppose that optical purity as determined from the above equation comes to be 60%.

$$\therefore \% \, (+) \text{ enantiomer} = 60 + \frac{100 - 60}{2} = 80\%$$

$$\text{and } \% \, (-) \text{ enantiomer} = \frac{100 - 60}{2} = 20\%$$

It should be understood that the difference between these two calculated values is equal to the optical purity or enantiomeric excess (ee).

Note

The specific rotation is defined as the number of degrees of rotation observed when light is passed through 1 decimetre (10 cm) of its solution having a concentration of 1 g per millilitre. The specific rotation is calculated by using the equation.

$$[\alpha]_D^{t^0} = \frac{\alpha_{aba}}{l \times c}$$

where $[\alpha]_D^{t^0}$ stands for specific rotation determined at t °C using D-line of sodium light, α_{abc} is the observed angle of rotation; l is the length of the solution in decimetres; and c is the concentration of the active compound in grams per millilitre.

Note

This is green synthesis, since no by-products which may cause pollution are formed.

3.7.3 Benzoin

It is well known that benzoin is obtained from benzaldehyde using cyanide ion, an inorganic reagent as catalyst.

A green synthesis of benzoin consists of the reaction of benzaldehyde with a biological coenzyme, thiamine hydrochloride as the catalyst (R. Breslow, J. Am. Chem. Soc., 1958, *80*, 3719) (Scheme 3.46).

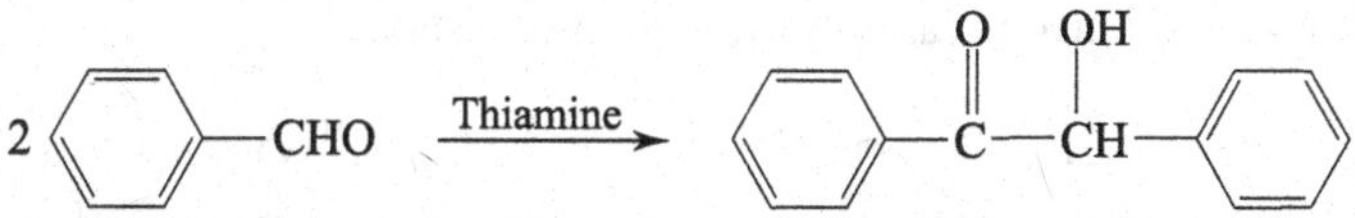

Benzaldehyde   Benzoin

Scheme 3.46 A green synthesis of benzoin

Mechanism

Thiamine loses a proton to the solvent (or to the enzyme) to give conjugate base of thiamine, which adds to benzaldehyde to give thiamine-benzaldehyde. The thiamine-benzaldehyde loses a proton to give a resonance-stabilized equivalent of acyl carbanion, which adds to a second molecule of benzaldehyde to give a product that eliminates thiamine to give benzoin and to regenerate the catalyst (Scheme 3.47).

Materials

Benzaldehyde	0.9 ml
Thiamine hydrochloride	0.3 g
Ethanol (95%)	3 ml
Sodium hydroxide solution	0.9 ml
(obtained by dissolving 2 g NaOH	
in 25 ml H_2O)	

Procedure

In an Erlenmeyer flask (25 ml capacity) dissolve thiamine hydrochloride (0.3 g) in water (0.5 ml). Add ethanol (3 ml), shake the container till all solid dissolves. To this, add sodium hydroxide solution (0.9 ml, 8%) and swirl the flask until the bright yellow colour fades to a pale yellow colour. Add benzaldehyde (0.9 ml). Swirl the flask until homogeneous solution is obtained. Stopper the flask and let it stay in dark place for 48 h.

Cool the flask (ice-bath), scratch the sides of the flask (to induce crystallization) and filter the formed benzoin. Crystallize from alcohol (using 0.8 ml alcohol/0.1 g of crude benzoin). Record the yield, m.p. 134–35 °C.

Notes

1. Pure benzaldehyde should be used. For use, it is best to shake benzaldehyde with an equal volume of 5% aq. Na_2CO_3 solution. Remove the lower layer of Na_2CO_3 solution, wash the aldehyde layer with H_2O and dry over anhydrous $CaCl_2$. The resulting purified benzaldehyde is suitable for the above preparation.
2. It is equally important to use pure thiamine hydrochloride, which should be stored in a refrigerator.
3. Thiamine serves as coenzyme for the following three types of enzymatic reactions

    (a)   Non-oxidative decarboxylation of α-ketoacids.

$$R-\overset{\overset{\displaystyle O}{\|}}{C}-COOH \xrightarrow{\ B_1\ } R-\overset{\overset{\displaystyle O}{\|}}{C}-H \ + \ CO_2$$

    (b)   Oxidative decarboxylation of α-ketoacids.

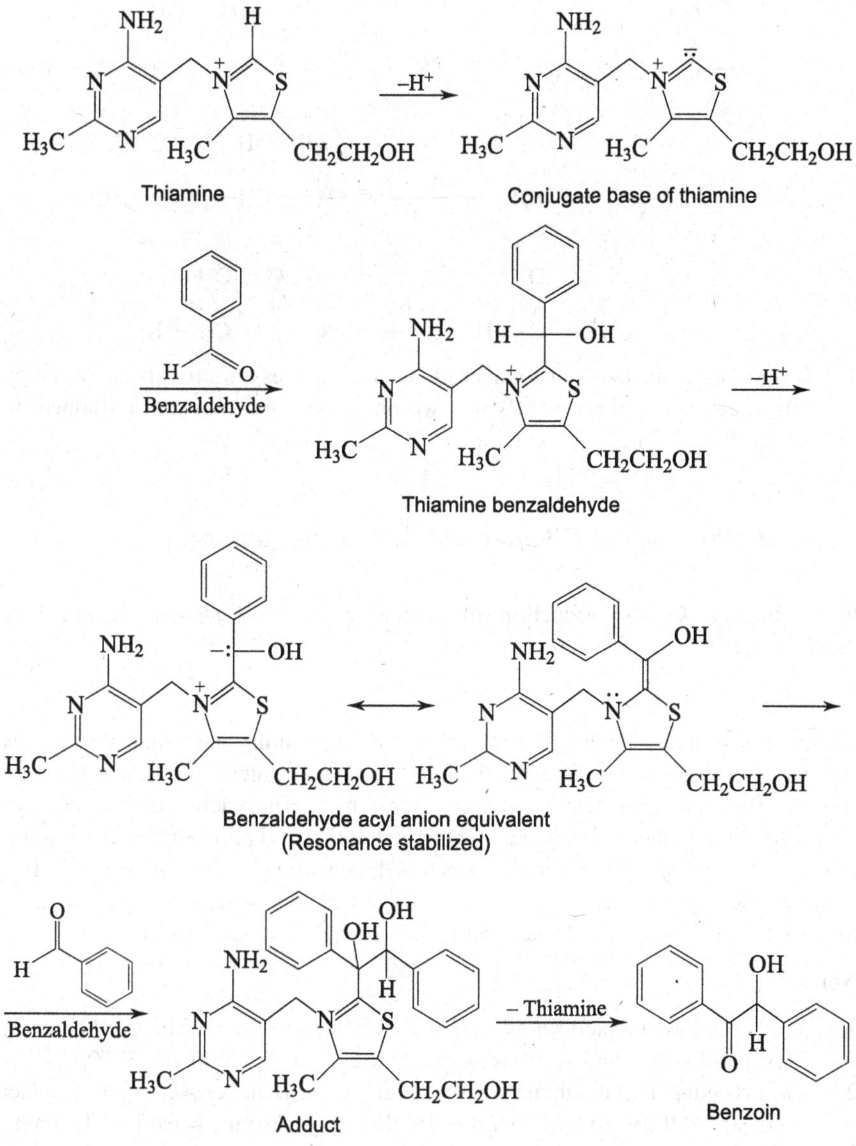

Scheme 3.47 Mechanism of the synthesis of benzoin using thiamine

$$R-\overset{\overset{\displaystyle O}{\|}}{C}-COOH \xrightarrow{B_1, O_2} R-\overset{\overset{\displaystyle O}{\|}}{C}-OH + CO_2$$

(c)   Formation of acyloins (α-hydroxy ketones).

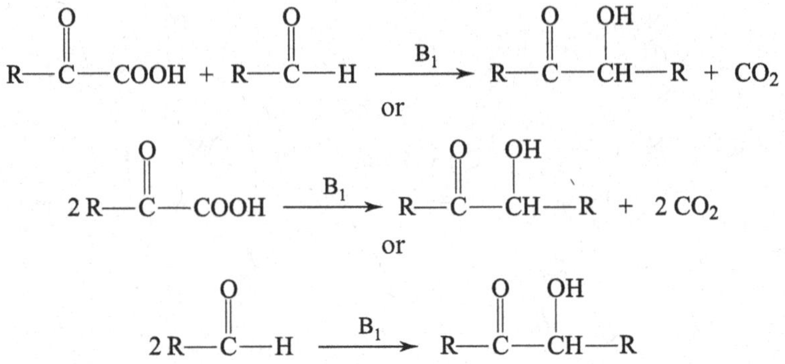

$$R-\overset{\overset{\displaystyle O}{\|}}{C}-COOH + R-\overset{\overset{\displaystyle O}{\|}}{C}-H \xrightarrow{B_1} R-\overset{\overset{\displaystyle O}{\|}}{C}-\overset{\overset{\displaystyle OH}{|}}{C}H-R + CO_2$$

or

$$2\,R-\overset{\overset{\displaystyle O}{\|}}{C}-COOH \xrightarrow{B_1} R-\overset{\overset{\displaystyle O}{\|}}{C}-\overset{\overset{\displaystyle OH}{|}}{C}H-R + 2\,CO_2$$

or

$$2\,R-\overset{\overset{\displaystyle O}{\|}}{C}-H \xrightarrow{B_1} R-\overset{\overset{\displaystyle O}{\|}}{C}-\overset{\overset{\displaystyle OH}{|}}{C}H-R$$

4. It is reported that benzoin condensations of aldehydes are strongly catalysed by quaternary ammonium cyanide in a two-phase system (J. Soludar, Tetradhedron Lett., 1971, 287).

3.7.4 1-Phenyl-(1S) Ethan-1-ol from Acetophenone

It is obtained by the reduction of acetophenone with *Daucus carota* root (Scheme 3.48).

Procedure

Remove the external layer of carrot and cut the remaining into small thin pieces (approx. 1 cm long slice). To acetophenone (1 ml) and water (20 ml), add slices of carrots (10 g). Shake the reaction mixture occasionally. The reaction takes about 40 h for completion. Isolate the product by ether extraction and purifying by chromatography over silica gel (200 mesh). Elution was done with ether–petroleum ether (1:2) mixture. Yield 73%. % ee 92. The alcohol obtained was 1-phenyl-(1S)-ethan-1-ol. Record its NMR spectra and rotation $[\alpha]_D^{25} = -39.1$ (C = 3.5, CHCl$_3$).

Notes

1. The procedure adopted for asymmetric reduction is of J.S. Yadav, S. Nanda, P. Thirupathi Reddy and A. Bhaskar Rau, J. Org. Chem., 2000, *67*, 3900.
2. The procedure is better than that uses baker's yeast as the recovery of the product is not straightforward (O.P. Ward and C.S. Young, Enzyme Microbiol Technol., 1998, *12*, 4822).

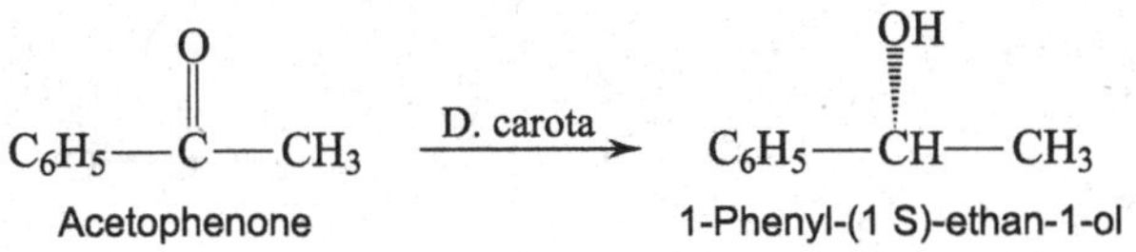

$$C_6H_5-\overset{\overset{\displaystyle O}{\|}}{C}-CH_3 \xrightarrow{D.\ carota} C_6H_5-\overset{\overset{\displaystyle OH}{|}}{C}H-CH_3$$

Acetophenone   1-Phenyl-(1 S)-ethan-1-ol

Scheme 3.48 Enzymatic synthesis of 1-phenyl-(1S) ethan-1-ol from acetophenone

3. The chiral alcohol obtained had S configuration, which is in perfect agreement with Prelog's rule.

4. Using the above procedure a number of ketones (viz., p–Cl, p–Br–, p–F–, p–NO$_2$–, p–CH$_3$, p–OCH$_3$–, p–OH–acetophenones could be reduced to the corresponding S alcohol with ee 90–98% and in 70–82% yield. The reduction was monitored by TLC carried out on 0.25 mm silica gel plates with UV light and 2.5% ethanolic anisaldehyde (with 1% CH$_3$COOH and 3% conc. H$_2$SO$_4$)–heat as developing agent.

5. The procedure can also be used for the reduction of β-ketoesters, cyclic ketones, azido ketones and open chain ketones like 2-butanone, 2-pentanone etc.

3.7.5 Deoximation of Oximes by Ultrasonically Stimulated Baker's Yeast

The enzymatic conversion of oximes by ultrasonically stimulated baker's yeast yields the corresponding aldehydes and ketones.

Materials

Oxime of p-methoxy benzaldehyde	0.7 g
Ethanol	15 ml
Baker's yeast	15 g

Procedure

To ultrasonically pretreated baker's yeast (15 g) in phosphate buffer (pH 7.2; 150 ml), add the oxime (0.7 g) in ethanol (15 ml).

The mixture was incubated at 37 °C for 2–3 days. It was filtered, and the filtrate was extracted with ethylacetate. The organic phase was dried and evaporated under reduced pressure. The residue obtained was purified by column chromatography to give p-methoxy benzaldehyde. b.p. 247 °C. Yield 98%.

Notes

1. The procedure followed was described by A. Kamal, M.V. Rao and H.M. Meshram, J. Chem. Soc. Perkin Trans I., 1991, 2056.

2. Oxime of p-methoxy benzaldehyde was prepared by the usual procedure.

3.8 Sonication Reactions

3.8.1 Butyraldehyde

It is obtained by **Bouveault reaction**[1] of 1-chlorobutane by ultrasonic irradiation with lithium and dimethyl formamide (Scheme 3.49).

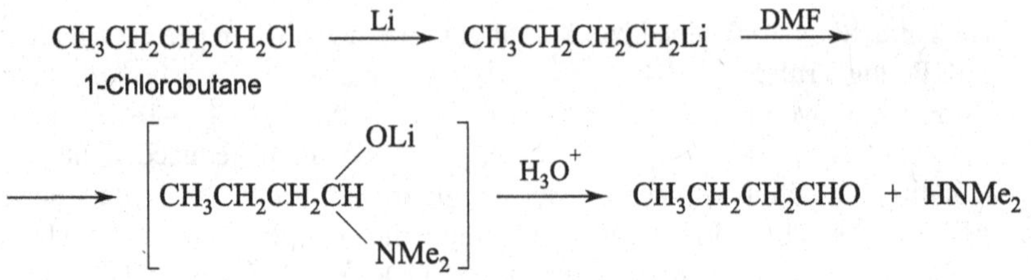

$$CH_3CH_2CH_2CH_2Cl \xrightarrow{\ Li\ } CH_3CH_2CH_2CH_2Li \xrightarrow{\ DMF\ }$$

1-Chlorobutane

$$\longrightarrow \left[CH_3CH_2CH_2\underset{\overset{|}{NMe_2}}{\overset{\overset{OLi}{|}}{CH}} \right] \xrightarrow{H_3O^+} CH_3CH_2CH_2CHO \ + \ HNMe_2$$

Scheme 3.49 Synthesis of butyraldehyde from 1-chlorobutane

Materials

1-Chlorobutane	1.85 g
Lithium	0.1 g
Dimethyl formamide	1.5 g

Procedure

A mixture of 1-chlorobutane (1.85 g), dry dimethyl formamide (1.5 g) and lithium sand (0.1 g in 4 ml dry THF) is sonicated in an ultrasound cleaner (96 w/l, 40 kHz). The reaction is complete at 10–20 °C in 15 min. Extract the mixture with ether (50 ml), wash ether with dil. HCl and sodium chloride solution and dry (anhydrous Na_2SO_4). Distillation of the ether solution gave n-butyraldehyde with 78% yield.

Notes

1. The procedure followed is that of C. Petrier, A.L. Gemal and J. L. Luche, Tetrahedron Lett., 1982, *23*, 3361.
2. Lithium metal is obtained as suspension in mineral oil. For using, it is washed with anhydrous THF under inert atmosphere.
3. Using appropriate halides (chlorides or bromides), a number of aldehydes can be prepared. For example, cyclohexyl halides give the corresponding aldehyde.

3.8.2   2-Chloro-N-Aryl Anthranilic Acid

2-Chloro-N-arylanthranilic acid is prepared by the **Ullmann condensation** of 2-chlorobenzoic acid with 2-choloraniline in the presence of copper powder and cuprous iodide in boiling DMF with ultrasonic irradiation (Scheme 3.50).

Materials

2-Chlorobenzoic acid	1.5 g
2-Chloroaniline	1.0 g
Copper powder	0.1 g
Cuprous iodide	0.1 g
Potassium carbonate	0.55 g

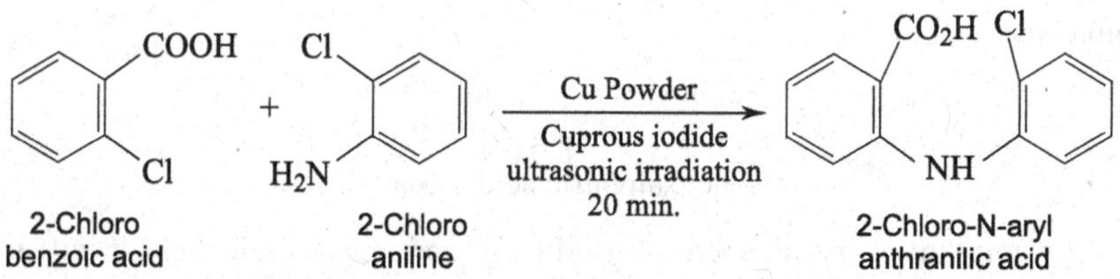

Scheme 3.50 Synthesis of 2-chloro-N-aryl anthranilic acid

Procedure

Heat a mixture of 2-chlorobenzoic acid (1.5 g), 2-chloroaniline (1.0 g), potassium carbonate (0.55 g), copper powder (0.1 g) and cuprous iodide (0.1 g) in boiling DMF (5 ml) for 20 min with ultrasonic irradiation (virsonic 300 cell disrupter at 20 kHz). Pour the reaction mixture onto water (20 ml). Filter the separated solid, treat with conc. K_2CO_3 solution. Acidify the clear filtrate with dilute acetic acid (1:3). The product separates at pH 6–8. Crystallize the product, m.p. 196–8 °C (75% yield).

Notes

1. The procedure described is that of R. Carrasco, R.F. Pellon, Jose' Elguero, Pilar Goya and Juan Antorio Paez, Synthetic Communications, 1989, *19* (*11* and *12*), 2077–2080.
2. The procedure using ultrasound is very convenient, shortens the reaction time, i.e., 20 min in comparison to 4–6 h by the usual procedure. Also, the reaction proceeds in a much cleaner way and affords higher purity of crude products.

3.9 Esterification

3.9.1 Benzocaine (Ethyl p-Amino Benzoate)

It is a local anaesthetic and is prepared by direct esterification of *p*-aminobenzoic acid with ethanol (Scheme 3.51).

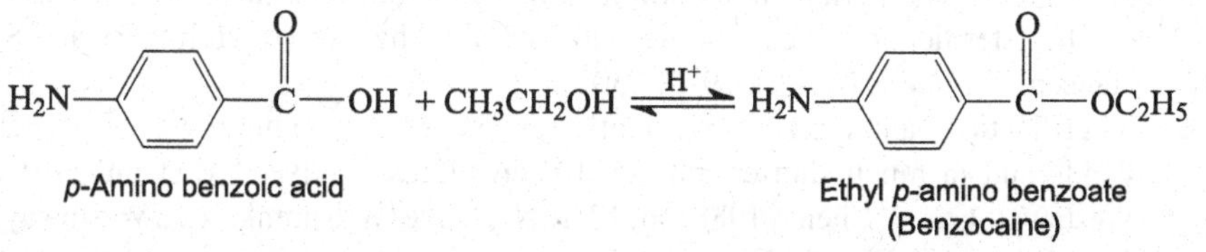

Scheme 3.51 Synthesis of benzocaine

Materials

> p-Aminobenzoic acid   4 g
> Absolute ethanol   40 ml
> Conc. sulphuric acid   3 ml

Stir a mixture of p-aminobenzoic acid (4 g) and absolute alcohol (40 ml) in a round-bottomed flask till complete solution results. To the stirred solution, add conc. sulphuric acid dropwise (3 ml). A precipitate forms, but it dissolves on stirring and subsequent refluxing. Reflux the mixture using a reflux condenser for 1–1.5 h at about 105 °C. Stir the mixture during this period.

Allow the reaction mixture to cool down to room temperature. Add water (30 ml) and then add sodium carbonate solution (10%) dropwise to neutralize the mixture. As the pH increases, a white precipitate of benzocaine is produced. Add more sodium carbonate solution until pH is about 8. Filter the separated benzocaine, wash with water and crystallize from methanol–water. Record the yield and m.p. The pure benzocaine melts at 92 °C.

Notes

1. You may test benzocaine for its anaesthetic action on frog's muscle.
2. Carboxylic acids can be conveniently esterified [H.E. Hennis, L.R. Thompson and J. P. Long, Ind. Eng. Chem. Prod. Res. Dev., 1968, 7, 96. H.E. Hennis, J. P. Esterly, L.R. Collins and L. R. Thompson, Ind. Eng. Chem. Prod Res. Dev., 1967, 6, 143] with alkyl halide and triethyl amine. In this case, PTC is generated in situ by the reaction of triethylamine and alkyl halide.

$$Et_3N \; + \; \underset{\substack{\text{Alkyl} \\ \text{halide}}}{RX} \; \longrightarrow \; \underset{\text{PTC}}{Et_3\overset{+}{N}RCl^-}$$

$$\underset{\substack{\text{Sodium salt of} \\ \text{carboxylic and} \\ (aq.\ \text{solution})}}{R'COONa} \; + \; \underset{\substack{\text{Alkyl} \\ \text{halide}}}{RX} \; \xrightarrow{\text{PTC}} \; \underset{\text{Ester}}{R'COOR} \; + \; NaX$$

In the above case, alkyl halide must be highly reactive, e.g., benzylchloride. Alternatively, quaternary ammonium or phosphonium salts can be directly used for the esterification of carboxylic acid with alkyl halides (R. Holmbery and S. Hansen, Tetrahedron Lett., 1975, 2307).

3. Esterification can be very conveniently carried out using a microwave in better yields and in much shorter time (5–10 min) (R.N. Gedye, F.E. Smith, K.C. Westaway, Can. J. Chem., 1988, 66, 17; R.N. Gedye, W. Ramnk, K.C. Westaway, Can. J. Chem., 1991, 69 706).

3.9.2 Isopentyl Acetate (Banana Oil)

The ester, isopentyl acetate is referred to as banana oil, because it has familiar odour of this fruit. It is prepared by esterification of acetic acid with isopentyl alcohol. Since the equilibrium does not favour the formation of the ester, it must be shifted to the right in order to get better yield of the ester. For this purpose, one of the starting materials is used in excess. Acetic acid being less expensive than isopentyl alcohol and more easily removed from the reaction mixture is used (Scheme 3.52).

The mechanism of the reaction is similar to that described for the preparation of methyl salicylate (see Sect. 3.9.3).

Materials

Isopentyl alcohol	12.5 ml
(Isoamyl alcohol)	
Acetic acid	18 ml
Conc. sulphuric acid	2.5 ml

Procedure

In a round-bottomed flask take isopentyl alcohol (12.5 ml) and glacial acetic acid (18 ml). To the clear solution add dropwise conc. sulphuric acid (2.5 ml) during shaking. Reflux the mixture for 1 h. To the cooled solution add water (25 ml). Remove the lower aqueous layer and discard (use a small separatory funnel). Extract the organic layer with aqueous sodium bicarbonate (2 × 5 ml) and then with water. Dry the organic layer over anhydrous sodium sulphate, filter and distil. Yield 10 g. b.p. 142 °C.

Notes

1. See note 2 in the preparation of benzocaine (Sect. 3.9.1).
2. See note 3 in the preparation is benzocaine (Sect. 3.9.1).

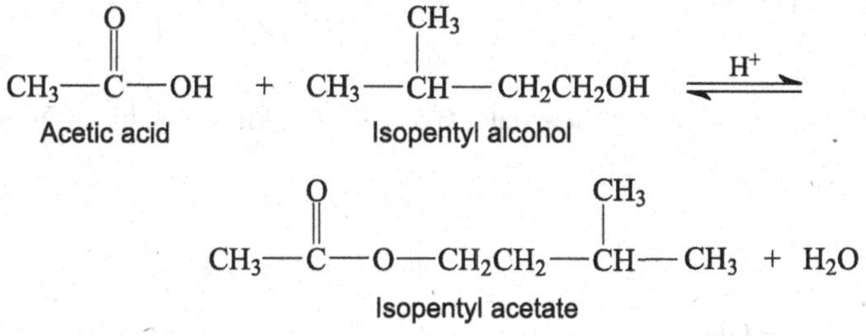

$$CH_3-\overset{\overset{\displaystyle O}{\|}}{C}-OH \ + \ CH_3-\overset{\overset{\displaystyle CH_3}{|}}{CH}-CH_2CH_2OH \ \underset{}{\overset{H^+}{\rightleftharpoons}}$$

Acetic acid   Isopentyl alcohol

$$CH_3-\overset{\overset{\displaystyle O}{\|}}{C}-O-CH_2CH_2-\overset{\overset{\displaystyle CH_3}{|}}{CH}-CH_3 \ + \ H_2O$$

Isopentyl acetate

Scheme 3.52 Synthesis of isopentyl acetate

3.9.3   Methyl Salicylate (Oil of Wintergreen)

Methyl salicylate was isolated as a familiar–smelling organic ester–oil of wintergreen from the wintergreen plant (*Gaultheria*) in 1843. It was found to have analgesic and antipyretic character almost identical to that of salicylic acid when taken internally. This medicinal characteristic is due to the ease with which methyl salicylate is hydrolysed to salicylic acid under alkaline conditions in the intestinal tract. Salicylic acid is known to have analgesic and antipyretic properties. Methyl salicylate can be taken internally or absorbed through the skin. It finds its application in liniment preparations. On applying to skin, it produces a mild tingling and soothing sensation. The ester is also used to a small extent as a flavouring principle due to its pleasant odour.

Methyl salicylate is obtained by the esterification of salicylic acid with methanol in the presence of acid. It is an equilibrium reaction. The equilibrium can shift to right, i.e., more product can be obtained by increasing the concentration of one of the reactants (methyl alcohol being cheaper is used in excess) (Scheme 3.53).

Mechanism (Scheme 3.54)

Materials

Salicylic acid	6.5 g
Methyl alcohol	20 ml
Conc. sulphuric acid	3 ml

Procedure

In a round-bottomed flask dissolve salicylic acid (6.5 g) into methanol (20 ml). To the clear solution, add conc. H_2SO_4 dropwise (3 ml) during shaking. Reflux the solution at 80 °C for 2.5 h. Remove excess methyl alcohol by distillation (steam bath) and pour the residual product into water (40 ml). Extract with ether (2 × 50 ml), wash ether extract with sodium bicarbonate solution (till free of acid) and finally with water. Dry ether extract over anhydrous sodium sulphate. Methyl salicylate is obtained as a colourless liquid with fragrant smell. b.p. 223–25 °C. Yield 6.5 g (85%).

Notes

1. In a similar way, methyl benzoate (b.p. 190 °C), known as oil of Niobe can be prepared.

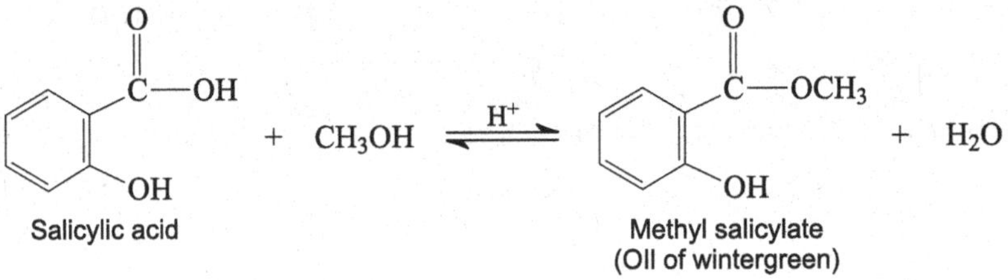

Scheme 3.53 Preparation of methyl salicylate

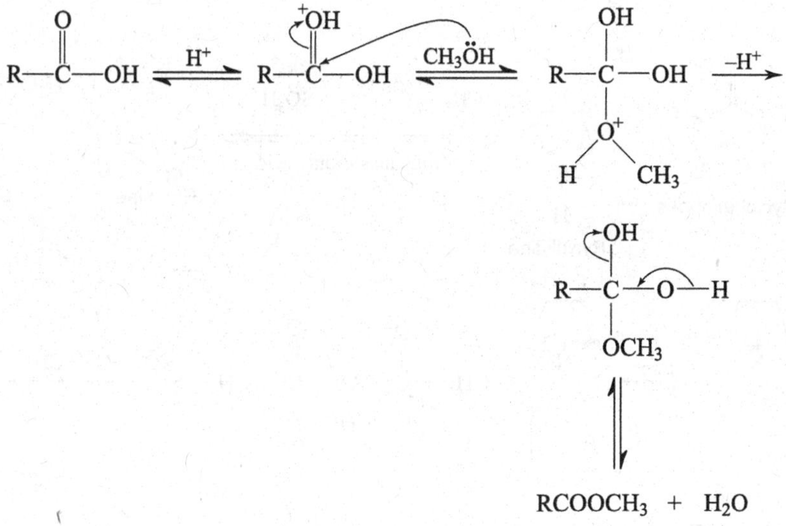

Scheme 3.54 Mechanism of esterification

2. See notes 2 and 3 in the preparation of benzocaine (Sect. 3.9.1).

3.10 Enamine Reaction

3.10.1 2–Acetyl Cyclohexanone

It is prepared by the reaction of enamine (obtained from cyclohexanone and pyrrolidine in the presence of p-toluene sulfonic acid) with acetic anhydride (Scheme 3.55).

Step 1: Preparation of enamine.

Materials

Cyclohexanone	1.92 ml
p-Toluene sulfonic acid	55 mg
Monohydrate	
Pyrrolidine	1.62 ml
Toluene	12 ml

Procedure

Take cyclohexanone (1.92 ml) in a round-bottomed flask (25 ml capacity). Add toluene (12 ml) and p-toluene sulfonic acid monohydrate (55 mg). To the mixture, add pyrrolidine (1.62 ml). Stir the mixture using a magnetic needle for 10–15 min under anhydrous conditions. While stirring, set up the distillation apparatus. Heat

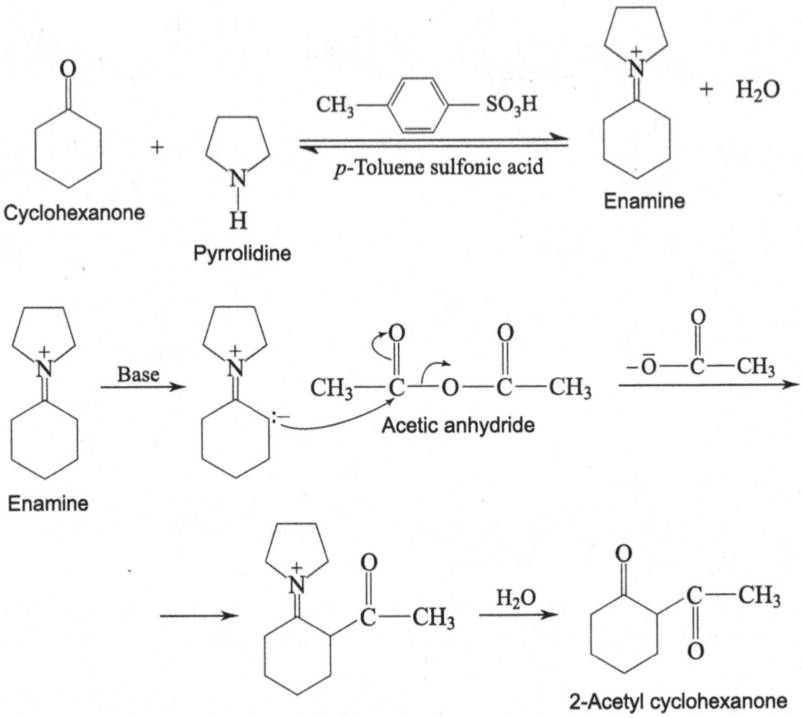

Scheme 3.55 Preparation of 2-acetyl cyclohexanone

the reaction mixture to 140–45 °C (sand bath). The rate of distillation should be controlled so that the distillate 11–12 ml gets collected in about 30 min. By this process, the water formed in the reaction is removed as an azeotrope. The flask contains the enamine and is used as such for the next step.

Step 2: 2-Acetyl cyclohexanone

To the enamine obtained in the step 1, add acetic anhydride (1.92 ml) dissolved in toluene (3 ml). Swirl the flask (which has been corked) for about 5 min at room temperature and allow the mixture to stand for 48 h.

After this period, add water (3 ml). Reflux the mixture (using a water-cooled condenser) at about 120 °C for 30 min. Cool the flask to room temperature, add water (3 ml) and separate the upper toluene layer using a small separatory funnel. Add to toluene layer (containing 2-acetyl cyclohexanone) hydrochloric acid (6 M, 6 ml). Shake the mixture. By this procedure, pyrrolidine is extracted by HCl. Discard the acid layer. Shake the organic layer with water (3 ml). Dry the solution (anhyd. Na_2SO_4) and evaporate the dried toluene solution in a water bath at 70 °C using a stream of dry air or nitrogen. When all the toluene is removed, the dry product (2-acetyl cyclohexanone) is purified by column chromatography using alumina as the absorbent and methylene chloride as the eluent.

2-Acetyl cyclohexanone is obtained as a yellow liquid. Determine its yield and record the NMR spectra.

Note

1. The above procedure can also be used for the preparation of 2-alkyl cyclohexanone by using alkyl halide in place of acetic anhydride.

3.11 Reactions in Ionic Liquids

3.11.1 1-Acetylnaphthalene

It is obtained by **Friedel–Crafts reaction** of naphthalene with acetyl chloride in the presence of ionic liquid, such as the [emin] Cl–AlCl$_3$ [emin = 1-methyl 3-ethylimidazolium cation] (Scheme 3.56).

Materials

Naphthalene	6.4 g
Acetyl chloride	3.9 g
[emin] Cl-AlCl$_3$	20 g

Step 1: 1-methyl-3-ethylimidazolium chloride. It is prepared by the procedure described by R.S. Verma, V.V. Namboodiri, Chem. Commun., 2001, 643.

A mixture of ethyl chloride (7.8 g) and 1-methyl-3-ethyl imidazole (11.2 g) were mixed thoroughly and the mixture heated in unmodified household microwave oven (240 W) for 1 min or till a clear single phase is obtained. Cool the resulting ionic liquid and wash with ether (3 × 20 ml) to remove unreacted starting material, and the product 1-methyl-3-ethylimidazolium chloride is obtained.

Step 2: 1-Acetylnaphthalene

A mixture of naphthalene (6.4 g), acetyl chloride (3.9 g) and the ionic liquid [emin]Cl–AlCl$_3$ [emin-1-methyl-3-ethyl-imidazolium cation] (20 g) [obtained by mixing equimolar amounts of 1-methyl-3-ethyl imidazolium chloride with anhyd. aluminium chloride] was allowed to react at 0 °C at room temperature for 1 h. The product obtained (after extraction with ether) was crystallized from alcohol to give 1-acetyl naphthalene with 89% yield. Record its NMR spectra.

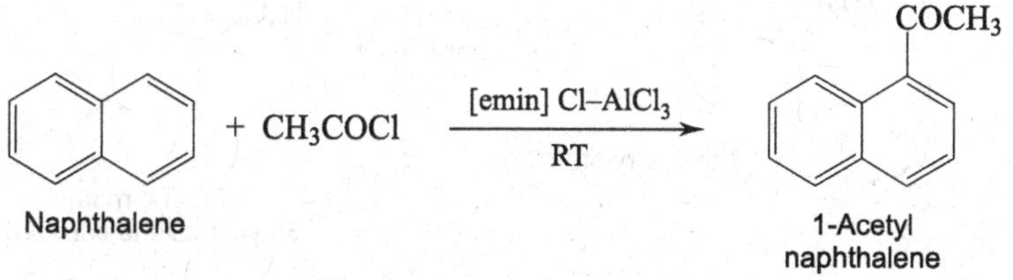

Scheme 3.56 Synthesis of 1-acetylnapthalene

Notes

1. The procedure described is adopted from C.J. Adams, M.J. Earle, G. Roberts and R. Seddon Chem. Common, 1998, 2097.
2. Friedel–Crafts acylation of naphthalene in ionic liquid gives the thermodynamically unfavourable 1-isomer. On the other hand, conventional Friedel–Crafts reaction gives the 2-isomer.
3. Using the above procedure, toluene, chlorobenzene and anisole give the corresponding 4-acetyl compounds with 97–99% yields.
4. The ionic liquid can be used again.

3.11.2   Ethyl 4-Methyl-3-Cyclohexene Carboxylate

It is prepared by **Diels–Alder reaction** of isoprene (dienophile) with ethyl acrylate (diene) in neutral ionic liquids, viz., 1-ethyl-3-methyl imidazolium tetrafluoroborate, [emin] BF_4 (Scheme 3.57).

Step 1: 1-Ethyl-3 methyl imidazolium tetrafluoroborate [emin] BF_4.

It is prepared by the procedure described by J.D. Holbreg and K.R. Seddon, J. Chem. Soc. Dalton Trans., 1999, 2133–2139.

Tetrafluoroboric acid (12.2 ml, 0.116 mol, 48% solution in water) was added slowly to a rapidly stirred slurry of silver (I) oxide (13.49 g, 0.058 mol) in water (50 ml) over 15 min. The reaction mixture was covered with aluminium foil to prevent photodegradation. The stirring was continued (1 h) until all the Ag(I) oxide had completely reacted to give a clear solution. A solution of 1-ethyl-3 methyl imidazolium bromide (22.24 g, 0.116 mol) in water (200 ml) was added to the reaction mixture and stirred at room temperature for 2 h. The resulting yellow precipitate of silver bromide was removed by filtration. The solvent was removed from the supernatant liquor by heating at 70 °C initially under reduced pressure and finally *in vacuo* to yield the tetrafluoro borate salt as a pale yellow liquid, with yield 21.36 g, 93%.

Step 2: Ethyl 4-methyl-3-cyclohexene carboxylate

The reaction of diene (isoprene) with dienophile (ethyl acrylate) in the binary liquid [b_{min}] BF_4] was performed in 1:5:1.0:1.0 molar ratio mixture of diene:dienophite:solvent at 70 °C for 24 h to give 97% yield of the required product (endo:exo ratio 4.0:1).

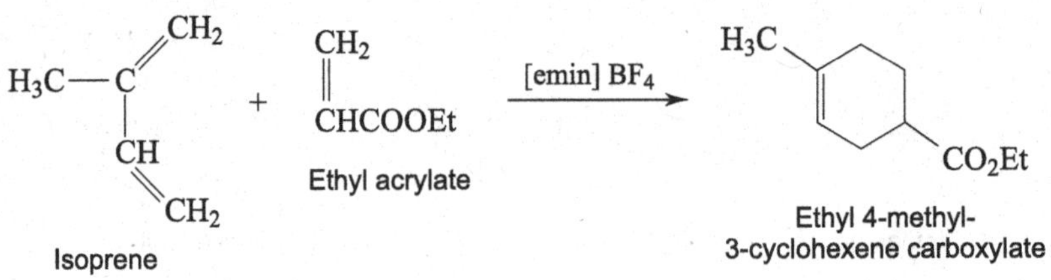

Scheme 3.57 Preparation of ethyl 4-methyl-3-cyclohexene carboxylate

Notes

1. The procedure described is of M.J. Earle, P.B. McCormac and K.R. Seddon, Green, Chemistry, 1999, 23.
2. Using isoprene (diene) and but-2-en-3-one (dienophile) in [b$_{min}$] [BF$_4$] at 20 °C, for 2 h gave 90% yield of 4-acetyl-1-methyl cyclohexene.
3. The addition of 5 mol per cent of zinc (II) iodide dramatically increases the rate of the reaction.
4. The final product was isolated by extraction with ether.
5. The binary liquid can be used again.

3.12   Green Preparation Using Renewable Resources

3.12.1   Biodiesel from Vegetable Oil

Biodiesel is a mixture of methyl esters of fatty acids and can be prepared from vegetable oil. The method involves mixing vegetable oil with methanol in the presence of catalytic amount of KOH. The products are obtained in two layers. The upper layer is the biodiesel. The reaction takes place by trans esterification (Scheme 3.58).

Procedure

Methanol (15 ml) is added carefully to vegetable oil (100 ml) taken in a 250 ml Erlenmeyer flask. Subsequently, a solution of KOH (1 ml, 9 M) is added slowly. The mixture is stirred for 10 min and poured in a separatory funnel. Two layers separate out. Drain out the lower layer and wash the remaining upper layer of biodiesel with distilled water (10 ml). The biodiesel is dried over anhydrous Na$_2$SO$_4$. The amount of biodiesel obtained is measured.

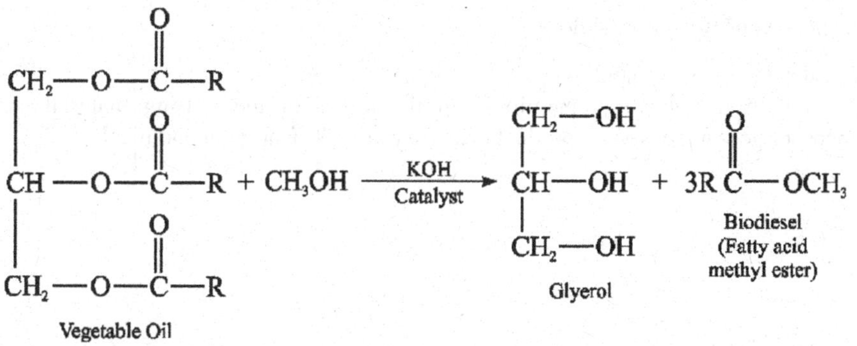

Scheme 3.58 Biodiesel from vegetable oil

Notes

1. A modified procedure is described in Introduction to Green Chemistry, American Chemical Society.
2. The biodiesel obtained from vegetable resources has negligible sulphur content and so does not produce SO_2 on ignition and thus help in reducing acid rain.
3. The IR spectra of biodiesel shows an absorption at about $1735 \, cm^{-1}$ due to the presence of ester group.
4. The pH of biodiesel is about 7. This can be observed by mixing 2–3 drops of diesel with 1 ml distilled water. The mixture is mixed and the pH of aqueous solution measured using a pH paper.
5. A number of other products can also be obtained by using renewable starting material. These include Ethanol (Sect. 3.7.1).

3.13   Reactions Using the Principles of Atom Economy (Avoiding Waste)

One of the important principles of green chemistry is designing synthetic method which should have maximum atom economy. It has been observed that rearrangement and addition reactions are 100% atom-economical while substitution reactions and elimination reactions are much less atom-economical.

3.13.1   Rearrangement Reactions

Synthesis of 2-alkyl phenol

For details see Sect. 3.6.1.

Synthesis of Benzopinacolone

For details see Sect. 3.6.2.

In both the above preparations, all the atoms of the starting material are incorporated in the formed products, i.e. they are 100% atom-economical.

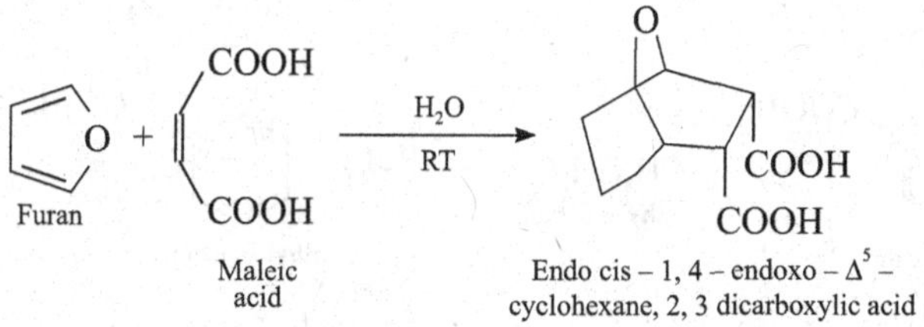

Scheme 3.59 Synthesis of maleic acid adduct

3.13.2 Addition Reactions

Synthesis of Endo-cis-1, 4-endoxo–Δ^5-cyclohexane-2, 3 dicarboxylic acid

It is obtained from Diels–Alder reaction of furan with maleic and in water at room temperature (Scheme 3.59).

Procedure

A mixture of maleic acid (5 g) and furan (2.5 g) in water (25 ml) is stirred at room temperature for 2–5 h. The separated adduct is filtered and washed with water. Yield 7.2 g. Record its m.p.

Notes

1. R.B. Wood ward and H. Baer, J. AM. Chem. Sec., 1948, 70, 1161; R. Breslow and D. Rideout, J. Am. Chem. soc. 1980, 102, 7816.
2. The above reaction is 100% atom-economical.
3. The reaction can also be satisfactorily performed under microwave irradiation. The reaction is complete in 90 s.
4. Another example of addition reaction is preparation of anthracene-maleic anhydride adduct (9,10-dihydroanthracene-endo-α, β succinic anhydride). For details, see Sect. 3.6.1.

3.13.3 Substitution Reactions

Ethyl Benzoate

It is prepared by the reaction of benzoic acid with ethyl alcohol in the presence of acid catalyst (Scheme 3.60).

In the above reaction, the OH group of benzoic acid and H atom of ethyl alcohol are not incorporated into the final product.

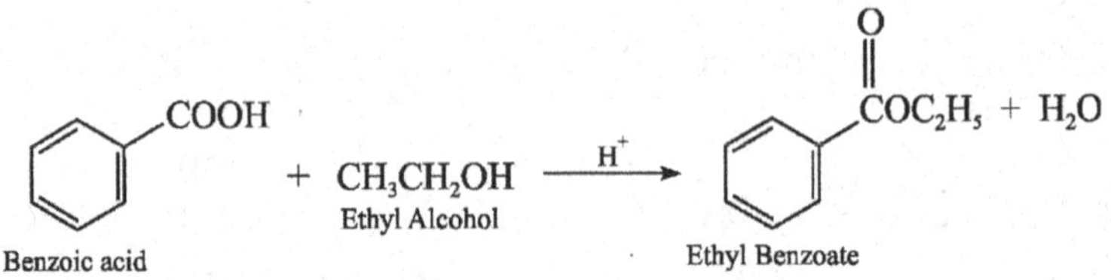

$$
\underset{\text{Benzoic acid}}{\text{C}_6\text{H}_5\text{-COOH}} + \underset{\substack{\text{CH}_3\text{CH}_2\text{OH} \\ \text{Ethyl Alcohol}}}{} \xrightarrow{\text{H}^+} \underset{\text{Ethyl Benzoate}}{\text{C}_6\text{H}_5\text{-COC}_2\text{H}_5} + \text{H}_2\text{O}
$$

Scheme 3.60 Synthesis of ethyl benzoate

Procedure

A mixture of benzoic acid (15 g, 0.123 mol), absolute ethyl alcohol (25 ml) and conc. H_2SO_4 (1 ml) is refluxed (using a water bath) for 4 h. The R.B. flask is fitted with a Dean and Stark distillation head so that the water formed can be removed. Finally, excess ethyl alcohol is distilled off (steam bath) and the residual mixture is poured into water. Excess of acid is neutralized by addition of solid $NaHCO_3$. The remaining liquid is extracted with ether (2 × 25 ml), ether extract dried (anhydrous K_2CO_3) and ether distilled yield. Yield 18 g. B.P. 213 °C.

Notes

1. In the above procedure, it is not esterified to use Dean–Stark distillation head, but the yields are low.
2. The carboxylic acid can also be esterified by heating with alkyl halides in the presence of triethylamine (H.E. Hennis, L.R. Thompson and J.P. Long, Ind. Eng. Chem. Prod. Res., Dev., 1968, I. 96; H.E. Hennis, J.P. Esterly, L.R. Collins and L.R. Thompson, Ind. Eng. Chem. Prod. Res. Dev. 1967, 6, 193). In this case the PTC is generated in situ by the reaction of ethyl amine and alkyl halide (Scheme 3.61).
The generated PTC catalyses the reaction between a carboxylic acid (as sodium salt) and alkyl halide to form an ester (Scheme 3.62).
In the above esterification reaction, the alkyl halide must be highly reactive, viz. benzyl chloride, which rapidly forms the quaternary salts (PTC). Alternatively, the quaternary ammonium or phosphonium salts can be directly used in the esterification of carboxylic acid with alkylhalide.

Scheme 3.61 In situ generation of PTC

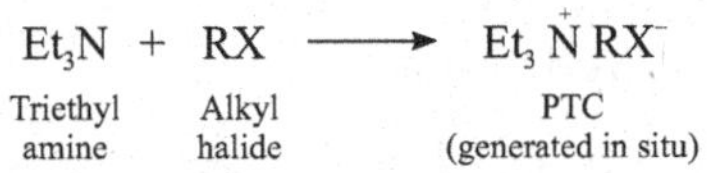

$$
\underset{\substack{\text{Triethyl} \\ \text{amine}}}{\text{Et}_3\text{N}} + \underset{\substack{\text{Alkyl} \\ \text{halide}}}{\text{RX}} \longrightarrow \underset{\substack{\text{PTC} \\ \text{(generated in situ)}}}{\text{Et}_3\overset{+}{\text{N}}\text{RX}^-}
$$

Scheme 3.62 Synthesis of esters

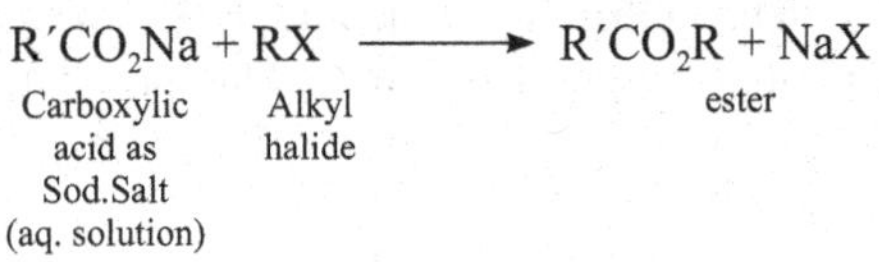

$$
\underset{\substack{\text{Carboxylic} \\ \text{acid as} \\ \text{Sod.Salt} \\ \text{(aq. solution)}}}{\text{R}'\text{CO}_2\text{Na}} + \underset{\substack{\text{Alkyl} \\ \text{halide}}}{\text{RX}} \longrightarrow \underset{\text{ester}}{\text{R}'\text{CO}_2\text{R}} + \text{NaX}
$$

$$RCOOH + R'OH \xrightarrow[\text{)))}]{\text{RT, conc} \cdot H_2SO_4 \text{ (catalyst)}} RCOOR'$$

Scheme 3.63 A simple process for esterification

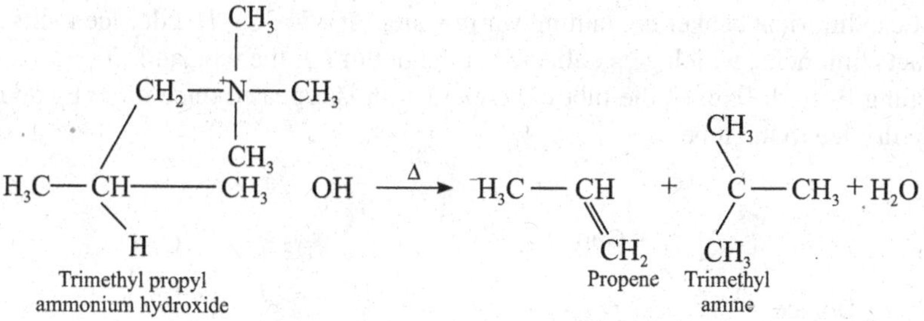

Trimethyl propyl ammonium hydroxide

Propene   Trimethyl amine

Scheme 3.64 An elimination reaction

3. A simple process for the esterification of carboxylic acid with alcohols in the presence of conc. H_2SO_4 can be easily carried out using ultrasound (J.M. Khurana, P.K. Sahoo and G.C. Mackap, Synth. Commun., 1990, 2267) (Scheme 3.63).
4. The percentage atom economy in the above esterification reaction is P 9.3.
5. For more details about esterification, see Sect. 3.9.

3.13.4 Elimination Reactions

In elimination reactions (e.g., Hofmann's elimination reaction), two atoms or group of atoms are lost from the reactant to form a π bond. Such reactions are not very atom-economical as the groups that are lost from the reactants are not incorporated into the final product. An example is given in Scheme 3.64.

The above reaction is found to be 35.30% atom-economical (see Sect. 1.7.2 subsection 'd'). Also see Sect. 2.33.

3.14 Extraction of D-Limonene from Orange Peels Using Liquid CO_2

d-Limonene occurs in citrus fruits like orange peels. It is used as a flavouring agent in food, beverages and chewing gums.

Carbon dioxide is normally a gas but at and over its critical pressure and temperature it becomes liquid which find its use for the extraction of natural products.

Procedure

A copper wire with a loop at one end (see Fig.) is inserted in a plastic centrifuge tube. A filter paper is placed on the loop of the copper wire. Grated orange peels (about 5 g) is inserted into the loop. The plastic centrifuge tube is filled with solid CO_2 (dry ice) and the top of the tube is closed with a cap. The tube is finally immersed in a plastic cylindrical beaker containing warm water (40–50 °C). The dry ice melts and extracts limonene, which gets collected at the bottom of the tube and CO_2 gas starts escaping from the cap of the tube. The extraction is repeated once more by adding more dry ice to the tube.

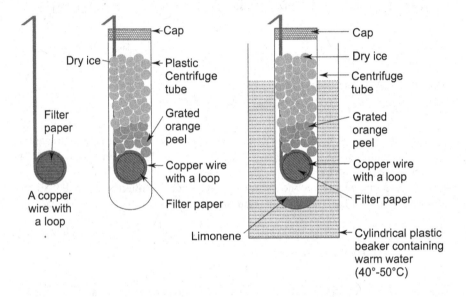

Extraction of Limonene from orange peel using liquid CO_2

Notes

1. a-Limonene has a density 0.8422. Its refractive index is 1.4743 and rotation + 123.8°.

2. Limonene can also be extracted from orange peels by using water as solvent. A mixture of blended orange peel (5–10 g) and water (about 150 ml) is blended in a blender (15–20 s). The blended mixture is distilled (using an RB flask). The distillate (25–30 ml) is collected and taken in a separatory funnel and allowed to stand. The lower aqueous layer is discarded. The clear limonene layer is dried in a vacuum desiccator.

Multiple Choice Questions

1. Which of the following is an obstacle in the pursuit of goals of green chemistry?
   (a) It is not possible to obtain starting materials from renewable resources.
   (b) It is not possible to avoid the formation of byproducts.
   (c) Use of benign solvent is not always possible.
   (d) Atom economy is not always possible.
   (e) All are the obstacles.

2. The most important principle of green chemistry is
   (a) Prevention of waste.
   (b) Maximum in corporation of all starting materials into the final product.
   (c) The requirement of energy should be minimum.
   (d) As far as possible catalysts be used.
   (e) All are equally important.

3. Which of the following reactions are 100 percent atom economical
   (a) Rearrangement reactions   (b) Addition reactions
   (c) Substitution reactions   (d) Elimination reactions

4. Which of the following is most harmful
   (a) Bhopal Gas Tragedy   (b) Flexborough disaster
   (c) Minimata disease   (d) Itai-Itai disease
   (e) Seveso Disaster   (f) Love Canal Disaster

5. The most important green solvent is
   (a) Water   (b) Ionic Liquid
   (c) Polythylene glycol   (d) Fluorous Solvents

6. The requirement of energy in a chemical reaction is kept minimum by
   (a) Sonication   (b) Microwave Irradiation
   (c) Photochemical activation   (d) all are equally important

7. The best result for the reaction is

$$RCH_2Cl + NaCN + H_2O \longrightarrow RCH_2CN$$

   (a) stirring the mixture for 5-6 hrs.
   (b) heating the mixture for 4-5 hrs.
   (c) carrying out the reaction in presence of a PTC at room temperature
   (d) The reaction is not possible.

8. For the conversion of benzaldehyde into benzene, the green process involves
   (a) Heating benzaldehyde with NaCN in H_2O/alcohol.
   (b) Stirring Benzaldehyde with NaCN in H_2O/alcohol
   (c) Carrying out the reaction of benzaldehyde in presence of a biological catalyst, thiamine hydrochloride
   (d) Any of the above process.

9. For the extraction of d-limonene from orange peels the green process is
   (a) Extraction of powered orange peels with liquid CO_2.
   (b) Extraction of powdered orange peels with H_2O followed by steam distillation.
   (c) Extraction is powdered orange peels with organic solvent.
   (d) Either process (a) or (b) is a green process.

10. The reaction of aromatic aldehydes having α-hydrogens with nitro alkanes is called
    (a) Aldol Condensation
    (b) Crossed aldol condensation
    (c) Henry Reaction
    (d) None of the above is correct

11. The reaction of Keto Oximes of its type $\begin{matrix} R \\ \quad \diagdown \\ \quad C{=}N \diagup ^{OH} \\ \quad \diagup \\ R' \end{matrix}$ with acidic

    reagent gives
    (a) RCONHR'   (b) R'CONHR
    (c) A mixture of (a) and (b)   (d) The reaction is not possible

12. Benzaldehyde can be best converted into benzyl alcohol by
    (a) Canizzaro reaction by treatment with concentrated aqueous alkali.
    (b) Ionication of benzaldehyde in alcohol in presence of $Ba(OH)_2$ for about 10 minutes.
    (c) Heating benzaldehyde and paraformaldehyde in presence of $Ba(OH)_2 . 8H_2O$ is oil bath or Microwave oven.
    (d) Any of the above process

13. Allyl phenyl elter on heating at 200°C gives

    (a) OH, $CH_2CH{=}CH_2$ (b) OH, $CH{=}CH{-}CH_2$

    (c) A mixture of (a) and (b)   (d) No reaction takes place

14. The reaction $C_6H_5CHO + CH_3CHO \xrightarrow{\text{NaOH}} C_6H_5CH{=}CHCHO$ is a type of

    (a) Claisen Reaction   (b) Aldol condensation
    (c) Claisen-Schmidt reaction   (d) Any of the above

15. The Carbonyl group of aldehydes and Ketones can be best converted with the corresponding hydrocarbon by

    (a) Reduction of carbonyl group by reflaxing with Zinc-Amalgam and hydrochloric acid.
    (b) Reduction of carbonyl group into the corresponding hydrazone by treatment with hydrazinc followed by heating with KOH.
    (c) Sonication of the carbonyl compound with Zn—$NiCl_2$ (9:1) in EOOH—H_2O (1:1) at room temperature for about 2 hrs.
    (d) Any of the above process.

16. p-hydroxy benzaldehyde can best be converted into quinol (p-hydroxy phenol) by

    (a) Dakin reaction with H_2O_2/NaOH followed by hydrolysis
    (b) Dakins reaction with sodium percarbonate in H_2O-THF under ultrasonic irradiation.
    (c) Heating and Stirring a mixture of hydroxy aldehyde with urea hydrogen peroxide adduct (in molar ratio 1:2) at 95°C for 20 min – 1.5 hr and isolation of the product by extraction
    (d) Any of the above process.

17. In the reaction $C_6H_5CH_2COCH_3 + ClCH_2CN \xrightarrow[\overset{+}{Q}X]{\text{aq} \cdot \text{NaOH}}$?, the product obtained is

    (a)
    $$\begin{array}{c} C_6H_5{-}CH_2 \\ \diagdown \\ \overset{H_3C}{\diagup}C{-}CH{-}CN \\ \diagdown \!\! / \\ O \end{array}$$

    (b)
    $$\begin{array}{c} H_3C \\ \diagdown \\ C{-}CH_2{-}CN \\ \diagup \diagdown\!\!/ \\ C_6H_5{-}H_2C \quad O \end{array}$$

    (c)
    $$\begin{array}{c} C_6H_5CHCOCH_3 \\ | \\ CH_2CN \end{array}$$

    (d)
    $$\begin{array}{c} CH_3CHCOCH_2C_6H_5 \\ | \\ CH_2CN \end{array}$$

18. Dieckmann condensation of ethyl adipate is best obtained by treatment with

    (a) Na or NaOEt
    (b) reaction in presence of K in toluene and sonication for 5 min
    (c) treatment with base followed by distillation
    (d) any of the above process

19. The reaction [⟨diagram⟩] + ‖ —Heat→ [⟨diagram⟩] is

(a) [4+2] cycloaddition reaction

(b) the reaction is 100% atom economical

(c) The reaction is a green reaction

(d) all the above are correct.

20. Friedal-craft alkylation of anthracene with CH_3 COCl using $AlCl_3$ in nitrobenzene gives

(a) 9-Acetylanthracene   (b) 1-Acetyl Anthracene

(c) 1:1 mixture of (a) and (b)   (d) A 1 : 9 mixture of (a) and (b)

21. Following photochemical cycloaddition reaction gives

$$R—\overset{\overset{O}{\|}}{C}—R' + \overset{\overset{H_3C \quad H}{\diagdown \diagup}}{\underset{\diagup \diagdown}{\underset{H_3C \quad H}{C}}} \xrightarrow{h\nu} ?$$

(a)

$$R—\overset{\overset{\overset{H}{|}}{O—C—CH_3}}{\underset{\underset{R'}{|}}{C}}—\overset{}{C}—CH_3$$

(b)

$$R—\overset{\overset{\overset{CH_3}{|}}{O—C—CH_3}}{\underset{\underset{R \quad CH_3}{|}}{C}}—\overset{}{C}—H$$

(c) a mixture of (a) and (b)   (d) There is no reaction

22. Reformatsky reaction can be best performed by

(a) Reacting an α-bromo ester (usually an α -bromoester) with an aldehyde or Ketone in presence of Zn metal in inert solvent to give hydroxy ester

(b) Performing the reaction under sonication at room temperature for 5-30 min in dioxane.

(c) Treating aromatic aldehyde with ethyl bromoacetate and Zn-NH_4Cl in solid state.

(d) Any of the above method

23. A Green Simmons-Smith reaction involves

(a) reaction of alkenes with CH_2I_2 and Zn-Cu couple in ether followed by treatment with dilute NaOH

(b) By generating carbene by the reaction of CH_2I_2 with Zn under sonication and reacting with alkene

    (c) any of the above methods

    (d) None of the above methods

24. The best method for the synthesis of unsymmetrical ether involves

    (a) reaction of an alkyl halide with sodium alkoxide

    (b) reaction of an alphol with alkyl halide in NaOH solution in presence a PTC.

    (c) Any of the above methods

    (d) Only by method (a)

25. A green procedure for the synthesis of disodium iminodiacetate involves

    (a) reaction of ammonia, formaldehyde, hydrogen cyanide and HCl

    (b) copper catalysed dehydrogenation of diethyl amine in presence dilute NaOH.

    (c) any of the above process

    (d) none of the above process is green.

Answers:

1. (e)	2. (e)	3. (a) and (b)	4. (a)	5. (a)
6. (d)	7. (c)	8. (c)	9. (d)	10. (c)
11. (a)	12. (b) or (c)	13. (a)	14. (c)	15. (c)
16. (c)	17. (c)	18. (b)	19. (d)	20. (d)
21. (c)	22. (b)	23. (b)	24. (b)	25. (b)

Short Answer Questions

1. What is Green Chemistry?

2. What is need for Green Chemistry?

3. What are the obstacles you face for pursuit in goals of Green Chemistry?

4. Give the Principles of Green Chemistry?

5. Discuss the twelve principles of green chemistry?

6. For planning a green synthesis what steps must be taken.

7. How can energy requirement be kept to a minimum for green synthesis.

8. How will you synthesise (give green synthesis only)

    (a) Adipic Acid

    (b) Catechol

    (c) Benzoic acid from methyl benzoate

9. Discuss the green conditions for the following reactions:
   (a) Baeyer-villiger oxidation   (b) Baylis-Hillman Reaction
   (c) Benzoin Condensation   (d) claisen-Schmrott Reaction
   (e) Dieckmann condensation   (f) Grignard Reaction
   (g) Reformatsky Reaction   (h) Simmons-Smith Reaction
   (i) Stecker Synthesis   (j) Wittig Reaction

10. How are following obtained?
    (a) Azomethines
    (b) Copper Phthalocyanine
    (c) Biodiesel
    (d) Ethyl Benzoate

Index